Practical and Theoretical Geoarchaeology

Dedicated by RIM to the late Peter Reynolds (Butser Ancient Farm, United Kingdom) and Roger Engelmark (University of Umeå, Sweden): pioneers of experimental farms (see Chapter 12)

From RIM: to Mum, Jill, and Flora
From PG: to my folks for sending me away to summer camp

Practical and Theoretical Geoarchaeology

Paul Goldberg and Richard I. Macphail
Department of Archaeology, Boston University and
Institute of Archaeology, University College London

With contributions from
Wendy Matthews
Department of Archaeology, The
University of Reading

Blackwell
Publishing

BLACKWELL PUBLISHING
350 Main Street, Malden, MA 02148-5020, USA
9600 Garsington Road, Oxford OX4 2DQ, UK
550 Swanston Street, Carlton, Victoria 3053, Australia

First published 2006 by Blackwell Science Ltd

5 2011

Library of Congress Cataloging-in-Publication Data

Goldberg, Paul.
Practical and theoretical geoarchaeology/Paul Goldberg and Richard I. Macphail.
p. cm.
Includes bibliographical references and index.
ISBN 978-0-632-06044-3 (pbk. : alk. paper) 1. Archaeological geology. I. Macphail, Richard. II. Title.

CC77.5.G65 2006
930.1′028–dc22

2005018433

A catalogue record for this title is available from the British Library.

Set in 10/12.5 pt Meridien
by Newgen Imaging Systems (P) Ltd, Chennai, India
Printed and bound in Singapore
by C.O.S. Printers Pte Ltd

The publisher's policy is to use permanent paper from mills that operate a sustainable forestry policy, and which has been manufactured from pulp processed using acid-free and elementary chlorine-free practices. Furthermore, the publisher ensures that the text paper and cover board used have met acceptable environmental accreditation standards.

For further information on
Blackwell Publishing, visit our website:
www.blackwellpublishing.com

The publishers wish to acknowledge that Wendy Matthews contributed to chapters 10,11, and 13 and holds copyright to the relevant sections of those chapters.

Contents

Color plate section between pp 212 and 213

Preface

Geoarchaeology within the past two decades has become a fundamental discipline whose value is recognized by everyone interested in past human history. First, archaeologists have become increasingly sensitive to the fact that sediments and stratigraphy provide the ultimate context for the artifacts and features that they excavate. Understanding this sedimentary context and its implications is a requisite for carrying out modern archaeology and for interpreting the archaeological record fully and accurately. Environmentalists (e.g. paleoecologists, soil scientists, sedimentologists, geologists, biologists, and climatologists) on the other hand, have turned to archaeology, because it provides a long-term view of human–environment interactions that have shaped both Quaternary and Holocene landscapes. Such knowledge can have a critical bearing on the future, for example, through identifying areas of sustainable landuse. Interpretation of such ecological information depends upon detailed understanding of the pedosedimentary and geomorphological context, as in archaeology. Geoarchaeology is an important discipline, because as shown recently, it increases our understanding of human impacts on the landscape through the study of ancient soils and occupation deposits. These investigations can provide detailed histories of human endeavors up to the present.

Although geoarchaeology – though not defined as such – had its roots at least in the eighteenth century (Rapp and Hill, 1998) it never came together as a subdiscipline until the appearance of the edited volume "*Geoarchaeology: Earth Science and the Past*" by Davidson and Shackley (1976). At this time Renfrew (1976) coined the term "geoarchaeology" in his Preface. Since then, a number of texts have appeared, the earliest ones (e.g. Rapp and Gifford, 1985) again were collections of papers, which were followed by single authored volumes (e.g. Waters, 1992; Rapp and Hill, 1998). In most cases however, the archaeological and human element as geoarchaeology was often underexplored in favor of landscape studies. This present volume, *Practical and Theoretical Geoarchaeology*, attempts to address this deficiency by providing what we feel is a more balanced view of the discipline by including detailed investigations of human activities as revealed by geoarchaeology. Potentially fruitful avenues of expansion of the discipline are highlighted, for example, in chapters on caves, experiments, occupation deposits, and forensic applications.

This book aims to be both an educational and practical tool, describing how geoarchaeology is carried out across a wide range of environments and periods, employing examples from many countries. It endeavors to teach readers both how to approach geoarchaeological problems theoretically and how to deal with them in practice. The topics covered range from regional-scale studies down to smaller, open area excavations and strata in past and present urban areas, such as Roman and Medieval London. The book presents numerous field and laboratory techniques, exposing readers to approaches suitable to a variety of site situations. Instructive guidelines and protocols are given to show the reader how to create integrated reports that include field evaluations, laboratory assessments, and archive and publication reports.

The book is designed primarily for undergraduate students in Archaeology and those Environmental and Geoscientists, who wish either to train in geoarchaeology or gain a background in this applied science. It is intended to serve as a basic text and an intermediary course in geoarchaeology. It also serves as a necessary text for advanced undergraduates and postgraduate students requiring access to geoarchaeological skills. In addition, it should act as a valuable resource for professionals in order to help develop their awareness of both field and laboratory methods and to identify the full potential of geoarchaeological investigations either for research or mitigation archaeology. Beginners in the subject may benefit from reading Chapters 1 to 3 first, which provide introductions to stratigraphy, sediments, and soils.

Acknowledgments

Those who know us are keenly aware how long it has taken us to get this book together. We thank them here at the outset for their patience and any grief we might have inflicted. There are too many people to thank individually, but a number of folks helped us considerably in writing the book.

Early discussions with Wendy Matthews were very stimulating and helped frame the focus of the book. Similarly, before we were too far along, David Sanger provided some fundamental insights about geoarchaeology and its practitioners. He reminded us that most geoarchaeology books are not written by archaeologists. Again, we hope that this helped us steer a more equitable course on the subject. Along the way we benefited from extensive and intensive collaboration and picked up valuable insights from Ofer Bar-Yosef, Steve Weiner, Takis Karkanas, Trina Arpin, Sarah Sherwood, Carolina Mallol, Arlene Rosen, Harold Dibble, Shannon McPherron, Steve Kuhn, Mary Stiner, Susan Mentzer, Lauren Sullivan, as well as Mike Allen, John Crowther, Gill Cruise, Johan Linderholm, the late Peter Reynolds, and Pat Wiltshire.

The staff at Blackwell has been exemplary, especially in light of the experiences that the authors have had with publishers. Delia Sandford and Rosie Hayden provided congenial help with a great deal of enthusiasm: a sheer pleasure. They also supplied a good deal of forgiveness and support along with Ian Francis, our editor. It was his questionable idea to contact us in the first place, and we are grateful for the opportunity to capitalize on it.

Students at Hebrew University, Boston University, the University of Tours, and the Institute of Archaeology, and members of the Archaeological Soil Micromorphology Working Group served as test subjects for various parts of this book over the years. Colleagues at HU (Na'ama Goren-Inbar, Nigel Goring-Morris, Anna Belfer-Cohen, and Erella Hovers) offered immense opportunities to observe exciting real-time geoarchaeology. Nigel Goring-Morris, simply one of the best field archaeologists around, furnished several examples of key sites that really helped build a geo-archaeological story. Colleagues at Texas (Mike Collins, Tom Hester, Britt Bousman, Lee Nordt, Charles Frederick) provided a different outlook and a growing experience, in spite of the accent. The years of field work at Dust Cave with Boyce Driskell revealed that anthropogenic deposits exist even for hunters and gatherers in the New World.

Over the years, the interaction with many geoarchaeologists similarly shaped our thinking. At the outset, the late Henri Laville served as an inspiration for cave sediments, along with Prof. F. Bordes.

Marie-Agnès Courty has carried on this tradition of scholarship and friendship and has set the bar for geoarchaeological standards. Our first collaboration with her was invigorating and subsequent geoarchaeological interactions have helped us all develop and profit. We continue to be indebted to her.

Reid Ferring and PG were graduate students working in the Negev in the early 1970s and the latter has gained lots of insights into New World and Old World, geology, archaeology, and geoarchaeology, particularly in the field, waiting for things to happen. Rolfe Mandel and Vance Holliday, are among the foremost geoarchaeology practitioners, and they were instrumental in providing support, knowledge, and insights over the years. Much of this was sharpened by sharing the editorial helm with Rolfe at *Geoarchaeology*. Many hours were spent talking about the state of the discipline and the people who practice it. Vance Holliday in particular, devoted a lot of his time to furnish us with timely, constructive comments that significantly improved our message and how we should get it across. Not only are all these folks helpful, but they are simply nice people and made collaboration a pleasure. We also specifically would like to thank our other reviewer (from the United Kingdom), and Gill Cruise who played the devil's advocate with parts of the text. Duncan FitzGerald provided valuable advice on several chapters.

Over the years, we have been encouraged by many other people and organizations who also supplied us with help, support, and information: Bud Albee, Nick Barton, Nick Bateman, Martin Bell, Francesco Berna, Sandra Bond, Mike Bridges, Barbara Brown, Brian Byrd, Dana Challinor, Nick Conard, Jane Corcoran, Andy Currant, John Darlymple, Roger Engelmark, John G. Evans, Nick Fedoroff, Charly French, Henri Galinié, Anne Gebhardt, Daffyd Griffiths, Ole Grøn, Simon Groom, Joachim Henning, John Hoffecker, Bruce Larson, Tom Levy, Elisabeth Lorans, Stephen Macphail, Curtis Marean, Roberto Maggi, Dominic Powlesland, Kevin Reeves, Mark Roberts, Thilo Rehren, Philippe Rentzell, Pete Rowsome, Tim Sara, Solveig Schiegl, Joe Schuldenrein, Astrid Schweizer, Christine Shaw, Jane Sidell, Michela Spataro, Julie Stein, Chris Stringer, Kathryn Stubbs, Ken Thomas, Peter Ucko, Brigitte Vliet-Lanöe, Steve Weiner, Alasdair Whittle, Tim Williams, Liz Wilson, Jamie Woodward, Elizabeth Zadora-Rio, the Universities of Frankfurt and Tours, the Institute of Archaeology (UCL), The Alexander von Humboldt Foundation, University of Tübingen, Albion Archaeology, Butser Ancient Farm, Corporation of London, English Heritage, Framework Archaeology, Museum of Londan Archaeology Service Norfolk Archaeological Unit, Suffolk Archaeological Unit, Oxford Archaeology, and Wessex Archaeology. We are grateful to all of you.

Introduction

Geoarchaeology has become highly prominent these days, and geoarchaeological themes appear in journal articles, monographs, and reports, either within a specific section of an article or as a stand-alone publication. Journals embrace not only mainstream "geoarchaeological" type of publications, such as *Geoarchaeology* or the *Journal of Archaeological Science*, but also other journals that touch on more mainstream archaeological, anthropological, or geological subjects: *Journal of Human Evolution, American Antiquity, Journal of Sedimentary Research*, and *Antiquity*. In the United States, annual meetings of both the Geological Society of America (GSA) and the Society for American Archaeology (SAA), generally have at least one session or poster session, in addition to society-sponsored symposia on the subject. The GSA has an Archaeological Geology Division, and the SAA has the Geoarchaeology Interest Group. The Association of American Geographers (AAG) commonly has geoarchaeology sessions at their annual meetings.

Moreover, the most exciting archaeological sites that one reads about today – either in the popular press or professional literature – commonly have a substantial geoarchaeological component. The reader has only to be reminded about the significance of the geoarchaeological aspect of sites that are concerned with major issues relating to human development and culture. Some high profile issues and sites include: the use and evidence of the controlled use of fire (Zhoukoudian, China); the sedimentary context and the origin of early hominins (Dmanisi, Republic of Georgia; Boxgrove, United Kingdom; Mediterranean and South African caves – Gorham's Cave, Gibraltar; Kebara Cave, Israel; Blombos Cave, South Africa); peopling of the New World (Meadowcroft Rockshelter United States; Monte Verde, Chile); Near Eastern urban cultures (Çatal Höyük, Turkey; Abu Salabikh and Tel Leilan, Syria), and early management of domestic animals (Arene Candide, Italy). Even previously excavated localities, such as the Folsom site, are being reinvestigated from a more sophisticated geoarchaeological perspective.

These well-known landmark sites have really drawn attention to the contribution that geoarchaeology can make to, and its necessity in, modern archaeological studies. This situation was not the case a few decades ago when only a handful of archaeological projects utilized the skills of the geoarchaeologist; one of the authors (PG) could enumerate the sparse number of geoarchaeological studies at the time he was a graduate student in the late 1960s. Still, the best results have come from highly focused geoarchaeological investigations that have employed the appropriate techniques, and that have been intimately linked to multidisciplinary studies that provide consensus interpretations.

This book is about how to approach geoarchaeology and use it effectively in the study of archaeological sites and contexts. We shall not enter into any detailed discussion of the origins and etymology of "Geoarchaeology" versus "Archaeological Geology." (Full discussions of this irrelevant debate can be found in Butzer, 1982; Courty *et al.*, 1989; Rapp, 1975; Rapp and Hill, 1998; Waters, 1992.) In a prescient, no-frills view of the subject Renfrew (1976: 2) summed it up concisely and provided these insights into the nature of geoarchaeology:

> *This discipline employs the skills of the geological scientist, using his concern for soils, sediments and landforms to focus these upon the archaeological "site," and to investigate the circumstances which governed its location, its formation as a deposit and its subsequent preservation and life history. This new discipline of geoarchaeology is primarily concerned with the context in which archaeological remains are found. And since archaeology, or at least prehistoric archaeology, recovers almost all its basic data by excavation, every archaeological problem starts as a problem in geoarchaeology.*

These issues of context, and what today would be called "Site Formation Processes" in its broadest sense, can and should be integrated regionally to assess concerns of site locations and distributions, and geomorphic filters that might have controlled them.

In this book – as will be seen throughout the volume – we take a decidedly pragmatic and functional approach. We do not see any need to differentiate between the two and consider geoarchaeology, archaeological geology, and geological archaeology to fall under the same rubric: any issue or subject that straddles the interface between archaeology and the earth sciences. Classifications – and in this case the distinctions between geoarchaeology and archaeological geology – are of value only if they are ultimately useful. Does it *really* matter how we categorize research that is aimed at studying postdepositional dissolution of bones at a site? According to the above this research would fall into both camps, but does it help us to know if we are doing geoarchaeology or geological archaeology or archaeological geology? For the sake of brevity, we employ the simple term, *Geoarchaeology*.

Geoarchaeology is practiced at different scales (Stein and Linse, 1993). Furthermore, its use and practice vary according to the training of the people involved and the goal of their study. For example, geologists and geographers may well emphasize the mapping of large-scale geological and geomorphological features – such as where a site may be situated within a drainage system – or other regional landscape features. This is a regional perspective scale that exists in three dimensions, with relative relief possibly being measured in thousands of meters, especially if working in the Alps and Andes. Much of the geoarchaeological research carried out in North America is focused at this landscape scale. Geologists would also be interested in the overall *stratigraphy* of site deposits and how these might interrelate to major landforms, such as stream terraces, glacial landforms, and loess plateaus. Pedologists, on the other hand, would be more concentrated on the parent materials, the surfaces upon which soils were formed, and how both have evolved in conjunction with the landscape; such materials can be buried by subsequent deposition or can be found on the present-day surface. In either case, pedologists' focus tends to be on the scale of the soil pit, that is, on the order of meters. Archaeologists themselves may want to focus geoarchaeological attention upon microscale, centimeter-thick occupation deposits: what they are, and how they reflect specific or generalized past human activities. In the case of rescue/mitigation archaeology – commonly termed Cultural Resource Management (CRM) in the United States – geoarchaeology is tailored to the nature of the "job specifications" proscribed by the developer under the guidelines of salvage operations. Finally, the geoarchaeologist may well be just one member of an environmental team whose task is to reconstruct the full biotic/geomorphic/pedologic character of a site and its setting, and how these environments interacted with past human occupations. All these approaches can be relevant depending on the research questions involved, and holistically they could be subsumed under the term, "site formation processes" (Schiffer, 1987).

Archaeologists come from a variety of backgrounds. As stated above, in North America, archaeology is taught predominantly in anthropology programs, although some universities (e.g. Boston University, United States, and Simon Fraser University in Canada) actually have archaeology departments; Classical and Near Eastern Archaeology programs are not rare, and these tend to emphasize written sources over excavation. In Europe, archaeology is included within programs, or in departments and institutes, and not necessarily as an extension of anthropology.

Although in the United Kingdom, geoarchaeology is taught in a number of archaeological departments, this is not always the case in Europe as a whole. In France, for example, this subject may only be taught to prehistorians and not to classical or medieval archaeologists. Commonly, even in the United Kingdom and elsewhere in the world, geoarchaeology is more likely to be seen as an *ad hoc* offshoot of geology and geography. In North America, it is not anchored in any particular department and may be cross-listed among Anthropology, Archaeology, Geology, and Geography. In spite of good intentions and good training, many geoscientists tend to be naïve in their approach to solving archaeological problems, and as a consequence they effectively reduce their potential in advancing this application of their science. This often diminishes or even negates their contributions to interdisciplinary projects. The opposite situation can be found, where an archaeologist does not even know what questions to ask (Goldberg, 1988; Thorson, 1990).

Thus, as Renfrew (1976) so cogently demonstrated, geoarchaeology provides the ultimate context for all aspects of archaeology from understanding the position of a site in a landscape setting to a comprehension of the context of individual finds and features. Without such knowledge, even the most sophisticated isotope study has limited meaning and interpretability. As banal as it might sound, the adage, "garbage-in, garbage-out" is wholly pertinent if the geoarchaeological aspects of a site are ignored.

In the past, geoarchaeology was carried out very much by individual innovators. In North America, the names Claude Albritton Jr., Kirk Bryan, E. Antevs, E.H. Sellards, and C. Vance Haynes immediately come to mind as the early and prominent leaders in incorporating the geosciences into the framework of archaeology (see Holliday, 1997 and Mandel, 2000a for details). In fact, Mandel aptly points out that for the Great Plains, geoarchaeology – or at least geological collaboration – locally constituted an active part of archaeological survey for several areas, although it was patchy in space and time. Much of the emphasis was focused on evaluating the context of Paleoindian sites and how these occurrences figured into the peopling of the New World (Mandel, 2000).

In Europe, during the period from the 1930s to the 1950s, Zeuner (1946, 1953, 1959) at the Institute of Archaeology (now part of University College London), developed worldwide expertise in the study of the geological settings of numerous Quaternary and Holocene sites that ranged from India to Gibraltar. After Kubiëna (1938, 1953, 1970) called the world's attention to soil micromorphology, Cornwall (1958) also at the Institute of Archaeology, applied this technique to archaeology for the first time (see below). At the same time, Dimbleby (1962) developed the link between archaeology and environmental studies, and produced one of the first detailed investigations of past vegetation and monument-buried soils for Bronze Age England Dimbleby (1962). Duchaufour (1982) in France also systematically studied environmental change and pedogenesis. In mainland Europe, the legendary French prehistorian François Bordes (1954) – whose doctorate in geology dealt with the study of loess, paleosols, and archaeological sites, principally in Northern

France – placed the French Palaeolithic within its geomorphologic setting. Vita-Finzi (1969), working in the Mediterranean Basin, used archaeological sites to suggest the chronology of Mediterranean valley fills, which he related to both climatic and anthropogenic factors. Cremaschi (1987) investigated paleosols and prehistoric archaeology in Italy.

Although some geoarchaeological research is funded by granting agencies (NSF, NGS, NERC, CNRS), much, if not most, of modern geoarchaeological work – in both the New and Old Worlds – is fostered and sponsored by CRM projects, ultimately related to human development throughout the world. Approaches and job specifications vary according to whether investigations are at one end of the spectrum, short-term one-off studies, or long-term research projects at the other. Geoarchaeological work can be done by single private contractors or by huge international teams, which may well include specialists who also act as private contractors. Nowadays, local authorities, government agencies (e.g. State Departments of Transportation in the United States) and national research funding agencies (e.g. NSF in the United States, AHRB and English Heritage in the United Kingdom, AFAN and the CNRS in France, and Nara National Institute in Japan) may all be involved in commissioning geoarchaeological investigations. It is currently a very flexible field. It is also one where there is an increasing need for formal training, but where relatively few practitioners have been in receipt of one.

Geoarchaeological work is now often broken up into several phases, with desktop investigations, fieldwork survey, excavation, sample assessment, and laboratory study, all being likely precursors to full analysis and final publication. This is all part of modern funding and operational procedures.

Single-job or site-specific studies may be as straightforward as finding out "What is this fill?" whereas problem-based research could involve the gathering of geoarchaeological data on the possible first controlled use of fire, as at Zhoukoudian, China (Goldberg *et al.*, 2001; Weiner, 1998). Sites are investigated at different scales and sometimes, for very different reasons. At one time "dark earth" – the dark colored Roman-medieval urban deposits found in urban sites across northern Europe – engaged the particular interest of geoarchaeologists because these enigmatic deposits commonly span the "Dark Ages," and human activities at this time were poorly understood (Macphail, 1994; Macphail *et al.*, 2003). Analysis of "dark earth" therefore, became a research-funded topic for urban development sites (CRM projects in urban areas) across Belgium, France, and the United Kingdom, for example. On the other hand, attention can be focused on individual middens and midden formation because they provide a wealth of material remains, particularly organic, that are normally poorly preserved and complex to understand and interpret (Stein, 1992). Regional studies of the intertidal zone, for example, may include the investigation of middens as one single component. The recent study of the intertidal deposits of the River Severn around Goldcliff, Wales, for example, involved analysis of sediment and drowned soils in order to investigate sea level changes and their effects on the populations living in the coastal zone during the Mesolithic through Iron Age periods (Bell *et al.*, 2000). Equally, studies of alluvial deposits and associated floodplains (Brown, 1997; French, 2003) have involved the search for buried sites, within the overall realm of evaluating the distribution of archaeological sites. The Po plain of Italy (Cremaschi, 1987) and the Yellow River of China (Jing *et al.*, 1995) both feature a series of late prehistoric settlements. Many of the most significant Paleoindian and Archaic sites in the United States are situated within alluvial sequences (Ferring, 1992, 1995; Mandel, 1995, 2000a).

Modern geoarchaeological research makes use of a vast number of techniques that either have been used in geology and pedology or have been developed or refined for geoarchaeological purposes. Early geoarchaeological research until the latter part of the last century, at least in North America, was predominantly field-based and made use of both natural exposures and excavated areas. More recently, field techniques have become more improved and technologically sophisticated (Hester *et al.*, 1997). Natural exposures can be supplemented with surface satellite remote sensing data (Scollar, 1990), as well as subsurface data derived from machine-cut backhoe trenches, augering, coring, and advanced geophysical techniques (e.g. magnetometry, electrical resistivity, and ground penetrating radar-Kvamme, 2001). Moreover, such data can be assembled and interrogated using Geographic Information Systems (GIS; Wheatley and Gillings, 2002) that can be used to generate and test hypotheses.

Laboratory techniques have similarly become more varied and sophisticated. At the outset, many geoarchaeological studies adopted techniques from geology and pedology that were aimed at sediment/soil characterization. Thus traditional techniques characteristically consisted of grain-size analysis (granulometry), coupled with other physical attributes (e.g. particle shape, bulk density, bulk mineralogy), as well as basic chemical analyses of organic matter, calcium carbonate content, extractable iron, and so on. The analysis of phosphate to elucidate activity areas or demarcate site limits has a longer history spanning over 70 years (Arrhenius, 1931, 1934; Parnell *et al.*, 2001). Conventional techniques with long historical pedigrees, such as x-ray diffraction (XRD), electron microprobe, x-ray fluorescence (XRF), instrumental neutron activation analysis (INAA), and atomic absorption (AA) have been enhanced by rapid chemical, elemental, and mineralogical analyses of samples through the use of Fourier transform infrared spectrometry (FTIR), Raman spectrometry, and inductively coupled plasma atomic emission spectrometry (ICP-AES) (Pollard and Heron, 1996).

In addition, a notable advance in geoarchaeology has been the application of soil micromorphology to illuminate a wide variety of geoarchaeological issues (Courty *et al.*, 1989; French, 2003). These issues range from the development of soil and landscape use (e.g. Ayala and French, 2005; French and Whitelaw, 1999; Romans and Robertson, 1983) and the formation of anthropogenic deposits (Macphail *et al.*, 1994; Matthews, 1995; Matthews *et al.*, 1997) to the evaluation of the first uses of fire (Goldberg *et al.*, 2001).

Finally, geoarchaeological research has been facilitated by the development of numerous dating techniques just within the past two to three decades. Now, sites within the span beyond the widely accessible limits of radiocarbon are potentially datable with techniques, such as thermoluminescence (TL), optically stimulated luminescence (OSL), and electron spin resonance (ESR) (Rink, 2001).

In this book, we aim to present a fundamental, broad-based perspective of the essentials of modern geoarchaeology in order to demonstrate the breadth of the approaches and the depth of problems that can be tackled. As such, it is also aimed to promote a basic line of communication and understanding among all multidisciplinarians. We cover a variety of topics that discuss thematic issues, as well as practical skills. The former encompasses such broad concepts as stratigraphy, Quaternary and environmental studies, sediments, and soils. We then provide a survey of some of the most common geological terrains that provide the natural settings for almost all archaeological sites. These are established geoarchaeological topics into which we have incorporated some new findings. Unlike previous books on geoarchaeology, we have dedicated a second major portion of the volume to new topics that are normally not entertained by previous geoarchaeological texts, such as "human impact," "experiments," and occupation deposits including Roman and medieval archaeology, and forensic applications. It is important also to obtain some insights into practical aspects of geoarchaeology, including how specifically geoarchaeologists should fit in to a project. Similarly, two chapters are devoted to a presentation of the methods – pragmatic and theoretical – currently used in geoarchaeology. These include not only field techniques (e.g. the use of aerial photos, how to describe a profile, and collect samples), but also those techniques that are used in the laboratory. Although we summarize the "what," and "how," we also try to emphasize the "why," and provide a number of example-based caveats for important techniques. A final facet deals with the practical aspects of reporting geoarchaeological results, keeping in mind that material presented in reports differs from that in articles. Reports essentially present the full database and arguments, whereas articles are commonly more thematic and focused, and by necessity are constrained to present results more concisely. Reports, which are seldom published in full, constitute the "gray literature" and make up an important part of the scientific database. They are too commonly overlooked, ignored, or simply are not readily accessible.

As a final point, we maintain that geoarchaeology in its broadest sense, must be made understandable to all players involved, be they archaeologists with strong training in anthropology, or the geophysicist, with minimal exposure to archaeological issues. All participants should have enough of a background to understand what each participant is doing, why they are doing it, and most importantly, what the implications of the geoarchaeological results are for all team members. Too often we hear about the geospecialist simply turning over results to the archaeologist, essentially being unaware of the archaeological problem(s), both during the planning stages and later during execution of the project. Hence they cannot correctly put their results to use. On the other hand, many archaeologists tacitly accept results produced by specialists with few notions on how to evaluate them. This book attempts to level the playing field by providing a cross-disciplinary background to both ends of the spectrum. Such basic material is needed to establish a dialogue among the participants so that problems can be mutually defined and mutually understood, regardless of whether you call yourself an archaeological geologist or geoarchaeologist.

Part I

Regional scale geoarchaeology

Introduction to Part I

Geoarchaeological endeavors operate at a variety of scales. These undertakings range from regional views of archaeological sites – their distributions and associated activities (e.g. cultivation, hunting ranges, trading routes, and networks) – to microscopic study of the deposits, artifacts, and features found within them. In the following chapters we examine the issues linked to landscape scale geoarchaeology, and describe the most salient aspects of some of the principal geological environments and processes that geoarchaeologists are likely to encounter. Although archaeological sites and associated remains can be found in most geological environments (including what is now marine; Faught and Donoghue, 1997), most human occupations, or at least their traces, are not evenly spread throughout these environments. Sites associated with temperate *fluvial* (stream) environments, for example, are considerably more abundant than those from desertic or glacial terrains. Thus, although we try to touch on all these situations, we may provide more detail for those with greater representation in the geoarchaeological record. Detailed treatises on geological environments can be found in many geomorphology and sedimentology texts, such as Boggs, 2001; Easterbrook, 1993; Reading, 1996; and Ritter *et al.*, 2002.

1

Sediments

1.1 Introduction

Sediments – those materials deposited at the earth's surface under low temperatures and pressures (Pettijohn, 1975) – constitute the backbone of geoarchaeology. The overwhelming majority of archaeological sites is found in sedimentary contexts, and the material that is excavated – whether geogenic or anthropogenic – is sedimentary in character. In this section we examine some of the basic characteristics of sediments, many of which can also be applied to soils (Chapter 3). We have two principal goals in mind. Since sediments are so ubiquitous in archaeological sites, it is necessary to have at least a working knowledge of some of these characteristics so as to be able to share this descriptive information with others. Essentially, these descriptive characteristics constitute a *lingua franca*: the term "sand" (Table 1.1), for example, corresponds to a defined range of sizes of grains, irrespective of composition. Second, and perhaps more important, is that many of the descriptive parameters that we observe in sediments commonly reflect – either individually or collectively – the history of the deposit, including its (1) origin, (2) transport, and (3) the nature of the locale where it was deposited, that is, its environment of deposition. Figuring out these three aspects of a sediment's history constitutes a subliminal mindset of sedimentologists, whether they are studying a 100 m thick sequence of Carboniferous sandstones in Pennsylvania or a 10 cm thick sandy layer within a Late Pleistocene cave in the Mediterranean. In sum, by observing and recording the lithological attributes of a sediment we not only provide an objective set of criteria to describe it, but also a means to get some insights into its history.

1.2 Types of sediments

Sediments can be classified into three basic types, clastic, chemical, and organic, of which the first two are generally the most pertinent to geoarchaeology. Clastic sediments are the most abundant type. They are composed of fragments of rock, other sediment, or soil material that reflect a history of erosion, transport, and deposition. Most clastic sediments are terrigenous and deposited by agents such as wind (e.g. sand dunes), running water (e.g. streams, beaches), and gravity (e.g. landslides, slumps, colluvium). Typical examples of clastic sediment (as based on decreasing sizes of the components) are sand, silt, and clay (Table 1.1). In the geological record, when such materials become lithified, the resulting rock types are sandstone, siltstone, and shale, respectively.

Volcaniclastic debris, consisting of volcanic ash, blocks, bombs, and pyroclastic flow debris are also considered as clastic sediments (Fisher

TABLE 1.1 Common grain size scales used in geology and pedology

Wentworth class (geology)[1]	Size range	Phi (Φ) units[2]	UK soil science class equivalent[3]	Size range	USA soil science class equivalent[4]
Boulder	>256 mm	−8	Boulders Very large stones	>600 mm 200–600 mm	
Cobble	64–256 mm	−6 to −8	Large stones	60–200 mm	
Pebble	4–64 mm	−2 to −6	Medium stones Small stones	20–60 mm 6–20 mm	
Granule	2–4 mm	−1 to −2	Very small stones	2–6 mm	
Very coarse sand	1–2 mm	0 – 1			1–2 mm
Coarse sand	0.5–1 mm	1–0	Coarse sand	0.6–2 mm	0.5–1 mm
Medium sand	250–500 μm	2–1	Medium sand	212–600 μm	250–500 μm
Fine sand	125–250 μm	3–2	Fine sand	63–212 μm	100–250 μm
Very fine sand	63–125 μm	4–3			50–100 μm
Coarse silt	31–63 μm	5–4	Coarse silt	20–63 μm	Silt = 2–50 μm
Medium silt	16–31 μm	6–5	Medium silt	6–20 μm	
Fine silt	8–16 μm	7–6	Fine silt	2–6 μm	
Very fine silt	4–8 μm	8–7			
Clay	<4 μm	>8	Clay	<2 μm	<2 μm

1. Modified from Nichols, 1999
2. $\Phi = -\log_2 d$ (d = grain diameter)
3. Avery, 1990; Hodgson, 1997
4. Soil Survey Staff, 1999

and Schmincke, 1984), Overall, they are relatively uncommon in geoarchaeological contexts as they are restricted to volcanic areas. Nevertheless, they constitute an important aspect in the stratigraphy and formation of some key archaeological sites. Pompeii, considered the type site for depicting instances in archaeological time (Binford, 1981), is covered by about 4 m of volcaniclastic debris (tephra), consisting of pumice, volcanic sand, lapilli (2–64 mm), and ash (<2 mm) (Giuntoli, 1994). The site of Ceren, in San Salvador, represents a similar type of setting, where structures and agricultural fields were buried under several meters of tephra (Conyers, 1995; Sheets, 1992). Volcaniclastic deposits in rift valleys play critical roles in the dating and stratigraphy of Pliocene and Pleistocene deposits from sites in East Africa, the Jordan Rift, Turkey, and Georgia. Sites such as Olduvai Gorge, Koobi Fora, Gesher Benot Ya'akov, and Dmanisi, are just a few of the sites where archaeological and hominin remains are intercalated with volcanic rocks and tephra (Ashley and Driese, 2000; Deocampo *et al.*, 2002; Gabunia *et al.*, 2000; Goren-Inbar *et al.*, 2000; Stern *et al.*, 2002).

Marine organisms such as mollusks and corals, for example, produce shells of calcium carbonate. In cases where these hard body parts are subjected to wave action, they can be broken into small centimeter to millimeter size fragments, resulting in the formation of a bioclastic limestone, for example (Table 1.2). Coquina is an example of a coarse bioclastic

Table 1.2 Types of sediments; consolidated (lithified), rock equivalents for clastic sediments are given in parenthesis. Note that *bioclastic* limestones, composed of biologically precipitated shell fragments (e.g. coquina), can be thought to be both clastic and biochemical in origin

Clastic and bioclastic		Nonclastic	
Volcaniclastic	**Terrigenous and marine**	**Chemical**	**Biological**
Lapilli, blocks, bombs	Cobbles, boulder; gravel (*conglomerate*)	Carbonates (*limestones*)	Peat (lignite, coal)
Ash (*welded tuff*)	sand (*sandstone*)	Evaporates (chlorides, sulphates, silicates; phosphates)	
	silt (*siltstone*)	Travertines and flowstones (cave and karst settings)	Algae, bacteria, diatoms, ostracods, foraminifera
	clay (*shale*)		
	Bioclastic: coarse (e.g. coquina); fine (chalk)		

sediment, whereas chalk is composed of silt and fine sand-size tests of marine organisms (foraminifera); in other cases organisms may have siliceous skeletons as is the case with diatoms, resulting in the formation of diatomite. Biological remains, such as ostracods, diatoms, and foraminifera, can be preserved within otherwise mostly mineralogenic deposits. The moat deposits from the Tower of London, for example, contained numerous diatoms that suggested shallow, turbid water in disturbed sediments (Keevill, 2004). These conditions were inferred to be a result of inputs into the moat of waste disposal, surface water, and water from the Thames and the City Ditch.

Chemical sediments are those produced by direct precipitation from solution. Lakes in semiarid areas with strong evaporation, for example, will exhibit a number of precipitated minerals, such as halite (table salt), gypsum (calcium sulphate), or calcite or aragonite (both forms of calcium carbonate). In cave environments, chemical sediments are widespread and typically produce sheets of calcium carbonate (e.g. travertine or flowstone), or ornaments such as stalactites and stalagmites. These are usually composed of calcite or aragonite, but other peculiar minerals can be composed of phosphates, nitrates, or sulphates (Hill and Forti, 1997). The third group, biological sediments, is composed mostly of organic materials, typically plant matter. Peats or organic-rich clays in swampy areas and depressions are characteristic examples.

1.2.1 Descriptive and interpretative characteristics of sediments

1.2.1.1 Clastic sediments

Clastic sediments display a number of properties that can be described and ultimately interpreted. These characteristics include composition, texture (grain size and shape), fabric, and sedimentary structures.

Composition. Sediments can exhibit a wide variety of composition of mineral and rock types, and normally this is a function of the source of the material (Tables 1.3 and 1.4). For this reason, geologists are able to reconstruct geological landscapes (e.g. former landmasses in Table 1.4) that have long since been eroded. In spite of their wide variety, certain rocks and minerals occur repeatedly in sediments ("major minerals" in Table 1.3). Their relative abundance in a sediment can vary with location and age. In the case of the latter, some minerals (e.g. olivine) are more susceptible to alteration/destruction than others, and thus they can be less persistent in older sediments. Furthermore, overall sediment composition can be influenced by secondary processes (e.g. weathering, soil formation, diagenesis) that result in the precipitation of minerals that either cement the skeletal grains of the sediment, or that precipitate as concentrations within the sedimentary mass (e.g. nodules and concretions). Secondary mineralization may involve carbonates (e.g. calcite, aragonite), silicates (e.g. opal, microcrystalline quartz/chert), sulfates (e.g. gypsum, barite), and iron oxides (e.g. limonite, goethite).

Texture involves attributes of individual grains themselves, and these like other traits, have both descriptive and interpretative value. One of the most basic and widespread attributes is that of grain size (Table 1.1), and it is one that both geoscientists and archaeologists use and intuitively understand: "this deposit is fine grained, while the one in the corner is quite gritty to sandy." Obviously we have to be more precise than this example, and both geologists and pedologists employ formal names

Table 1.3 Common minerals and rock fragments in sediments (modified from Boggs, 2001)

Major minerals

- Quartz
- Potassium and plagioclase feldspars
- Clay minerals

Accessory minerals

Micas: muscovite and biotite

Heavy minerals (those with specific gravity >2.9):

- Zircon, tourmaline, rutile
- Amphiboles, pyroxenes, chlorite, garnet, epidote, olivine
- Iron oxides: Hematite, limonite, magnetite

Rock fragments

- Igneous
- Metamorphic
- Sedimentary

Table 1.4 Heavy mineral associations and related geological sources (modified from Pettijohn *et al.*, 1973)

Mineral association	Typical geological source
Apatite, biotite, brookite, hornblende, monazite, muscovite, rutile, titanite, tourmaline, zircon	Acid igneous rocks
Augite, chromite, diopside, hypersthene, ilmenite, magnetite, olivine	Basic igneous rocks
Andalusite, corundum, garnet, phlogopite, staurolite, topaz, vesuvianite, wollastonite, zoisite	Contact metamorphic rocks
Andalusite, chloritioid, epidote, garnet, glaucophane, kyanite, sillimanite, staurolite, titanite, zoisite-clinozoisite	Dynamothermal metamorphic rocks
Barite, iron ores, leucoxene, rutile, tourmaline (rounded grains), garnet, illmanite, magnetite, zircon (rounded grains)	Reworked sediments

and size limits to describe the classes of particle sizes that range from fine, micron-size (1 μm = 0.001 mm) grains of dust, up to large boulders several meters across.

The grain size scale commonly used by geologists in the United States is that developed by Wentworth (1922) (Table 1.1); note that this scale is a geometric grade scale with the limits between classes having a constant ratio of 1/2 (Krumbein and Sloss, 1963). Geologists have later modified this by introducing the phi (Φ) scale, which is a logarithmic transformation of the grain size to the base 2 ($\Phi = -\log_2 d$, where d is the grain size in millimeter). These arithmetic intervals are convenient for the graphical presentation of grain size data, such as histograms (see Chapter 16), and for performing statistical analyses. In any case, as can be seen in Table 1.1, the limit between silt and clay is 3.9 μm in the Wentworth scale, and the limit between silt and sand is 62.5 μm. Furthermore, it is important to note that soil scientists (in both the United States and United Kingdom) employ different limits between sand/silt and silt/clay: for soils in the United States, silt includes material between 50 and 2 μm, whereas in the United Kingdom, it is 63–2 μm (Table 1.1). These differences are particularly significant while evaluating data in reports and maps, and whether the descriptions have been done by a geologist or pedologist: the same sediment can have different percentages of silt or sand, depending upon who it was analyzed by a geologist or a pedologist.

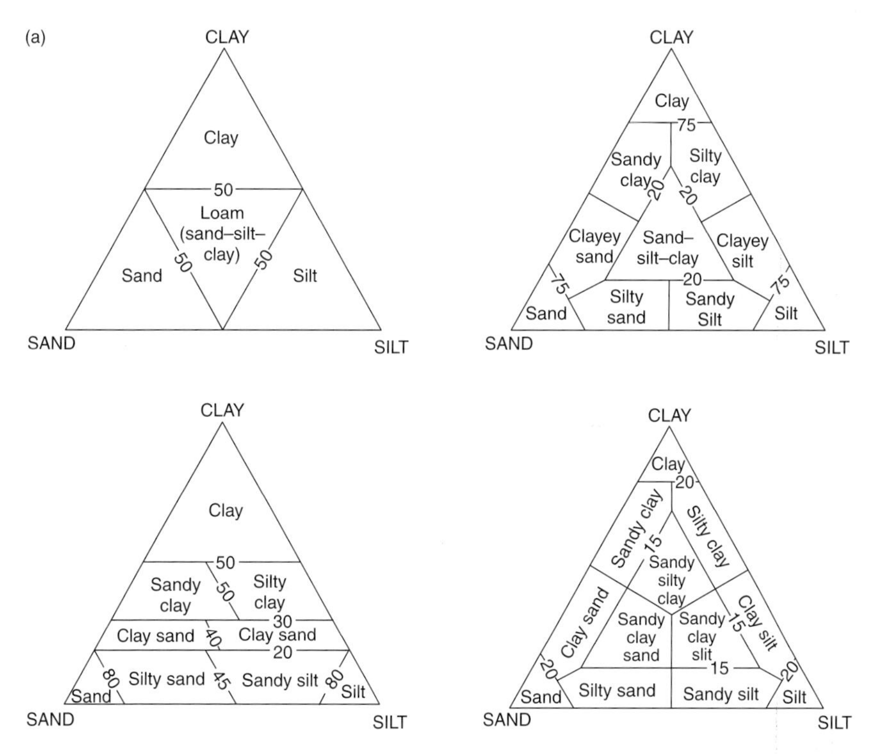

FIGURE 1.1 Triangular grain size diagrams with sand, silt, and clay end-members. (a) These diagrams illustrate the large differences and variability in class names and boundaries as used by American geologists. (b) A similar type of triangular diagram as employed by American and British soil scientists. Note the widespread occurrence of the term "loam" and its variants (e.g. sandy loam) to describe soil textures. Furthermore, compare the size of the "Clay" field in this soil diagram with that of "Clay" as used by geologists in (a) (1.1a, modified from Pettijohn, 1975).

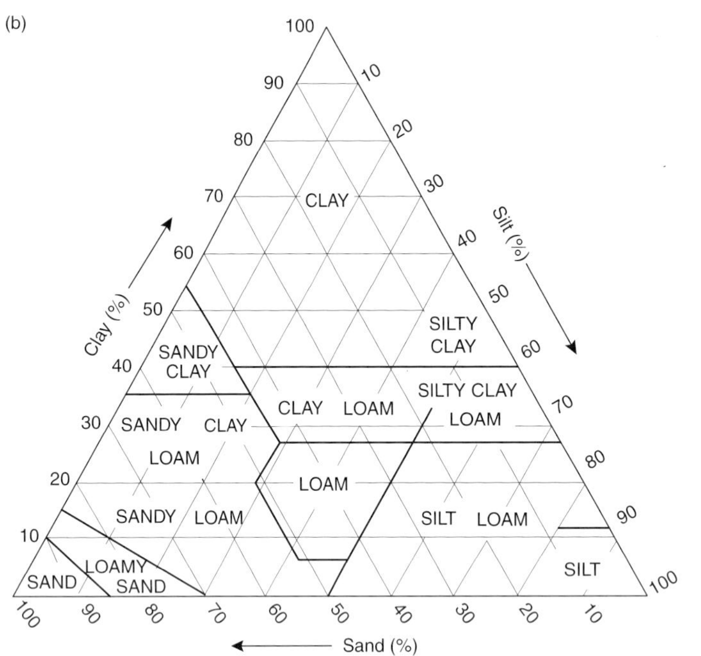

FIGURE 1.1 *(cont.)*

Although the methods for grain size analysis are discussed in Chapter 16, we point out here that sediments are commonly mixtures of different sizes of particles. The proportions of sand–silt–clay, for example, in sediments and soils are commonly presented in terms of three major end-members, as shown in the triangular diagrams of Figure 1.1. Not surprisingly, different mixtures of these end-members have different names, such as "sandy clay" or "silty sand" that vary from author to author, and from discipline to discipline (again, geology versus pedology – cf. Figs 1.1a versus 1.1b). Figure 1.1a, for example, shows four triangular diagrams with different limits between different classes. In addition, soil scientists use the additional term loam, which is defined by the USDA to be composed of 7–27% clay, 28–50% silt, and <52% sand; many soils fall into the "loamy" category, being sandy, silty, or clayey loams (Soil Survey Staff, 1999). A silty loam equivalent for a geological deposit could fall into the "silty sand" or "silty clayey sand" category depending upon which sedimentary classification scheme is used. The point is that, again, one has to be aware of the background of the author of the grain size analysis: whether the classification system used is geological or pedological and from which country they are from, since class limits vary from country to country. Thus, the term loam, should more properly be used to describe soils and not sediments.

Sorting is a term applied to the proportion and number of different size classes comprising the grain populations. In particular, it relates to the statistical distribution of sizes around the mean (the standard deviation; see Tucker, 1988 for ways to measure and evaluate it). Sorting can be readily visualized in Figure 1.2. The predominance of one particle size indicates a well-sorted sediment. Beach and dune sands, for example, are characteristically well sorted, as are wind-blown dust deposits, known as loess. A poorly sorted sediment consists of numerous amounts of different particle sizes. Slope deposits, where a mass of sediment has been moved downhill (the process of colluviation) and glacial till are typically poorly sorted deposits.

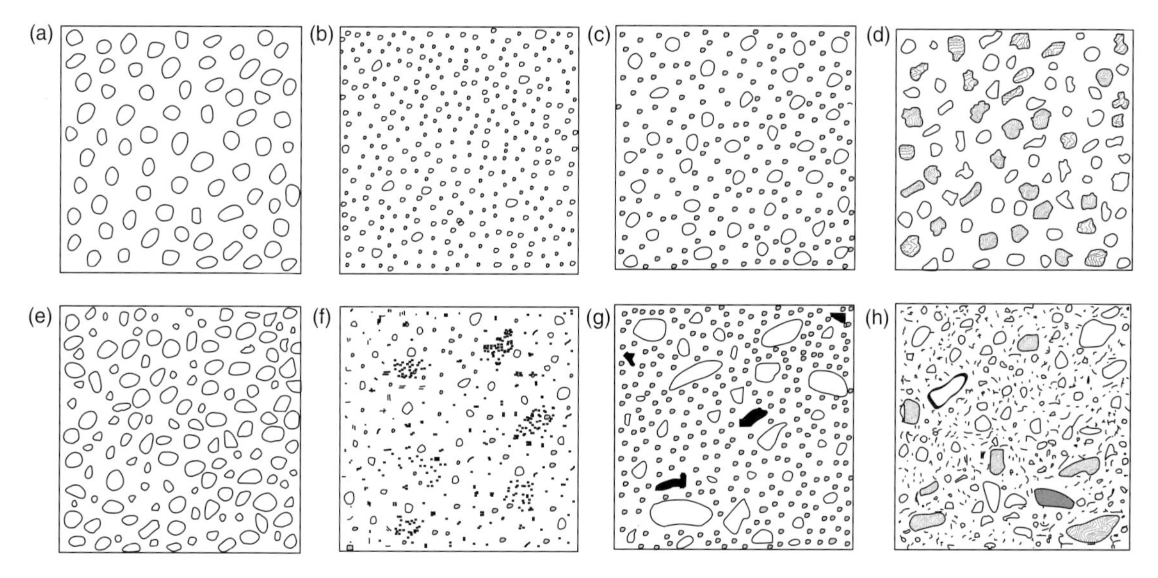

FIGURE 1.2 Illustration of the concept of sorting. (a) well-sorted sand; (b) well-sorted silt; (c) bimodal: well-sorted silt *and* sand; (d) well-sorted sand of varying composition; (e) moderately sorted sand; (f) poorly sorted silt; (g) bimodal: poorly sorted sand in a well-sorted silt; (h) unsorted (modified from Courty *et al.*, 1989).

Particle shape is usually considered for pebbles and sand-size grains. It is another descriptive parameter and an indicator of grain and sediment history. Three related features of shape are generally considered. *Form* refers to the general outline of the grain and ranges from equant grains (with roughly equal length, width, and thickness dimensions approaching the shape of a sphere), to platy or disc-shaped grains, in which the thickness is markedly less than the length or width (see Boggs, 2001; Mueller, 1967 for methodological details to measure shape). *Roundness,* on the other hand, relates to the angularity of a grain and is concerned with the jaggedness of the edges (Figs 1.3a,b). Finally, *surface texture* refers to the microtopographic features of the grain, such as pitting, etch marks, and micro fractures; grains may vary from pitted to smooth.

The importance of form is not universally agreed upon, and its measurement should be viewed in light of other sedimentary parameters (Boggs, 2001). Form is more pertinent to the coarser, pebble fraction, as the finer sand fraction (generally quartz) seems to be only slightly modified if at all by transport. The most noteworthy example comes from beach pebbles and cobbles, which tend to be flattened, although the reasons for this are not clear. In fluvial environments, shapes can be associated with ease of transport, so that spherical and prolate shapes tend to be more readily transportable than are blady particles.

Grain roundness is a function of mineralogy, size, transport, and distance of transport (Boggs, 2001). For sand-size quartz grains, for example, increased roundness is commonly associated with aeolian transport, which is more effective in rounding grains than is water. For larger pebbles, composition plays an increasingly important role. Limestone pebbles, for example, are much easier to round in fluvial environments than are those from cherts, which tend to fracture before they become rounded. On the other hand, the presence of

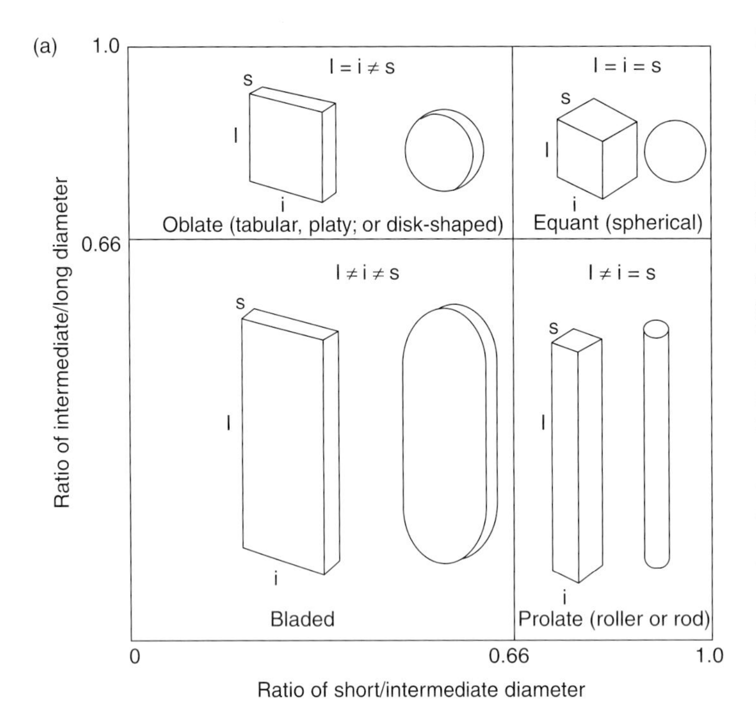

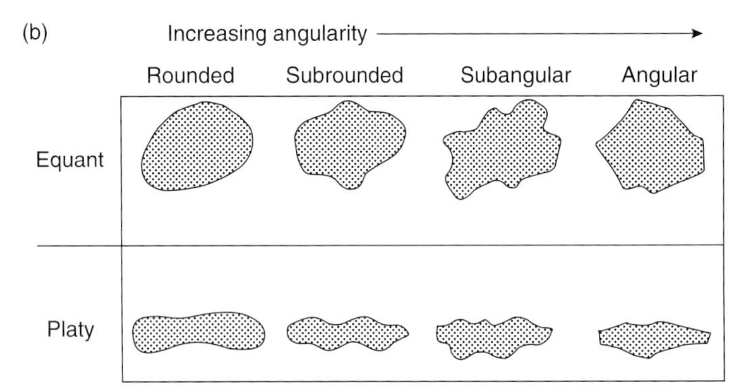

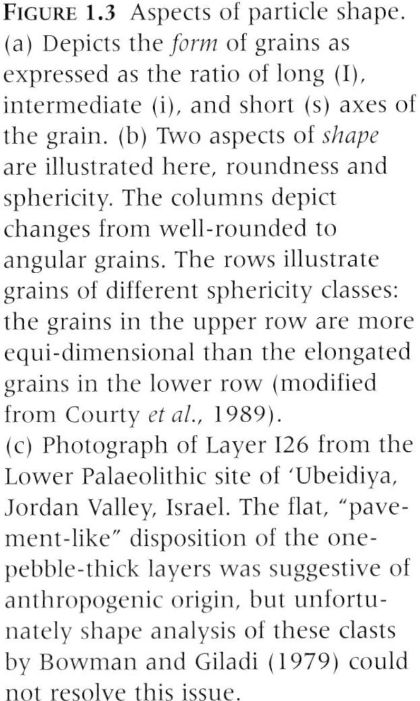

FIGURE 1.3 Aspects of particle shape. (a) Depicts the *form* of grains as expressed as the ratio of long (I), intermediate (i), and short (s) axes of the grain. (b) Two aspects of *shape* are illustrated here, roundness and sphericity. The columns depict changes from well-rounded to angular grains. The rows illustrate grains of different sphericity classes: the grains in the upper row are more equi-dimensional than the elongated grains in the lower row (modified from Courty *et al.*, 1989). (c) Photograph of Layer I26 from the Lower Palaeolithic site of 'Ubeidiya, Jordan Valley, Israel. The flat, "pavement-like" disposition of the one-pebble-thick layers was suggestive of anthropogenic origin, but unfortunately shape analysis of these clasts by Bowman and Giladi (1979) could not resolve this issue.

well-rounded chert pebbles is commonly indicative of several cycles of reworking and points to relatively greater age than fresh angular chunks. In any case, well-rounded pebbles point to fluvial transport, where rounding tends to take place relatively rapidly after a clast enters the fluvial system.

Shape analysis was applied to the study of pebbles and cobbles from Layer I26 at the Lower Palaeolithic site of 'Ubeidiya in the Jordan Valley (Fig. 1.3c) (Bar-Yosef and Tchernov, 1972; Bar-Yosef and Goren-Inbar, 1993). Here, it was not clear whether a pebble "pavement" represented the substrate of living floors or natural accumulations. Unfortunately, morphometric analysis (size, roundness, shape, and sphericity) and comparison to modern beach pebbles from the Sea of Galilee (Bowman and Giladi, 1979) revealed ambiguous results, and it was not possible to differentiate whether these gravels were of fluvial, beach, or human origin. The uniform thinness of the gravel layer and lack of imbrication, however, was suggestive of anthropogenic influence.

(c)

FIGURE 1.3 (*cont.*)

Studies of surface texture usually involve sand-size particles. Signs of fracture and abrasion produced during transport, as well as diagenetic changes expressed as etching and secondary precipitation can be observed under the optical (binocular and petrographic) and electronic (scanning electron – SEM) microscopes (see Chapter 16). Study of scores of quartz grains under the SEM grains from modern environments has revealed that different sedimentary environments impart different assemblages of surface signatures on the quartz grains (Krinsley and Doornkamp, 1973; Le Ribault, 1977; Smart and Tovey, 1981, 1982). Surface analysis of quartz sand grains from the Lower and Middle Palaeolithic site of et-Tabun Cave (Fig. 1.4), for example, was made by Bull and Goldberg (1985). They found that the basal deposits (Layers F and G of Garrod; Lower Palaeolithic – LP) show aeolian and diagenetic surface characteristics. The middle unit, Layer E (LP) is principally aeolian but with signs of marine alteration. Quartz grains in Layers D, C, and B (Middle Palaeolithic) in the upper part, show only aeolian modifications.

Fabric. Sedimentologists and pedologists use the term fabric in different ways. For sedimentologists (e.g. Boggs, 2001; Tucker, 1981), fabric relates to the orientation and packing of

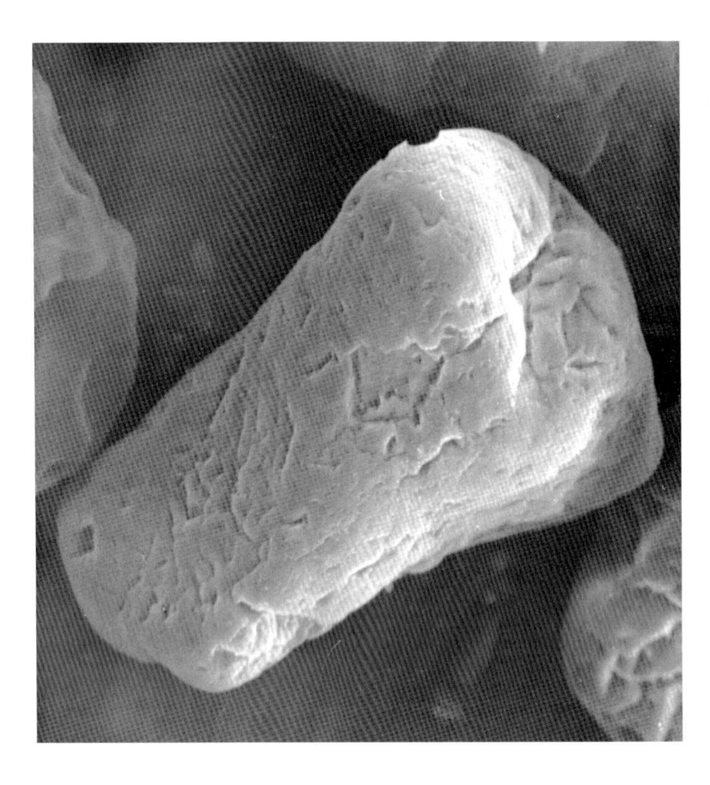

FIGURE 1.4 SEM photo of quartz grain from Layer E (Lower Palaeolithic) in Tabun Cave, Israel showing chemical etching of the grain (Bull and Goldberg, 1985). Magnification of grain is 340×. (Photo courtesy of Darwin Spearing.)

grains, which is commonly a function of flow direction. Elongated pebbles, for example, can be oriented in the same direction, either parallel or perpendicular to the flow, depending on the hydraulic characteristics in the environment of deposition. Sand-size particles will commonly be oriented parallel to the flow direction. Another sedimentological concept is packing, which describes the contacts between grains; it is associated with porosity and permeability. Sedimentologists differentiate grain-supported fabric, in which grains are in contact with adjacent ones, from matrix-supported fabric, where coarser clasts (e.g. sand, pebbles) are enclosed within a finer matrix (Fig. 1.5).

On the other hand, soil scientists – particularly micromorphologists – have a slightly broader and more nuanced view of fabric (Bullock *et al.*, 1985; Courty *et al.*, 1989; FitzPatrick, 1993). The most generalized and encompassing viewpoint is that used in the *Handbook for Soil Thin Section Description,* which we endorse: "Soil fabric deals with the total organisation of a soil, expressed by the spatial arrangements of the soil constituents (sold, liquid and gaseous), their shape, size and frequency, considered from a configurational, functional and genetic viewpoint" (Bullock *et al.*, 1985: 17). This approach to fabric encourages detailed observation of all the components of a deposit, enabling the deconstruction of a soil or sediment into its essential primary (depositional) and secondary (postdepositional) elements (see Chapter 16, lab techniques).

Bedding, Bedforms, and Sedimentary structures. Bedding is an important sedimentary feature, as it reveals information about the environment of deposition as well as postdepositional changes, such as bioturbation, that may erase original traces of bedding. Although in theory, an individual bed accumulated "under constant physical conditions," (Reineck and Singh, 1980: 95), it is commonly difficult or impossible to recognize such individual events and conditions. A single bed is distinguished from adjoining ones by surfaces generally called

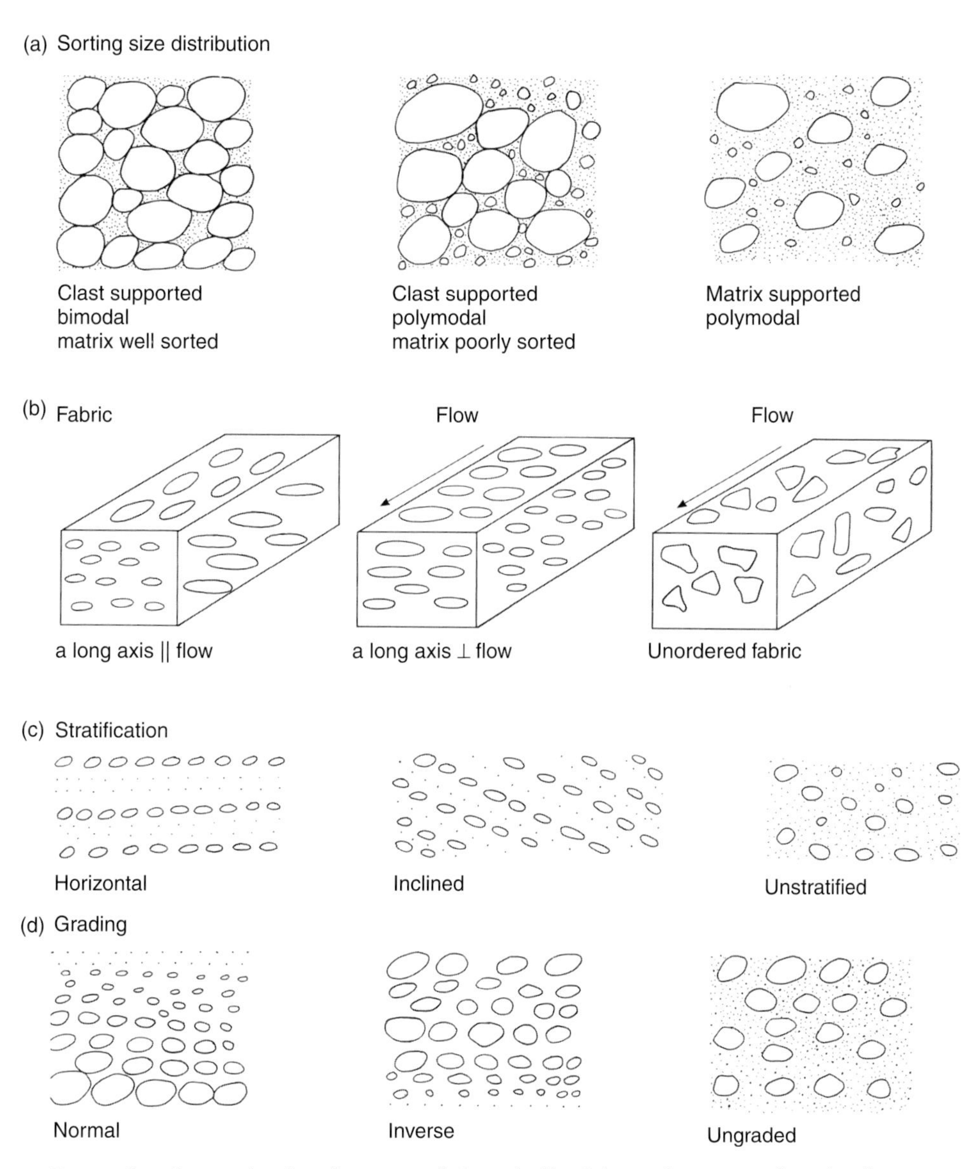

FIGURE 1.5 Textural and organizational aspects of clasts in fluvial conglomerate, showing features such as sorting and internal organization, including fabric, stratification, and grading. This visualization can be equally applied to other types of deposits (e.g. slope deposits) (modified from Graham, 1988, figure 2.1).

bedding planes, which delineate beds of different composition of texture, for example Fig. 1.6. Bedding can be described by a number of criteria, including thickness (Tables 1.4a,b) and shape (Fig. 1.7). A lamina is essentially a thin bed, and generally has uniform composition and smaller areal extent than a bed; it results from a minor fluctuation in flow conditions rather than representing constant physical condition. Individual lamina are bounded by laminar surfaces (Reineck and Singh, 1980).

Bedforms are surface morphological features that are produced by the interaction of flow

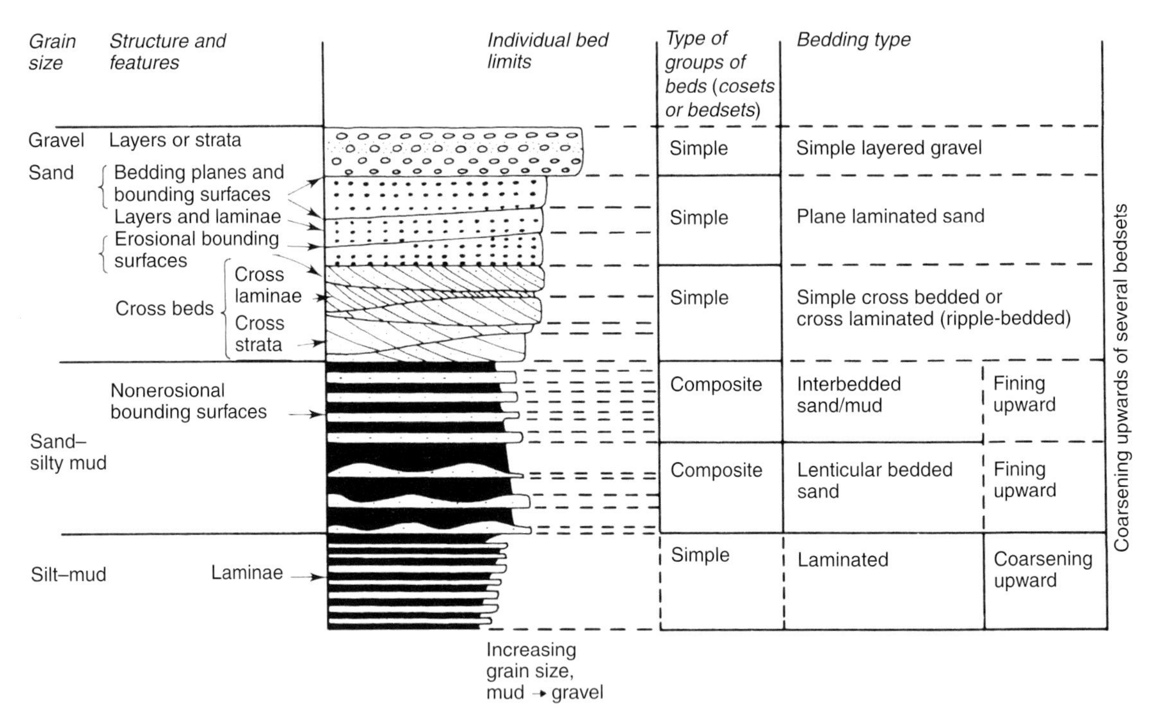

Figure 1.6 Grouping and subdivision of sedimentary beds according to grain size and sedimentary structure (modified from Collinson and Thompson, 1989, figure 2.2).

Table 1.4 (a) Nomenclature used to describe bedding types (adapted from Collinson and Thompson, 1989)

Name	Sedimentary unit	Comments
Bed	>1 cm thick	• Upper and lower surfaces called bedding planes • Subdivided into *"layers"* or *"strata"*
Cosets, bedsets	Groups of beds	*Cross-bedding/cross lamination* = Bedding inclined to depositional surface
Laminae	<1 cm thick	

(b) Nomenclature applied to describe thickness characteristics of beds and laminae (adapted from Boggs, 2001)

Beds		Laminae	
Thickness (mm)	**Name**	**Thickness (mm)**	**Name**
>1000	Very thick	>30	Very thick
300–1000	Thick	10–30	Thick
100–300	Medium	3–10	Medium
10–100	Thin	1–3	Thin
<10	Very thin	<1	Very thin

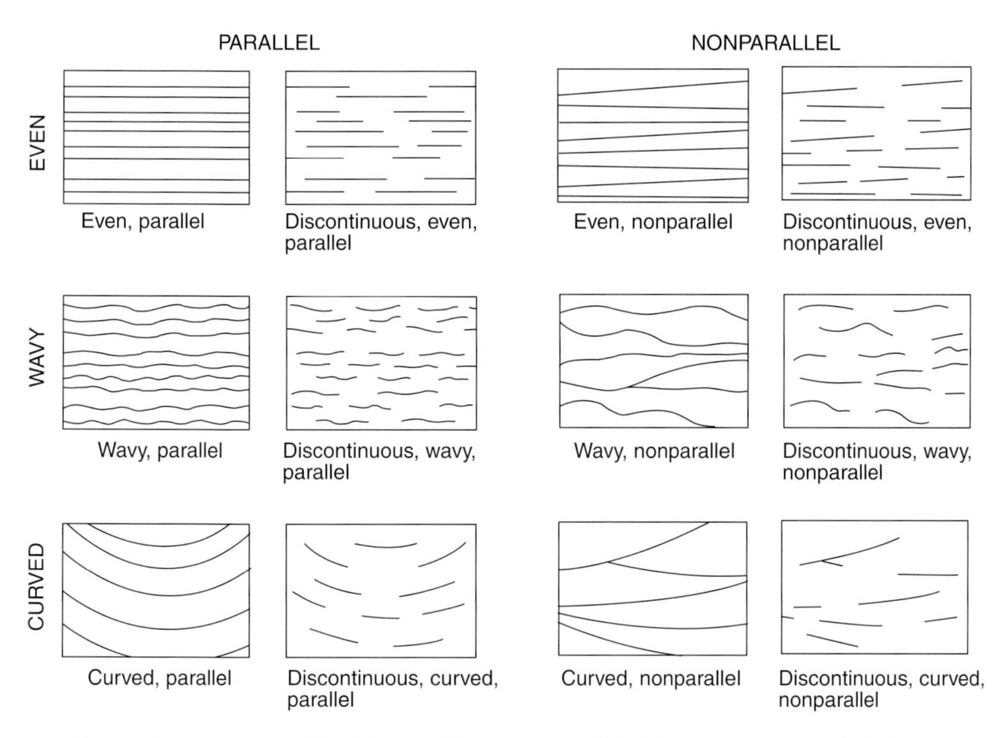

FIGURE 1.7 Shape characteristics of bedding and laminae (modified from Reineck and Singh, 1980, figure 152).

(water or air) and sediment on a bed (Nichols, 1999). Familiar examples are ripple marks as seen on beaches and in streams (see Chapter 5), and dunes (see Chapter 6). Bedding represents sedimentary structures associated with these surface bedforms, which are higher order arrangements or organizations of groups of particles within a sediment. As with grain size, both the bedforms and associated structures have descriptive value, and also furnish information about the environment of deposition, including direction, depth, and intensity of flow. Systematic treatments of sedimentary structures can be found in Boggs (2001); Collinson and Thompson (1989), Reineck and Singh (1980), and only a few will be mentioned here.

Current ripple marks, with flow from one direction, are small-scale bedforms, on the order of centimeters. In plan view, they can have different morphologies that include parallel or subparallel, and sinuous to lunate shaped (Fig. 1.8). In cross section, they may be symmetrical or asymmetrical, with crests that range from sharp to flattened (Reineck and Singh, 1980). Internally, stratification is expressed by various forms of cross-bedding that are divided into two basic types, tabular and trough cross-bedding as exposed in three-dimensional view (Fig. 1.8). Moreover, cross-bedding can be produced in a number of different ways in different types of sedimentary environments, such as channels, point bars in meandering streams (see Chapter 5), beaches (see Chapter 7), and sand dunes (see Chapter 6).

Cross-bedding in fine laminae is exemplified by ripple cross lamination. In this case, if the

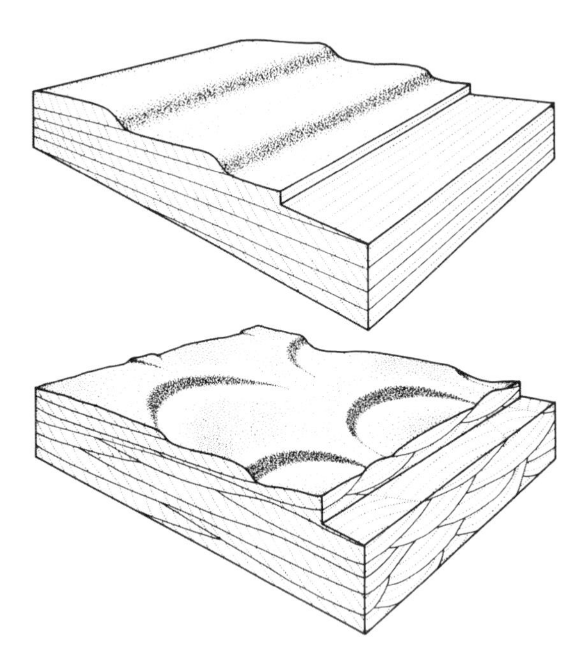

FIGURE 1.8 Ripple marks. Block diagrams showing two types of surface ripple bedforms. The underlying sediments are organized into cross beds, of which two types are illustrated here. The upper block illustrates tabular cross-bedding, which is produced by the migration of ripples with straight crests; in the lower block, troughs are formed by the migration of sinuous ripples (Tucker, 1981) (modified from Tucker, 1981, figure 2.21).

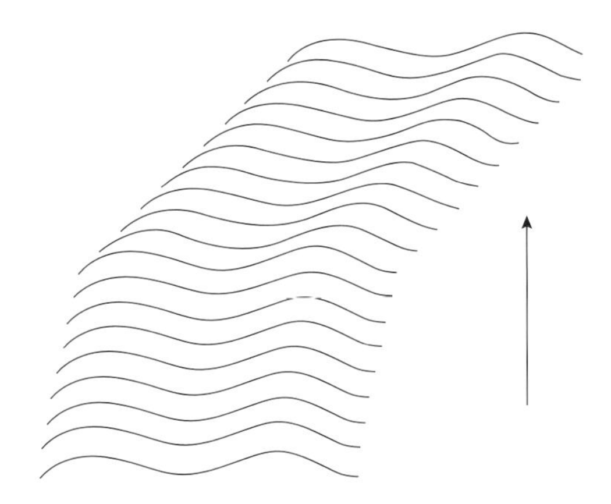

FIGURE 1.9 Diagram illustrating climbing ripple cross lamination. The arrow point to the change from in-phase forms at the bottom to climbing forms at the top, which represents decreasing rates of sediment depositon, greater migration, and a reduction of water depth. These features can be produced in the waning stages of deposition on a floodplain.

rate of sand addition is larger than the rate of migration of the ripple, elevation of the ripples rises. This process results in the formation of climbing ripples, which point to rapid sedimentation and high sediment input (Nichols, 1999) (Fig. 1.9). Cross-bedding occurs at a variety of scales that range from millimeters to meters. In small-scale forms, cross-beds range from millimeter up to ca 4 cm thick and are usually trough shaped. They are produced by migrating current and wave ripples (Reineck and Singh, 1980). Larger-scale cross-bedding has bedding units that range from >4 mm up to 1 to 2 m in thickness, and can be both planar and trough shaped. They can be found in different environments (e.g. dunes, longshore bars).

Other structures found within sediments and soils can be of nondepositional origin and produced by physical, chemical, or biological agents, some partially synchronous (penecontemporaneous) with the original deposition. They include features, such as those produced by freeze-thaw (e.g. ice cracking and frost wedging (Fig. 1.10a)), and desiccation, bioturbation (e.g. burrowing by rodents and soil insect fauna – Fig. 1.10b), and deformation, convolution structures, and load structures (Figs 1.10c,d).

1.2.1.2 Chemical sediments (or nonclastic sediments)

Chemical sediments are those precipitated from solution, and in certain geoarchaeological contexts (e.g. caves), they can form a significant and critical part of the deposits and geoarchaeological story.

In the open-air context, where rates of evapotranspiration are high, chemical sediments can be related to regional-scale features, such as old lake basins and sediments (e.g. Great Salt

Figure 1.10 Examples of sedimentary structures. (a) Ice wedge formed in bedded and iron stained silts from Normandy, northern France; (b) burrows produced from small rodents at the site of Hayonim Cave, Israel. In addition, the finely bedded and laminated nature of these silty and clay deposits points to deposition in pools of standing water near the entrance; (c) convolution structures in the lighter area of Layers 3/4b/4u at Boxgrove, United Kingdom in which calcareous pond deposits were deformed while they were still wet and undergoing dewatering; (d) soft sediment deformation of finely laminated lake sediments from the Pleistocene Lisan Lake, Dead Sea Region, Jordan Valley; these deformation features are related to earthquakes within the rift valley (Enzel *et al.*, 2000).

Lake and other lakes in the Great Basin, United States; Dead Sea). Many lacustrine sediments in these arid areas commonly tend to be rich in carbonates [calcite – $CaCO_3$, magnesite – $MgCO_3$, and dolomite – $CaMg(CO_3)_2$], as well as chlorides (halite – NaCl), and sulfates (gypsum – $CaSO_4 \cdot 2H_2O$; anhydrite – $CaSO_4$). The Dead Sea Scrolls, for example, were found in caves developed within Late Pleistocene lacustrine calcareous and gypsiferous marls (Bartov *et al.*, 2002; Begin *et al.*, 2002) deposited in Lake Lisan, the precursor to the modern Dead Sea (Fig. 1.10d). Similar types of deposits in archaeological contexts are not uncommon, and are known from the American High Plains (e.g. Holliday and Johnson, 1989) and Australian Lake Mungo (Bowler *et al.*, 2003) among others.

Many of these same secondarily precipitated salts and minerals also occur in the soil environment, either on the surface or at shallow depths. Calcite, gypsum, and halite are all common minerals in arid and semiarid environments, and it is not uncommon to find ceramic sherds with a thin carbonate crust on their undersides. The formation of iron and manganese oxides is also widespread in waterlogged situations, such as Boxgrove, United Kingdom (Collcutt, 1999; Macphail, 1999). Widespread carbonate accumulations occur in association with open-air spring deposits, such as the Transvaal Caves (Brain, 1981; Butzer, 1976b). The site of Bilzingsleben in Germany is noted for its detailed Middle Pleistocene sequence, including its collection of *Homo erectus* remains. The site formed as a combined result of lacustrine carbonate deposition and extensive travertine formation, which has sealed much of the site (Mania, 1995).

Caves are the focus of an astonishing variety of chemical sediments that comprise most of the major mineral groups, including oxides, sulfides, nitrates, halides, carbonates, phosphates, and silicates (Ford and Williams, 1989; Ford and Cullingford, 1976; Gillieson, 1996; Hill and Forti, 1997) (see Chapter 8). Since the majority of prehistoric and archaeological caves are in karstic terrains, the most common minerals are the carbonates. These tend to occur in a variety of well-known geometric forms, including stalactites, stalagmites, and massive bedded accumulations (flowstones).

Other minerals, such as phosphates, can be quite frequent in many cave sites, particularly those from Mediterranean climates but less so from temperate areas. These phosphates are derived from bat and bird guano, large vertebrate excreta (e.g. hyaena), bones, vegetal materials, and ashes. The minerals range from apatite (calcium phosphate and related minerals) to more complex aluminum, iron, and calcium phosphates (e.g. crandallite, montgomeryite, leucophosphite) produced by the reaction of siliceous clay with guano-derived phosphate (Hill and Forti, 1997; Karkanas *et al.*, 1999, 2000; Macphail and Goldberg, 2000; Weiner *et al.*, 1993).

1.2.2 Organic matter and sediments

Organic matter and sediments are generally localized in subaerial or subaqueous environments. In subaerial environments, such as A horizons in soils, organic matter occurs in various forms, depending on the degree of breakdown. As a result of decay, organic matter decomposes into humus, a product of plant decomposition aided by microbial activity. It is "composed of large molecules of heterogeneous polymers formed by the interaction of polyphenols, amino acids, polysaccharides and other substances" (FitzPatrick, 1986: 110–111).

In more poorly drained situations, the presence of water and anaerobic conditions hamper the breakdown of organic matter by bacteria and oxidation, and thick accumulations can occur in the form of peats. Because of the associated acidity and anaerobic conditions, bones are less well preserved than plant remains (including pollen) or soft animal tissues. Some spectacular remains have been recovered from peat bogs including the Lindow Man in England (Stead *et al.*, 1986), Tollund Man in Denmark (Glob, 1970), and intact brain tissue from the Windover site in Florida

(Doran and Dickel, 1988). Although these are Holocene examples, we note the remarkably well-preserved, worked wooden spears from the Lower Palaeolithic site of Schöningen in Germany, which are about 400,000 years old (Thieme, 1997).

1.2.3 Archaeological sediments

One of the chief concerns of this volume is the consideration of archaeological sediments. These deposits – a main subject area in geoarchaeological research – are little dealt with in geoarchaological textbooks apart from Courty *et al.* (1989). For this reason and because any sediment that may have a cultural connection can be termed an archaeological sediment, we devote two chapters to the subject (Chapters 10 and 11).

1.3 Conclusions: sediments versus soils

In closing, we wish to reiterate the differences between sediments and soils, discussed in Chapter 3 (see also Mandel and Bettis, 2001). A good part of sediments in archaeological contexts are clastic in nature, with the remainder being composed of organic-rich deposits, and in special environments (particularly caves), chemical sediments. As such, these clastic sediments typically embody the concepts of source, transport, and ultimately deposition. As discussed in chapter 3, soils form under a different set of conditions, which are mostly ones that take place in a single, essentially stable location. Whereas detrital components, such as silt and clay may be translocated or moved within the soil profile, and salts may be leached or precipitated in these same horizons, the dynamic is one of vertical movements or displacements generally on the order of centimeters or decimeters. Thus, one should be cautious in the casual use of phrases such as "soil samples," "archaeological soils," or "compacted soil" when actually the term "sediments" should be applied. This point might appear to be overly picky, but as discussed in many places in the book the dynamics of formation of sediments are very different from those of soils, and so are the interpretations.

2

Stratigraphy

2.1 Introduction

Geoarchaeology, like geology and archaeology, is a field-based endeavor that relies on empirical data. No matter how much we might like to theorize in either discipline, the bottom line is that we are constrained by the observations we make first in the field and later in the laboratory. This notion, along with the reality that geology supplies the ultimate context for archaeological sites and their contents, underscores the need for people working in the field to have at least a basic understanding – even if a rudimentary one – of the materials and landscapes that articulate with the archaeological record and sites being investigated.

These materials commonly represent a complex, interwoven concoction of geological and human accumulations that normally have been modified after their deposition by natural soil-forming processes (pedogenesis), or by human activities, such as cooking, trampling, digging, or by building activities (Fig. 2.1). In order to carry out accurate geoarchaeological and archaeological research, we must be cognizant of such processes. Moreover, we must be able to record them in a standardized and systematic way, which can portray the information objectively and enable others to interpret them in their own way.

The purpose of this chapter is to set out some of the ground rules and guidelines for observing and recording field data related to soils and sediments, both of which have complex and interrelated origins in archaeological settings. We begin with a general discussion of stratigraphy and how it is structured to examine temporal and spatial relationships among sediments and soils. Most of the previously published geoarchaeological books and articles have emphasized geogenic deposits and their context to the relative exclusion of anthropogenic accumulations and modifications. An extraordinary amount of hitherto untapped information – often of quality similar to or higher than traditional data derived from artifacts or faunal remains – resides in these deposits, and the reader should be aware of their nature and usefulness.

2.2 Stratigraphy and stratigraphic principles

Stratigraphy has an intuitive meaning to many people, even if specific definitions vary among authors and disciplines. In the more classical geological literature, stratigraphy tends to be viewed at the regional scale: "The crux of much of stratigraphy is the spatial relationships of

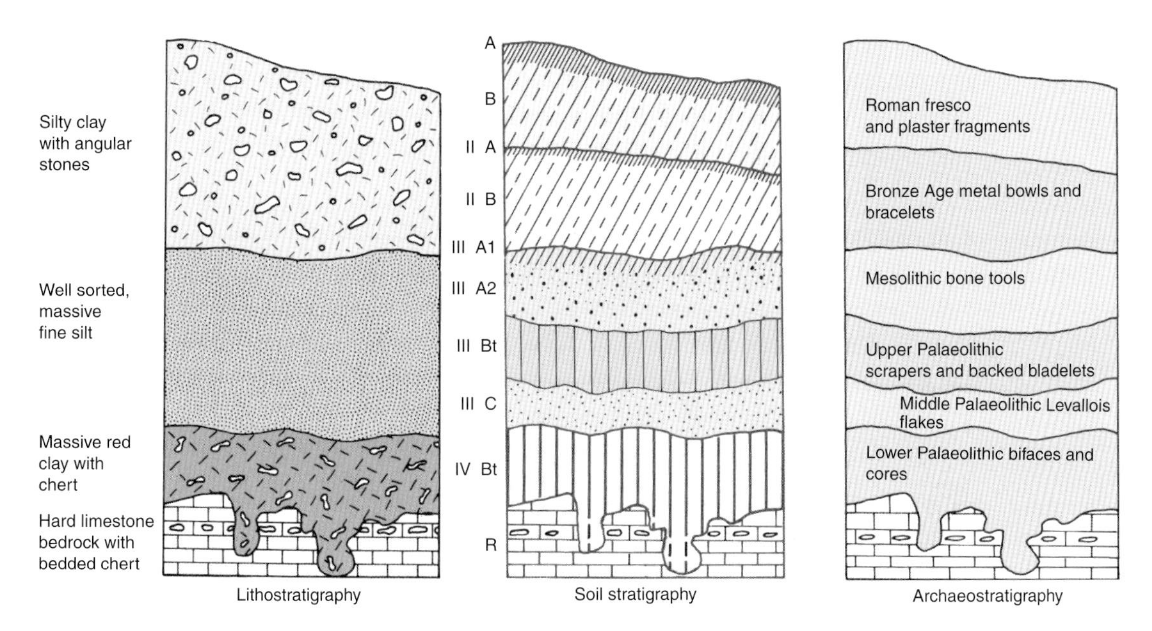

FIGURE 2.1 Hypothetical triptych illustrating the complexities involved in archaeological deposits. Shown here is the same profile as viewed from the eyes of the geologist (lithostratigrapher), pedologist (soil stratigrapher), and archaeologist (archaeostratigrapher). In the first panel, lithostratigraphy is presented on the basis of the characteristics of the deposits, such as composition, texture, and bedding. The second panel illustrates division of the profile on the basis of soil forming events as expressed by different soil horizons (see Chapter 3); these features represent post-depositional modifications of the original primary deposits that took place on stable or semi-stable surfaces. The third panel depicts the profile from the standpoint of the archaeological materials it contains, such as artifacts or features and architecture. Thus the stratigraphy from an archaeological site can represent the combined effects of depositional and post-depositional processes, coupled with human activities, such as flint knapping, pottery manufacture, and discard, as well as construction activities associated with buildings. The trick is to isolate these factors. The first stage to accomplish this is the detailed recording of the stratigraphy (modified from Courty *et al.*, 1989, figure 3.3).

rocks over geographic areas" (Schoch, 1989). For F. J. Pettijohn, a renowned sedimentologist, it also encompasses physical characteristics of sediments that include notions of space and time:

> *stratigraphy in the broadest sense is the science dealing with strata and could be construed to cover all aspects – including textures, structures, and composition. . . . Stratigraphers are mainly concerned with the stratigraphic order and the construction of the geologic column. Hence the central problems of stratigraphy are temporal and involve the local succession of beds (order of superposition), the correlation of local sections, and the formulation of a column of worldwide validity. Although these are the objectives of stratigraphy, the measurement of thickness and description of gross lithology are commonly considered a part of the stratigrapher's task.* (Pettijohn, 1975: 1)

Archaeologists are used to viewing stratigraphy, deposits, features, and artifacts at the scale of the site and perhaps its immediate surroundings (Stein and Linse, 1993). For them, stratigraphy can be simply, "the natural and cultural layering of the soil at a site" (Feder, 1997) (intuitively correct despite the incorrect use of "soil"). D.H. Thomas comes closer to a more holistic definition of stratigraphy: "An analytical interpretation of the structure produced by the deposition of geological and/or cultural sediments into layers, or strata" (Thomas, 1998: 205). He rightly includes the interpretative aspects, since stratigraphy does not exist by itself. For better or worse, it is something that is recognized somewhat subjectively, and different persons – depending on experience and background (archaeologist, geologist, geographer, pedologist, and other environmentalist) might construct different stratigraphic sequences for the same physical stratigraphic profile.

In the geoarchaeological sector, Waters takes a traditional geological, "depositional" viewpoint, emphasizing deposits and soils that are observed principally on a geomorphologic scale: "Stratigraphy is the study of the spatial and temporal relationships between sediments and soils. Stratigraphic sequences are created because depositional environments are dynamic and constantly changing" (Waters, 1992: 60). On this regional scale, stratigraphic study is useful to organize sediments and soils into objectively identifiable packages that can be arranged in some kind of chronological order and absolute age on the basis of temporal markers, such as soil formation, or perhaps erosion. "Parcels" of soils/sediments can be physically (spatially) or temporally linked and integrated over an area, either on the scale of meters or kilometers (correlation; see below).

In this book, we take a scale-neutral, and broad approach to stratigraphy and characterize it as the three-dimensional organization in space and time of geological layers, soils, archaeological features, and artifacts. In a sense, it embodies the internal fabric of these entities, whereby stratigraphy and the stratigraphic record represent encoded pages of geoarchaeological history, including the gaps (blank pages) where nothing apparently took place. By being able to view these pages within a structure – the stratigraphic framework – we are better able to decode these pages and reconstruct the depositional, pedological, and geomorphological history of the site and its surroundings, including the activities of the people who acted within and around the site. It also enables us to evaluate the relative timings of human and geological events, which can be extended to, and correlated with, other sites and regions.

During the latter part of the nineteenth century, classical geologists realized that it was necessary to separate the physical characteristics of a sedimentary rock (i.e. color, texture, and composition) from the time in which it formed. This came about because they realized from their fossil content that a geological unit can accumulate over long period of time and be older in one place than in another (diachronous or "time-transgressive" units; Krumbein and Sloss, 1963; Vita-Finzi, 1973). Deltaic deposits, for example, become progressively younger seaward, even though the overall "deltaic lithologies" do not change radically and horizontally within the area of the delta. To overcome these issues of time and space, stratigraphey recognized a number of different types of stratigraphers units, which at present are embodied in the *International Stratigraphic Guide* (Hedberg, 1976; Schoch, 1989) and the *North American Stratigraphic Code* (NASCN, 1983).

Table 2.1 summarizes the most common types of stratigraphic units. Note that the two major groups of units are delineated on the basis of content or physical limits (Group I), as opposed to those units in Group II, which are defined solely on the basis of time. Thus, the units in Group I are derived, for example, on the basis of their mineral composition or texture, biological/fossil constituents, soil characteristics, or stratigraphic boundaries or gaps marked by lapses in time (unconformities).

Lithostratigraphic units are the most basic, ubiquitous, and relevant to the majority of geoarchaeological situations. They are denoted on the basis of lithological characteristics, such as color, texture, composition, thickness, upper, and lower boundaries. They do not imply any notions of time, just the descriptive aspects of the sedimentary bodies. In regional-scale geological contexts, the primary lithostratigraphic unit is the formation, which is one that can be

TABLE 2.1 Types of stratigraphic units (modified from NASCN, 1983; table 1 and Fritz and Moore, 1988; table 1.3)

Stratigraphic unit	Example of unit name	Real-life geological example
I. Units based on content or physical characteristics		
Lithostratigraphic	• Regional scale: Formation • Site scale: Bed	• Regional scale: "Ubeidiya Formation, Israel (interbedded layers of clay, marl, and gravel with intercalated Lower Palaeolithic site) (Bar-Yosef and Tchernov, 1972) • Site scale: band of yellow silt over layer of clay (Stratigraphy Box Fig. 2.4)
Biostratigraphic	Biozone	Younger dryas (pollen zone), northwest Europe (Faegri and Iverson, 1989)
Magnetopolarity	Polarity zone	Reverse polarity (e.g. Unit B2 at Dmanisi, Georgian Republic) (Gabunia *et al.*, 2000)
Pedostratigraphic	Geosol	Sangamon Geosol (United States); Barham soil (United Kingdom)
Allostratigraphic	Alloformation	Unit Q2 (Strat Box Fig. 2.3)
II. Units based on age time units		
Geochronologic	Period	Quaternary Period, ca 1.8 to 0 my
Polarity chronologic	Polarity Chron	Brunhes Chron of "normal" polarity (see Table 2.2)

mapped over a region. It can be represented by a 20-m thick accumulation of alternating beds of clay and silt or by a 2-m thick massive layer of limestone. Two or more formations that form a consistently uniform lithological package can be combined into a group. On the other hand, formations can be subdivided into smaller units, such as a member and even finer, bed. The latter is the smallest unit of the lithostratigraphic units, and is lithologically distinct from units above and below it.

On an archaeological scale, which can be on the order of tens (e.g. building) to thousands of square meters (e.g. tell or other types of mound), a lithostratigraphic unit can take the form of a centimeter-thick band of red clay underlying a plastered floor from a Bronze Age house. In caves, a layer of rock fall mixed with clay would also be a common type of lithostratigraphic unit (see Section 2.3 for a discussion of facies).

Biostratigraphic units are delineated on the basis of the fossils that they contain, including their appearance, disappearance, or relative abundance. The fossils can be either plants or animals, marine or terrestrial. For example, the Tertiary Period was subdivided by Lyell on the basis of the abundance of fossil molluscan species that are living today. Thus, the Eocene, for example, has only 3.5% of living forms whereas the Pleistocene has about 90% of modern species (Farrand, 1990).

Magnetic polarity units represent certain remnant magnetic properties of a sediment (e.g. dipole–field pole position, intensity of the magnetic field, etc.) that differ from those above and below it (NASCN, 1983). The most often cited example of magnetic polarity units is based on reversals in the earth's magnetic field, whereby rocks or sediments could possess either "normal" or "reversed" polarity. At the important site of Dmanisi, Georgian Republic, remains of hominids, animals, and lithics were found in the reversely magnetized Unit B (Gabunia *et al.*, 2000).

Related to these are polarity chronologic units, which refer to the specific *time* interval over which a state of magnetic polarity exists. Both basalts and water-deposited sediments are typically measured, as they possess magnetic polarity properties of the earth's magnetic field at the moment that the basalt crystallized or the sediment accumulated (NASCN, 1983) (Table 2.2). Major polarity periods (Chrons) exist to the order of millions of years, whereas shorter-term polarity events take place to the order of 10^4–10^5 years (Bradley, 1999). For example, the Matuyama Reversed Chron corresponds to a period between 2.581 and 0.78 million years (my) ago, and deposits at this time display polarity opposite to the present day. Subchrons represent shorter intervals or "events" within longer magnetic polarity periods. Thus, at the Lower Palaeolithic site of Olduvai Gorge, Tanzania, basalts dated from about 1.95 to 1.77 my (the Olduvai subchron) display normal polarity within the longer Matuyama Reversed Chron (Table 2.2).

Pedostratigraphic units represent whole or part of a buried soil, which exhibits one or more soil horizons (see Chapter 3) that are preserved in a rock or sediment, and are traceable over an extended area (Schoch, 1989; NASCN, 1983). These types of units, the foremost of which is the geosol, represent relatively short periods of geological time. Hence, they serve as temporal pegs or "marker horizons" that temporally place the relative ages of deposits and events that overlie and underlie the soil (Holliday, 1990). Geosols also have palaeoenvironmental significance (Fedoroff and Goldberg, 1982). The interglacial Sangamon soil in the midwest United States has been extensively used as a stratigraphic tool (Birkeland, 1999), and similar such soils have been documented in the United Kingdom by Kemp (1986, 1999). The Barham Soil in eastern England, for example, is a clear stratigraphic marker horizon that denotes a major landscape change of marked climatic degradation (Rose *et al.*, 1985).

Table 2.2 Palaeomagnetic polarity timescale for the last 3.5 my (modified from Bradley, 1999).

Age (my)	Chrons (Epochs)	Subchrons (Events)	Age range (my)
0–0.78	Brunhes		
		Laschamp excursion	ca *0.041*
0.78	*Matuyama*		
		Jaramillo	0.99–1.07
		Cobb Mt.	1.201–1.211
		Olduvai	1.77–1.95
		Réunion	2.14–2.15
2.581			
	Gauss		
		Kaena	*3.04–3.11*
		Mammoth	*3.22–3.33*
3.58			

Normal polarity is represented in roman and reversed polarity in italic

Allostratigraphic units are rock or sediment bodies that are overlain and underlain by temporal discontinuities (unconformities). A prevalent case in point are stream terrace deposits produced by successive episodes of alluvial deposition and erosion. Allostratigraphic units are a convenient means to map widespread fluvial deposits, for example, and constitute a basic component in documenting the Holocene fluvial geoarchaeological history of much of the mid-continental United States (Brown, 1997; Ferring, 2001; Goldberg, 1986; Holliday, 1990; Mandel, 1995). Such fluvial deposits commonly contain archaeological sites and can be traced over several kilometers within the same allostratigraphic unit (see Box 5.1 on fluvial and Box 2.1 on facies and stratigraphy).

Geochronologic units are those that are defined solely on the basis of the time interval that they encompass. As such, they do not represent actual specific rocks but are more conceptual, representing divisions of time. These units are differentiated on the basis of radiometric dating, such as potassium/argon ($^{40}K/^{40}Ar$) or

Box 2.1 Facies and stratigraphy: The Paleoindian-Archaic site of Wilson–Leonard, Texas

Facies and microfacies are important facets of geoarchaeology. Understanding lateral and vertical changes in the lithological aspects of deposits over a landscape and within a site are vital for comprehending fully the deposits and stratigraphy, and integrating them into the reconstruction of site settlement and sediment history. The Paleoindian-Archaic site of Wilson–Leonard in central Texas illustrates some of these concepts.

The Wilson–Leonard site contains one of the most complete records of occupations in the Southern Plains of the United States (Bousman *et al.*, 2002; Collins *et al.*, 1993), spanning Paleoindian through Late Prehistoric times, roughly from about 11,400 radiocarbon years ago up to about 4,000 years ago. The site is located about 33 km north–northwest of Austin, Texas within the valley of Brushy Creek (Fig. 2.2), a tributary of the Brazos River (Q-2 in Fig. 2.3; Collins and Mear, 1998). This valley displays several Quaternary fluvial terraces, which, following the custom in North America, are labeled Q1, Q2, Q3, etc., from youngest to the oldest. Excavations in the 1980s and early 1990s, in the Q2 terrace fill, revealed a sequence of deposits ranging from gravels at the base to organic-rich silts at the top that are marked by the presence of numerous remains of

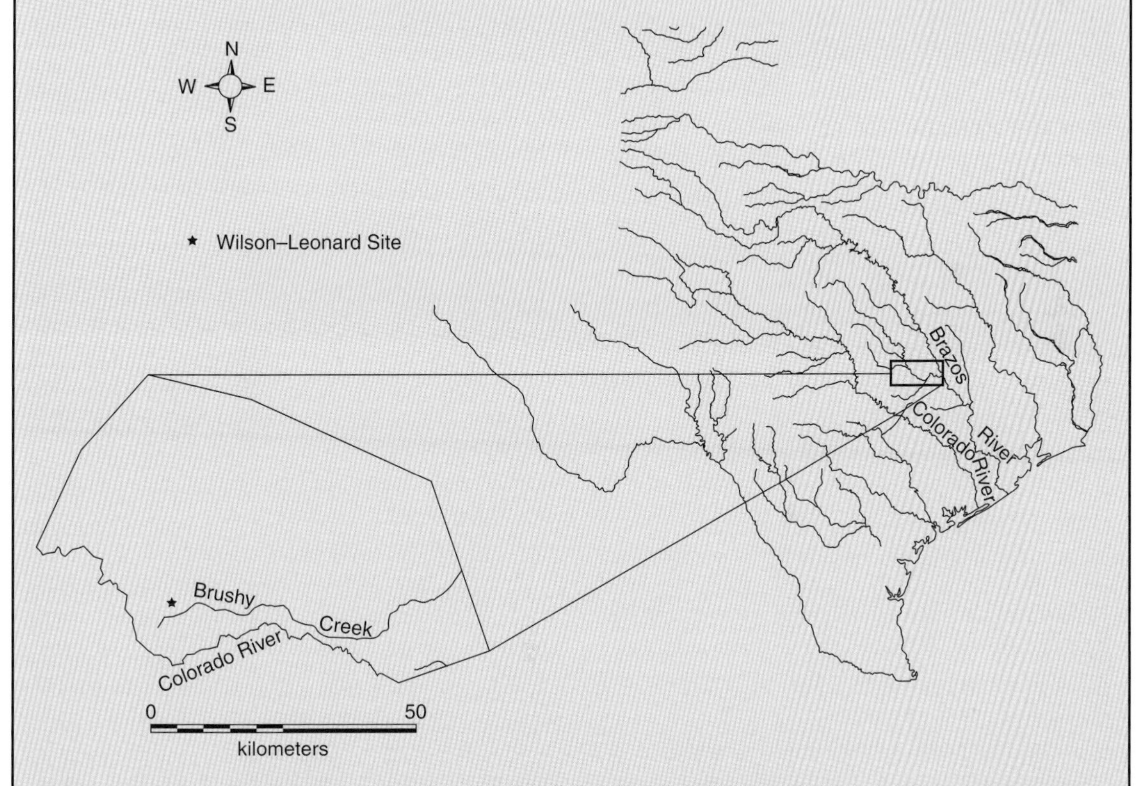

Figure 2.2 Location of Wilson–Leonard site, Central Texas marked with * (from Collins and Mear, 1998).

(cont.)

Box 2.1 *(cont.)*

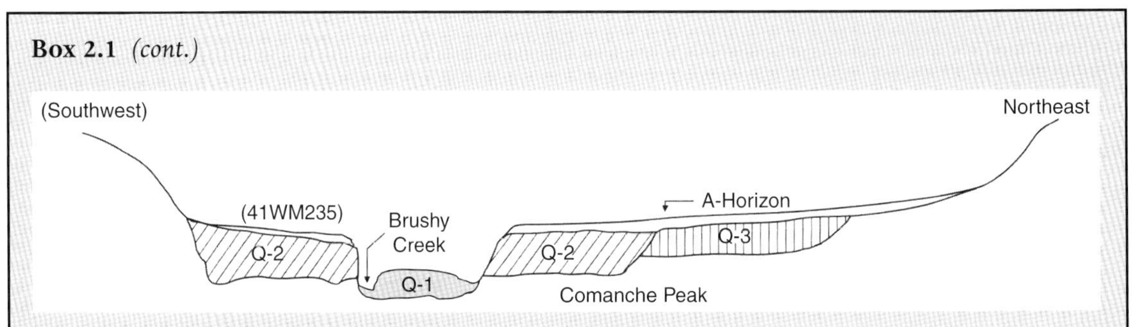

FIGURE 2.3 Schematic cross-section through Brushy Creek, showing the position of the Wilson–Leonard site (41WM 235) in relation to the Q-2 fluvial terrace, which can be seen on both sides of the valley (from Collins and Mear, 1998). Q-3 is an older Pleistocene terrace, whereas Q-1 is modern gravelly alluvium.

burnt rock ovens. The accompanying schematic cross-section and photograph of the Western profile of the site (Fig. 2.4) illustrate a number of facies that are characteristic of a typical Late Quaternary fluvial setting in this part of Texas (Goldberg and Holliday, 1998), where rivers and creeks drain the Edwards Plateau and flow into the Gulf of Mexico. The following discussion describes some the different types of lithostratigraphic units at the site; these are keyed to a depiction of the historical evolution of the site and its local landscape (Fig. 2.5).

Occurring at the base of the section is Unit Ig, which consists of medium to coarse, well-rounded gravel that laterally interfingers with Unit Isi. The gravels are interpreted to be channel deposits that accumulated when Brushy Creek was flowing at the location of the site (Fig. 2.5, #1). The silty sediments of Unit Isi that overlie and interfinger with the gravels of Unit Ig are overbank deposits that are partly contemporary with Unit Ig as shown by their interfingering. Unit Icl is an organic silty clayey lens that overlies Ig but also interfingers with Unit Isi. Unit Id is a thin, discontinuous lens of silt that is draped on the underlying deposits of Isi. Unit II consists of fine gravelly silt that overlies Isi and Id with a sharp, erosional contact, thus representing a different facies and depositional regime. It, in turn, is overlain by thick, darker brown, locally, organic-rich accumulations of gravelly silts in the upper part of the profile. Intercalated within these upper deposits of Unit III are numerous burned rock features with fire-cracked rock that represent clear anthropogenic features. This uppermost Unit III was further divided in to subunits on the basis of color and texture, and which reflect the relative contributions and rates of sedimentation from different sources. The darker, organic matter-rich components are derived from fire-related anthropogenic activities, whereas the lighter-colored silts indicate relatively more geogenically derived calcareous alluvium and colluvium, the latter originating from the slopes behind the site. Thus, the darker color of deposits from Subunits IIIa and IIIc result from high inputs of organic matter, fuel, and plant residues associated with numerous burned rock oven features (Goldberg and Guy, 1996; the base of Unit III in Fig. 3b).

These descriptive lithostratigraphic units and facies relate to specific geogenic and anthropogenic processes acting at the Wilson–Leonard site. They include calcareous fluvial deposits associated with Brushy Creek (e.g. Units Ig and Isi) and stony colluvial material derived from the slopes behind the site to the west. By being able to recognize these different facies and their lateral and vertical

(cont.)

Box 2.1 *(cont.)*

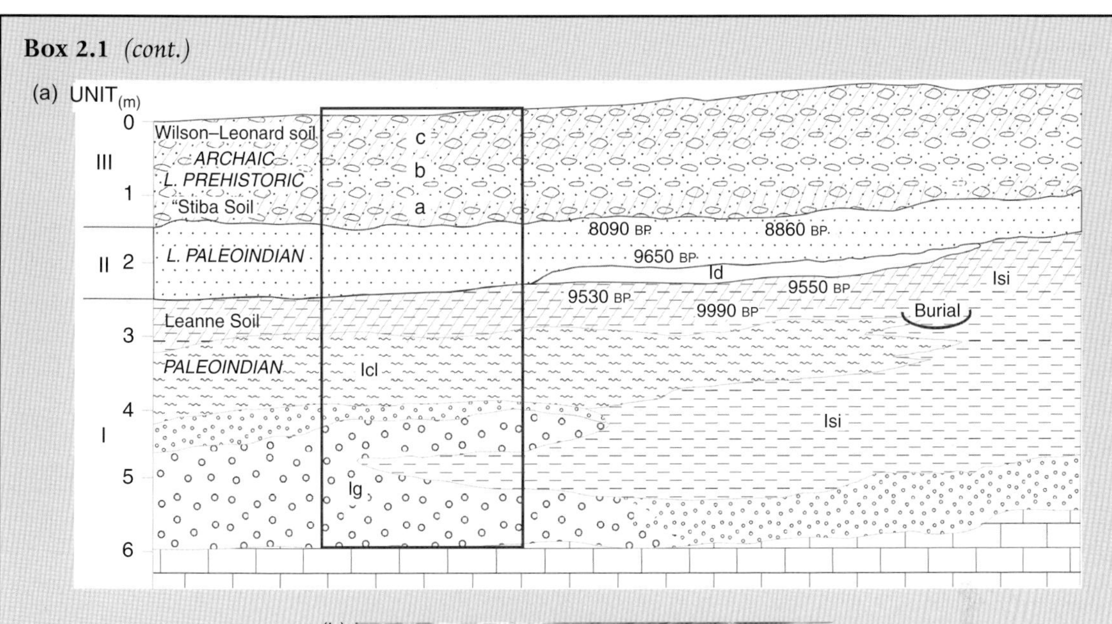

Figure 2.4 (a) Schematic view of major lithostratigraphic units and their lateral and vertical variations as shown by the interfingering of some of the units. Unit Icl, for example (a clayey cienega – depression deposit) grades laterally into massive alluvial silt of Unit Isi. The profile is approximately 8 m wide. (b) Photograph of the area outlined in (a) with some of the labeled stratigraphic units (Geologist T. Stafford can be seen taking notes).

(cont.)

Box 2.1 *(cont.)*

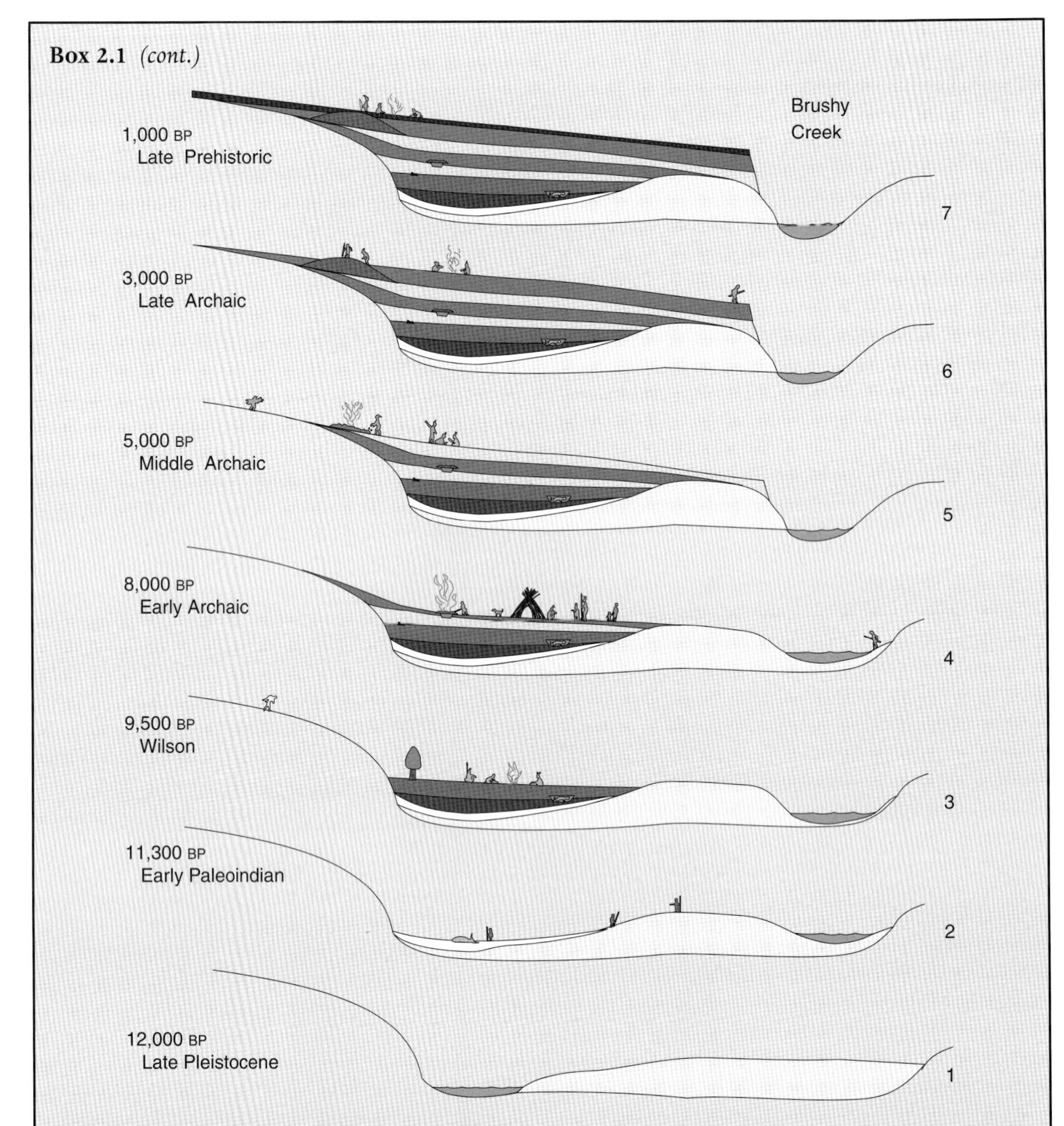

FIGURE 2.5 Geological evolution of Wilson–Leonard site (modified from Bousman and Goldberg, 2001).

changes, it is possible to piece together the evolution of the geological environment of the site during the Late Pleistocene–Holocene (Fig. 2.5; numbers on the right side of the figure refer to individual stages of depositional history). Thus, Unit Ig represents a gravel bar that formed within the channel of Brushy Creek over 11,000 years ago (#1). When this channel was suddenly abandoned by avulsion, and the stream began to flow within a new channel (right side of figure), it left behind a closed, elongated depression (*cienega*) that was

(cont.)

Box 2.1 *(cont.)*

successively filled in with organic-rich clays (Unit Icl) and overbank silts (Unit Isi; #2, 3) that accumulated during flooding events. As this mass of sediment continued to accumulate, or aggrade, the influence of stony colluviation began to be more pronounced with the accumulation of Unit II (#4–7); fluvial material originating from Brushy Creek concomitantly became less important, and likely covered the area of occupation only during times of major floods. During Early Archaic times (#4), the intensity and area of occupation increased, along with the use of numerous burned rock ovens. These factors led to the addition of anthropogenically derived organic matter-rich fine detritus mixed with burned stones and colluvially derived stony silts (Units IIIa and IIIb).

This example from the Wilson–Leonard site illustrates that lateral facies shifts in lithologies from gravels to silts can take place even within the boundaries of an excavated area of only 8 × 8 m; in fact, the geoarchaeological picture of the site was initially most confusing during the reexcavation of the site during the 1990s. Thus, understanding these lateral and vertical facies changes is crucial to understanding the overall stratigraphy at the site, the geological context from which the artifacts come, the local environment in which the former inhabitants lived, and how all these changed through time.

uranium-series techniques; they can contain mixtures of different types of fossils and lithologies. The Pleistocene, which includes much of the geoarchaeological record, represents the last 1.8 my.

The system of stratigraphic units described above, evolved within the geological community over many decades as a practical means to facilitate the description and interpretation of the rock record. More recently, the archaeological community has expressed concerns about the lack of attention being paid to archaeological sediments and stratigraphy, and attempts have been made to develop a system of nomenclature geared to the recording of archaeostratigraphy (Gasche and Tunca, 1983; Stein, 1990). The first two authors formally proposed, for example, the term "layer" for the fundamental lithological unit, and "phase" for chronostratigraphic unit.

Both these and similar concepts appear to have little sensible merit, as the foundations for the objective description of soils and sediments (whether of geological or human origin) exists within these disciplines, and there is no need to add another level of descriptive complexity. A "4-cm thick lenticular mass of crumbly organic-rich sand" can be simply described as that. Moreover, the use of a descriptive term, such as "layer," adds a clearly genetic flavor to the description. In the Earth Sciences, small-scale stratigraphic subdivisions, such as beds, are expected to represent individual natural depositional episodes. In archaeological settings, the sedimentary dynamic is not always apparent (geogenic versus anthropogenic causes), so it is generally not realistic to attempt to identify a distinct layer that indicates a specific "geoarchaeological" event. In addition, the isolation of "layers" may likely be unwarranted and unwanted, since it supposes that such a layer is solely anthropogenic in nature. It would be at odds with a situation that involves reworking of a previously deposited anthropogenic accumulation. As is discussed in Chapter 15, field descriptions have to remain as free as possible from genetic overtones. This is especially true in light of the fact that field observations provide only partial insights into the history of a deposit; analytical data from the laboratory are normally needed to fully determine the complete history of a deposit or stratigraphic sequence.

Gasche and Tunca (1983) also make the case for the "ethnostratigraphic unit" to depict deposits based on their archaeological contents. Again, while we clearly need to observe a stratigraphic profile from the standpoint of the artifacts found within the deposits (cf. Fig. 2.1), we strongly question the need to define a formal "ethnostratigraphic unit." It is a far less restrictive practice, and more pragmatic to be able to say that a certain layer contains Roman lamps, than to have to "define" the layer as a "lamp layer." We stress on a practical approach throughout this book, and practical needs should guide classification systems – it should not be the other way around. The "need" to fit classifications is a nuisance that should be avoided to ensure complete operational flexibility as new discoveries are made every day.

2.3 Facies and microfacies

Geologists realized that at a given time, different types of sediments can be accumulating within the different depositional environments. So, for example, at the coast sandy sediments accumulate at the beach zone, but further offshore toward deeper water, the sediments often become increasingly fine grained and "muddier." Lateral landward change in lithologies might include sand deposits of aeolian (wind blown) origin or organic-rich, sandy, and sand silty deposits found in coastal marshes. In solely continental (i.e. terrestrial) environments, we can see coarse gravel deposits accumulate within a river channel, which grade laterally into silt and clay deposits of the floodplain (see Chapter 5). These laterally equivalent lithological bodies that accumulate at the same time are called *facies*, and the term permits us to visualize how sediments/rocks of different compositions can be of the same age.

Over decades, the term facies has been used in a variety of ways (Boggs, 2001). The most appropriate use is descriptive and includes terms such as "sandy" facies, and "clayey organic" facies. Moreover, different sublithofacies can be recognized on the basis of lithological characteristics (e.g. massive sand versus finely bedded sand), or biofacies based on fossil content (e.g. gastropod-rich clay versus clay with bivalves) in order to provide more flexibility (Fig. 2.6). Less desirable practices involve names that have a genetic flavor, such as "shallow marine" facies or "intertidal" facies. This practice of using facies, names that connote interpretation or genesis should be avoided, as they can be commonly subjective or wrong. In sum, facies must be identified on the basis of objective criteria, such as lithology or fossil content.

The facies concept is a very valuable and functional approach in geoarchaeology, both at regional scales, as above, and at scales related to individual stratigraphic units exposed in an excavated balk. At the regional level, recognition of different facies enables detection of different parts of former landscapes and geological environments; it also links up with questions, such as site location, site burial and erosion, and resource availability.

At the site-specific level, where individual deposits are concerned, the concept of *microfacies* has been developed. It involves the recognition of facies at the microscopic level using petrographic thin sections or polished slabs of impregnated blocks of sediment. The use of the microfacies concept is critically important in deciphering the formation of archaeological sediments, site-specific activities of past inhabitants, and the integrity of the archaeological record (Courty, 2001: 229) ". . . the ultimate goal of archaeological facies analysis is to restore the three-dimensional image of a human-related space at a give time and to describe its evolution." Thus, in order to be able to recognize penecontemporaneous (approximately the same time) human activities within different areas at a site (e.g. cooking area versus storage area) it is necessary to recognize

(a)

(b)

FAN

SAND FLAT

MARSH

Water table

Facies / FEATURES	Sand–silt	Calcareous sand–silt	Calcareous muds
SEDIMENTARY STRUCTURES	Well-preserved, lower flow regime planar beds, and small-ripple cross-beds	Well-preserved, mainly lower flow regime planar beds	Rare laminae
BIOTURBATION	Dispersed: Small insect burrows and vertical roots	Moderate: Cicada burrows and vertical roots	Extensive: *Typha* and other roots, small insects (?)
SECONDARY CARBONATE	Absent except where pedogenic	Common: Platy carbonate, cicada nodules, minor mottled carbonate	Extensive: Mottled most common; rod-shaped rhizo concretions
FAUNA	None	None	Aquatic molluscs, abraided megafaunal remains on margins
TEXTURE	Sand = silt, rare gravel	Silt > sand, no gravel	Clayey silt > sand

FIGURE 2.6 Facies. (a) Photograph of alluvial fan in the Ka Valley of Western Sinai, Egypt. From the base of the mountain at the head of the fan, down the fan, toward the toe of the fan, the fan surface flattens and grades topographically into a plane. (b) A stratigraphic cross section through such a fan sequence shows the lateral lithostratigraphic changes in facies from gravels to sands and ultimately calcareous muds. Other lithostratigraphic characteristics are also shown (modified from Ashley, 2001).

and be able to trace small-scale "microfacies" across the site and see where one type of microfacies grades into another. So, for example, a layer of ashy silts in an area can be resolved microscopically into individual ash accumulations resulting from *in situ* burning in one area; laterally however, these ashy deposits change into loose, organic-rich ashy silts produced by trampling and of the dumping of cleaned-out hearth materials that is mixed with organic refuse and discarded.

2.4 Correlation

A related concept to facies is correlation. As used in an everyday sense, correlation implies a certain degree of bringing together two similar things. In geology, the idea of correlation, like facies, is used in different ways (Boggs, 2001). Two are most prevalent. One approach refers to a demonstration of equivalence of time, such that two bodies of rock were formed at the same time. The alluvial facies example discussed above (channel gravels versus floodplain silts; up fan–down fan changes), illustrates this concept. Another viewpoint is more encompassing: equivalence is not strictly based on time equivalence, but on lithologic, palaeontologic, and chronologic grounds. Thus in the case of *lithocorrelation*, relationships are established on the basis of similar lithologies or stratigraphic position. Two clay deposits, for example, may look similar and be "correlatable," but they may not necessarily be the same age. *Biocorrelation* is effected on the basis of fossil makeup or by biostratigraphic comparisons. In *chronocorrelation*, comparisons are made by means of chronological age. Whereas with older geological periods it is critical to keep these different types of correlation separated, this practice is less critical in younger, geoarchaeological settings because stratigraphic and temporal resolution is greater and lithologic, palaeontologic, and chronostratigraphic boundaries tend to overlap, or coincide.

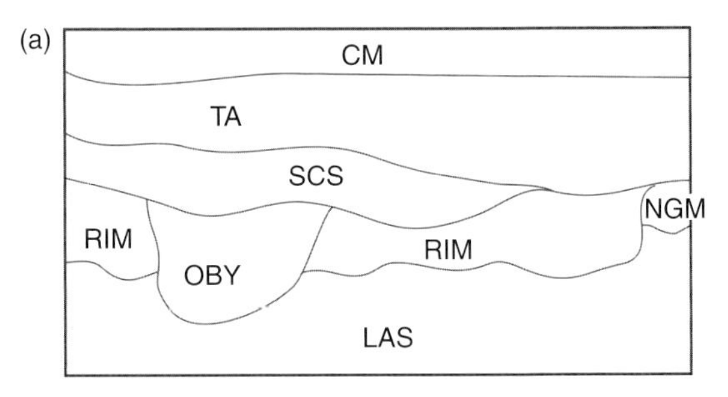

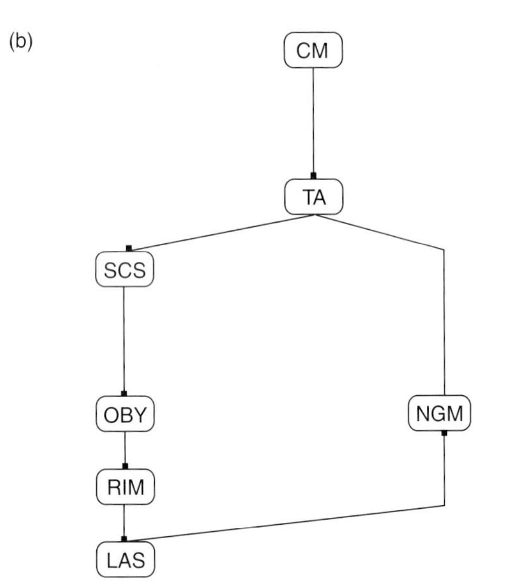

FIGURE 2.7 Example of the Harris Matrix. (a) A hypothetic stratigraphic section. Correlating such units and microfacies can be very difficult to do in the field and also after excavation. This is particularly true in excavations of larger surface areas (*décapage*) where the depth/time dimension is not conserved or is difficult to follow. (b) A simplified Harris Matrix made with the ArchEd program (Version 1.4.1) (http://www.ads.tuwien.ac.at/ArchEd/).

2.5 Keeping track: the Harris Matrix

One of the thorny, operational issues that arise during the investigation of archaeological deposits, particularly those from sites with complex stratigraphic sequences, is how to organize all the stratigraphic data (features, stratigraphic units, etc.) and how to correlate stratigraphic units across space. This is less of an issue with most open-air hunter and gatherer sites, where more homogeneous, stratigraphic units can be recognized in the field and which are laterally and vertically extensive. But when an area is intensely occupied, such as in a tell or in a cave habitation site, layers may extend only for a short distance, and there can be intersecting pits and complex architectural features.

A technique that has received much attention is that of the Harris Matrix (Harris, 1979, 1989; http://www.harrismatrix.com). Essentially, this is a house-keeping strategy that permits one to visualize the stratigraphic relationships among stratigraphic units and features in diagrammatic form (Fig. 2.7). Harris stressed the recognition of not only the stratigraphic units but also the contacts between them, which reflect specific human actions (e.g. clearing and pit filling). Whereas there are some controversies associated with the technique (Barham, 1995; Farrand, 1984; Stein, 1987), it nevertheless serves as a convenient form of shorthand, if it can be implemented with caution in the proper context. In many cases it is quite difficult to produce a realistically accurate and understandable diagram because of the complexities and ambiguities of the stratigraphic deposits and their relationships. Computer programs (e.g. www.stratify.org; WINBASP www.uni-koeln.de/~a1001/basp) can be helpful in construction of Harris Matrices, but their implementation is not so straightforward. These programs should not be considered as a black box, and they still require considerable attention to details of data input. and a complete understanding of all the stratigraphic relationships at the site.

2.6 Conclusions

The techniques and strategies used to recognize and record stratigraphic information in the field are cousidered in Chapter 15 (field techniques). In this chapter we have discussed the means to distinguish characteristics produced by primary (i.e. depositional) processes from those attributes produced by secondary processes, such as soil formation or diagenesis. In conclusion, we would like to stress the point that stratigraphy can be simple, complex interesting, dull, or difficult to interpret. However, there is no such thing as "good" or "bad" stratigraphy. It can be either correctly conceptualized or poorly studied, but these deficiencies are human aspects and not geological ones.

3

Soils

"The soil is a natural body of animal, mineral and organic constituents differentiated into horizons of variable depth which differ from the material below in morphology, physical make-up, chemical properties and composition, and biological characteristics."

(Joffe, 1949)

3.1 Introduction

This chapter introduces the importance of soils in geoarchaeology, and why it is essential that soils be clearly differentiated from natural and archaeological sediments. In order that both environmental and human influences on soil formation are understood, the concept of the *five soil forming factors* is introduced. This provides the basis for analysis and reconstruction of past soil landscapes, and shows how soil classification and mapping are essential tools in this endeavor. Basic soil types and the horizons that make up different soil profiles are also given, and some important soil forming processes are described.

An understanding of soils is fundamental to archaeology. Apart from soil being an essential component of any environmental reconstruction and background to any human activity, they provide an absolutely essential resource to humans. For example, the type of woodland, richness of pasture, and character of agricultural undertaking, are all governed by soil conditions. Moreover, the nature of man's impact on soils, through erosion, acidification, desertification, and any management practices can contribute to getting information on the nature of contemporary human societies.

At the broad scale, soils were, and in many cases, still are, one of the most important factors affecting human occupation and development in the Holocene, whether in China, Mesoamerica, or Europe. For example, fertile soils, perhaps as a response to regular rainfall, warm temperatures, and amenable geology, were often the foci for early arable activity (Limbrey, 1990). It was in the area of the "Fertile Crescent" that the earliest arable activity was recorded in cultures that were aceramic. At the detailed scale, many sites are situated on soils, and these soils normally make up most of the archaeological deposits and the archaeological stratigraphy. Thus the soils that make up archaeological deposits, which also form the substrate upon which occupation takes place, and which may themselves be involved in the preservation of a site as overburden, can all provide crucial information about site history (see Boxes 3.1 and 3.2). The study of paleosols (generic term for ancient soils) and associated hominid/human remains, both in the New and Old Worlds have provided clear clues to past environmental conditions (Catt and Bronger, 1998; Holliday, 1992, 2004, Chapter 8; Kemp, 1986) (see Box 3.2). Moreover, when a landscape is affected by human activity local soils may have been utilized in the manufacture of

Box 3.1 The Five Factors of Soil Formation and Bronze Age Brean Down, United Kingdom

FACTORS OF SOIL FORMATION (modified from Jenny, 1941)

Jenny (1941) enumerated five factors that affect the formation of soils:

$$s = f'(\text{Cl, O, R, P, and T}),$$

where

- *s* Soil properties are dependant on the function (f') of soil forming factors
- **Cl** Climate (arctic conditions, for example, result in slight soil formation; high effective precipitation can lead to leaching of soils and peat; high evapotranspiration can lead to secondary precipitation of carbonates, gypsum, halite, and other salts)
- **O** Organisms (Both large and small: plants – mosses to mangroves; animals – ants to antelopes)
- **R** Relief (droughty ridges and uplands, eroding slopes, colluvial footslopes, and boggy valleys)
- **P** Parent material (hard igneous rocks may produce thin Ranker soils, soft clays are easily eroded, but are poorly drained)
- **T** Time (permits the increased development of the soil profile horizonation – A, B, C horizons. Changes relating to time: age of landscape, with older exposed landscapes having more strongly developed horizons)

The Five Soil Forming Factors at Bronze Age Brean Down, United Kingdom

These soil forming factors can be observed at the multiperiod site of Brean Down (Somerset coast of Bristol Channel). Natural and excavated exposures revealed Pleistocene slope deposits and Holocene soil formation; Beaker (Early Bronze Age) through Iron Age occupations and colluvium, punctuated by blown beach sand episodes (Bell, 1990). The Local geology is composed of Carboniferous Limestone, and the talus slope deposits that constitute the Brean Down headland form the basal deposits at the site. Pleistocene Alfisols (argillic brown earths/luvisols) formed on the limestone (and probably on the fine drift cover). Here and on other limestones, such as the Jurassic Oolitic Limestone (e.g. Cotswolds, Gloucestershire), slow weathering of the bedrock produces a red-colored clay rich *β* horizon between the overlying Bt horizon and underlying weathered C horizon (limestone substrate) (Catt, 1986) (Fig. 3.1). This soil profile shows the main results of *time*

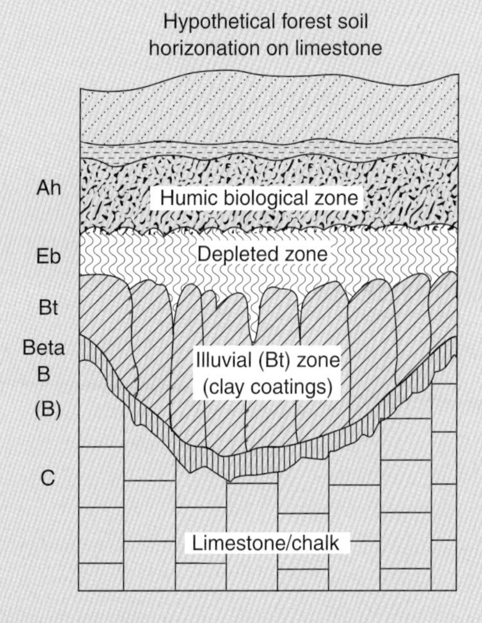

FIGURE 3.1 Alfisol formed on limestone under woodland in a humid western temperate climate, long weathering developing a reddish clay rich β horizon (cf. *terra fusca* and red Mediterranean soil).

(cont.)

Box 3.1 *(cont.)*

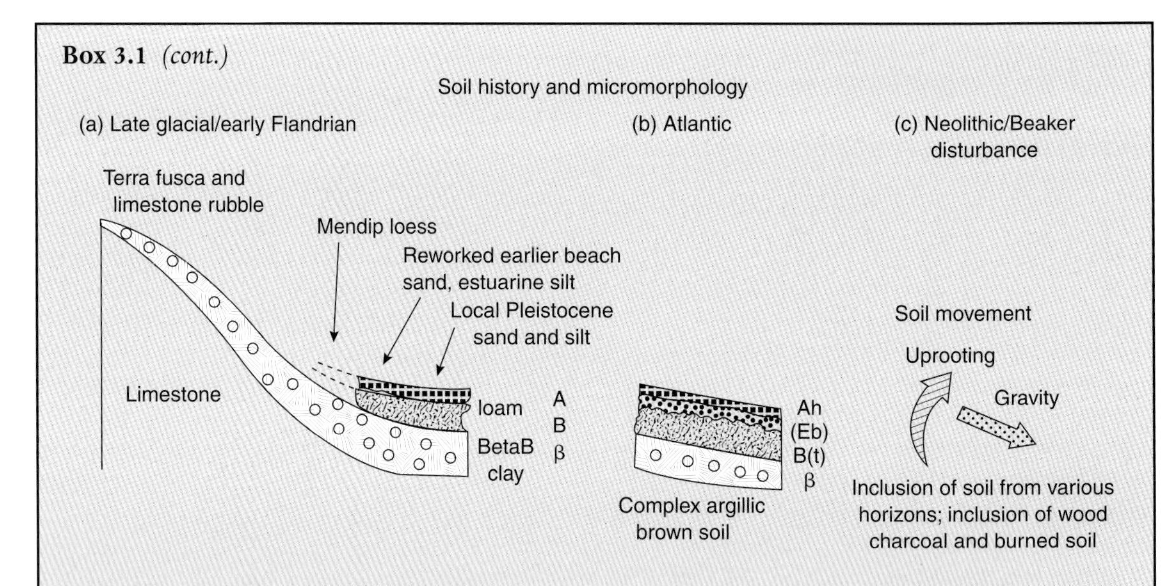

FIGURE 3.2 Evolutionary development of soils at Brean Down from Late Glacial through Neolithic/Bronze Age times. (a) Catenary development of alfisol (A, B, and β horizons) formed along the toe slope where *terra fusca* and limestone rubble interfinger with beach/estuarine sand; (b) during the Atlantic period (ca 5,000 BP) this profile naturally evolved into [Ah, A2(Eb), Bt, and β horizons], which eventually were ultimately modified by Neolithic/Beaker period disturbance which mixed horizons (see below) (Macphail, 1990).

(well-developed ABC profile forming since the Late Glacial), *parent material* (limestone), climate (western humid temperate – inducing leaching and clay translocation), and *organisms* (under woodland).

At Brean Down, however, the soil forming factors are complex and less straightforward:

- *topography* and *relief* play a role as the site is located on a slope toward the base of the catena, and soil materials are affected by gravity;
- multiple *parent materials* exist, including limestone talus, previously formed red β soils (terra fusca), "Mendip loess," local Pleistocene sand and silt, and reworked beach sand and estuarine silt;
- *time* is represented by some 5,000 years from the Late Glacial to the Atlantic Period;
- *organisms* here involve not only natural soil fauna, but also human activity, such as uprooting and clearance of trees (Macphail, 1990) (Fig. 3.2). The results of all these *Factors* are revealed in the soil micromorphology (and grain size and chemistry) of the Beaker palaeosol (buried by Bronze Age colluvium and blown sand) (Bell, 1990) (Box 3.1; Fig. 3.3).

These features are manifestations of the following Soil Forming Factors at the site: *Climate* (Mid-Holocene humid western temperate climate), *Organisms* (human induced plough erosion and colluviation; inclusion of burned and charcoal-rich soils [relict from clearance], *in situ* cultivation disturbance; *in situ* earthworm working during postcultivation pasture stage), *Relief* (footslope position resulting in

(cont.)

Box 3.1 *(cont.)*

colluvial additions to the soil and overthickening of the profile), *Parent Material* (as Fig. 3.2, with inclusion of various soil fragments from the relict *terra fusca* [β horizon], Holocene profile and burned soil, which, for example, affect measurements of soil organic matter, grain size, magnetic susceptibility, and phosphate) and *Time:* here very old (Late Glacial), earlier (clearance, ploughsoil colluvium), and contemporary (Beaker Period pasture topsoil) are all present; Fig. 3.3) (from Macphail, 1990).

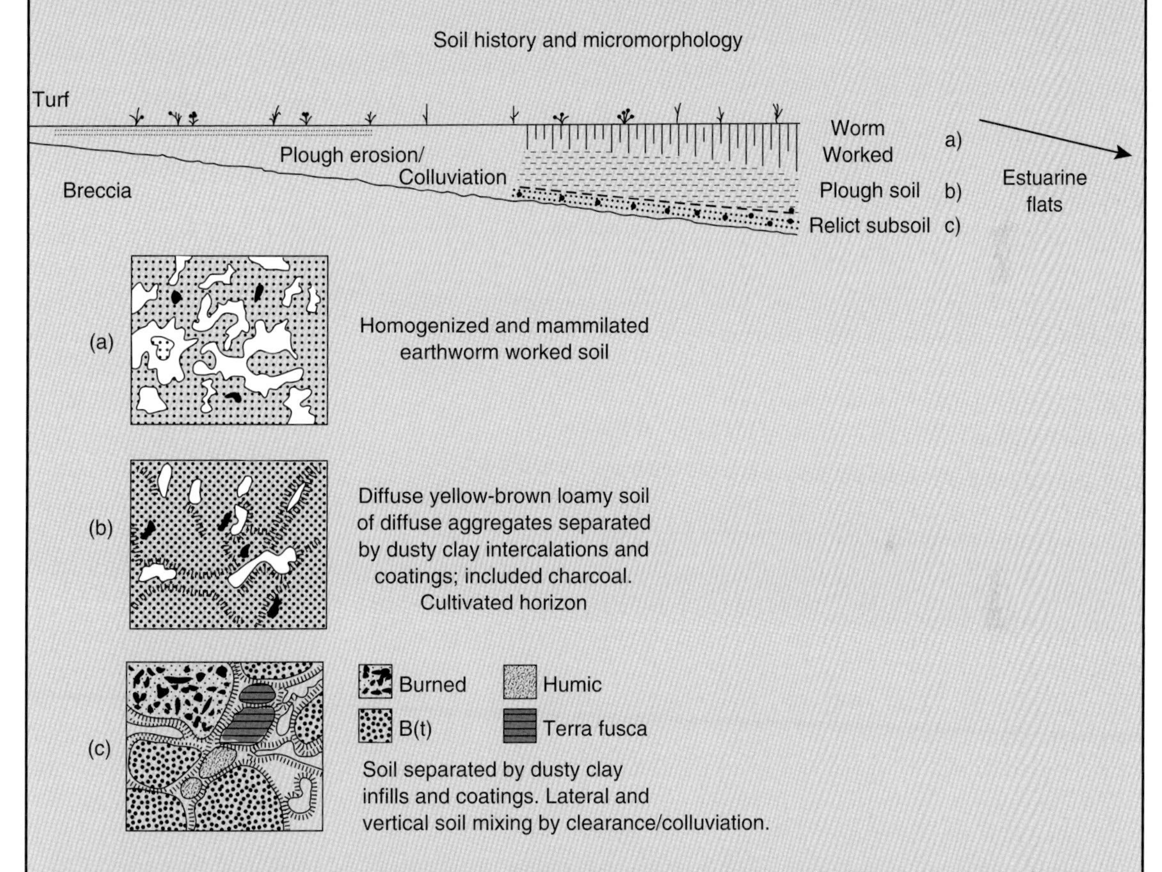

FIGURE 3.3 Details of Box 3.1, Figure 3.2, showing Neolithic/Beaker soil development: Upper figure – three layer (a, b, c) slope soil profile development buried by blown sand, and how this slope soil eventually interdigitates with estuarine sediments; Lower figure – (a) uppermost earthworm-worked and homogenized pasture soil formed in colluvium, (b) buried ploughsoil (cultivated) colluvium that contains fragmented soil and textural pedofeatures indicative of soil disturbance and mixing, and (c) lowermost subsoil formed and deposited over the Pleistocene talus substrate and composed of fragments of red *terra fusca,* and clasts of the Holocene Bt horizon, Ah topsoil, and burned soil – the last containing charcoal and probably relict of clearance.

daub and mud brick (Chapter 13), and served as the medium for arable activity (Chapter 9). Finally, the site itself may be buried under slope deposits that result from soil erosion (Chapter 4). In the last case, conditions of burial relate directly to soil conditions, such as vegetation cover, aridity, waterlogging, acid or alkaline influences, which in turn affect conservation of archaeological materials and environmental reconstructions (Table 3.1). Of greater importance to ancient populations, however, are soil-workability and soil stability/erodibility, because these factors will govern how a landscape can be exploited for agriculture.

It is important to understand the essential differences between a soil, a sediment/rock, and an archaeological deposit. A good, pragmatic, and generalized definition of soil is: "soil is a natural body of animal, mineral and organic constituents differentiated into horizons of variable depth which differ from the material below in morphology, physical make-up, chemical properties and composition, and biological characteristics" (Joffe, 1949). The various soil forming processes (e.g. leaching, translocation) normally take place on an extant, exposed substrate. Sediments are generally little affected by the *five soil forming factors* described below, which transform sediments into soils (Box 3.1). During soil formation *(pedogenesis)*, minerals are subjected to weathering, elements may become hydrated or leached, and biological activity, whether involving bacteria or larger plants and soil animals, mixes organic matter with mineral material; the sum total is the initiation of soil horizon formation.

Archaeological deposits themselves may have accumulated through sedimentary processes, some or all of which are purely anthropogenic in origin. Furthermore, they themselves may be affected by pedological or geogenic postdepositional processes (e.g. cryoturbation), and this may destroy original layering and transform or completely remove some easily weatherable materials, such as wood ash (Weiner *et al.*, 1995). Extreme pedological alteration of archaeological deposits results in the formation of *dark earth* as found in Roman/Early Medieval Europe (Macphail, 1981; Macphail *et al.*, 2003; Yule, 1990) and South and Mesoamerica where it is more commonly known as *terra preta* (Woods and McCann, 1999) (see Chapter 13).

3.1.2 Differentiating soils and sediments

There are clear differences between soils and sediments (Mandel and Bettis, 2001). A sediment has a dynamic history, which encompasses erosion, transport, and deposition over a landscape or area (e.g. glacial till, aeolian loess, beach sand) (see Chapter 1, Sediments). In contrast, a soil is static in that it has formed *in situ* through various weathering and biological processes that are described below. This soil forming episode is often termed a period of "stability" ("stasis"). An archaeological deposit is therefore clearly a sediment and not a soil, with a source (e.g. a combustion feature) and a mode of deposition (e.g. dumping, accumulation of stabling waste). Like any sediment, an archaeological deposit can undergo postdepositional effects that are geogenic and/or pedogenic in character. When these three aspects are understood, the site formation processes involved in any one archaeological context can be identified. As this chapter is focused on soils, postdepositional *soil forming processes* are discussed in some detail (see Section 3.3).

Another important soil/sediment is colluvium, which is a downslope accumulation of eroded rock, soil, and in some cases, archaeological construction materials (e.g. mud brick). Although detailed aspects are considered in Chapters 1, 4, and 10, we note here that even in the case of "soil" colluvium or "hill-wash," these slope deposits accumulate through sedimentary processes. Thus colluvium can be described both as a sediment and a soil. Alluvial soils (soils formed in alluvium)

Table 3.1 Soil characteristics and preservation of artifacts and ecofacts in soils

Conditions	Parent material	Soil type	Humus form	Soil fauna	Preservation potential	Comments/other
Acid pH <3.5–6.5	Quartz sand; schist	Ranker; Podzol (e.g. Entisols and Spodosols)	Mor and Moder	Mites, Enchytraeids, and non-burrowing earthworms	Pollen; macrobotanical material; phytoliths, soil diatoms	Loss of bone; possible damage to pottery; corrosion of iron (loss of magnetic susceptibility signal)
Neutral pH 6.5–7.5	Loams	Brown earth, Luvisols (e.g. Mollisols and Alfisols)	Mull	Collembola, Arionids (slugs) and burrowing earthworms (Lumbricids)	Bone, molluscs, phytoliths, soil diatoms; some macrobotanical material	Pollen is oxidized; large-scale mixing by meso and macrofauna; charcoal can become fragmented
Alkaline pH >7.5	Shell sand, chalk, salt lake sediments	Rendzina; Solonetz (e.g. Xerolls and Xeralfs)	Mull	As above but becoming restricted	Molluscs; bone; phytoliths (except for extreme pH > 8)	Pollen is oxidized; salt, carbonate, and gypsum crusts on pottery and other artifacts; large-scale mixing by meso and macrofauna, and fracture by secondary crystal growth
Waterlogged	Peat, estuarine sediments	Peat (Histosols)	Peat	None	Pollen, molluscs, diatoms, most organic materials, including fragile insects, skin, leather	Corrosion (loss of magnetic susceptibility signal from iron slags when fluctuating oxidizing/reducing conditions); soft-tissue "pickling" (bog man); secondary mineral formation when water tables fluctuate (gypsum, jarosite, pyrite, siderite, vivianite)

Note: Caveats

1. Fluctuating water tables (i.e. oxidation/reduction cycles) destroy pollen and "rust" iron slags
2. High biological activity mixes stratigraphy and displaces artifacts
3. Possible serendipitous preservation of strongly residual iron slag, burnt soil, fused fly-ash, phosphatized materials

(Note: gypsum ($CaSO_4 \cdot H_2O$), jarosite ($KFe_3(OH_6)_6(SO_4)_2$), pyrite (FeS_2), siderite ($FeCO_3$), vivianite [$Fe_3(PO_4)_2 \cdot 8H_2O$], see Bullock *et al.*, 1985, Table 6.2)

TABLE 3.2 Soil horizons – with suggested associated major soil processes and international terms (possible associated USDA diagnostic horizons[a] are also given)

Common horizon term	Europe (FAO/UNESCO)	UK	USA (USDA[1])	Common soil name Europe/UK/USA	Major soil process
Superficial organic horizons					
L (litter)	Litter	L	~Oi	Mull/O1/cf. Oi	High biological activity; only accumulation of plant fragments (L)
L/F (fermentation)	Fermentation	F	~Oe	Moder/O2/cf. Oe	Moderately low biological activity; accumulation of excrements of soil fauna and decomposing plant fragments (F) below L
L/FH (humus)	Humus	H	~Oa	Mor/O2/cf. Oa	Very low biological activity; accumulation of amorphous organic matter termed humus (H) below L and F
Peat	Histic H	Peat O horizon	~Oa (Histic epipedon)	Peat/O	Dominant accumulation of organic matter because waterlogging inhibits biological breakdown of organic matter
Topsoil horizons (Epipedons)					
Humic topsoil	e.g. Mollic A, Umbric A	Ah horizon	A (e.g. Mollic epipedon, Umbric epipedon)	Humic topsoil	Accumulation of organic matter in the mineral soil (along with associated nutrients of N, K, and P); focus of biological activity and organic matter turnover (oxidation and alteration). Mollisols include grassland prairie soils and chernozems
Plaggen	Plaggen (Cultosol)	Cultosol (Ap)	Ap (Plaggen epipedon; Anthropic epipedon[b])	Plaggen/Anthrosol	Over-thickened (0.40–1.0 m) humic topsoil developed through additions of manure, turf, household waste (e.g. since AD 1000 in Holland)
Ploughsoil	Cultivated A	Ap	Ap (in the United States includes both ploughed soils and pasture)	Ploughsoil/Arable topsoil	Topsoil, mechanically homogenized to depth of plough share (ca. 0.40 m) or ard (e.g. 0.06 m); liable to loss of organic matter through oxidation; arable soils ameliorated by additions of organic manures, possibly since Neolithic
Upper subsoil horizons					
Leached/eluviated upper subsoils	Albic E (A2)	Eb horizon	E (Albic)	Alfisol/Argillic/Luvisol	Eluviation of clay (along with cations, including iron; organic matter and phosphorus)

Leached/eluviated upper subsoils	Albic E (A2)	Ea horizon	E (Albic)	Podzol/Spodosol/podzol	Eluviation of iron and aluminum (sesquioxides) (after acid breakdown of clay and mobilization by plant chelates)
Subsoils					
Cambic B	Cambic B	Bw horizon	Bw (Cambic)	Cambisol/Brown soil/Cambisol	General weathering of minerals and clay formation
Argillic Bt	Argillic B	Argillic Bt	Bt (Argillic)	Luvisol/Argillic brown soil/Alfisol	As above, but in some cases, with illuviation of clay from overlying A2 horizon (clay translocation)
Spodic Bs/Bh/Bhs	Spodic B	Podzol Bs/Bh/Bhs	Bh/Bs/Bhs (Spodic)	Podzol/podzol/spodosol	Illuviation of sesquioxides (Fe and Al) often along with humus
Calcic B and K horizon	–	–	Alkaline Bk	Calcisols/calcic brown soils/aridisols	Illuviation of alkaline earths, especially calcium carbonate; cementation by calcium carbonate (K horizon/calcrete)
Gypsic horizon	n.a.	Gypsol	By (Gypsic)	N.a./gypsic/gypsid	Concentration/cementation by gypsum
Gleyed (Ag, Bg, G) horizon	Gleyic G (Bg)	Gleyic G (Bg)	Bg	Gleys/Gleys/Aquents, Aquepts	Gleying or hydromorphic process of reducing iron (Fe^{+++} to Fe^{++}); pale (G) to mottled (Bg) colors
Ironpan	Ironpan Bf	Placic Bf		Ironpan/ironpan/iron pan, plinthic	Cementation by iron (usually under gleyed conditions)
Oxic B	n.a.	Ferrasols	Bo (Oxic)	N.a./ferrasols/Oxisols	Strong (tropical) weathering and formation of iron-rich laterite (hematite and goethite)
Rubefied argillic B	n.a.	Planosols	B (Argillic)	N.a./acrisols/Ultisols	Extreme (tropical) illuviation of clay from E horizon
Fragipan	Fragipan	Fragipan	Bx (Fragic)	?/cryosols/Gelisols	Compaction of soil material associated with freezing and thawing
Andic horizon	n.a.	Andic horizon	B (Andic)	n.a./andosols/Andepts	Organic matter accumulation and weathering of volcanic ash (e.g. Tephra) to produce allophanes
Natric and salic horizons	–	–	Btn (Natric) Bz (Salic)	Saline soils/Salorthids, Natrustalfs, Natrargids	Concentration of salts, especially sodium salts (halite) in Natric horizons

TABLE 3.2 (*Contd.*)

Common horizon term	Europe (FAO/UNESCO)	UK	USA (USDA[1])	Common soil name Europe/UK/USA	Major soil process
Parent material					
C horizon	C horizon	C	C (Cox, Cr, Cu)	n.a.	Partial weathering of substrate (Poorly consolidated or unconsolidated regolith or surficial sediment)
R horizon	R horizon	R	R	n.a.	Unweathered hard bedrock

[a]In contrast to field-identified A, B, and C horizons, diagnostic horizons of the USDA's *Soil Taxonomy* (1999) are rigidly defined according to criteria measured in the field *and* laboratory

[b]Anthropic epipedon requires >250 ppm citrate-extractable P_2O_5 (A mollic epipedon has < 250 ppm citrate-extractable P_2O_5)

Caveat: workers should check their classifications (and possible revised versions) with the appropriate texts for their regions. Some good examples of UK soils classified according to the US classification system can be found in Avery (1990).

(Birkeland, 1999; Bridges, 1990; FAO-UNESCO, 1988; Hodgson, 1997; Holliday, 2004; Pape, 1970; Soil Survey Staff, 1999; van de Westeringh, 1988)

n.a. – not applicable

develop in a similar way, with most or all evidence of their sedimentary history also being lost through biological working. Before, however, detailing some of the soil forming factors and processes that contribute to soil profile formation, which at one time were linked to geographical regions (i.e. *Zonal Soils*, Section 3.2.2) – and various human effects on them – some fundamentals are given on how soils are described in the pedological literature (Tables 3.2 and 3.3).

Table 3.3 Soil classification (Soil Survey Staff, 1999)

Soil order	Main characteristics
Entisol	No or weak evidence for soil horizon formation
Inceptisol	Humid region soils with soil horizon formation, loss of bases but weatherable minerals still present
Aridisol	Soils formed with little/rare precipitation
Mollisol	"Grassland" (prairie/steppe/pampa) base rich soil
Alfisol	"Forest" base rich soil with diagnostic argillic subsoil horizon
Spodosol	Leached soil with diagnostic spodic subsoil horizon, variously enriched in Fe, Al, and humus
Gelisol	Cold climate permafrost soils
Ultisol	Warm climate, low base status soils with diagnostic argillic subsoil horizon
Oxisol	Tropical and subtropical soils often with reddish iron oxide rich horizons of great age
Andisol	Often warm, humid climate soils formed on volcanic substrates
Vertisol	Humid climate soils, characterized by seasonal swelling and deep cracking
Histosol	Peats

3.2 Soil profiles and soil properties

Essentially, soil profiles were first categorized into soil horizons using the ABCD system of horizon designation (Bridges, 1990):

1 the A or humose topsoil horizon
2 the B or pedologically formed subsoil horizon (Fig. 3.18)
3 the C horizon or weathered parent material and
4 the D or more commonly the R horizon of consolidated bedrock.

A, B, and C horizons are often given a generic modifier (Table 3.2). For example, humose A1 topsoils can also be termed Ah horizons, whereas subsoil B horizons enriched with *sesquioxides* (aluminum and iron) are called Bs horizons (Bridges, 1990). In some soils the upper subsoil becomes leached and is commonly termed an A2, E, or Albic horizon. Those soils that have horizons with the subscript "g" (Ag, Bg, Cg), indicates that soils are mottled due to the effect of gleying ("hydromorphism"), and ochreous, grey to blue-green colors may be encountered in the field.

Different countries have their own classification systems. For instance, in the United Kingdom, upper subsoils that have lost clay are called Eb horizons, whereas those depleted in sesquioxides are known as Ea horizons (Avery, 1990). Fortunately now, there is broad agreement between Europe and North American soil classification systems. These have *diagnostic horizons*, a term first employed in the United States, but now used internationally although with some differences (Avery, 1990; Birkeland, 1999; Soil Conservation Service, 1994; Soil Survey Staff, 1999) (see Table 3.2). Diagnostic horizons, however, are *identified* only after both field and laboratory measurements have been carried out. For example, subsoils enriched in translocated clay are termed Bt ("argillic") horizons and relate to the translocation or "washing" of clay from the A horizons (including the A2 horizon) into the subsoil. Soils with a *diagnostic* argillic horizon are variously called Alfisols (Soil Survey Staff), Luvisols (FAO, Europe), or argillic brown soils (Soil Survey of England and Wales) (Avery, 1980; Bridges, 1990; FAO-UNESCO, 1988; Soil Survey Staff, 1999). A further example is the poorly fertile, acid soil commonly called a podzol. This has an Ah horizon over a leached upper subsoil (A2/E/Ea) horizon that can also be termed an albic horizon. This is present over a subsoil enriched in humus and/or sesquioxides (Bh/Bs) – termed a spodic diagnostic horizon. Hence these podzols are known as Spodosols in countries employing the American system (Soil Survey Staff, 1999). Often soil mapping is carried out using the identification of horizons although such identifications of diagnostic horizons is normally verified later through laboratory analyses (Chapters 15 and 16). For example, soil color, soil structure (peds/aggregates), amounts and type of roots, and approximate grain size (through "finger texturing") can be noted in the field. Later in the laboratory, pH, organic matter status (through organic carbon and loss-on-ignition – LOI analyses), fertility (Cation Exchange Capacity [CEC], N [nitrogen], K [potassium], and P [phosphorus]) and accumulation of sesquioxides (Fe_2O_3 and Al_2O_3), can all be measured. An important characteristic of soils, is their *base status*, as discussed later, and essentially refers to whether a soil is *acid* ($pH < 6.5$) or base rich – alkaline ($pH > 7.5$), which again effects their fertility and the type of humus they develop (*Mull*, *Moder* or *Mor* – see Section 3.3.2) (Table 3.2).

Lastly, we note that different types of A and B horizons (see Table 3.2) are formed by *soil forming processes* (pedogenesis) (see Fig. 3.19). Researchers in northwest Europe, for example, should be aware that podzols/Spodosols are developed by *podzolization*, whereas Alfisols/Luvisols/argillic soils are formed through clay translocation that sometimes is still known by its French name for "washing" – *lessivage* – French-speaking soil scientists describing their soils as *sols lessivés* (Duchaufour, 1982; Soil Survey Staff, 1999) (Fig. 3.8).

3.3 The five soil forming factors

Jenny (1941) formulated an expression to account for the various types of soils, the so-called five soil forming factors (Box 3.1):

$$s = f'(\text{Cl, O, R, P, and T}).$$

Briefly, these are:

- Cl Climate (influences of temperature, precipitation, and seasonality)
- O Organisms (effects of plants and animals, including humans)
- R Relief (topography, and its control on drainage – see *catena*)
- P Parent material (rocks and archaeological deposits that provide a substrate for pedological activity)
- T Time (length of time over which soil formation has taken place governs degree and maturity of soil formation)

More recently, Johnson (2002) has stressed that the role of biological activity has been severely underestimated as a soil forming factor. Nevertheless, the relative weight of each factor in pedogenesis varies from soil to soil, with for example, the oldest (t) most weathered soils being Ultisols (Table 3.3); whereas poorly fertile substrates (p) and a climate (c) that encourages leaching, give rise to acid unproductive Spodosols (Birkeland, 1999) (see Box 3.1).

3.3.1 Climate

Very simply, this factor can be first understood from the examples of arctic and arid environments, in which little soil is produced. Freezing

Box 3.2 Cold Climate Soils

It is important to be able to recognize the effect of cold on soils, sediments, and archaeological deposits and materials. In the first place, cold climate processes, such as frost wedge infilling, hummock and hollow, and stone stripe patterned ground formation, can all produce natural features that archaeologists should not confuse with anthropogenic cut features. At the microscale, textural pedofeatures such as dusty and impure void coatings formed by freezing and thawing, must not be mistaken for similar features resulting from cultivation and trampling. Finally, fragmentation and mixing of archaeological deposits by cryoturbation may give the impression that a dump is present (Fig 3.4).

Fortunately, there have been numerous studies of cold climate geomorphic field and microscopic features, from Scandinavia and Canada (Courty *et al.*, 1994; McKeague *et al.*, 1973; Van Vliet-Lanöe, 1985, 1998). The field handbooks by Catt are also important in the investigation of cold climate deposits within the Quaternary (Catt, 1986, 1990). Unfortunately, this volume can hardly cover the mass of literature on Quaternary soils and associated cold soil-sediment studies that have been carried out. Many relate to the deposition of Pleistocene loess (Kemp *et al.*, 1994; Mücher and de Ploey, 1977; Mücher and Vreeken, 1981; Mücher *et al.*, 1981), cave sediment formation (Macphail and Goldberg, 1999, 2003) (Chapter 8), the disruption of occupation surfaces (Macphail *et al.*, 1994), the movement of artifacts by cryoturbation (Bertran, 1993, 1994; Hilton, 2002), and paleosol formation (Fedoroff *et al.*, 1990; Fedoroff and Goldberg, 1982; Kemp, 1986). Cold climate indicators have been particularly well studied from European caves in archaelogy (e.g. Bertran and Texier, 1999; Courty, 1989; Goldberg *et al.*, 2003). As examples, the Mid-Pleistocene Early Palaeolithic site of Boxgrove, United Kingdom

FIGURE 3.4 White Horse Stone, Kent: scanned thin section of valley bottom "Allerød soil," which is in fact a probable Younger Dryas solifluction sediment of reworked humic Allerød soil fragments (ASF) intercalated with a chalky mud flow soliflual layer (SL). Length of thin section is ~75 mm.

(see Chapter 7) and some Late Glacial Upper Palaeolithic occupations, are cited.

In the United Kingdom an Anglian Period (Isotope stage 12), cold stage soil developed under very cold and dry conditions, as characterized by blown sand infilled ice wedges at Barham, East Anglia (Kemp, 1985, 1986). At Boxgrove, West Sussex the cold climate terrestrial slope deposits that bury the early hominid (*Homo heidelbergensis*) coastal landsurface are formed out of periglacial flint and

(cont.)

Box 3.2 *(cont.)*

chalk pellet gravels and brickearth deposits (Roberts *et al.*, 1994; Stringer *et al.*, 1998). These deposits were thought to date broadly to the Anglian Cold Stage, but this "dating" correlation is hindered by the fact that the "soils" within these sediments show deposition under soliflual – cold and *wet* conditions, not cold and dry ones (Macphail, 1999). Conditions in the Anglian were also supposed to be so harsh that all hominid (*Homo erectus*) populations went south. At Boxgrove, however, refitting flints occur within these deposits, in horizons where ephemeral temperature amelioration (temperate interstadials) led to biological activity – rooting, and soil mixing by earthworms – showing that human populations had not simply disappeared (Roberts and Parfitt, 1999). This site thus records cold climate slope deposit formation where high-energy gravels are interbedded with silty soil-sediments formed under lower energy deposition (Figs 3.12–Plate 3.1); earlier sediments were fragmented into granules by cryoturbation, and ensuing meltwater caused the inwash of clay into voids. Mass-movement chalk gravels (Unit 8) also surprisingly contain an *in situ* refitting flint scatter, the surrounding soil matrix recording both the temperate conditions of an Interstadial and renewed cold climate effects (Figs 3.13–3.16). Such microfabrics may also show seasonal activity within an Interstadial.

Soils and deposits that date to the Lateglacial can be associated with Final Palaeolithic artefacts across Europe. Animal bones of lemmings and reindeer indicate cold climate conditions, whereas the presence of red deer has been used to identify a temperate Interstadial. Small mammal bones from the regurgitation pellets of raptors (birds of prey) have also been extensively used to reconstruct palaeoclimates (Fig. 3.9) (Andrews, 1990). In many open-air sites, the nature of the soil-sediments containing artefacts can be enigmatic because of postdepositional Holocene biological working and homogenization. Hence, Final Upper Palaeolithic artefacts are found to span enigmatically the Pleistocene-Holocene Boundary.

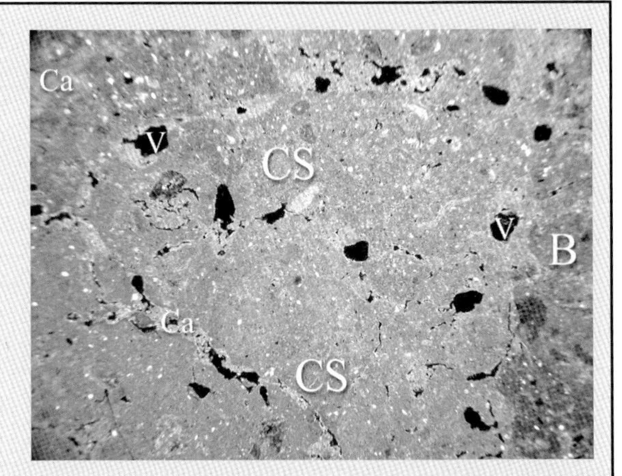

Figure 3.5 As in Figure 3.4, showing compact older Dryas solifluction deposit of eroded nonhumic chalky subsoil clasts (CS) – possibly rooted (voids – V) by seasonal plant growth and now showing cementation by micritic calcite (Ca). Note burrow (B) of humic soil on the right, of likely Allerød age. Cross Polaraized Light (XPL), frame width is ~7 mm.

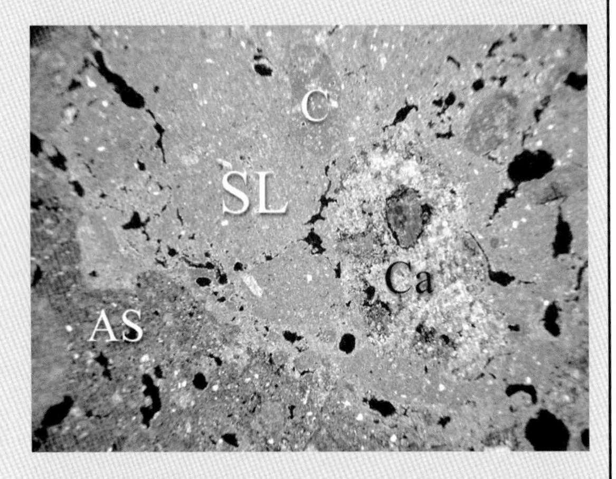

Figure 3.6 As in Figure 3.4, the junction of reworked humic Allerød soil fragments (ASF) and the overlying chalky mud flow soliflual layer (SL) that contains chalk (C), and where voids have become infilled with secondary micrite (Ca). XPL, frame width is ~7 mm.

(cont.)

Box 3.2 *(cont.)*

In southern United Kingdom the mollusc, insect, and soil-sediment characteristics from a number of lower slope and valley bottom Allerød soils have been studied during deep excavations for the Channel Tunnel (Preece and Bridgland, 1998). Such Allerød (11,000–13,000 yrs bp) soils were recognized in the field from their dark humic character, but when workers began studying them at the microscale they discovered that these "soils" resulted from very "disturbed" conditions that had led to mass-movement (Preece *et al.*, 1995; Van Vliet-Lanoë *et al.*, 1992). A series of illustrations from White Horse Stone, Kent, United Kingdom – a site excavated along the Channel Tunnel Link to London – shows the bedded character of the Allerød soil (Fig. 3.4), the typical nature of soliflucted (Older Dryas; 13,000–16,000 yrs bp) deposits (Fig. 3.5), a soliflual chalky mud flow layer *within* the Allerød soil (Fig. 3.6) and the character of the Allerød soil, as chaotic mixture of reworked humic (rendzina) soil clasts (Fig. 3.7). The Allerød soil

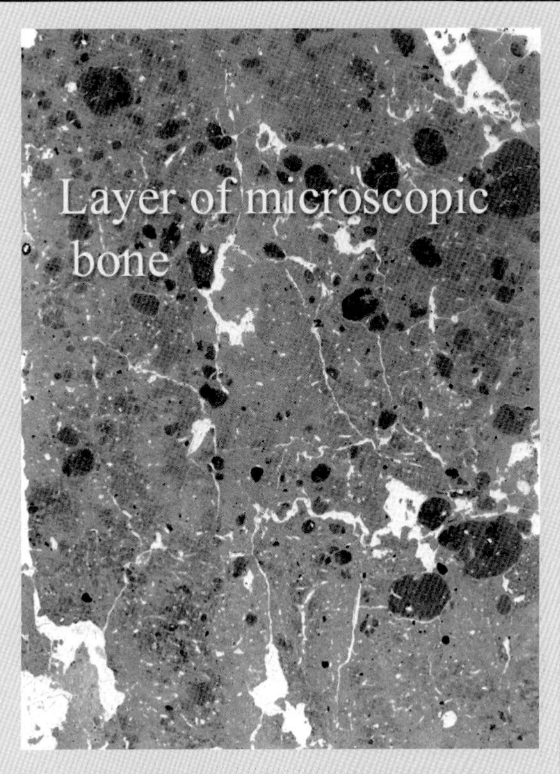

FIGURE 3.8 Underdown Lane, Kent; Scan of 2317-2, illustrating location of some 70 microscopic bone fragments, and ferruginized relict cold climate soil fragments. Width ~55 mm.

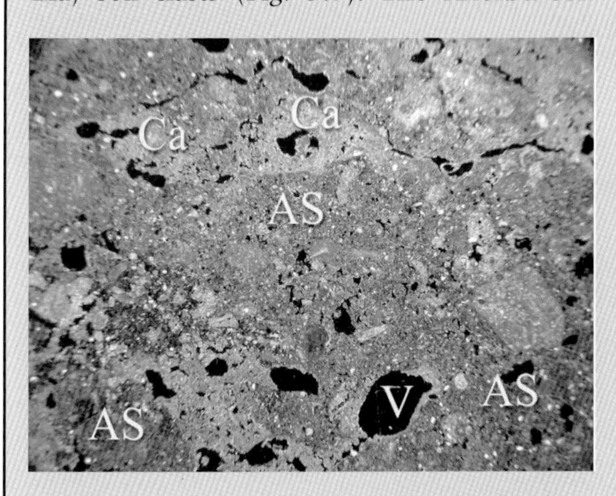

FIGURE 3.7 As in Figure 3.4, detailing chaotic sedimentary mixed nature of the Allerød soil layer (AS). Voids (V) have become partly coated and infilled with calcitic inwash (Ca), where secondary micrite coatings have also often formed. XPL, frame width is ~7 mm.

FIGURE 3.9 2317-2; microphotograph of ferruginized bone cluster (BC), voids (V), the iron depleted fabric (IDF) and intercalatory textural Pedofeatures (I) surrounding the "embedded" bone cluster. Plane polarized light (PPL), width is ~1.75 mm.

(cont.)

Box 3.2 *(cont.)*

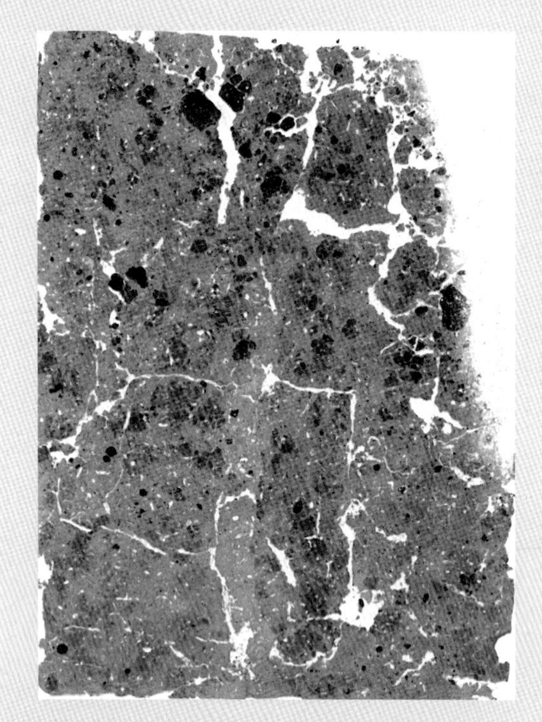

Figure 3.10 Scan of 2317-1 showing iron-depleted matrix and included iron and manganese stained relict (Allerød) topsoil clasts. Width ~55 mm.

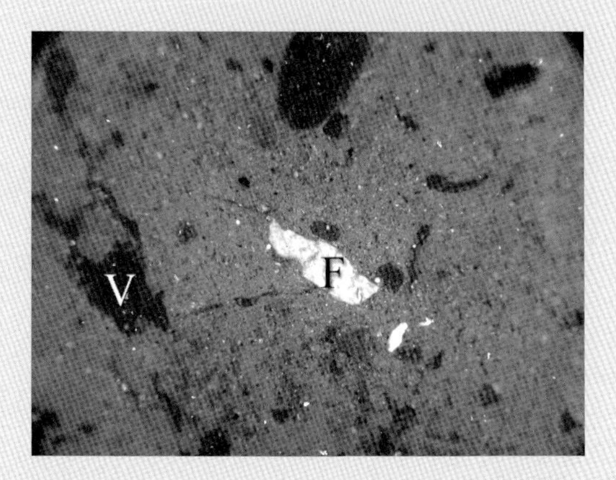

Figure 3.11 2317-1; microphotograph of soil containing a burned flint flake (F) and voids (V). Oblique incident light (OIL), width is ~7 mm.

here is thus probably an (Younger Dryas 10,000–11,000 yrs bp), *sediment*. On the other hand, *in situ* Allerød soils have been recorded as thin entisols (young soil's) in England, for example, as a humic ranker (Westhamptnett, West Sussex) and as a pararendzina on Older Dryas scree at King Arthur's Cave, Herefordshire; in both cases microscopic bone and fine charcoal are present (Macphail, 1995; Macphail *et al.*, 1999). King Arthur's Cave recorded an important Final Palaeolithic occupation contemporary with the soil (Barton, 1997). Similarly at Underdown Lane, Kent (Macphail, 2004), a Final Upper Palaeolithic artifact assemblage is associated with very wet and disturbed Lateglacial soil-sediments formed out of Late Pleistocene silty brickearth, where bone

Figure 3.12 Boxgrove: Unit 11, Middle Silt Bed and overlying gravels; part of the terrestrial slope solifluction deposits burying the early hominid coastal landscape, but containing refitting flints that record "occupation."

(cont.)

Box 3.2 *(cont.)*

Figure 3.13 Boxgrove, GTP 25; terrestrial slope solifluction deposits (here the "chalk pellet gravel" – Unit 8) burying the early hominid coastal landscape (here the beach); top of the ladder marks the *in situ* flint scatter.

Figure 3.14 Boxgrove, GTP 25; *in situ* refitting flints within the chalk pellet gravel.

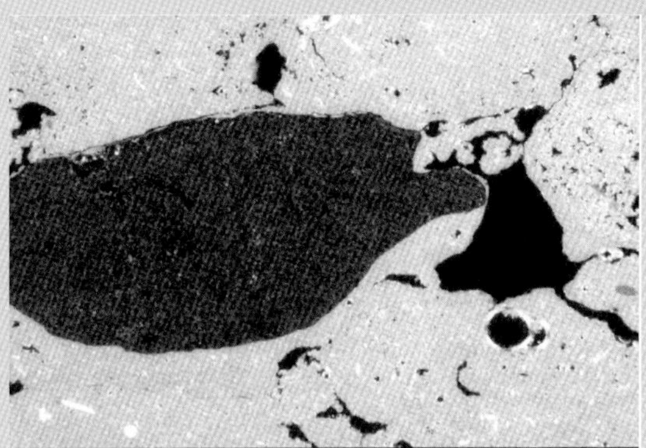

Figure 3.15 Boxgrove, GTP 25; terrestrial slope solifluction deposits; flint artifact within the "chalk pellet gravel" (Unit 8). The *in situ* flint scatter is set in a pedological microfabric composed of chalk and micritic chalk soil. A later phase of freezing and thawing has induced depositional inwash of colloidal chalk, coating some pores and forming a compact soil-sediment. XPL, frame is ~5.5 mm.

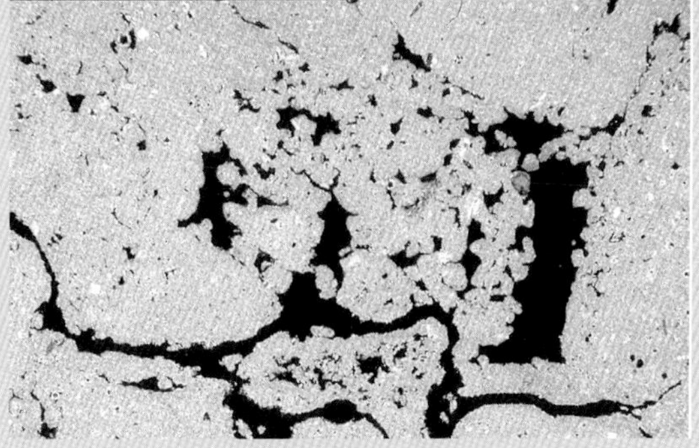

Figure 3.16 Boxgrove, GTP 25; terrestrial slope solifluction deposits; as in Figure 3.13, showing probable mammilated earthworm excrements demonstrating biological activity that was likely to be contemporary with occupation at this location. This microfabric probably marks an Interstadial during the formation of the generally cold climate Unit 8. XPL, frame is ~5.5 mm.

clusters possibly testify to the earlier presence of raptors and where burned flint was moved alongside the fragmentation and redeposition of probable humic ranker topsoils (Figs 3.10 and 3.11).

and thawing generate coarse sediments such as scree, and rockfall in caves, but little chemically weathered fine soil because there is a lack of biological activity and production of organic matter, and chemical processes are inhibited. Equally, cool conditions permit only modest breakdown of organic matter and peats (Histosols) may form. On the other hand, arid conditions lead to little plant growth, but high evapotranspiration can encourage secondary precipitation of salts, such as carbonates, gypsum, and salt (halite). The presence of such neo-formed minerals in ancient deposits has in fact led to interpretations of past desert conditions (Courty *et al.*, 1987). High temperatures under humid conditions give rise to the deep, highly weathered soils of regions such as those found in Australia, the Congo, and the Amazon, especially if formed in very ancient (pre-Pleistocene) landscapes where *time* has been influential. Traditionally, soil types that are governed by climate are called zonal soils.

3.3.2 Zonal and intrazonal soils

Some of the earliest classifications of soils employed the concept of zonality. *Zonal* soils are soils that simply reflect broad environmental conditions, such as podzols (Spodosols) of the boreal climatic/vegetation zone (coniferous forests of the mid-high latitudes); this concept stems from the work done by Dochuchayev at the end of the nineteenth century. Another example is the grassland soil of the dry mid-latitudes, an environment known as prairie, pampas, or steppe, and which typically features chernozems (Mollisols). Intrazonal soils, on the other hand, are those soil types that are governed by local conditions, such as poor drainage (gleys/hydromorphic soils) or their parent material (e.g. Andosols). Soils developed during one set of soil forming conditions are termed monogenetic soils, whereas polygenetic soils represent more than one, that is, superimposed, soil forming events over time. Good examples of monogenetic soils would be rendzinas (A/C horizons on limestone/chalk) or other Mollisols developed under grassland during the Holocene. Pleistocene and pre-Pleistocene paleosols are on the other hand much more likely to be polygenetic in character (Retallack, 2001; Wright, 1986). For instance, a Pleistocene soil could have developed as an Alfisol during an interglacial, but formed a fragipan and became gleyed as cold conditions of a glacial event ensued (Fedoroff and Goldberg, 1982; Kemp, 1985). As described earlier, some European podzols are polygenetic having formed as Alfisols under woodland before Bronze Age clearances led to podzol formation under heath.

3.3.3 Organisms

The uppermost soil horizon is the most organic because it is the focus of plant growth and associated biological activity (A1/Ah horizon). Fungae, mosses, and lichens, invade sediments that are newly exposed to subaerial weathering, and these are soon followed by higher plants, such as grasses, shrubs, and trees, which thrive when deeper, rootable soils begin to form. Close inspection of ash layers produced by burning on archaeological sites often reveals working by soil fauna and growth of lichens and mosses, which initiate weathering of such materials. Topsoils or A horizons develop through the accumulation and mixing of organic matter by soil fauna. Organic matter is essentially formed of carbon (C) and nitrogen (N), and organic matter breakdown (e.g. humification) is accelerated by the presence of nitrogen, so that plant materials like the grasses, for example, with the greatest amounts of nitrogen (low C : N ratio of <25) break down more rapidly than oak and beech forest litter (C : N of 30–40) and pine and Ericaceous litter (C : N of >60) (Duchaufour, 1982: 44–45). These differences are reflected in decreasing levels of biological activity present in Mull→Moder→Mor horizons (see Table 3.2).

An understanding of A horizon formation is also important in archaeology. Artifacts and traces of human activity, such as hearths and charcoal, are jumbled and scattered by soil mixing. This can occur through root growth, earthworm and insect burrowing, or more destructively by mammals such as moles, gophers, and rabbits. In earthworm-worked soils that are also burrowed by small mammals (voles and moles), artifacts become both buried by earthworm casting and mixed throughout the topmost 20–40 cm of the soil, after only a few decades. This was first demonstrated by Darwin but has been further elucidated (Armour-Chelu and Andrews, 1994; Atkinson, 1957; Balek, 2002; Darwin, 1888; Johnson, 2002; Van Nest, 2002). The effect of biota on soils and archaeological deposits are homogenization and porosity formation. Size, character, and abundance of excrements produced by soil invertebrates, their burrows (sometimes termed biogalleries), and channels formed by roots, reflect types of biota. The analysis of these contributes to the identification of past environments – both local and regional – and may also indicate modern disturbance. An important example of understanding sedimentation versus bioturbation can be cited from the Lower Palaeolithic site of Boxgrove, United Kingdom (see Fig. 7.1). In Unit IVb, artifacts left on ephemeral landsurfaces formed on lagoonal mudflats and buried by sedimentation, remain *in situ* (Roberts and Parfitt, 1999). In contrast, flints deposited later (Unit IVc) on a more mature terrestrial land surface (paleosol) that underwent bioturbation by roots, invertebrates, and small mammals spread both vertically and horizontally. The recognition of bioturbation on a site is therefore essential if dateable materials (e.g. by ^{14}C) and artifacts are to be interpreted correctly, including correlation within and among sites (Stiner *et al.*, 2001). Entire stratigraphic columns can be modified by bioturbation, and if this is not recognized, nonsensical analyses and interpretations ensue (Johnson, 2002). Lastly, an understanding of the factor "organisms" is also important in cultural and environmental archaeology. It is often an essential first step to calculate the natural resources, in terms of vegetation type and fauna, that are available to human populations in a given soil landscape. For example, grasslands provide grazing for both stock and wild herbivores, whereas "managed" woodlands supply fuel, wild fruit, nuts, as well as game (Evans, 1975). Neolithic to modern herders have also used "leaf hay" from trees to provide fodder for their animals. Equally, coasts can supply seaweed for manuring.

3.3.4 Relief

Soils are influenced by topography. Steeply sloping ridges are generally eroded and produce thin, droughty (excessively drained) soils. Eroded materials accumulate in valley bottoms producing thick, cumulative soils. Relative relief, that is, the difference in altitude between the highest ground and the lowest, and steepness of slope, have implications for both the soil and human landscape. In areas of steep slopes, such as 30–40° and marked relative relief, soils of different kinds will be closely juxtaposed, and humans will have access to the varied ecosystems present. In Mediterranean and arid areas, droughty south-facing slopes provide grazing land, whereas north-facing slopes are more moisture retentive and more likely to carry scrub and woodland. The constraints of slope and aspect also govern land suitable for agriculture and pasture in the high latitudes, because they receive more sunlight compared to less sunny slopes, which are best left to forestry. Sloping ground may also be managed by terracing to produce agricultural land (Sandor, 1992). Low slope areas, such as alluvial plains may have lower soil variability but supply large areas of flat, easily worked Class 1 land for cultivation, where there are very minor or no physical limitations to use (see land capability classification; Chapter 15) (Bibby and Mackney, 1972;

Klingebiel and Montgomery, 1961). Moreover, access to rivers will permit the gathering of wetland resources.

Relief also strongly governs drainage and depth to groundwater, and their control over soil type. This important concept has been modeled in soil science as the *catena*: the lateral variation in soil profile types across a crest-hollow transect, for a given geological strata (Fig. 3.18). Groundwater, which respects the slope of the catena, and which varies seasonally, reaches the surface in the valley bottom where it is expressed as a river, swamps/wetlands. Where permeable strata overlie impermeable rock or where the slope profile may coincide with the groundwater profile, a spring line or boggy ground may occur. Archaeological mapping may detect a coincidence of settlements and specific soil types (Ampe and Langohr, 1996), and a spring line, whereas the existence of boggy ground may promote organic preservation (Table 3.1).

3.3.5 Parent material

Rock/sediment type, and even the kind of archaeological deposit present on a site, governs the way pedological processes act. First, some parent materials, such as hard and poorly weatherable acid igneous rocks, or sediments that are derived from deposits that have already been strongly weathered (e.g. glaciofluvial sands) tend to produce acid soils (e.g. pH 3.5–5.5). On the other hand, soils on soft calcareous rocks like chalk give rise to base-rich alkaline (pH 7.5–8.5) soils, as these contain basic cations (Ca^{2+}, Mg^{2+}, K^+, and Na^+). These two extremes produce soils with different humus forms, levels of biological activity, and the way cultural and environmental materials may be preserved (see Table 3.1; Chapter 12). Soft clays are easily eroded, but commonly produce poorly drained ground.

Soil mineralogy is also an important consideration. Much of the coarse (sand and silt) fraction is composed of highly resistant quartz (SiO_2), with small amounts of feldspars and mica that are ultimately derived from igneous rocks. Sedimentary minerals like glauconite are more easily weathered and yield iron into the soil. Also present are the more rare "heavy" minerals (specific gravity >2.9), which can be separated from quartz (specific gravity 2.65) and feldspar (specific gravity 2.5–2.76) by heavy liquid or magnetic separation, or by centrifuging (see Chapter 16). Heavy minerals include iron oxides, sometimes termed opaque minerals because they are not translucent under the petrological microscope. Some examples are the magnetic mineral magnetite, hematite, and pyrite, although pyrite has often oxidized into limonite. There are over 30 translucent minerals, heavy minerals, such as the micas (biotite and muscovite), amphiboles, and pyroxenes, as well as the so-called semiprecious minerals that include garnet and topaz (Chapter 1). Such mineral suites are used to assess the provenance of deposits, and degree of weathering can be appreciated from the tourmaline : zircon ratio (Krumbein and Pettijohn, 1938). Minerals are now commonly identified on the basis of their chemical composition through SEM/EDAX techniques. Layers of tephra or volcanic glass have been crucial in relating sediment, including peat to sequences, and human events, such as the occupation of Iceland to specific historically dated volcanic activity (Gilbertson, 1995).

Clay minerals play an important role in governing the structure and fertility of the soil. They initially form as hydrated silicates of aluminum, iron, and magnesium, by rock weathering, and are termed phyllosilicates (Birkeland, 1999; Delvigne and Stoops, 1990). These are layered minerals that are divided into single (1 : 1) layered (kaolinite–serpentinite group) and double (2 : 1) layered (e.g. smectite, vermiculite, mica, and chlorite groups). 1 : 1 clays are less active chemically than 2 : 1 clays. In addition, 2 : 1 clay types such as illite and montmorillonite also absorb more water and are known as swelling clays in Vertisols. The deep (>1 m) cracking of such soils have the

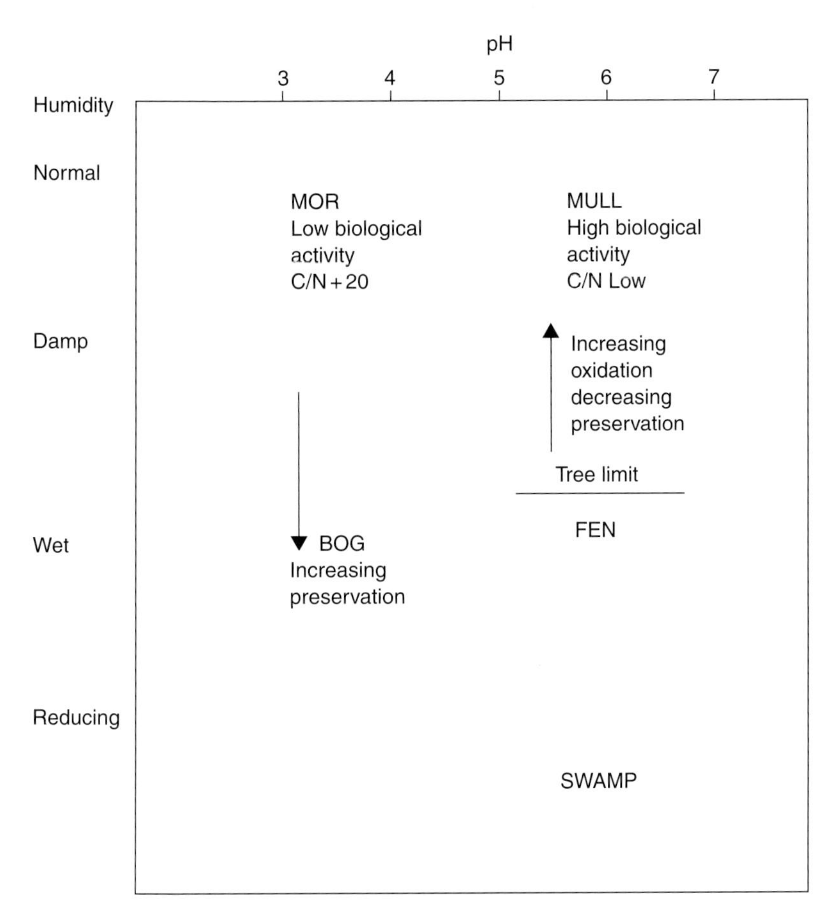

FIGURE 3.17 Generalized relations between soil pH, humidity, and humus types (modified from Pearsall, 1952; Macphail, 1987).

potential of markedly affecting vertical distributions of artifacts (Butzer, 1982). Clay types also reflect environment and the weathering/pedological history of landscapes. For example, kaolinite is associated with most weathered tropical soils – Ultisols, whereas the fibrous clay mineral palygorskite forms under extreme arid conditions, and at Boxgrove (see Chapter 7) occurred as an inherited mineral from the Cretaceous chalk geology (Birkeland, 1999; Roberts and Parfitt, 1999). Allophane, an amorphous clay mineral, is typically formed in volcanic Andosols (Tan, 1984).

Light sandy soils are easily worked and may have been the first to be cultivated in a region. Over the long term, however, they are often not suitable for sustainable agriculture without manuring. An example of this fertility problem is found in central Sweden, where Iron Age and later migrating populations needed to employ mixed farming. The stock (cattle) provided the manure necessary to grow barley (Engelmark, 1992; Viklund *et al.*, 1998). Another example is the Neolithic exploitation of the loess belt across northwest Europe (Ampe and Langohr, 1996). The poorly drained heavy clay soils of Europe could not be cultivated until the Romans introduced ox ploughing using a metal ploughshare (Barker, 1985); the introduction of the efficient "swivel plough" likely spread from the Roman Rhinelands (Germany) to Early Medieval (tenth–eleventh century) England (Henning, 1996). In contrast, in the Far East, areas of poorly drained soils were purposely landscaped in order to grow rice using wet cultivation of paddy fields (as opposed to dry rice cultivation) (see Chapter 9). Additionally, Limbrey (1990) identifies the self-mulching characteristics of Vertisols as a factor encouraging the earliest arable activity in southwest Asia.

3.3.6 Time

The length of time that a sediment or archaeological deposit is exposed to subaerial weathering controls the degree of alteration and the development of soil formation that can be expected. Time in itself is not a soil forming process, but the passage of time permits the other factors to drive soil forming processes. Estimating the numerical or chronometric age of a soil is notoriously difficult, but *relative dating* has been identified from some famous *chronosequences* where progressively older soils have been studied from marine and river terrace sequences and where glacial ice progressively retreated, for example, in North America (Birkeland, 1992). The major effect of time can be identified from soil maturity expressed in the formation of soil horizons. Short-lived weathering, termed "soil ripening" on reclaimed polders of Holland (Bal, 1982), produces immature A horizons (Entisols), whereas fully developed soils are recognized by their A, B, C horizons (Fig. 3.18). As we have mentioned, soil classification is in fact based upon the identification of a number of well-defined or diagnostic soil horizons (Soil Conservation Service, 1994; Soil Survey Staff, 1999) (see Table 3.2).

There are a several archaeological examples of attempts to "date" cultural phases and specific human activities by the maturity or immaturity of buried soils. A classic example of this is the analysis of an immature buried soil that marked a short-lived standstill phase in the construction of Caesar's Camp, Keston, Kent. The ephemeral surface became humus stained in consequence of its being briefly vegetated (during a period of stasis) before the next construction episode. Perhaps of wider importance is the ^{14}C dating of humus present in the subsoils of podzols in Belgium and France, which effectively increased the age estimate of some podzols from an expected ~3,500 years BP (Bronze Age) to ~6,000–7,000 years BP-Neolithic (Mesolithic) (Dimbleby, 1962; Guillet, 1982).

Certain soil indices that reflect the age of soils have been suggested by Jennifer Harden (Harden, 1982) and many others. These measurements include surface organic matter accumulation, leaching of $CaCO_3$, Bt (argillic horizon) development, a changing clay mineralogy, and red color (Birkeland, 1992). More recently, the study of experimental earthworks has shown just how rapidly changes to soils occur after burial, within the first 32 years on both base rich (chalk rendzinas) and acid (sandy podzols) soils (Bell *et al.*, 1996; Crowther *et al.*, 1996; Macphail *et al.*, 2003)(see Chapter 12). When the fills of ditches, graves, and pits are investigated, soil analyses can be applied to determine the rate of infill. Both slow infilling, that is, "ditch silting" and rapid dumping to fill pits or graves, can be identified. Likewise, thorough working by biological activity of once heterogeneous deposits may mark a standstill phase or stasis (Box 3.1). When investigating buried soils, one must strive to answer crucial questions: "how long after a certain activity occurred was the site buried?; days, weeks, years?"; "was the site: (1) cleared by fire, (2) ploughed, (3) trampled, or (4) razed (burned down)?" These questions underscore the importance of studying modifications to a site through time after it has ceased to be occupied (Chapter 12).

3.3.7 The interaction of humans and the soil forming factors

The above simple models are a first step in understanding soils and establishing the applicability of soil science to both cultural and environmental archaeology. This initial comprehension of pedogenesis is essential, but workers in geoarchaeology and forensic archaeology (Chapter 14) must become more aware of the complicating and commonly interactive environmental factors. Examples of such polygenetic soils are given below.

Montane soils are influenced by climate, itself a reflection of altitude, relief, and vegetation (Legros, 1992). Humans, however, have been successful in adapting to these environments: the construction of terraces offset problems of steep slope and their naturally shallow soils (Chapter 4). Such terracing controls soil loss by erosion, leaching of nutrients and low moisture content, and thus maintains soil depths sufficient for agronomy (Sandor, 1992). In the French and Italian Alps the ameliorating activities of terracing and stock management (pastures) by Chalcolithic (Copper Age) and later humans, brought about the development of brown soils at altitudes (1300–1900 m); this area naturally exhibited *zonal* podzols that formed under larch woodland (Courty *et al.*, 1989). In this case, exploitation by humans represents an example of the major influence that organisms play on soil formation.

It may seem strange that large populations are attracted to the potentially life-threatening areas around volcanoes. Here, however, *parent material* is a potent factor, because soils formed in volcanic ashes (andosols) are easily worked and naturally fertile, as their high organic matter content encourages high biological activity (Tan, 1984). An example from both prehistory and the classical world is Vesuvius, where arable soils were repeatedly buried under volcanic ash since the Bronze Age, long before the city and gardens of Pompeii were buried in AD 79 (Fulford and Wallace-Hadrill, 1995–6)

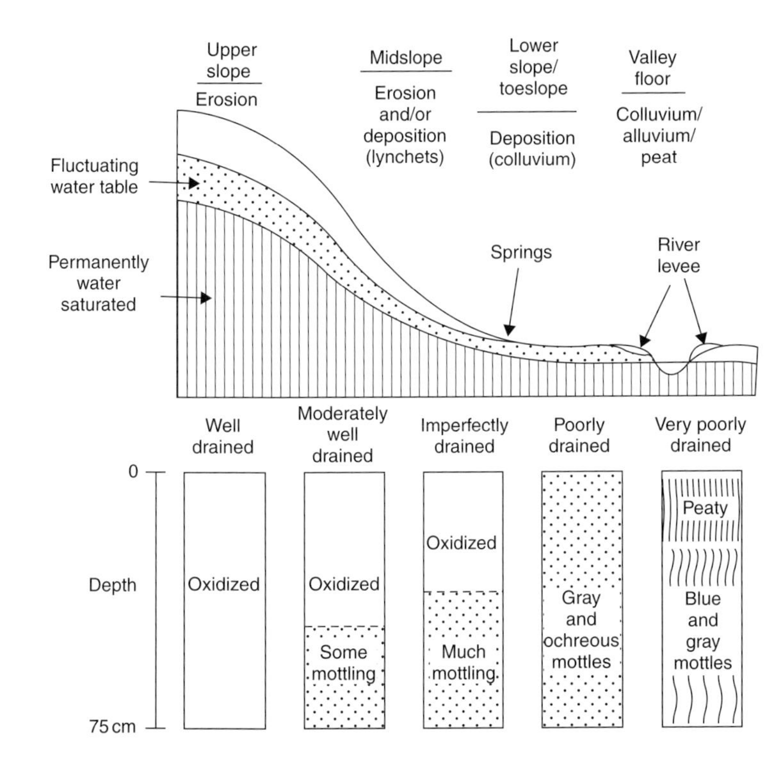

FIGURE 3.18 The soil *catena*; soil drainage classes in a moist temperate environment (after Avery, 1990); also showing generalized slope units, areas of predominant soil erosion or deposition and location of springs and river levees (see Table 4.1).

(see Chapter 2). Equally, the early hominid remains at Dmanisi (Gabunia *et al.*, 2000) that date to 1.7 my ago occur within a sequence of basalts and volcanic ash, the latter in part contributing to the good preservation of the human and other animal remains. It can also be noted that in the Auvergne (Massif Centrale, France) and in the Garroxta (northeast Spain), Pleistocene basalt flows provide upstanding topography that were often utilized by humans who built fortified villages and towns.

It is not only Roman Pompeii that was buried by volcanic ash. Buried Early Bronze Age (ca 2300–1700 BC) field systems have been found in Campania (southern Italy) below pyroclastic deposits from the Somma–Vesuvius and Flegrean Fields volcanic complexes. Elsewhere, Edo Period arable fields (Japan) and ochre covered floors (Alaska), have been preserved by volcanic ash falls (Hilton, 2002; Sunaga *et al.*, 2003). The site of Ceren, El Salvador, exhibits remarkable views of buildings and fields that were buried beneath several meters of ash (Sheets, 1992).

3.4 Important soil forming processes

The recognition of soil types in the field and from laboratory data is essential if correct soil classifications and accurate interpretations of landscape history are to be made. More importantly, the soil forming processes that govern soil type have to be understood if we are going to be able to comprehend past human environments; plant, animal, and human interactions, site formation processes, and preservation of sites and their artifacts.

3.4.1 Weathering

Soil formation and the development of horizons begin with weathering of the parent material. It is well known that weathering is influenced by temperature, and by warm conditions favoring chemical weathering (see Fig. 3.17).

Physical weathering can take place under conditions of heating and cooling, or salt crystal growth in hot, arid conditions, or freezing and thawing of water in rocks, in humid conditions. In the latter case, temperatures alternate around the freezing point, and rocks broken up by splitting form scree deposits that accumulate along talus slopes (scree) (see Box 3.2; Cold Climate Soils). Cryoturbation involves physical mixing of the upper regolith (e.g. rotation, horizontal, and vertical movement) by differential freezing and thawing, commonly in the seasonally freezing and thawing "active" zone above the frozen substrate or permafrost zone (Gelisols; Table 3.3; Box 3.2). Commonly on slopes, material in this active zone moves down slope as gravity flows, sometimes as turf-confined solifluction lobes. Where both solifluction and meltwater act, the deposits can be termed as soliflual (Catt, 1986, 1990). At both open-air and cave Palaeolithic sites, deposits (including hearths) can be either weakly or strongly disrupted by cryoturbation (Bertran, 1994; Goldberg, 1979b; Goldberg *et al.*, 2003; Macphail *et al.*, 1994; Texier *et al.*, 1998). Commonly, rock-falls in caves (*éboulis*) reflect major cold episodes in the Quaternary (Courty, 1989; Laville *et al.*, 1980; Macphail and Goldberg, 1999).

Chemical weathering includes the decay and alteration of minerals, as well as the breakdown of soil organic matter (see below). Typical geo-chemical processes include hydrolysis, related oxidation and reduction, and solution. To quote Duchaufour (1982: 7), "hydrolysis, or the effect of water containing such entities as Hydrogen ions (H^+) is the most important process of weathering" on minerals. For example, additions of hydrogen ions and water with feldspars is to produce clay minerals through hydrolysis (Birkeland, 1999). H^+ can be supplied from atmospheric sources ($CO_2 + H_2O \rightarrow H_2CO_3$) and organic acids from plants. In waterlogged/hydromorphic conditions, iron as Fe^{2+}

can be lost in solution, and this process is called reduction. The formation of the diagnostic cambic Bw horizon (weathered B horizon of a Cambisol) (Soil Conservation Service, 1994) (see Table 3.4) is expressed by the subsurface neoformation of clays, the replacement geological bedding by soil structures, removal of carbonates, and development of redder colors than the parent material because of the formation of hydrated iron oxides.

3.4.2 Leaching and clay eluviation

Weak carbonic acid forms from the mixing of rainwater and dissolved CO_2: $CaCO_3 + CO_2 + H_2O \leftrightarrow Ca^{2+} + 2HCO_3$. (This, along with organic acids derived from plants can weather calcareous materials, as well as human-made lime-based plaster and mortar. They also can leach base cations Ca^{2+}, K^+, Mg^{2+}, and Na^+ by replacing them with H^+. Soil reaction becomes acid ($pH < 6.5$). Soil that was well flocculated because clay was bonded by Ca^{2+} becomes more easily disaggregated and the process of pedogenic clay translocation can be induced. Pedogenic clay translocation is much less likely in calcareous, base-rich regimes (pH 7.5–8.0), and should not be confused with the physical inwash of calcareous fine soil induced by sedimentation and trampling, for example. These nonpedogenic clays are usually coarser, more poorly sorted, and can have fine micro-charcoal inclusions.

In addition to measuring pH, base-rich versus acid status of a soil can be deduced from its vegetation, surficial humus type, and associated mesofauna. For instance, a Mull topsoil has only a litter (L) layer (Table 3.2), because biological breakdown of organic matter produced by plants, such as leaf litter, is rapid (C : N of 5–15) (see Section 3.4.3). Such soils often have a neutral pH (6.5–7.5), and in western temperate countries earthworms such as *Lumbricus terrestris* make up the main part of the large invertebrate biomass in such soils. Termites and millipedes also thrive in neutral to base-rich soils. Plants requiring Ca, termed calciphyles, for example, Garrigue in the Mediterranean, and specific chalk and limestone grassland communities in the United Kingdom, are typically present (Polunin and Smythies, 1973; Rodwell, 1992). Mollisols, the soils of grassland and prairies, have characteristically high levels of biological activity. This is recorded at the microscale by what is termed a "total excremental or biological fabric," where millimeter-sized mammilated organo-mineral excrements dominate. This Mull humus (L) topsoil may also carry a population of moles who feed on earthworms; voles and prairie dogs eat grass roots in Mollisols (Table 3.3). Large grazing animals fertilize these soils with their dung, and this activity may be recorded by the presence of dung beetles and their burrows; hence the importance of insect studies in archaeology to infer pastoralism (Robinson, 1991).

It can also be noted that Collembola (Springtails) are more numerous above pH 5.4. If pH decreases, earthworms – which are strong mixers of soil and decomposers of organic matter in and on the soil – die out, and Enchytraeid worms become more common below pH 4.8 (Mücher, 1997). Mineral soil becomes much less mixed upward from less weathered subsoil horizons, because Enchytraeid worms are concentrated in the topmost few centimeters of the soil. Organic matter starts to accumulate on the soil surface, and thus more plant acids become available to leach out bases. Overall leaching is thus exacerbated, and bases are lost to ground water.

The surface organic matter thickens with the development of a fermentation (F) layer that is composed of mesofauna excrements and comminuted plant remains. This LF horizon is known as a Moder humus and is associated with Enchytraeids, insects, and their larvae, and is most commonly linked to broadleaved forests with a pH of about 4.5 and a C : N of 15–25 (Babel, 1975). The most commonly associated soils are Alfisols (Fig. 3.18; Table 3.3), which are characterized by pedogenic clay eluviation or clay washing (*lessivage*) from the E

FIGURE 3.19 Examples of soil types: (a) Field photo of typical Spodosol (podzol) formed under an acid lowland heath vegetation that probably dates from the Bronze Age in this area (Wareham, Dorset, United Kingdom; one of the control soil profiles at the Wareham Experimental Earthwork, 1996; Macphail *et al.*, 2003); note – from the top downward – the superficial organic litter (L) and fermentation (F) layers, and the very dark grey Ah, pale grey to white A2, and blackish Bh horizons formed in iron-poor Tertiary sands; (b) Field photo of typical deep Alfisol (Luvisol, argillic brown earth) formed in fine sandy silt loam brickearth (Pleistocene river terrace drift, here formed out of loess and alluvial sand); note that at this Roman site (Leadenhall St, City of London, United Kingdom), the Ah horizon has been truncated. The soil displays a deeply leached A2 (Eb) upper subsoil horizon, which had been extending into the A2&Bt and Bt horizons. These soils were extensively quarried from Roman times onward for use as a building "clay" (See Table 13.4). Plate 3.2a: Photomicrograph of diagnostic argillic horizon of an Alfisol. Here fluvial brickearth drift on the Chalk at Balksbury Iron Age Camp (Hillfort), Hampshire, United Kingdom (Ellis and Rawlings, 2001; Wainright and Davies, 1995), had developed into an Alfisol; note microlaminated clay coatings on ped surfaces and coating voids – although such coatings are often "dated" to the Atlantic and Sub-Boreal Periods in Europe, it is quite likely that some of the clay translocation relates to human impact (see Chapter 9). PPL, frame length is ~5.5 mm; Plate 3.2b: As Figure 3.3c, but under crossed polarized light (XPL) – note high interference colors of oriented clay in the void coatings.

(United States) or Eb (United Kingdom) or A2 horizon, with clay translocation into the underlying clay-enriched subsoil Bt horizon. This then constitutes the "diagnostic" argillic horizon. This horizon exhibits the presence of translocated clay in the form of clay coatings (McKeague, 1983) that can be seen in the field but are more clearly visible in thin section. Past forest soils, either from the Holocene or from interglacial (temperate) episodes within the Pleistocene, have been identified by the presence of such argillic horizons (Kemp, 1986, 1999; Kemp *et al.*, 1998). All horizons that seem to be argillic in character, however, are not necessarily relict of a forest soil development (Goldberg and Arpin, 2001; Goldberg *et al.*, 2001). Pre-Quaternary, Pleistocene, and Holocene (e.g. archaeologically buried) paleosols

TABLE 3.4 Selected types of soil horizons (simplified) (more detailed explanations are found in Bridges (1970, 1990) and Birkeland (1999))

a Basic characteristics of the A, B, C, and R soil horizon system

General process	Chief characteristic and result	Horizon
Homogenization and weathering (sometimes with leaching or accumulation)	Biological activity, organic matter accumulation, weathering sometimes with mineral accumulation or loss	A
Homogenization, weathering, mineral formation (neoformation) (sometimes with leaching or accumulation)	Often breakdown of primary minerals and formation of new ones; biological mixing; vertical additions from other horizons and/or losses to other horizons; vertical or lateral additions or subtractions by groundwater or throughflow	B
Weathering (see Table 3.4b)	Weathered, poorly to unconsolidated parent material (sometimes with minor losses and additions – as B horizon above)	C
	Unweathered hard Bedrock	R

b Weathering

Some general processes	Some characteristics and results	Horizon
Ripening (oxidation of organic matter and minor weathering; e.g. decalcification, iron mobilization) through to homogenization and mineral breakdown	Minor loss of carbonate, iron staining, incorporation of humus through to total biological mixing (biological microfabrics), organic matter incorporation and mineral weathering	A
Mineral breakdown and neoformation	Organic matter incorporation, clay formation, release of iron etc., biological mixing (biological microfabrics) (haploidization)	Bw

c Leaching, clay translocation, and podzolization

Process	Chief characteristic and result	Horizon
Leaching ▼▼▼▼	↓ ↓ ↓ Leaching and translocation of clay/humus/ sesquioxides (Fe_2O_3–Al_2O_3)	A2
Illuviation ⊥⊥⊥⊥	• Clay enriched • Humus enriched • Sesquioxide enriched	• Bt • Bh • Bs

d Hydromorphism and gleying

Process	Some characteristics and results	Horizons
Leaching (surface water stagnation) ▼▼▼▼ Leaching and precipitation (water table fluctuations) ▼▲▼▲	↓ ↓ ↓ Anaerobic leaching and translocation of iron and manganese and/or →↓↑→ reprecipitation of iron (Fe^{+++}) and manganese (Mn^{+++})	Ag and Bg
Leaching and loss from parent material ►►►►	Anaerobic reduction of iron (Fe^{++}) and manganese (Mn^{++}) and/or their removal →	(BG and CG)

TABLE 3.4 (*Cont.*)

e Calcification and alkalization

Process	Some characteristics and results	Horizons
Calcium carbonate movement (vertical) ▼▼▼▼ (throughflow) ▶▶▶▶ and accumulation ⊥⊥⊥⊥	Precipitation of calcium carbonate in voids, impregnating the soil and as root pseudomorphs (rhizoliths) (calcretes)	Calcic Bk and Km
Sulfate (SO_4^-) movement (vertical) ▼▼▼▼ (throughflow) ▶▶▶▶ and accumulation ⊥⊥⊥⊥	Precipitation of gypsum ($CaSO_4$) often as lozenge shaped crystals within voids or within the soil matrix, which then can become fragmented	Gypsic By
Vertical movement and precipitation of Cl^- and Na^+ ▲▲▲▲ (throughflow) ▶▶▶▶ and accumulation	Surface salt crystal (halite) efflorescence (salt crust) and subsoil accumulations (halite is not normally preserved in thin sections)	Natric Btn Salic Bz

have been studied (see Section 3.1), but it is obvious that long burial leads to diagenetic transformations to original soil features. Experimental archaeology has in fact demonstrated how quickly some changes occur in buried soils (see Chapter 12).

3.4.3 Podzolization

On substrates that contain few weatherable minerals and that are freely draining, such as glacial outwash or decalcified dune sands, leaching can rapidly cause the development of acid soils. At increasingly acid pHs clay is destroyed and Al (aluminum) is released. In such environments (pH 3–3.5) acidophyle plants dominate and commonly include *Ericaceae* (heath plants); in boreal conifer forests more rapid podzolization can take place: iron pan formation, for example, can occur in a decade in newly planted areas with high rainfalls. These types of plants produce leachates (e.g. chelates) that chemically combine organic matter and sesquioxides (Al_2O_3 and Fe_2O_3), thus making them water soluble and mobile. This process of destruction of clay and translocation of sesquioxides and humus, from a leached Albic E horizon to a diagnostic spodic (Bh/Bs) subsoil horizon, is termed podzolization (Fig. 3.2). In thin section, the E horizon is purely made up of quartz sand with very small amounts of organic matter. The spodic horizon is composed of two microfabrics, recently formed polymorphic (pellety) material and very much older monomorphic (grain coatings and cement) humic gels of Fe and Al

(De Coninck, 1980). The hard, cemented spodic horizons of podzols are therefore easy to recognize both in the field and under the microscope. Sometimes cementation causes localized drainage impedance and waterlogging (hydromorphism) causes an ironpan (Bf) to develop because Fe^{3+} is converted to mobile Fe^{2+}. Soil phosphorus also accumulates in spodic horizons, and phosphate surveys usually sample this horizon to examine ancient patterns of occupation in present-day areas of podzols (i.e. patterns of accumulated P that have been leached from the soil surface).

The form of organic matter accumulation that develops on podzols is termed a Mor humus and comprises L, F, and H (humus) layers (see Table 3.2). The H layer is typically composed of amorphous organic matter, and takes a longer time to form compared to the L and F horizons, because all recognizable plant material and excrements have been decomposed by bacteria. Under normal conditions of high acidity, for example, pH 3.5, and high C : N ratio (>25), this form of surface humus accumulation may include organic matter that is over a thousand years old and extant Holocene spodic horizons yield dates as old as 3,000 years that include both ancient and modern humus, providing what are termed "mean residence dates" (Guillet, 1982).

Podzols are typical of heath and coniferous woodland on poor substrates of the mid-latitudes, and across much of the boreal and taiga regions. When buried, for example, by Bronze Age barrows, these soils provide useful stratified pollen sequences because of their acidity and slow organic matter breakdown. Dimbleby and others (Behre, 1986; Dimbleby, 1962, 1985; Scaife and Macphail, 1983) showed, how heathlands and moorlands formed in once-wooded landscapes. Such change is demonstrated by the presence of a relict argillic (Bt) horizons overprinted by spodic horizons. This soil degradation sequence, forest Alfisol → heathland Spodosol demonstrates an important soil transformation (Gebhardt, 1993; Macphail, 1992). Equally, because of very low levels of biological mixing, pollen studies of surface horizons in the present-day coniferous woodlands of central Sweden, revealed evidence of cereal pollen from fields that had not been cultivated for 1,000 years (Sergeström, 1991).

3.4.4 Calcification, salinization, and solodization

In warm and dry regions, or where seasonal precipitation is strongly outweighed by evapotranspiration, soluble salts, such as Ca^{2+}, Mg^{2+}, Na^{+}, and K^{+} that are lost by leaching from more humid soil areas, become concentrated; they occur in a group of soils called Aridisols. Such soils are typified by the sparse growth of plants that are drought resistant and sometimes salt tolerant, and include grasses and cacti, with mesquite, creosote bush, yucca, and sagebrush in southwest United States and tamarisk in Asia (Eyre, 1968) (see Chapter 7). Levels of organic matter and biological activity reflect precipitation and can be very low, although a mollic epipedon (dark colored humic and bioactive topsoil) can be present in prairie areas (Table 3.2). Here, for example, calcic (Bk) and petrocalcic (cemented Km) diagnostic horizons form through the precipitation of calcium carbonate in subsoils; the depth of the $CaCO_3$ enriched horizon increases with increasing precipitation, resulting in the formation of calcretes (Raghavan *et al.*, 1991). Even in temperate areas, groundwater may supply ions that are drawn up-profile by plant roots, and that are precipitated as root pseudomorphs or rhizoliths, or as nodules within the soil. Past climatic fluctuations, effecting major fluctuations in ground water, for example, have been revealed from the investigation of calcite nodules in semiarid regions of India (Courty and Fedoroff, 1985). In the Negev and Sinai desert, secondary calcium carbonate precipitation, sometimes associated with archaeological sites, has proven a useful resource for dating and climatic inference (Goldberg, 1977; Magaritz *et al.*, 1981, 1987; Wieder and Gvirtzman,

1999). Gypsum (Table 3.1) accumulations (gypsic horizon [By]) have proven to be both a useful indicator of past arid conditions and very damaging to archaeological stratigraphy and materials. In the latter case, the growth of gypsum crystals can substantially fragment mudbricks (Goldberg, 1979a). Gypsum has also been shown to be an important constituent of the soils present in the raised fields of Maya sites in Mesoamerica, and the process of gypsum crystal growth has been cited as a pedological mechanism contributing to the present height of these raised fields (Jacob, 1995; Pohl *et al.*, 1990).

Saline soils, *sensu lato*, are formed mainly in arid steppe and semidesert areas where soluble salts are transported from high ground or in ground water from a source of sodium salts, and which are then concentrated in low-lying ground. Amounts of salt are measured in terms of the ratio between Na^+ and total exchange capacity. A natric horizon, for example, contains more than 15% of its CEC (Cation Exchange Capacity) as Na^+ (Duchaufour, 1982). A salic horizon is simply one that is enriched with salts (e.g. halite – NaCl) more soluble than gypsum. Saline soils normally have a pH below 8.5, whereas a natric horizon may develop a pH above 8.5, and such alkaline soils may attain a pH as high as 10. In addition to pH, the presence of salts is measured through conductivity. Saline soils as identified in the United Kingdom have a conductivity of the saturation extract of more than 4.0 S m^{-1} at 20° (Avery, 1990). The presence of any salts in groundwater is most important in the preservation or corrosion of artifacts (Goren-Inbar *et al.*, 1992). Also when sodium ions are introduced in large amounts to stable soils, such as during marine inundation, they have an effect of dispersing clay and forming unstable soils (see below and Chapter 7) (French, 2003; Macphail, 1994).

In saline soils salt in solution is drawn to the surface by evaporation, and forms a white salt crust or salic horizon, but because the topsoil is still dominated by Ca^{2+}, Mg^{2+}, and K^+, the soil remains stable; such solonchak soils were once known as white alkali soils. Where, however, the presence of Na^+ becomes completely dominant, perhaps as high as 70% of CEC, solodization or alkalization takes place, and all clay and humus is dispersed, all aggregate stability is lost, and a nonaerated massive structure forms (Duchaufour, 1982: 434). During humid periods, dispersed humus and clay are translocated down-profile and coat the natric horizon in what have been called black alkali or solonetz soils.

Salts such as sodium carbonate ($NaCO_3$) can also form in saline soils, and can be found in environments as different as playas and coastal muds. In the latter case salt tolerant vegetation can be found, because of the influence of saline tidewater. In estuaries, saltmarsh plants such as glasswort grow on saline Entisols (recently formed soils; Table 3.3) formed in marine alluvium (Avery, 1990: 324).

3.4.5 Tropical (and Mediterranean) soils

The two main types of soils formed in warm and humid regions are Oxisols and Ultisols, and develop through what was once known as laterite soil formation, for example, through the leaching of silica. Other primary pedogenic processes are, weathering of primary minerals, rapid biodegradation of the near surface organic matter, deep leaching, and the concentration of free oxides (Fe^{3+} and Al^{3+}) as gibbsite ($Al[OH]_3$), ochreous goethite ($\alpha FeO[OH]$), or reddish haematite (αFe_2O_3) (Duchaufour, 1982). Leaching of silica can lead to neoformation of 2 : 1 lattice clays (e.g. illite) rich in Si. Rubefication (reddening) of the subsoil may also occur if there is a marked dry season. A special case of rubefication of soils on limestone is the formation of the well-known Mediterranean reddish Ultisol called *terra rosa* (Table 3.3). Here, a typical brown Alfisol type known as *terra fusca* has developed in to a *terra rosa* from clays inherited from the rock after it has been weathered. In this case, rubefication is caused by the presence of some of the iron oxide becoming recrystallized as haematite (Lelong and Souchier, 1982). This process is important because many reddish palaeosols have been termed climatic indicators of Pleistocene interglacials (Kemp, 1986); such red soils have proven to be an ideal raw material for ochre production that was utilized from the Palaeolithic onward (Ferraris, 1997).

The mobility of iron in tropical soils can lead to down slope secondary hydromorphic ferric iron (Fe^{3+}) accumulations; in such cases ferrous iron (Fe^{2+}) in the groundwater precipitates to produce an "iron cuirasse" (Duchaufour, 1982: 405). Another form of iron and clay cemented tropical subsoil is the diagnostic plinthic horizon ("laterite"), that irreversibly hardens after repeated wetting and drying. A typical Ultisol develops under a forest cover when two combined processes occur: strong weathering produces a kaolinite- (1 : 1 lattice clay) dominated upper profile, whereas clay is translocated down-profile into the argillic subsoil. Such strongly leached soils have been known as planosols and yellow-brown podzolics. They are differentiated from Oxisols by the presence of clay translocation.

3.5 Conclusions

Pedogenic processes affect most archaeological sites. These may be short-term or long-term effects (e.g. prehistoric Raunds in the United Kingdom-see Chapters 16 and 17-and the Clovis sites of Arizona, United States-Holliday, 2004, Fig. 9.3). Basic soil science can also be applied to establishing amount and type of any postdepositional processes that may have affected a site, as well as providing snapshots of past environments through the study of buried soil profiles (Chapter 12). Armed with these essentials the geoarchaeologist can apply soil science techniques, as detailed in Chapter 16, in a focused way to tease out data that can then be applied independently to reconstructing past natural and cultural landscapes.

4

Hydrological systems I: slopes and slope deposits

4.1 Introduction

Slopes, streams, lakes, and wetlands are the most widespread and among the most important of geoarchaeological environments. They encompass a variety of readily available natural resources, and potentially furnish the right types of conditions for the preservation of archaeological sites and contexts. These environments are also the most written about of all the geoarchaeological situations (Blum and Valastro, 1992; Brown, 1997; Ferring, 1992; Frederick, 2001; Gladfelter, 2001; Howard *et al.*, 2003; Needham and Macklin, 1992; Mandel, 1992; Waters, 1986, 1992). The focal point of this environment revolves around water, which is not only necessary to sustain life but also serves to attract game, and furnishes water transport and raw materials, including wood, reeds, and sediments. These latter items serve a variety of functions and uses, including fuel, matting, basketry, pottery, and associated construction materials (principally wattle, mudbrick/adobe, or daub).

As is discussed in Chapter 5, fluvial systems also tend to preserve the context of archaeological sites and remains (Brown, 1997; Ferring, 1986, 2001; Howard *et al.*, 2003 Needham and Macklin, 1992) (The archaeology of estuarine environments is discussed in Chapter 7). Low energy inundations, of flood plains away from the higher energy channels, provide a favorable potential for preserving artifacts and features near to their original depositional contexts. The French sites of La Verberie (Audouze and Enloe, 1997) and Pincevent (Leroi-Gourhan and Brézillon, 1972) are noted for their evident preservation of the physical integrity and palaeoanthropological fidelity of the remains. The greatest concentrations of hand axes occur at the probable spring fed "waterhole" at the early hominid site of Boxgrove, United Kingdom (Roberts *et al.*, 1994, forthcoming).

In order to be able to gain an understanding of past human interactions with the hydrological system, and how geoarchaeologists can go about recognizing and interpreting them, it is first necessary to be aware of some of the aspects of how they operate. Alluvial environments can be viewed from any one of a number of standpoints, and here, we attempt to link geomorphic aspects with sedimentological ones, emphasizing how they interface with geoarchaeological considerations. Many basic works (e.g. Boggs, 2001; Bridge, 2003; Chorley *et al.*, 1984; Collinson, 1996; Marzo and Puigdefábregas, 1993; Miall, 1996; Reineck and Singh, 1980) can be consulted for additional details.

A prerequisite for investigating any landscape is the ability to identify areas of erosion, deposition (alluvium and colluvium), and locations of zero erosion or deposition. This chapter focuses upon landscapes with climates, where flowing water on slopes is the major mechanism for erosion and colluviation. Whereas sloping ground can be affected by water erosion, flat ground can suffer wind erosion (see Chapter 6). This present chapter spotlights slopes and the type of processes prevalent on them, as well as the possible associated archaeology. The susceptibility of certain soil types and their horizons to erosional processes are also outlined. Lastly, ways to recognize colluvium in the archaeological record are given, although types of human impact that initiate or encourage erosion are described later in Chapter 9.

4.2 Water movement on slopes

Water movement, and associated drainage channels, begins on upland positions of the watersheds in the landscape. On these slopes, precipitation falling as rain or snow leads to water flow that follows a number of options (Fig. 4.1) (Butzer, 1976). Initially water can infiltrate into the soil or joints or pores within the bedrock. With time – and depending on the environment and local conditions – such water can percolate laterally (*throughflow*) or vertically downward. In the latter case, it eventually reaches the groundwater table where it flows laterally; on the surface such water flow is expressed in the form of springs or as direct inputs into streams.

A number of factors influence the degree to which water is absorbed into the soil. Some of these factors include (Butzer, 1976a):

1 Vegetation cover (Fig. 4.2) and the way it intercepts and absorbs water: In contrast, bare slopes promote surface water flow that results in various forms of erosion, such as rilling, gullying, and sheet erosion (see below).
2 Soil texture, structure, and bedrock: water tends to percolate rapidly through sandy and open-structured soils, for example, and is impeded by dense, clayey soils or subsoils where void space has been infilled by clay, iron, or secondary carbonates. Jointed bedrock also promotes percolation.
3 Soil moisture is also a factor: Moist soils, which are limited in their ability to absorb additional water, will quickly become saturated, and any further water will accumulate on the surface or flow overland downslope.

Water that does not infiltrate into the subsurface may either rest in place, and in cases where there is no relief or there are horizontal surfaces or depressions it can remain in the form of ponds, or bogs, for example. The dynamics of water movement in the soil catena – that is, from plateau to river valley – was investigated by Dalrymple *et al.* (1968) who developed a nine unit landsurface soil water gravity model that helps account for numerous soil features (see Chapter 3). Water, standing on the surface can play a major role in soil formation (*pedogenesis*), resulting in the formation of *surface gley soils* with associated oxidation/reduction reactions that produce soil mottling overall, as is typical of bogs, there is a decrease in biological activity (see Chapter 3); in calcareous environments carbonate-rich spring deposits can form (e.g. tufa). Alternatively, water can move down slope. As shown in Figure 4.2, this situation is particularly true for bare slopes that are common in arid and semi-arid areas.

Down slope movements take on a number of forms that relate to the concentration of water and whether water flow is confined or not. Near the crest of slopes, water that just begins to flow typically takes the form of a millimeter-thick irregular sheets of water (*sheetwash* or *sheet flow*) that can transport dissolved and fine-grained load (Butzer, 1976; Chorley *et al.*, 1984).

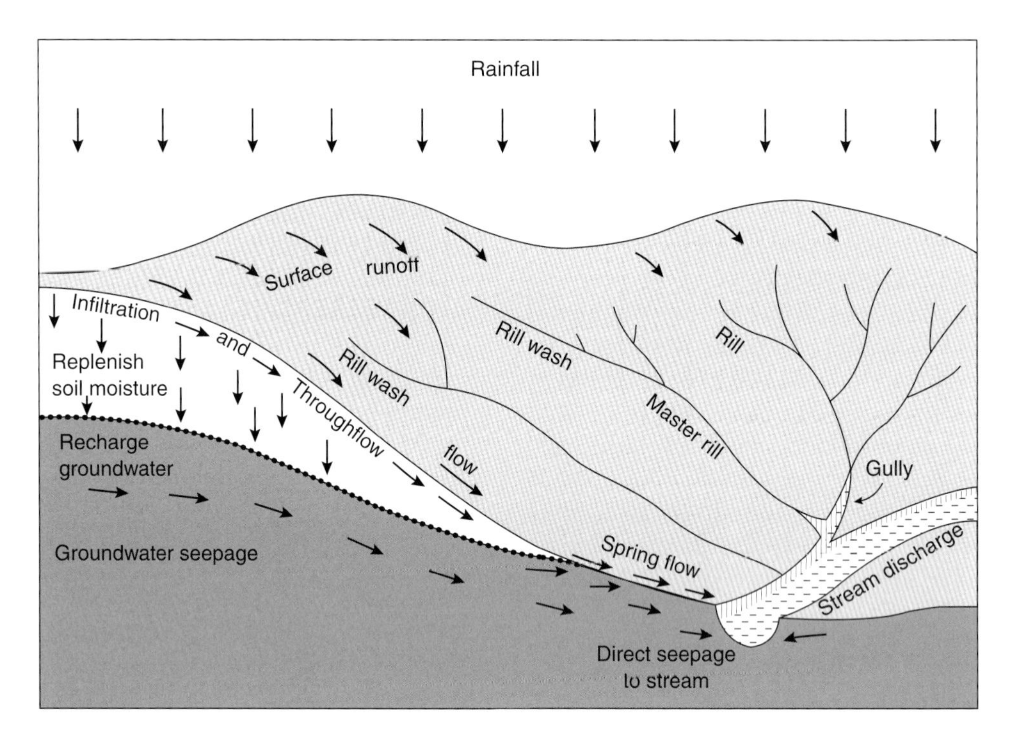

FIGURE 4.1 Pathways of water movement along slopes, and within soils and bedrock (modified from Butzer, 1976a, figure 6.1: 99).

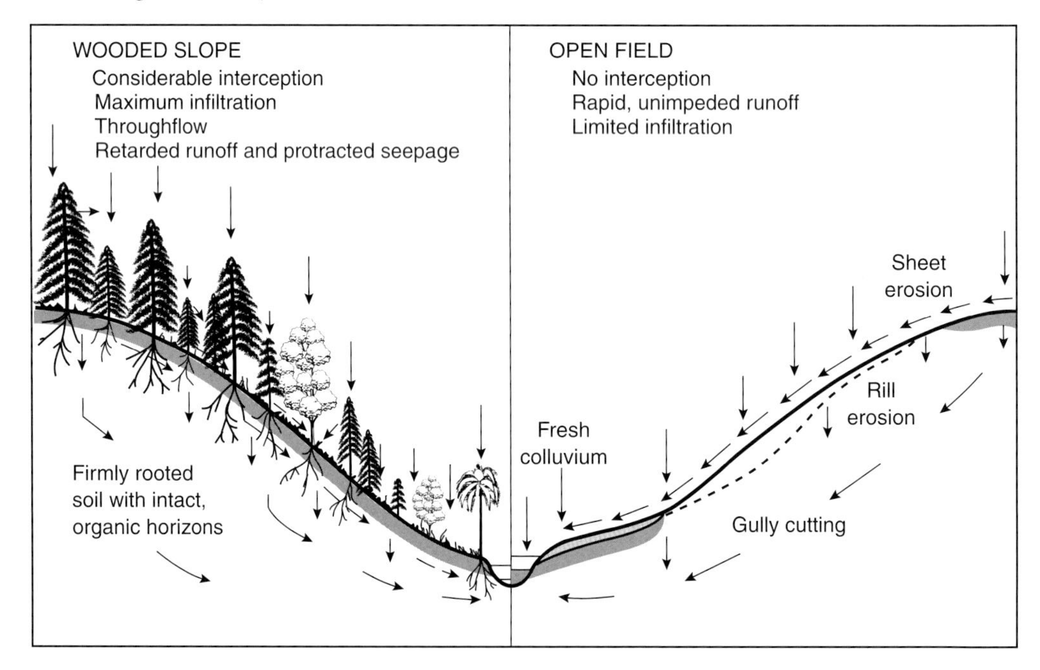

FIGURE 4.2 A generalized depiction of the effect of vegetation on runoff and infiltration along slopes. Not all slopes have behaved this way in antiquity, as for example, only very minor downslope soil movement has occurred under woodlands in Ardennes, Belgium (Imeson *et al.*, 1980). Commonly, pollen-rich bog sediments are concealed beneath the colluvium (Burrin and Scaife, 1984) (modified from Butzer, 1976a, figure. 6.2: 103).

Movement of particles is also encouraged by the action of rainsplash, which can dislodge and facilitate the transport of silt- and sand-size particles. The effects of rainsplash have been well documented by field studies and by numerous laboratory experiments (Chorley *et al.*, 1984; de Ploey *et al.*, 1981; Dijkmans and Mücher, 1989; Mücher and de Ploey, 1977; Mücher and Vreeken, 1981; Mücher *et al.*, 1981) (see below).

Thus, it is when the intensity of rainfall exceeds the infiltration capacity of the soil or substrate that overland flow commences, enabling erosion to take place (Ritter *et al.*, 2002). Irregularities on the surface bring about a transition to conditions, where sheet flow no longer operates, and water is carried along preferred lines of drainage. The smallest size feature, the *rill*, typically forms on fine-grained material and is a few centimeter deep and wide (Ritter *et al.*, 2002). Rills can be rather ephemeral, becoming obliterated by shrinking and swelling or freezing of the soils, or can be destroyed by rainfall itself; since they are in active environments and are mobile substrates they can reform and shift positions.

Gullies are larger scale features that are in the order of centimetres to metres in depth and width, and can easily grow to become tens of metres long. Both rills and gullies overlap in size and potentially occur in the same area, particularly in places that are undergoing rapid erosion, such as *badlands*. Gullies function ephemerally during rainfall events. Gully erosion advances by channel deepening on the slope and by headward erosion upslope (Butzer, 1976). On the surface, gullies and rills usually develop a dendritic pattern, which is partly controlled by the uniform (isotropic) nature of the substrate (e.g. silt, clay, and marl). Subsurface channels or *pipes* serve as conduits to water and sediment and with continued subsurface flow they become enlarged, often to the extent that their roofs collapse. This leads to lateral or headward extension of the gully.

Gullies and pipes, by reason of their relatively large size, are very important in geoarchaeological studies, since deeply buried archaeological sites that are normally undetectable can be readily revealed in gully exposures or collapsed pipes. One of the few sites that appears to span the typological transition from Middle Palaeolithic to Upper Palaeolithic, Boker Tachtit (Negev, Israel; Fluvial Box 5.1) was found by one of the authors (PG), while carrying out a foot survey in the Nahal Zin Valley, focusing on exposures in gullies and in eroded banks. In this case, charcoal was found eroding out of a gully that cut through an alluvial deposit containing the occupations (Goldberg, 1983; Goldberg and Brimer, 1983).

An additional noteworthy instance is illustrated from the Lower Palaeolithic site of Dmanisi, Republic of Georgia. Here, spectacular faunal and hominin remains (Gabunia *et al.*, 2001; Vekua *et al.*, 2002) owe their survival and remarkable preservation in part to their recovery from gullies and pipes. These features not only formed rapidly, but also were infilled quickly (personal observation and communication with C. Reid Ferring (Gabunia *et al.*, 2000)).

Along the upper part of slopes and along slope crests, overland erosion and gullying is relatively minor (Fig. 4.1), and gravitational movements remove volumetrically more material than by processes of running water (Boardman and Evans, 1997). On the other hand, gullying and rilling are more concentrated along the midslope positions. Materials derived from along the slopes accumulate at the base of the slope (the *footslope*; Fig. 4.2; see also Fig. 3.18 on *catena* and soils along slopes). The generic term for slope deposits is *colluvium*, which is generally massively bedded, and poorly sorted. Colluvium *sensu lato* includes both Quaternary and Holocene deposits. In the United Kingdom, in particular, water worked Pleistocene slope deposits are termed *soliflual* deposits (and are associated with solifluction sediments; see Box 3.2), whereas "colluvium" normally refers to fine and well sorted *hillwash* and in many cases relate to a human impact on the landscape, such as cultivation (Bell and Boardman, 1992; Catt, 1986, 1990) (Chapter 9). In much of the Mediterranean, however, colluvium is mainly composed of both eroded soils and sediments, and consists of heterogeneous, poorly sorted debris, with angular clasts. Elsewhere, such as in the United States, colluvium would include any type of unsorted slope mantle. Therefore, the nature of colluvia varies according to geographical region, and local and regional processes of deposition. Since the term is all encompassing, it is a useful term in the sense of its being specifically vague, especially in the field. On the other hand, slope deposits can be studied in detail to reveal the types of processes and activities operating on a slope – periglacial activity, effects of gravity, ratio of water to sediment, landuse practices, and management (e.g. clearance, cultivation, and terracing, see below). Hence, both human and climatic information can be extracted from their study. In any case, since such sediments occur in the lower part of slopes, colluvially derived slope deposits commonly grade into or interfinger with fluvial sediments deposited in the main fluvial system (Fig. 4.2).

4.3 Erosion, movement, and deposition on slopes

4.3.1 Slopes

The ability to recognize the relationship between sloping ground and colluvium is important, because such landscapes may provide the best examples of buried land surfaces and stratigraphic records of human activity, where erosion has been active. The landscape model of the *catena* (see Chapter 3) is significant here. Slope unit types and potential past land uses can be linker by applying modern

FIGURE 4.3 Gullies formed on soft, easily erodible Pleistocene/Holocene lacustrine sediments from Owens Valley, California. Many prehistoric lithic scatters and concentrations were found on these eroded deposits.

land capability classification. For example, it can be simply recognized that steeply sloping (16–25°/28–47%) land is difficult to plough, and pasture is a more likely land use (without terracing), with the extra constraint that soils are becoming more liable to the risk of erosion (Bibby and Mackney, 1972; Klingebiel and Montgomery, 1961). GIS (see Chapter 17) is now a common method for integrating this kind of survey work into archaeological models.

As a note of caution, it must be remembered that modern slopes are in fact *modern* and may bear little relationship to landscapes of the past. Massive Roman to medieval colluviation may have completely changed soil thicknesses, with once-concave slopes becoming convex as they became infilled. Sometimes very recent (twentieth century) landscaping to improve field size and slopes, is not that unusual, while humans have been causing soils to erode since the Mesolithic (see Chapter 9).

Even gently sloping (2–3°, 3–5%) plateau edge, as well as moderately (4–7°, 6–12%) to strongly sloping (8–11°, 13–20%) upper slopes may show evidence of erosion. On the other hand, mid and lower slopes, and the valley bottom themselves are the loci of colluvial deposition (see Chapter 2) (Table 4.1). This slope situation produces important archaeological stratigraphy, which on chalk downlands have been studied through analysis of molluscs (Bell, 1983, 1992). Bare arable ground is highly susceptible to erosion by rain splash and ensuing formation of rills and gullies. In fact sheetwash is now thought to be of negligible importance (Boardman, 1992) (see above). Slope bottom colluvium is deposited, at least in the first instance, as a laminated waterlain sediment (Farres *et al.*, 1992). Infrequent, but archaeologically significant storms can generate sufficient energy to erode and transport stone-sized material producing gravel fans that can include lithics (Allen, 1988, 1995).

The colluvial footslope/valley bottom interface is characterized by interdigitation of colluvial and alluvial deposits, for example, where poorly sorted silts may interfinger with organic-rich (peaty) clays (Brown, 1997). In chalk areas of Europe, such as the Chilterns and Yorkshire Wolds of the United Kingdom and the Pays de Calais, France, some valleys are termed "dry valleys," which were formed in the Pleistocene when frozen ground promoted overland flow even on normally porous rocks. These valleys can contain both periglacial mass-movement deposits as well as overlying Holocene colluvium (Van Vliet-Lanoë *et al.*, 1992). Toward the upper part of the Pleistocene deposits across Europe, a humic Lateglacial soil marks a temperate Interstadial at ca. 11,000–13,000 calendar years BP. This interval, termed the Windermere Interstadial in the United Kingdom, and in mainland Europe, includes the Bølling, Older Dryas, and Allerød climatic episodes (Barton, 1997) (Box 3.2).

Mediterranean valley fills and associated eroded landscapes have been increasingly studied in detail, as for example in Greece and the Levant (Bintliff, 1992; Goldberg, 1986; van Andel *et al.*, 1990; Vita-Finzi, 1969; Zangger, 1992). Such research, has compared the roles of humans and climate in erosion. Climatic change was first identified as a trigger for massive erosion, but as more archaeological sites and their environments were investigated, landuse, expanding populations, local soils, and geology were increasingly cited as factors influencing erosion. More commonly now, rare events, such as intensive rainstorms, are seen as the most important cause of major erosional and depositional events. On the other hand, landscape instability has been triggered by clearance and development of bare arable ground (see below) (Bell, 1983; Macphail, 1992; Zangger, 1992).

Lynchets usually occur at slight changes or dips in dip slope, through the combined effects of erosion and colluviation (Fowler and Evans,

TABLE 4.1 Generalized characteristics of slope deposits (see Figs 3.18)

Position on slope	Sedimentary/ erosional processes	Sedimentary attributes – macro	Soil characteristics	Micromorphology and other environmental indicators	Archaeology
Upper slopes	Rainsplash and rilling; sheetflow (not significant)	Aerobic decomposition Loose Stony	Bare rock and shallow soils • eroded topsoil • exposed subsoil Droughty soils	Possible identification of poorly humic, newly formed topsoils and truncated subsoils; negative features and subsoil hollows may contain molluscs	Eroded features Displaced artifacts Common location of negative features (e.g. pits, ditches, and post holes)
Midslopes	As above, with gullying; tilling effects and creep (gravity movement) Eroded and over thickened sediment/soil profiles	Cut and fill features; coarse (stones and sand) and fine (silts and clay) laminae	Bedrock exposed in gullies Eroded and overthickened sediment/soil profiles. Well-drained soils	Possible identification of charcoal and burned soil (clearance), lithorelics, and soil fragments of earlier landscape – for example, from A horizons and subsoil argillic Bt horizons, *terra rosa*; pollen and molluscs form feature-buried surfaces	Buried or eroded features and artifacts; common location of lynchets, walls, and terraces
Lower slopes/ Toeslopes	Minor rilling and gullying with more dominant accretionary soils (colluvium); possible springhead sapping	stone-free and stony layers; downslope-oriented finger-shaped stone (gravel) fans; possible spring associated	Homogeneous colluvial soils with relict/ephemeral stabilized topsoils and stony horizons; cumulic soils with overthickened A horizons; generally	Possible identification of soil fragments from eroded "natural" soils, and tillage soils (crust fragments from furrows); slaked soils relating to	Increased stratigraphic resolution due to rapid sedimentation; potential location of buried occupations,

		tufa and deposition of "salts"	well-drained soils, but possible seasonal affects of heightened groundwater (groundwater gleys), possible springs, and peats	mass movement; possible relict, sedimentary features; features commonly worked or partially worked by soil fauna; potential of well stratified molluscan sequencese	sites of wells and water holes utilizing springs; paddy fields
Valley floor	Interfingering of colluvium and alluvium/peat; potential combination of local valley and distant – up-river – deposits; likelihood of deep erosion of valley floor sediments by migrating meanders	Colluvial loams intercalated with silts, clay, and organic sediments; more likely tufa and deposition of "salts," as spring lines/water table meet	Cumulic soils show evidence of occasional poor drainage; alluvial soils display seasonal high water tables (ground water gleys); intercalated peats result from major changes to base level induced by climate/sea-level changes and human activities (clearance)	Possible identification of biologically homogenized colluvium and alluvium, and ground water effects (secondary calcium carbonate, e.g. tufa), iron and manganese mottling/panning, pyrite formation; salt pans and gypsic horizons; high potential for well stratified sequences of terrestrial and aquatic molluscs, and pollen and macrofossils (plants and insects),	Increased stratigraphic resolution due to rapid sedimentation; possible waterlogging and preservation of organic remains; wells and waterlogged pits; potential for preserved routeways – bridges, causeways, fords; Industrial activity – water mills, artificial channels, and ponds; paddy field

1967; Macnab, 1965). Erosion leads to a negative lynchet down slope, while colluvial deposition forms a positive lynchet upslope (Fig. 4.4). Occasionally, the positive lynchet may itself be formed on eroded ground, and slope soils may be stabilized by wall building along lynchets. At Chysauster, Cornwall, erosion of the negative lynchet reached an estimated depth of 0.50–70 m. In many societies, where both flat ground is rare and where erosion is a constant threat, labor-intensive terracing has been introduced. Examples can be cited from both the Old and New Worlds (French and Whitelaw, 1999; Sandor, 1992; Wagstaff, 1992; Wilkinson, 1997) (see Chapter 9).

4.3.2 Soil stability and erosion

Soil stability and erosion have been studied in great detail throughout the world and especially for Holocene geoarchaeological landscapes (Kwaad and Mücher, 1979; Macphail, 1992; van Andel *et al.*, 1990). The major mechanisms of erosion dealt with here are somewhat simplified but can be ascribed to: (1) erodibility and resistance to erosion by different soil types; (2) rainsplash impact; (3) water flow; and (4) effects of ploughing and cultivation.

Different soil materials erode more easily than others, as for example, wind blown sand (dunes) versus silt (loess) (Evans, 1992). In general, silty soils and sands – because of their inherent poor soil structure – are more erodible than soils with a high clay content. Moreover, soil organic matter, landuse practices, and general soil type also have an effect on soil erodibility. Although the interaction of these factors is complex, some useful generalizations can be made. Topsoils, for example, have greater aggregate stability than subsoils because of their greater humus content, and in general decreasing aggregate stability follows the trend:

grassland → woodland → arable land (Imeson and Jungerius, 1976).

Extant "natural" topsoils also have good structure and open porosity that allows them to drain well; consequently they are only slightly

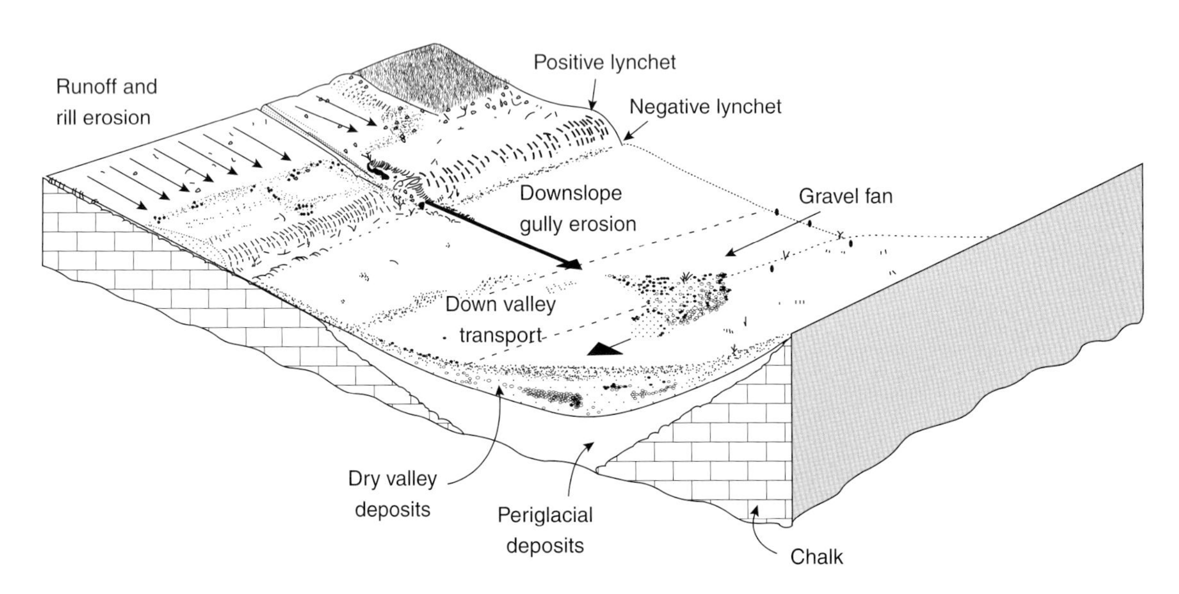

FIGURE 4.4 Colluvial soil movement down slope and down valley (after Allen, 1988); hypothetical soil erosion and deposition on the chalklands of southern England (see Table 4.1; Allen, 1992; Bell, 1983; Boardman, 1992).

influenced by surface water flow and erosion. On the other hand, research from the southern midwest of the United States shows that bare arable topsoils may suffer badly: rainsplash may break up soil structures causing surface crusts that lead to overland flow (McIntyre, 1958a,b). The resulting slaked soil is then transported along rills and gullies (West *et al.*, 1990). On flat ground, on the other hand, silty clay and sandy laminated surface crusts are formed as these constituents become redistributed. Crusts on arable ground have been the subject of much study (e.g. Boiffin and Bresson, 1987).

Leached upper subsoil Albic (A2; Ea in United Kingdom) horizons of podzols, are comprised of sand with little organic matter and therefore have weak structures and are in theory easily eroded; the presence of fungal hyphae, however, can actually make the sands water repellent. This not only may aid resistance to erosion, but also make such sandy soils droughty because of poor infiltration. Generally, however, when the protective and water absorbent Ah horizons of heath are breached by human and animal activity, such as by formation of trails, massive sand colluviation can take place (e.g. West Heath, Fig. 4.5a). The Bronze Age barrows at West Heath, West Sussex (UK), are commonly buried podzols with markedly truncated Ah and A2 horizons (ca 20 cm). Such erosion was likely induced by the gathering of local turf to construct eight barrows – earth mounds (Drewett, 1989; Scaife and Macphail, 1983) (Fig. 4.5b c). On the lower midslope and valley bottom sites, colluviation may produce over-thickened Ah and A2 horizons of 1.20 m. In many cases, on sloping areas of Spodosol landscapes, erosion and colluvial activity mix the humic Ah and leached A2 horizon material producing a dark gray colluvium. At the small Roman town of Scole, Norfolk, one of the successes of the geoarchaeological investigation was to be able to rapidly differentiate such dark gray colluvium from the cultural dark earth *sensu stricto* (Chapter 11) through auger and soil profile survey (Macphail, 1994; Macphail *et al.*, 2000). This study therefore saved much time and money for the excavator, who could then concentrate on the archaeologically more interesting dark earth and associated occupation deposits.

On argillic brown earths, the upper subsoil horizon (A2; Eb in United Kingdom) can become rather compact and poorly structured because of loss of clay and its translocation into the underlying Bt (argillic) horizon. Consequently this A2 horizon may permit only slow drainage, and is thus very susceptible to surface flow, subsurface "piping", and erosion (Dalrymple *et al.*, 1968). The underlying Bt horizon may also develop slow or impeded drainage because soil voids become increasingly coated or even sealed by translocated clay, forming a Btg horizon. If soil water stagnates in this horizon there is also the possibility that reducing conditions will lead to the breakdown of this translocated clay. In tropical soils, such as Ultisols and Oxisols, topsoils and upper subsoils may become eroded on slopes uncovering an iron and clay enriched subsoil, which on exposure irreversibly hardens (see Chapter 3) becoming a cemented *ferricrete* or *cuirasse* subsoil horizon that is strongly resistant to erosion.

4.3.3 Colluvium (hillwash)

The interpretation of colluvium is not always easy. Mücher (1974) showed that soil micromorphology could aid the identification of soil components composing colluvium. He also classified the types of processes that could be identified in thin section, such as the structural instability of medieval arable soils in northern Luxembourg (Kwaad and Mücher, 1979). At Ashcombe Bottom, West Sussex, Beaker colluvium (hillwash) was associated with arable activity that produced surface slaking crusts, and eroded fragments of these became embedded into the colluvium (Fig. 4.6) (Allen, 1988; Macphail, 1992).

FIGURE 4.5 (a) Colluvial podzol downslope of Bronze Age barrow cemetery, West Heath, West Sussex, United Kingdom – note horizontal gray colluvial layers over a thin buried black Ah and white Albic A2 horizon; (b) Barrow turf stack (earth mound) – best preserved turves are present at the base of the mound below (rabbit) burrowed layers; (c) Truncated buried Bronze Age Spodosol (podzol), showing junction of basal turf-stack and buried Ah, a thin Albic A2 and subsoil Bh, Bhs/Bf, and Bs horizons. A 1-kg sample of the buried Bh horizon produced a mean residence C14 date of 3770 ± 150 years (Drewett, 1976, 1989; Scaife and Macphail, 1983).

Often colluvium can display sequences of stone-free and stony horizons. One explanation for this phenomenon was that stones brought to the surface by tillage were periodically concentrated some 20–30 cm below stone-free soil, by earthworm working on ephemerally stabilized colluvial soils (e.g. during periods of fallow and/or abandonment).

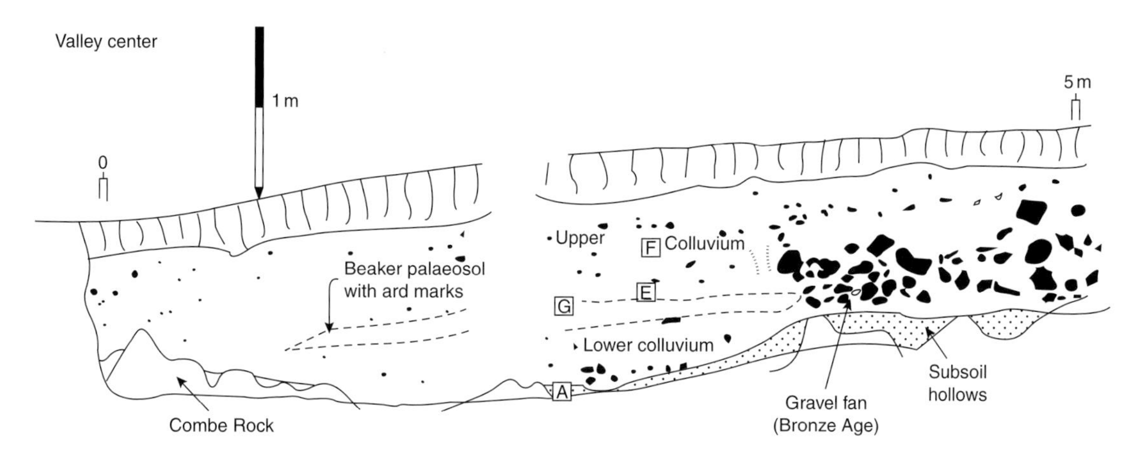

FIGURE 4.6 Ashcombe Bottom section drawing (courtesy of Mike Allen) showing gravel fan colluvial deposits, and location of thin section boxes and Beaker paleosol (modified from Macphail, 1992) (see figure 9.12a).

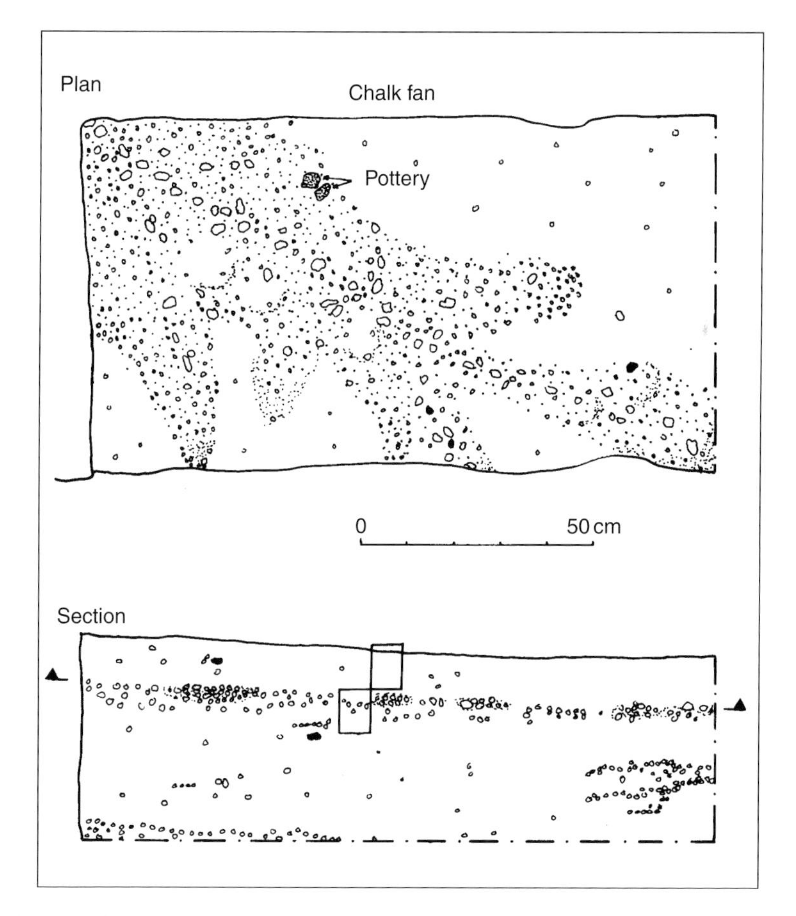

FIGURE 4.7 Strawberry Hill, Wiltshire, England; plan and section of colluvial fan (modified from Allen, 1994); rectangular boxes are thin section samples studied by R I Macphail.

The identification of high intensity rainstorm events in the present day record (Boardman, 1992), however, showed that rill and gully erosion could erode large numbers of stones, which were then deposited downslope as colluvial gravel fans. A study by Allen (1992), employed three-dimensional excavation, molluscan analyses, and soil micromorphology at Strawberry Hill, Salisbury Plain, Wiltshire (Fig. 4.7). He found that Iron Age rill erosion on the chalk produced spreading "fingers" of gravel (fan). The study of mollusks indicated the presence of large amounts of arable ground, whereas soil micromorphology showed that chalk stones had been deposited within a slurry of local chalk soil material (rendzinas); the underlying stone-free horizon was similarly deposited as a slurry. The stone-free horizons here had not resulted from earthworm working. Deposits from mass movement are expected in Pleistocene sites, but are perhaps less archaeologically acceptable in Holocene sequences. Mass movement produces specific types of microfabrics (Courty *et al.*, 1989), but such interpretations may be more readily taken on board if accompanied by flattened walls with a down slope orientation, or if they occur in well-known tectonically unstable areas (Joelle Burnouf, personal communication).

4.4 Conclusions

In this introduction to slopes, slope processes, and colluvium, the importance of recognizing all these at both regional and site scales has been noted. These findings contribute to arguments that reason for past land use patterns of perhaps arable versus pasture, or the location of wells, in addition to more obvious considerations of patterns of field/lynchet boundaries or the presence of terraces. They also contribute to the desktop assessment of where trenches may be best cut to find environmental information (Bell, 1983) (see Table 4.1). Chapter 5 examines the hydrological system as it manifests itself as rivers and lakes.

5

Hydrological systems II: rivers and lakes

5.1 Introduction

We have already seen how a fluvial system can be traced from the watershed, via the slopes to the valley bottom (Fig. 4.1). On the slopes there is both subsurface groundwater flow, which can affect both soils and the underlying geology, and overland flow, which can produce rill and gully features and slope deposits. In the case of slopes, different areas of the slope are more prone to erosion than others, while other portions of the slope may be wetter (e.g. springs) and/or more likely to accumulate colluvium (Table 4.1 and 5.1). The lowermost valley slope environment, however, is commonly the wettest and here colluvium may interfinger with fluvial sediments, producing interdigitated colluvium and alluvium that may well be influenced by groundwater. In this chapter we focus on the valley bottom channel and floodplain environment. Since many streams flow into closed depressions forming wetland and lakes, we also discuss lacustrine sediments (Table 5.2).

Fluvial environments are distributed throughout the globe from the subarctic to the tropics, reflecting a wide range of variability as a function of local environmental conditions, such as moisture, temperature, seasonality, and other climatic variables. Arid fluviatile systems are different from those in more humid areas, for example, in their lower frequency but potentially greater intensity of streamflow, and the geometry of the channel patterns (Huckleberry, 2001). In these chapters we examine some of the most important aspects of the fluvial system that would be useful to the geoarchaeologist in understanding why and where sites might be located, eroded, or buried in a given area. In Chapter 4, we deal with the watershed, fluvial systems on slopes and slope deposits, and associated archaeological features such as terraces, lynchets, and human impact, for example clearance (see Chapter 9), while in Chapter 5 the focus is rivers, and wetland, and their associated deposits.

5.2 Stream erosion, transport, and deposition

Water and sediment from the slopes eventually arrive at the valley floor where they enter the channel system. There, water and sediment inputs reflect physical characteristics of drainage basin (Ritter, 1986: 204). Furthermore, clastic materials in fluvial systems, as with slopes, can be further subjected to erosion, transportation, and deposition.

Rivers flow at various volumes during the course of the year. During periods of low flow (*base flow*), in which most of the water entering the stream is from springs or groundwater, volumes of water and *discharge* (the volume of

TABLE 5.1 Characteristics of valley sediments (modified from Summerfield, 1991)

Locus of deposition	Type	Characteristics
Channel	Transitory channel deposits	Mostly bedload temporarily at rest; may partially be preserved in more long-lasting channel fills or lateral accretions
	Lag deposits	Segregations of larger or heavier particles, more persistent than transitory channel deposits
	Channel fills	Accumulation in abandoned or aggrading channel segments; range from coarse bedload to fine-grained oxbow lake sediments
Channel margin	Lateral accretion deposits	Point and marginal bars resulting from channel migration
Overbank floodplain	Vertical accretion deposits	Fine-grained suspended load of overbank flood water; includes natural levee and backswamp deposits
	Splays	Localized bedload deposits spread from channel onto adjacent floodplain
Valley margin	Colluvium	Slope deposits, poorly to moderately sorted, stony to fine grained material, accumulated by slopewash and gravity (e.g. creep); commonly interfingers with channel margin and floodplain deposits
	Mass movement deposits	Debris avalanche and landslide deposits intermixed with colluvium; mudflows generally in channels but spill over banks

water carried per unit time, e.g. $m^3\ s^{-1}$) are relatively low; little change in the stream morphology occurs during these periods of base flow (Butzer, 1976). In northern temperate areas, base flow is particularly low in summer months; in drier areas, flow ceases for much of the year except for the rainy season, when it essentially results from contributions by runoff (*ephemeral* flow). However, in periods of flood, such as in spring with melting of snow or during major storms (e.g. hurricanes), associated with major precipitation events, discharges can increase dramatically resulting in the condition when the stream channel is full (*bankfull* stage), even overtopping its banks and resulting in *overbank* flow. Most fluvial changes, however, take place with frequent and continual flow when at $\frac{1}{2}$ to $\frac{3}{4}$ of the bankfull level (Butzer, 1976).

Within the stream channel, water flows at different velocities, depending on location. The maximum velocity is near the surface of the water, above the deepest part of the channel. However, next to the channel walls and along the channel bed, where friction is greatest, flow velocities are much lower. Thus (all things being equal), greater velocities occur in deeper, narrower channels than in shallow, broad ones, which have a proportionately larger wetted perimeter and a greater amount of external friction along the channel.

A stream transports different material in different ways, and these substances are known as *load. Dissolved load* refers to various materials like salts that are transported in solution, such as carbonates, sulfates, nitrates, and oxides. These are mixed throughout the water column. The *suspended load* is composed of finer particles (generally silt, clay, colloids, and organic matter) that are kept in suspension by turbulence. Coarser materials are carried as part of the *bedload* along or close to the bottom of the channel. These sand-size and larger particles move by bouncing (*saltation*), rolling, or sliding. Associated with this movement is the overall organization of the material being transported, and we see that the movement of sand-size grains is associated with different types of *bedforms* (see Chapter 2) that change according to flow velocity and mean sediment size (Fig. 5.1).

The load that a stream can actually transport is expressed in two ways. *Capacity* is the total amount of material that a stream can transport

TABLE 5.2 Types of river patterns and characteristics (modified from Morisawa, 1985)

Type	Morphology	Load type	Width/depth ratio	Erosion	Deposition
Straight	Single channel with pools and riffles; meandering talweg	suspension-mixed load or bedload	<40	Minor channel widening and incision	shoals
Sinuous	Single channel with pools and riffles; meandering talweg	Mixed	<40	Increased channel widening and incision	shoals
Meandering	Single channels	Suspension or mixed load	<40	Channel incision, meander widening	Point bar formation
Braided	Two or more channels with bars and small islands	Bedload	>40	Channel widening	Channel aggradation, mid-channel bar formation
Anastomosing	Two or more channels with large, stable islands	Suspension load	>10	Slow meander widening	Slow bank accretion

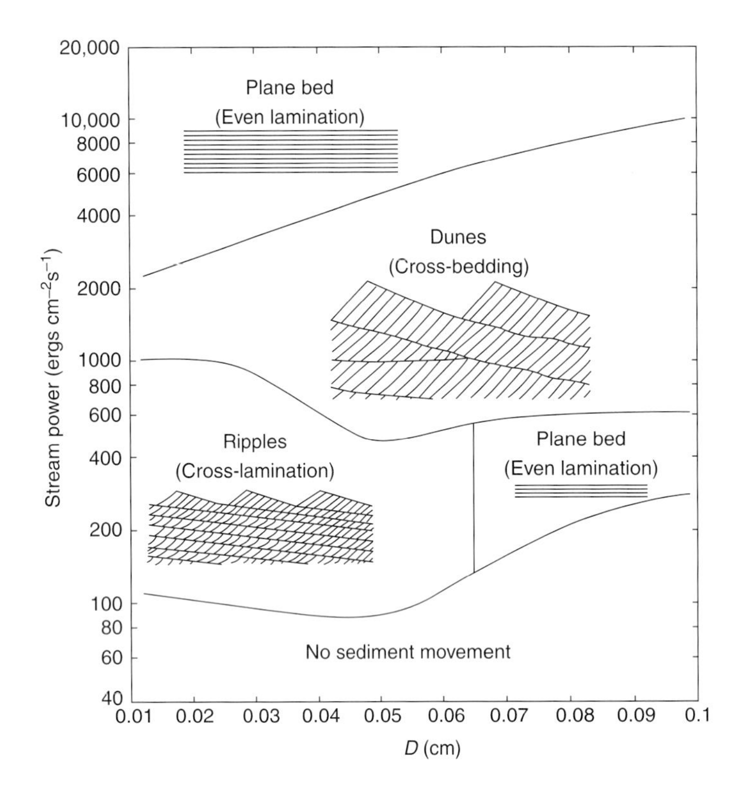

FIGURE 5.1 Bedforms in relation to grain size (mostly sand) and stream power (modified from Allen, 1971).

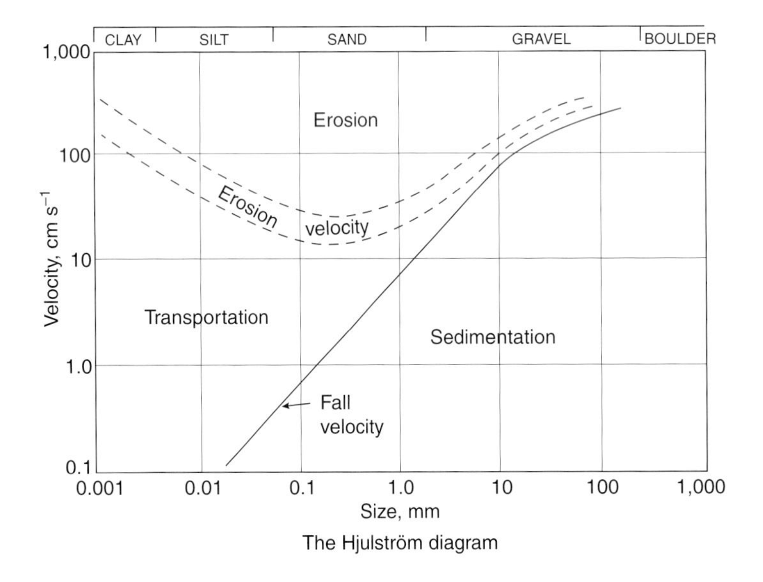

FIGURE 5.2 Hjulström's diagram showing velocity associated with the erosion, transport, and deposition of different sizes of particles.

and depends on the current velocity and the discharge. *Competence* refers to the size of the largest particle that a stream can carry under existing conditions. It depends mostly on velocity.

The proportion of the types of load that a stream carries varies from year to year, and within the year; it is also a function of climate. For example, streams in arid climates transport little dissolved load, but much bed load, especially during flooding; dissolved load is obviously more important in humid climates. Thus, the ability of a stream to erode and transport sediment is related to velocity and discharge; it is evident that most erosion and transport is carried out during major flow events, such as spring flooding or during exceptional events such as hurricanes and other storms.

A classical study (Hjulström, 1939), related erosion, transportation, and deposition to particle size and mean velocity (Fig. 5.2). This figure shows that coarse sand-size grain (ca 0.5 to 1 mm) is the most readily eroded sediment by water, requiring more energy for both coarser and finer sediment; the latter is due to the cohesion of finer particles requiring higher energies to entrain them and lower drag force. The diagram also shows the relatively low velocities required to keep finer material entrained and transported.

5.3 Stream deposits and channel patterns

Deposition takes place predominantly along the bottom where discharges fluctuate in response to inputs of water or sediment. In the case of point bars, for example (see below), which are built vertically during bank full and high stage events, deposition takes place along channel sides. In addition, different types of deposits occur in response to certain related conditions, such as energy of the stream (e.g. low flow versus flooding conditions), whether flow is confined to the channel or outside the channel, and extraneous inputs from the valley walls (Table 5.1).

The type of sediment being transported is generally linked to the morphology of the stream, which in turn is related to the flow conditions; both can be ultimately conditioned by climate (e.g. precipitation regime, vegetation). Geomorphologists and sedimentologists recognize several types of stream patterns (Fig. 5.3). Familiarity with these types of patterns and associated deposits is important because it can aid in the geoarchaeological interpretation of specific fluvial deposits and facies (see Chapter 2) that can then be valuable in looking for (or avoiding) certain loci for past human settlement. Low energy deposits occurring outside the channel, for example, are more likely to preserve *in situ* archaeological material compared to those from high energy gravel bars that are found within active channels (Figs 5.4–5.6). Moreover, an understanding of stream morphology and processes, including associated rates of change, may help explain an absence of surface sites in some areas (Guccione *et al.*, 1998).

Channels can be characterized as single or divided, with shapes that vary from straight, sinuous, meandering, braided, and anastomosing (anabranching) (Fig. 5.3; Table 5.2). These different types represent an average condition as channel morphology can change from season to season and year to year, and often grade into each other (Bridge, 2003; Ritter *et al.*, 2002). Such short-term changes, however, are not normally visible on the geoarchaeological scale, especially with older, Lower Palaeolithic sites, in East Africa for example, (Rogers *et al.*, 1994; Stern *et al.*, 2002).

Straight, single channels, tend to be rare and carry a mixture of suspended and bedload materials. The latter commonly accumulates on opposite sides of the channel (Ritter *et al.*, 2002), called *alternate bars*; shallow zones are called *riffles*, whereas deeper areas are called *pools*. The path that connects the deepest parts of the channel is called the *thalweg*. Erosion is relatively minor, with slight lateral widening or vertical incision. Deposition occurs along bars during the flood stage.

In *braided systems* rivers tend to be straight and many small channels are split off from the main channel; separating the channels are raised portions (bars), which are covered during periods of high water. Braided channels are characteristically found in arid and semi-arid areas (Fig. 5.4a,b), especially in alluvial fans. (see below), and in areas of glacial outwash. Braiding occurs in such locales where there are rapid shifts in water discharge, in the supply of sediment (generally coarse), and the stream banks are easily erodible (Boggs, 2001); the ratio of bedload to suspended load is high. Deposition of coarse material results in the formation of mid

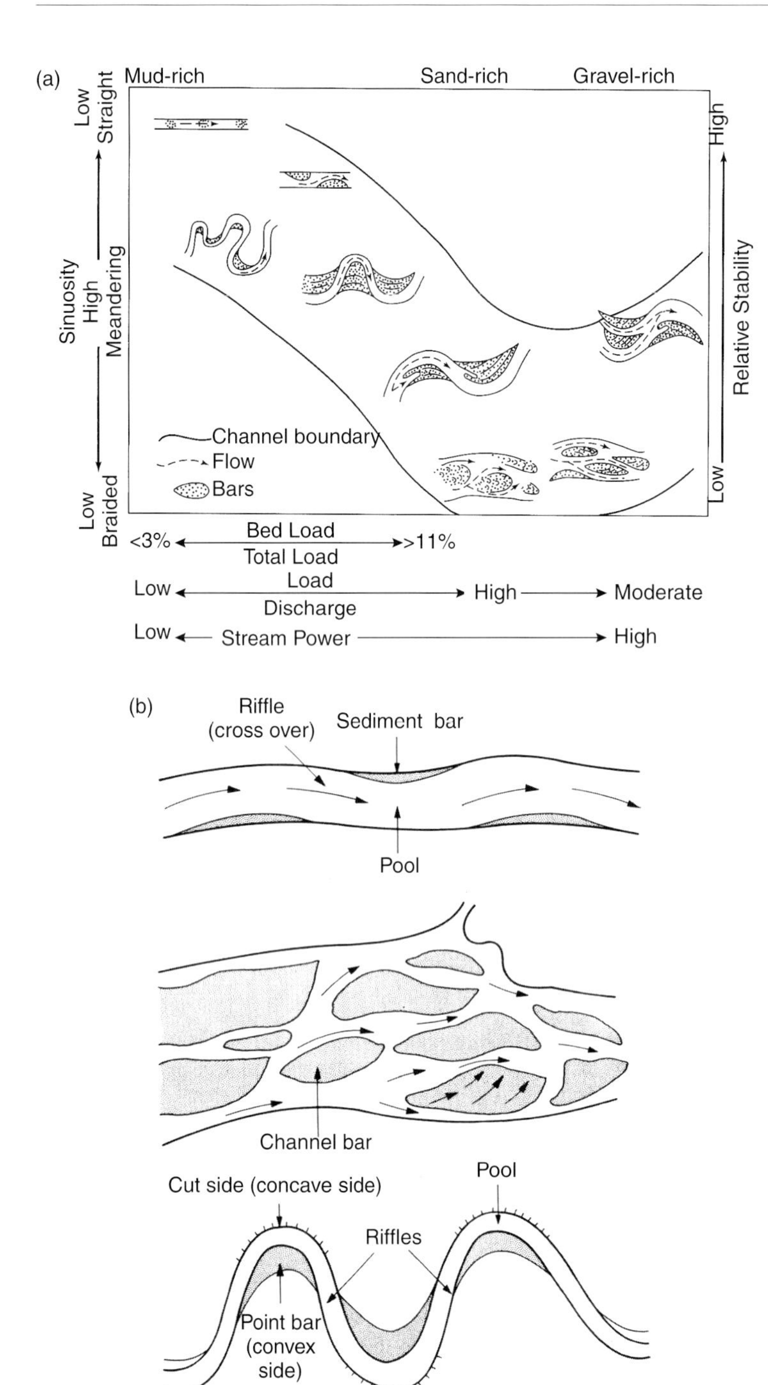

FIGURE 5.3 (a) Different fluvial channel pattern types showing the interrelation between channel form, sediment load, stream power, and stability (modified from Boggs, 2001. (b) Detailed view of straight, braided and meandering patterns (modified from Reineck and Singh, 1980, figure 372).

channel bars, which direct the flow around them during high water discharges. Through time, valley floors can accumulate several meters of braided stream deposits, which appear as nested, lens-like masses of gravels (Fig. 5.5). It is not surprising that in light of the periodically high discharges, the likelihood of finding intact archaeological sites within such deposits is low. Nevertheless, archaeological sites can be found on *abandoned* braided stream deposits, such as those on old alluvial fan surfaces. In Sinai, for example, deflated remains of coarse bouldery deposits (*lags*) are commonly exploited as large building stones; these can be difficult to distinguish from the natural bouldery surface cover (Fig. 5.6).

In *Anastomosing channels*, in contrast to braided channels, flow is around bars that are relatively stable and not readily eroded. In the Welland Valley area of eastern Britain, anastomosing channels are inclined to represent cold, periglacial conditions between 10,900 and 10,00 years ago (French, 2003). However, they are relatively infrequent in the archaeological record.

Meandering channels and systems are widespread in geoarchaeological settings and many sites are associated with them. Unlike braided streams, in meandering ones, flow is within a single channel. They also differ in having lower sinuosity, shallower gradients, and finer sediment loads (Boggs, 2001). The principal aspects of a meandering river system are illustrated in Figure 5.7.

In a meandering system, coarse, gravelly material is transported within the channel, generally during times of flood; under average conditions, sandy bedload is transported (Walker and Cant, 1984) (Table 5.1). Erosion takes place along the outer reaches of meander beds where velocities are higher. In contrast, deposition occurs in the inner part of the meander loop, leading to the formation of the point bar (Figs 5.3 and 5.7). Thus, through time the point bar can be seen to shift laterally across and downstream along the valley bottom and floodplain. This process of *lateral accretion* leads to the formation of a package of sediment that fines upward from gravel, sand, and finer silts and clay.

Increased erosion along the outer cut banks can lead to meanders being cut off and their abandonment, ultimately leading to the formation of an oxbow lake (Fig. 5.7). The sediment in these depressions consists of fine silt and clay, and is typically organic rich; they may contain diatoms, molluscs, and ostracods. Since they are away from the channel and contain water before being silted up, they are attractive to vegetation, game, and human occupation. A similar type of depression and infilling can be formed through *avulsion*, in which a channel breaks through its levee and abandons its channel. Such abandoned and isolated basins are not uncommon in the rivers that drain eastern Texas upto the Gulf of Mexico. A considerable part of the Late Paleoindian occupation at the Wilson–Leonard site, for example, took place next to an *avulsed channel* (see Box Fig. 2.5). A banana-shaped accumulation of organic silty clays within the avulsed channel was accompanied by bison kills (Bousman *et al.*, 2002; Goldberg and Holliday, 1998). At the Aubrey Clovis site on the Trinity River in north-central Texas, a cut-off channel was the scene of a groundwater spring-fed pond which slightly predates and is partly contemporaneous with the Clovis occupation (Humphrey and Ferring, 1994).

5.4 Floodplains

Floodplains, flat to gently sloped areas adjacent to stream channels, are notable in geoarchaeology, for not only are they one of the most widespread of fluvial landscapes – past and present – but they served as significant loci for past occupations, which can be well preserved. They are dynamic landscapes that exhibit a variety of local sedimentary environments and processes (Brown, 1997) which change over time as the floodplain evolves (Fig. 5.8). Close to the channel, for example (Fig. 5.9), sand and silt are found in raised, natural levees. These deposits grade laterally into finer silts and clays that accumulate in lower, backswamp areas where drainage is poor and only finer material accumulates during a flooding event. Consequently, although the backswamp might appear as an attractive locale for game, actual habitation is more likely to occur closer toward the levee, where the deposits are coarser and drainage is better. On the other hand, the position of these environments continually changes as the meanders sweep across the floodplain,

(a)

(b)

FIGURE 5.4 (a) Braided channel during a flood in the Gebel Katarina area, south-central Sinai, 1979. Note the gravel bar between two branches of the channel. (b) Braided channel from Qadesh Barnea area, Sinai. The braided channels are shown in the central part of the photograph with the main channel carrying water, and abandoned channels adjacent to it. The entire area between the vegetated areas on the right and left is covered during occasional flooding events that occur in winter; this area currently receives <100 mm of rainfall per year. On the left is a terrace riser (MP) composed of gravel and associated with Middle Palaeolithic artifacts. The height of the terrace is ca 19 m above the present valley floor and indicates that during the Middle Palaeolithic, the valley was filled with gravel to this height (Goldberg, 1984).

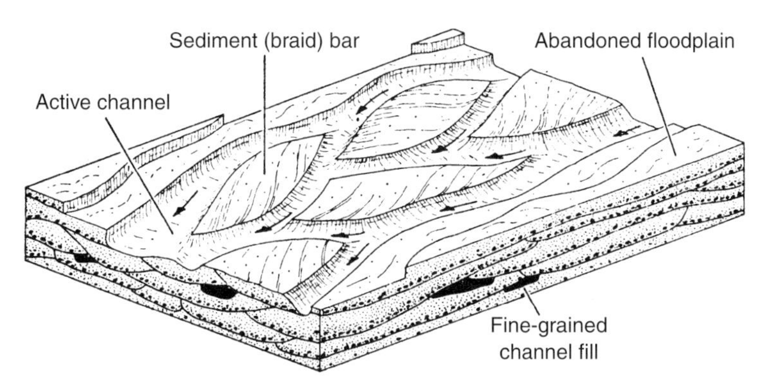

FIGURE 5.5 Block diagram showing surface morphology of a braided stream with active channels and bars. The depth dimension shows deposition of sediments within lenticular bodies, beginning with gravels that grade upward to finer sand and gravel. An abandoned floodplain on the right exists above the braided channel area The preservation potential of intact sites within such deposits is low. (modified from Allen, 1970, figure 4.7).

FIGURE 5.6 Coarse alluvium exposed at the surface of an alluvial fan in southwestern Sinai. This coarse lag of boulders and gravels was originally deposited by braided channels of the fan when it was active. O. Bar-Yosef and N. Goren-Inbar examine the remains of a Bronze Age building whose inhabitants exploited these boulders for construction.

(a)

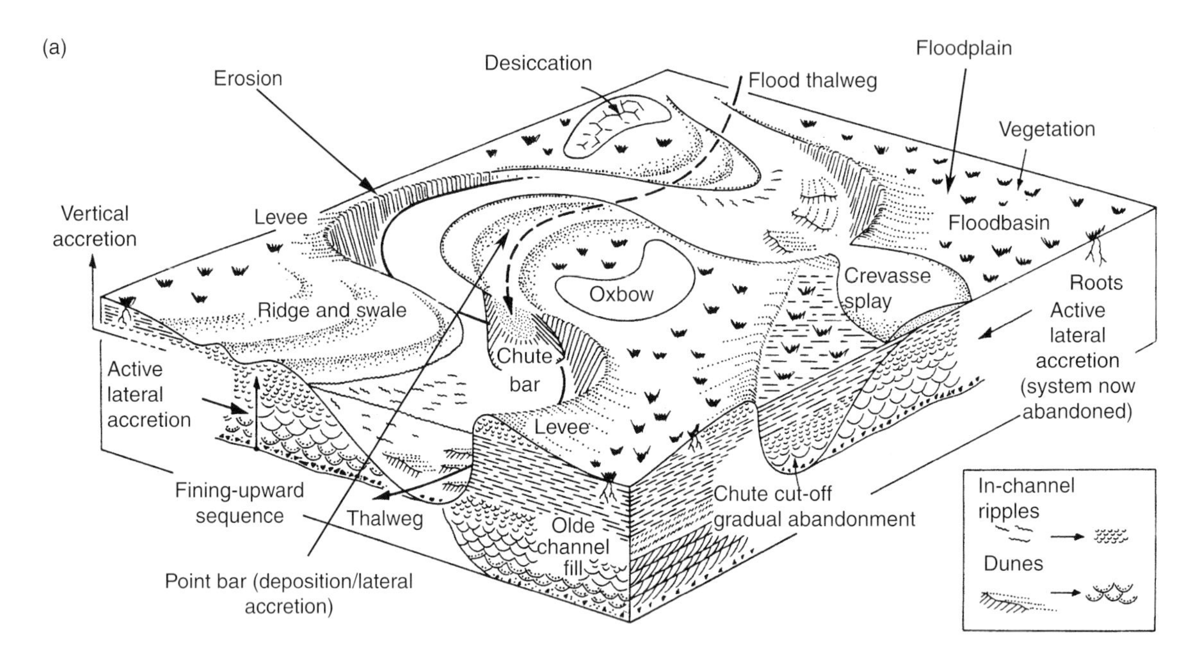

(b)

FIGURE 5.7 (a) Block diagram illustrating the major features of a meandering river system (modified from Walker and Cant, 1984 figure 1). (b) Aerial view of meandering river system showing flood basin (FB), present channel (Ch), and abandoned meander (MS) and associated ridge and swale topography along the point bar.

covering or erasing previous deposits. Along the meanders of the Red River, Arkansas, for example (Guccione *et al.*, 1998) found sites of different ages associated with different positions of the meander. The most recent meander belt (limits of meanders within the valley bottom) is about 200 to 300 years old and has covered many previous artifacts or sites with 1 to 2 m of alluvium. An older prehistoric site, on the other hand, is found on the surface associated with a 500 to 1,000 year old abandoned meander belt. Finally, they indicate:

> *At locations proximal to the river, the site may be buried by overbank sediment 0.4 m thick, but at more distant locations the site is at the surface or only buried by thin overbank sediment because of low sedimentation rates (0.04 cm yr^{-1}) over the span of a millennium. Sites, such as 3MI3/30, [Fig. 5.9] that are occupied contemporaneous with overbank sedimentation may be stratified; however, localized erosion and removal of some archeological material may occur where channelized flow crosses the natural levee. (Guccione et al., 1998: 475)*

The study of Guccione *et al.* (1998) elegantly demonstrates the active nature of meander belts, even within the last few centuries. Furthermore, it establishes that one must take into account issues, such as the vigor of the meander belt when working in such geomorphologically dynamic areas. Many of the techniques discussed in Chapter 15 (e.g. aerial and satellite photos, trenching, soil surveys) are helpful in evaluating the age of a fluvial landscape and rates of change, and whether sites of a given age will be likely preserved, if at all, either on the surface or beneath it.

5.4.1 Soils

Soils play an important part in fluvial geoarchaeology. They provide a window into understanding the relationships between sedimentation, pedogenesis, and archaeological site formation processes (Ferring, 1992). Furthermore, their formation can reflect regional landscape and climatic conditions, which in turn can be used as time parallel (roughly synchronic) stratigraphic marker horizons (see Chapter 2).

Soil visibility on floodplains involves a balance between rate of pedogenesis and geogenic sedimentation. At one extreme, in cases with continuous deposition in the same location, there is little chance for a mature profile to form: horizons (perhaps as defined in the strictest

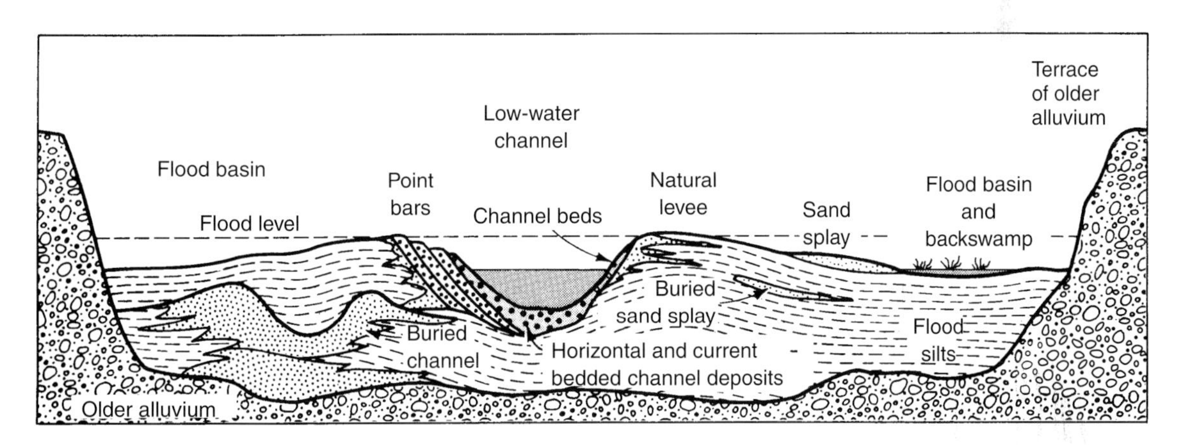

FIGURE 5.8 Lateral variations in deposits and facies are clearly shown in this schematic view of a floodplain. Buried and active channel gravels interfinger with finer grained flood silts, some associated with point bar deposits, which here are migrating from left to right. In this example, the entire floodplain sequence is set into an eroded valley consisting of older, gravelly alluvium, which forms a terrace on the valley flanks (modified from Butzer, 1976a, figure 8.3).

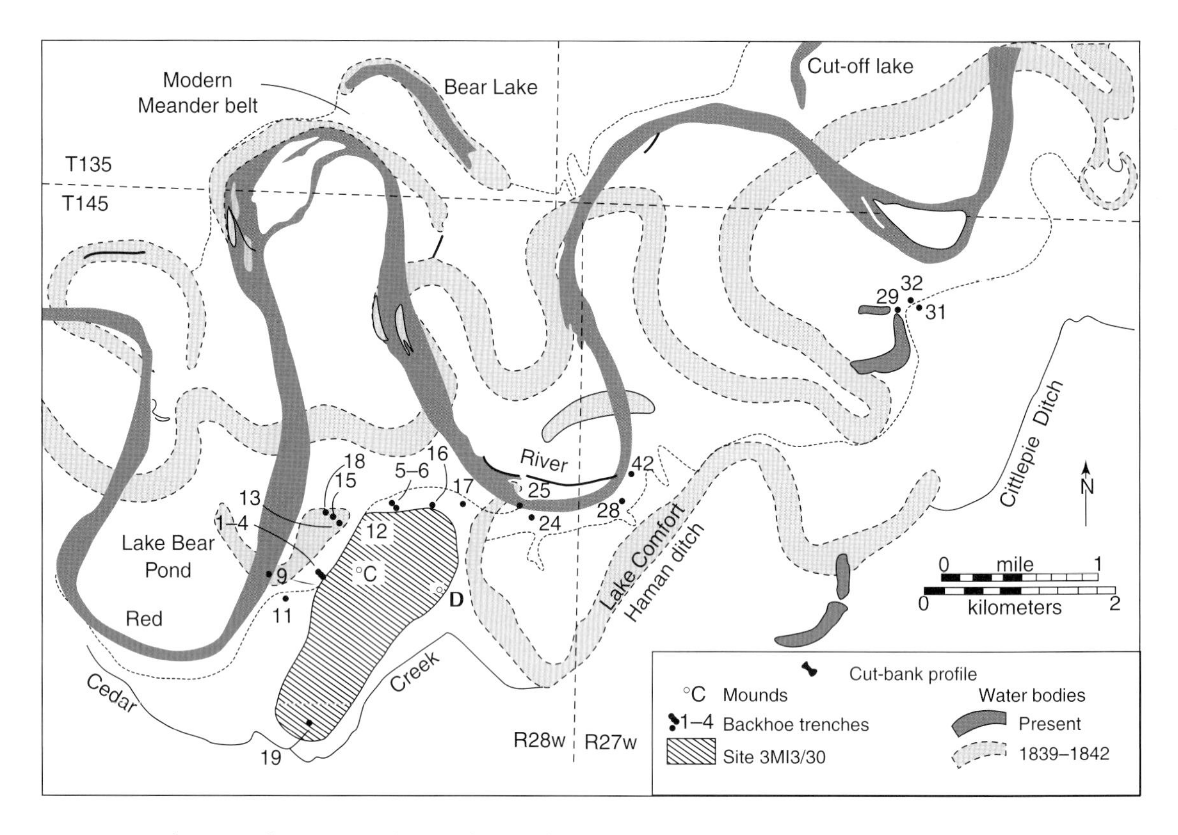

FIGURE 5.9 Changes of positions of meanders within the Red River during the past 160 years as determined by trenching and excavation. It is clear that any site situated next to the older channel (e.g. 3MI3/30, a mound site) could be buried under point bar deposits of the present meander (modified from Guccione *et al.*, 1998, figure 3).

sense) are continually forming with each subsequent flooding event, whether this interval is weekly, monthly, or annually. Such types of alluvial soils or *cumulic soils* straddle the area between sediments and soils (Fig. 5.10). In such cases of relatively rapid rates of deposition, any archaeological occupation has a reasonably good chance of being buried, with the net result that several discrete occupations can be observed throughout this overthickened A horizon.

At the other extreme, where there is little or no sedimentation, repeated occupations occur on top of each other, resulting in the formation of *palimpsests* (Ferring, 1986). As a result, it is virtually impossible to isolate individual occupations. Furthermore, inasmuch as the archaeological material lies about the surface, it can be subjected to modification and disturbance by a host of postdepositional processes and pedogenesis, such as displacement by water, (bioturbation) burrowing and gnawing by animals, chemical weathering, and pedoturbation (e.g. shrink-swell) (Limbrey, 1992). If the surface is exposed for a long enough period of time (depending on factors such as climate, parent material; see Chapter 3) soil horizon development will take place. If such a soil (now paleosol) profile ultimately becomes preserved and protected by renewed deposition, future excavation will reveal archaeological material occurring within the soil profile. Finally, since soils can be faithful recorders of local or regional (climatic) conditions (see Chapter 3) the types of soils can therefore provide palaeoenvironmental information during or close to the time of one or more of the

FIGURE 5.10 Photograph of late Holocene floodplain Ford alluvium in Cowhouse Creek, central Texas, United States. Shown here is sequence of buried A and A-Bw horizons (Entisols) within vertical accretion flood plain deposits consisting of silts and clays (see Nordt, 2004).

occupations. Such is the case for a calcic horizon, which is developed on red clayey alluvium in the Jordan Valley and which is universally associated with Epi-Palaeolithic (Geometric Kebaran; ca 14.5 to 13.7 ky uncal.) sites (Bar-Yosef, 1974; Maher, 2004; Schuldenrein and Goldberg, 1981).

Another indicator of landscape stability on fine-grained flood plain deposits is the formation of surface gleys. These result from poor drainage and associated oxidation/reduction conditions with the accumulation of organic matter. In addition, under wet conditions as such formerly humic topsoils and turbated topsoil material become preferentially impregnated with iron and manganese compounds (mottling), and commonly show the occurrence of pyrite or pyrite pseudomorphs (g horizon) (Miedema *et al.*, 1974; Wiltshire *et al.*, 1994).

Paleosols such as those described above also serve as temporal markers because they may have formed at roughly the same time interval in many places. In the Southern Plains, for example, soil formation took place during a period of flood plain stability between 2,000 and 1,000 BP (Ferring, 1992). Although the paleosol goes by different names in different places (e.g. Caddo in Oklahoma, Quitaque in Texas), it serves as a useful stratigraphic marker for placing Late Archaic, Plains Woodland, and Late Prehistoric sites in a firm stratigraphic and temporal framework (Ferring, 1992).

Alluvial archaeology is equally well established in Europe, particularly during the 1980s when sand and gravel extraction increased exponentially the number of CRM/"Rescue" sites (Brown, 1997; Howard *et al.*, 2003; Needham and Macklin, 1992). These studies provided insights into landscape and climatic changes. For instance, in the Fens (East Anglia), and along major rivers such as the Nene (Northamptonshire), Ouse (Bedfordshire), and Upper Thames (Oxfordshire), entire prehistoric landscapes have been buried by fine alluvium dating to Iron Age through to Medieval times (Robinson, 1992) (Fig. 5.11). The original landscape consisted of coarse silty Pleistocene/early Holocene river terrace deposits in which the soils had developed by mid–Holocene times. Ritual and burial monuments and landscape divisions (e.g. avenues, barrows, cursuses) exhibited treethrow features that could date back as early as the Late Mesolithic–Early Neolithic transition, some 6,000 years ago (French, 2003; Lambrick, 1992; Macphail and Goldberg, 1990).

Such sites thus provide the opportunity to investigate the archaeology and environment over long periods. Treethrows provide windows into the formation and state of prehistoric landscapes, such as at Raunds, Northamptonshire (Early Neolithic and Early Bronze Age monuments; Healy and Harding, Forthcoming), Roman and Saxon settlements, and at Drayton Cursus (Upper Thames). In the latter, the Neolithic Cursus seals treethrow hollows, and subsequent prehistoric soils are buried by Roman and Saxon alluvium (Barclay *et al.*, 2003). At both sites the alluvium buried early to mid-Holocene Alfisols, which provide "control soils" that can be used to compare soil development in later prehistory and historic times (Macphail and Goldberg, 1990). At Raunds, the consensus interpretation of the environmental data is for a Neolithic and Bronze Age grazed floodplain landscape on which stock concentrations produced trampled and phosphate-enriched soils that were sealed by numerous earth monuments (Macphail and Linderholm, 2004). Textural pedofeatures associated with this soil disturbance were readily differentiated from "Saxon and Medieval" clay inwash induced by alluviation (Brammer, 1971) (see Chapter 9; Fig. 16.1b, Table 16.1b).

A number of other riverine sites suffered the effects of increased wetness during the Holocene. For example, at Three Ways Wharf, Uxbridge Upper Palaeolithic artifacts (ca 10,000 BP) were biologically worked throughout the earliest Holocene soil, whereas flints from the Early Mesolithic (ca 9,000–8,000 BP) occupation (ca 1,000 years later) remained near or on the surface because they were sealed by a fine charcoal-rich humic clay (and peat) associated with wetter conditions. Similar peaty deposits formed throughout the Colne River valley during the early Boreal period (Lewis *et al.*, 1992). This wetness essentially stopped bioturbation of the Mesolithic artifacts at this site. The Upper Palaeolithic site was located on a low bar of a braided stream and functioned as a likely ambush camp at a river crossing. In contrast, by early Mesolithic times the site was swampy and humans were seemingly affecting the coniferous landscape by the use of fire.

The Upper Palaeolithic site of Pincevent in France, famous for the preservation of its living floors and activity areas (Leroi-Gourhan, 1984)

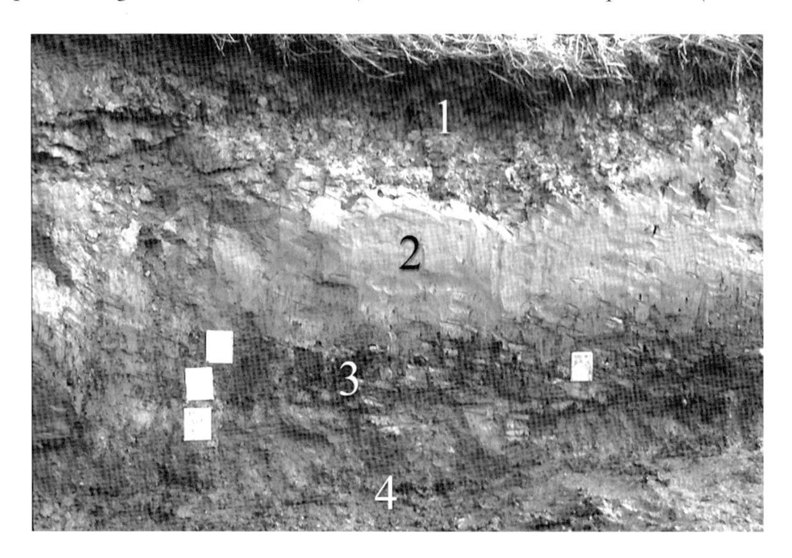

FIGURE 5.11 Section through charcoal and burned soil-rich remains of prehistoric treethrow hole fill buried by a meter of Saxon and Medieval fine alluvium at Raunds, Northamptonshire, United Kingdom. 4 – subsoils and gravels of Holocene Alfisols formed in Nene river terrace sandy silty sands, sands, and gravels; 3 – truncated burned tree hollow with burned soils with high magnetic susceptibility values (e.g. χ = 894 10^{-8} SI kg^{-1}); 2 – Saxon-medieval fine alluvium (e.g. clay loam); 1 – modern pasture soil with grass rooting and earthworm burrowing into top of fine alluvium.

exhibits exceptional preservation, due in large part to its location on the floodplain of the Seine. Here, low energy overbank flooding led to interment of the occupational remains, without movement of the artifacts. Lastly, it can be recorded that alluviation of the Lower Thames at London led to the deposition of highly polluted (P, Pb, Cu, and Zn) moat deposits at the Tower of London (Macphail and Crowther, 2004).

5.5 Stream terraces

As discussed earlier, channels constantly shift their position, either on a seasonal or yearly basis. In meandering channels, the stream sweeps across the valley floor resulting in the lateral accumulation of point bar deposits that over time can lead to vertical net accumulations. Similar net accumulation of deposits can occur in braided systems.

5.5.1 Characteristics of terraces

In many environments and locales, this net accumulation is interrupted by a marked change in hydraulic regime, resulting in the vertical incision (*entrenchment*) by the stream into previously deposited alluvium, or even bedrock. Such incision often leaves behind a geomorphic feature, a *terrace*, which consists of a flat portion (the "tread") and a steeper sloping surface (the "riser") that connects the tread to the level of the new floodplain or a lower terrace surface (Fig. 5.4; 5.12). Thus, in many cases, terraces represent abandoned floodplains that existed when the river flowed at this higher elevation. However, it should be kept in mind that terraces are essentially a geomorphic/topographic feature that can form on bedrock, or previously deposited alluvium or other sediment(s) (Fig. 5.12).

Two types of terraces are commonly recognized: erosional and depositional. Erosional terraces are produced by downcutting associated with lateral erosion, producing stepped surfaces as shown in Figure 5.12. If the underlying material is bedrock, they are commonly called *strath terraces*. With depositional terraces the tread and riser are composed of valley alluvium (Fig. 5.13).

Although processes involved in the formation of terraces are complex, varied, and often a product of combined factors, a phase of valley accumulation (*aggradation*), must be followed by a phase of downcutting (entrenchment). Similarly, aggradation occurs when the supply of material is greater than the ability of the system to remove it. This in turn can be brought about by (1) inputs of glacial outwash, (2) changes in local or regional base level caused by changes in sea level (brought about by eustatic or tectonic causes), (3) inputs of coarse sediment related to tectonic uplift, or (4) climate change (Ritter *et al.*, 2002) (see Box 5.1). Entrenchment arises from similar causes, principally tectonism or climate change. In valleys adjacent to the Dead Sea, for example, terraces were formed as the streams incised their channels in response to a lowered base level resulting from the shrinking of the Pleistocene Lake Lisan, the precursor to the modern Dead Sea (Fig. 5.12) (Bartov *et al.*, 2002; Bowman, 1997; Niemi, 1997; Schuldenrein and Goldberg, 1981).

5.5.2 Archaeological sites in terrace contexts

Depositional terraces provide a wealth of stratigraphic, palaeoenvironmental, and geoarchaeological information as they represent fossilized loci of human activities focussed on alluvial environments. Furthermore prehistoric sites associated with terraces provide materials (organic matter/charcoal/hearths) that are datable by ^{14}C or cultural remains that can be temporally bracketed. Within recent years the success and refinement of OSL/TL techniques shows that they can be used to date the sediments directly (Fuchs and Lang, 2001), thus

FIGURE 5.12 Photo of strath terraces from Nahal Ze'elim, western Jordan Valley. This ephemeral stream flows into the Dead Sea from the Judean Mountains during major rainfall events. During Pleistocene times a large lake, Lake Lisan, over 200 m higher than the present day Dead Sea (ca −405 m below sea level), existed in the Jordan Valley. When this lake dried up at the end of the Pleistocene (Bartov *et al.*, 2002), the base level lowered rapidly, and streams such as this one incised into their channels relatively rapidly as they moved across the valley floor. The result is this flight of terraces developed on bedrock, and fluvial and lacustrine deposits.

avoiding the problems of the charcoal being reworked or derived from old wood.

5.5.2.1 New World sites

Sites found in terrace contexts exist over the entire globe and many have been extensively studied. In central North America, terrace sequences are common along major and minor drainages, and many contain prehistoric sites with among the earliest Paleoindian sites (Blum *et al.*, 1992; Blum and Valastro, 1992; Ferring, 1986, 1992; Mandel, 1995; Nordt, 1995). The common practice is to label the lowest and youngest terrace, that is, the one closest to the level of the modern channel, T_0 and successively older and higher terraces, T_1, T_2, etc.; this practice of terrace designation is not widespread outside North America.

Geoarchaeological research in the American southwest has been intensive and extensive for over 65 years with the early work of Kirk Bryan and E. Antevs (Waters, 2000), followed by the dedicated work of Haynes (1991, 1995) and his student, Waters (Waters, 1992; Waters and Ravesloot, 2000). Waters' lifetime work in the area has sought to address a number of issues relating to alluvial stratigraphic sequences, how they reflect upon palaeoenvironments and the completeness of the archaeological record (Waters, 2000). For example, the occurrence of Clovis sites in the San Pedro Valley of southern Arizona is not matched in neighboring valleys where Clovis remains are absent. The question, of course, is: does this distribution represent a cultural preference or another, perhaps geological control of site distribution?

Detailed geoarchaeological work in these valleys is shown in Figure 5.20, which reveals some interesting patterns. It is clear that the absence of sites in the Santa Cruz River and Tonto Basin, for example, is related to the absence of sedimentary traps to store these sites; even if they were there, they are long gone. Secondly, even where deposits do exist

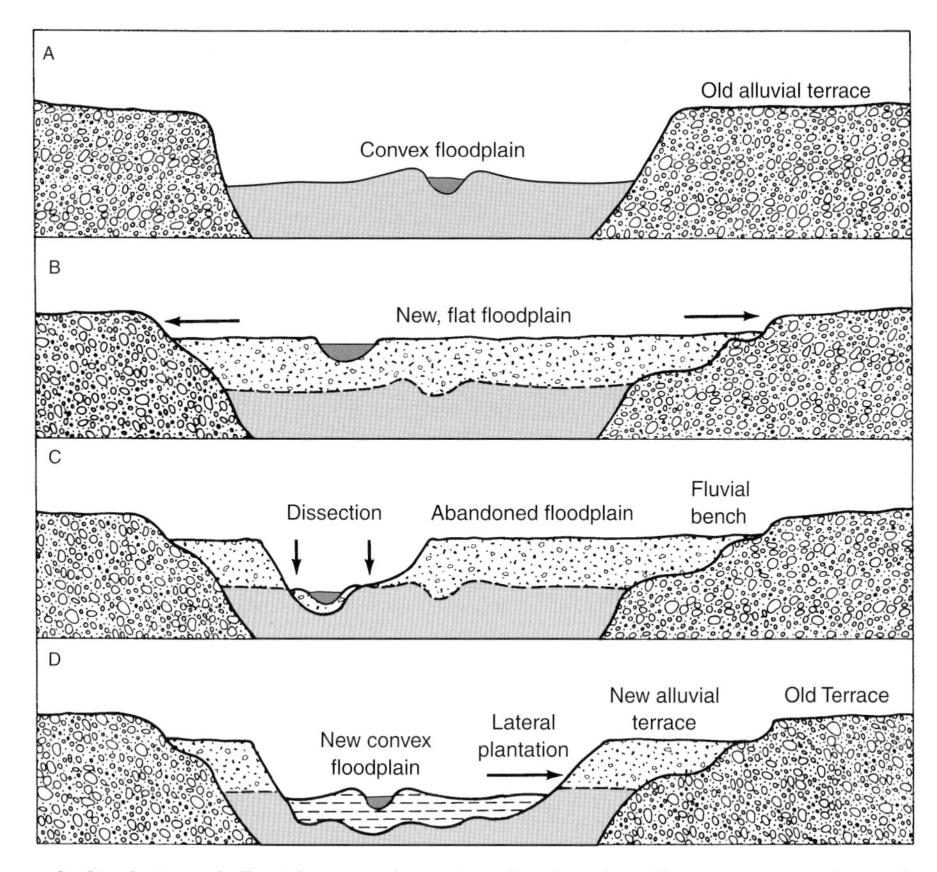

FIGURE 5.13 Idealized view of alluvial terrace formation developed in alluvium. (A) As shown here, this first stage depicts build-up (aggradation) of fine-grained alluvium within an eroded valley that formed by the downcutting through previously deposited gravelly alluvium; (B) Coarser alluvium (braided stream, perhaps dryer conditions) covers these initial deposits, enlarging the floodplain by lateral planation; (C) initial incision and downcutting (degradation) takes place due to factors such as climatic change or base level lowering, leaving behind an abandoned floodplain surface; (D) downcutting ceases and renewed alluviation takes place that is characterized by fine-grained sediments (as in (A)), with lateral planation resulting in the widening of the floodplain and removal of some of the previous alluvium from stages (A) and (B); as a consequence, a second alluvial terrace is formed at a lower elevation than the first, but it is composed of two different types of deposits (Modified from Butzer, 1976, figure 8.13, p. 170).

(e.g. Whitewater Draw, Gila River) the braided stream hydrology would result in cultural material being eroded and reworked throughout the gravelly deposits. Finally, we see that valley deposition and erosion in the Late Pleistocene and Early Holocene was not uniform throughout the area. This pattern would be expected if each valley had specific and different geomorphic variables (e.g. rock type, slopes), or if there were climatic differences among valleys, or the climate was the same but each valley reacted differently to climate changes (Waters, 2000). In any case, this research shows that a geological filter (viz. erosion) is responsible for the apparent absence of sites beyond the San Pedro River.

In the same area, Waters evaluated the effects of fluvial landscape change on Hohokam prehistoric agriculturalists who occupied the Gila and Salt River valleys from AD 100 to 1,450 (Waters, 2000). The major questions involved the change in pattern in public areas around

Box 5.1 Upper and Middle Palaeolithic sites of Nahal Zin, Central Negev, Israel

Fluvial environments contain rich records of archaeological sites Arid environments, are in particular, effective in preserving and displaying sites in fluvial contexts because the stratigraphic profiles are well exposed. They are also abundant. Such conditions enable us to observe and recognize buried sites that are commonly concealed in more humid environments where vegetation cover is more extensive. An area that has yielded a wealth of prehistoric sites is the Nahal Zin area of the Central Negev in Israel (Fig. 5.14) (Marks, 1983). Sites ranging from Lower Palaeolithic through Chalcolithic Periods were uncovered during the extensive surveys and excavations made in the 1960s and 1970s. Most of all, numerous Middle and Upper Palaeolithic

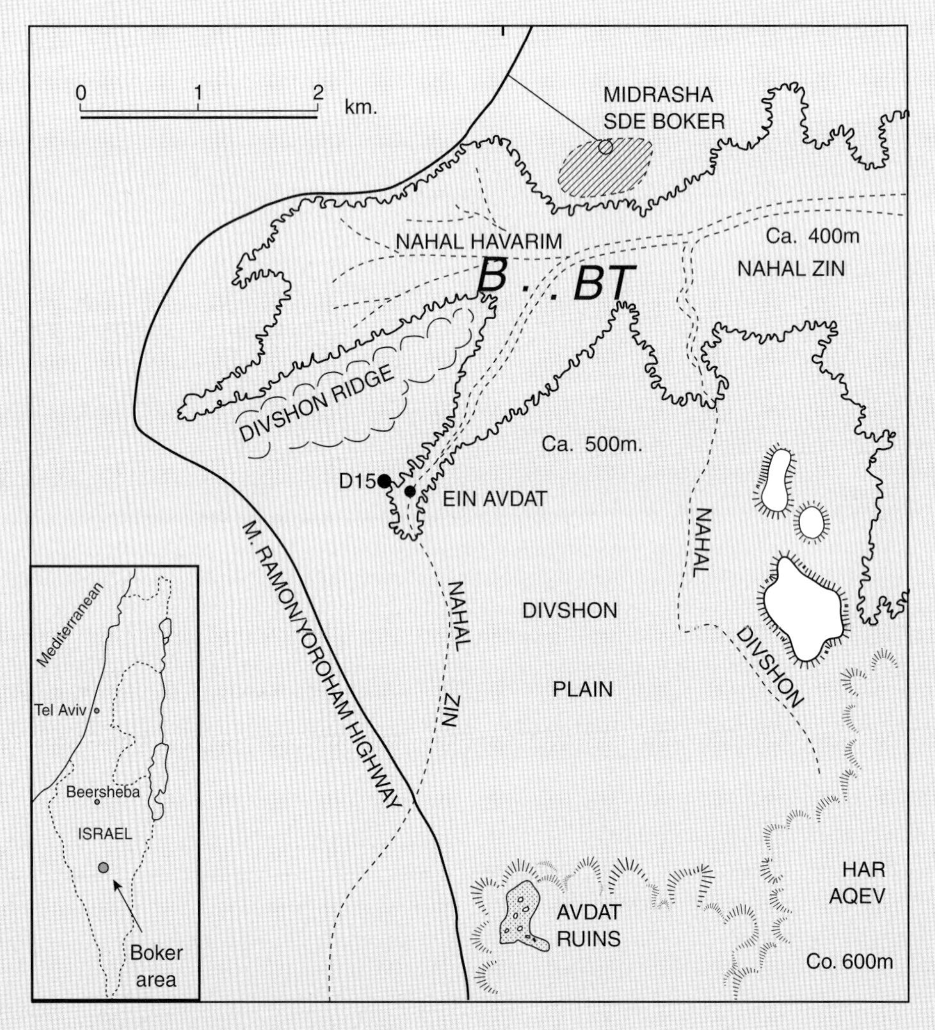

FIGURE 5.14 Nahal Zin area in the Central Negev, Israel. The Upper Palaeolithic site of Boker (B) and Middle/Upper Palaeolithic site of Boker Tachtit (BT) are shown (modified from Goldberg, 1976).

(cont.)

Box 5.1 (*contd.*)

sites were found within discreet fluvial deposits representing succinct *aggradational* (depositional) events, separated by erosion and vertical downcutting. Younger Epipalaeolithic and Neolithic sites tend to occur on the surfaces of these older deposits.

The Nahal Zin (*nahal* is Hebrew for ephemeral stream) drainage originates at about 1,000 m in elevation, in the upper part of the Central Negev Highlands, a plateau composed of Eocene chalk and limestone with abundant layers of chert. Along the northern edge of the plateau, in the area of Sede Boker (Fig. 5.14), the stream has incised into the channel forming a canyon and a series of *knickpoints* that are associated with springs (e.g. Ein Avdat) along the ca 100 m of vertical drop from the plateau down to its lower course. At this lower elevation, the stream flows on softer marl and chalk, and continues to flow northwest for about 1.75 km, after which it turns east and flows eventually into the Dead Sea. Along this northeast stretch two important Middle and Upper Palaeolithic sites were found, Boker Tachtit (**BT**) and Boker (**B**) (Figs 5.14 and 5.15).

Both sites are contained within a sequence of fluvial deposits that include soft, marly and chalky silts, gravels, and colluvial clay. The site of Boker Tachtit is the older of the two sites, and is situated on the east side of the

FIGURE 5.15 Nahal Zin valley looking southwest toward the Central Negev Plateau from the area of Sede Boker (cf. Fig. 5.14). The site of Boker Tachtit (**BT**) is situated within terrace deposits adjacent to the road, about 7 m above the floor of the channel. The flat surface of the terrace continues along to the southwest and northeast. The site of Boker (**B**) sits across the channel from it. The hill immediately above and to the right (west) of Boker is a remnant of a landscape surface and gravelly deposits that contain the remains of an undated Middle Palaeolithic site (D6). During the Middle Palaeolithic occupation of the site, Nahal Zin was flowing about 60 m above the present day channel and was more continuous with Nahal Havarim, which flows in from the west (Goldberg and Brimer, 1983). Compare with cross section in Figure 5.16.

(*cont.*)

Box 5.1 (*contd.*)

valley (Figs 5.14 and 5.15) within the lower part of a terrace remnant. The archaeological material, which spans the Middle Palaeolithic/Upper Palaeolithic boundary, is composed of predominantly fine marly silts, with lenses and bands of gravel that accumulated as overbank deposits not far from the channel (Fig. 5.16). These deposits became somewhat consolidated before they were eroded by gravels that truncate (erode) them with a sharp contact. At this time, the channel – and associated higher energy gravels – had shifted to a new position directly over the former finer-grained flood bank. Occupation at this location, after the shift of channel, was not very likely as it would have lined up with the center of the gravel-laden stream channel.

The site of Boker consists of a number of individual occupations that are found at different elevations and locales within a similar terrace vestige of similar height on the west side of the valley (Figs. 5.14, 5.15, and 5.18a). The deposits that constitute the terrace are exposed over a length of about 85 m and vary from 3 to 9 m in thickness. The sites are enclosed within interbedded sands, silts, and clays (Figs 5.18b and Plate 5.1), and locally some gravel lenses. The gravels and sands are predominately of alluvial origin from Nahal Zin. On the other hand, the silts are both alluvial and aeolian in origin. Windblown silts are common in the Negev and recognizable by their good sorting and massive character. Hard, grey clay lenses are colluvially derived from the surrounding grey Palaeogene marls into which the terrace sediments are set (Fig. 5.18a,b; Plate 5.1).

The causes of such periods or cycles of downcutting (incision) and infilling (aggradation) for arid areas such as these has been much discussed over several decades. Changes in base level, climate change, tectonic movements, and intrinsic changes in the fluvial system are commonly cited to produce such changes in these fluvial landscapes. In the Nahal Zin area, fluctuations in Pleistocene climate – mostly precipitation – appear to be the most reasonable cause, although in light of the complexity of alluvial environments it is difficult to reconstruct specific processes and responses. Other data from the southern Levant (e.g. similar

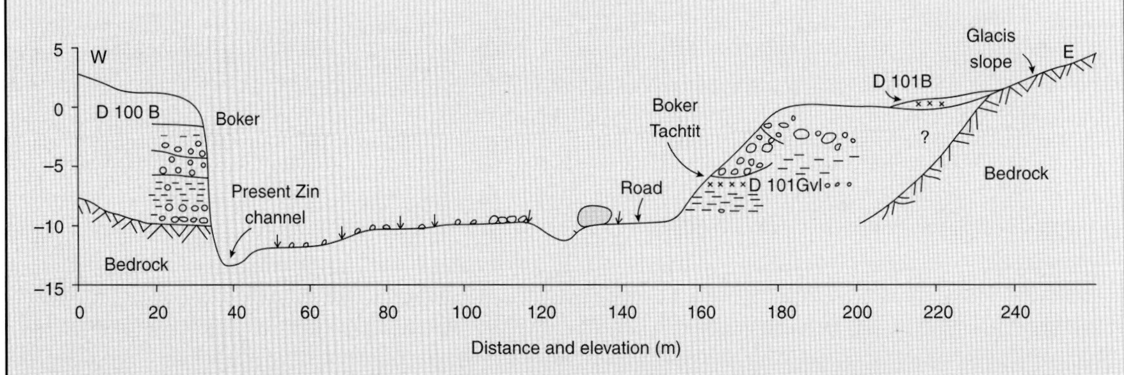

FIGURE 5.16 Schematic west–east cross-section across Nahal Zin showing the topographic and stratigraphic relationships between the sites of Boker and Boker Tachtit Note the well-preserved form of the terraces, particularly Boker Tachtit, which shows the classic tread and riser morphology. The Epipalaeolithic site of D101B rests within a colluvial and aeolian loess mantle on top of the Boker Tachtit tread surface and is shown in Figure 5.17a.

(*cont.*)

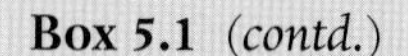

Box 5.1 *(contd.)*

(a)

(b)

FIGURE 5.17 (a) The site of Boker Tachtit as seen from the site of Boker. The two major sedimentary units exposed during the excavation can be seen within the terrace. The upper surface of the terrace (the tread) is actually covered with a thin veneer of colluvial and aeolian tan silts (arrow) that contain Epi-palaeolithic artifacts that date to about 13 ka. (b) The terrace sediments of Boker Tachtit are composed of two major fluvial units: (1) a basal light-colorer chalky silt with lenses (stringers) of gravel, which are marked by a erosional contact of (2) the overlying silty gravel. The sharp and undulating contact clearly indicates that the lower deposits were somewhat consolidated before the gravels truncated them. The transition between the Middle and Upper Palaeolithic, which occurred about 40 to 45,000 years ago, takes place within the lower, chalky unit. Scale is 50 cm.

(cont.)

Box 5.1 *(contd.)*

geomorphological and stratigraphic sequences over large regions – draining into both the Mediterranean and Dead Sea Basins – and distribution of archaeological sites) lend support to the influence of climatic fluctuations (Goldberg, 1986; Goldberg and Bar-Yosef, 1995).

FIGURE 5.18 (a) The site complex of Boker is situated on the west bank of Nahal Zin across from Boker Tachtit. The terrace sediments are set into eroded soft grey chalks and marls which can be seen in the foreground and immediately to the right of the site. Eroded and colluvially reworked portions of these marls make up the grayish layers in the terrace sediments shown in the excavated areas of the site (Area BE of the original excavation; Fig. 5.18b and Plate 5.1). The terrace form here is less pronounced and somewhat more eroded than at Boker Tachtit. Its continuation to the south is visible in the upper part of the photograph where a well-defined remnant can be seen sloping down to the east. (b) Area BE of the original excavation. Layers of pale brown-colored sand and silt alternate with wedges of lighter, grayish clay. The sand is alluvial in origin with some aeolian loess layers. The grayish clay layers are colluvially derived marls from the surrounding hills (cf. Fig. 5.18a). In this regard note that the clays thicken in the colluvial source to the right, whereas the pale brown-colored alluvium thickens to the left, toward the direction of the Nahal Zin. The black streaks are thin lenses of charcoal, representing dispersed hearth components. The height of the profile in the center of the photograph is about 110 cm.

(cont.)

Box 5.1 *(contd.)*

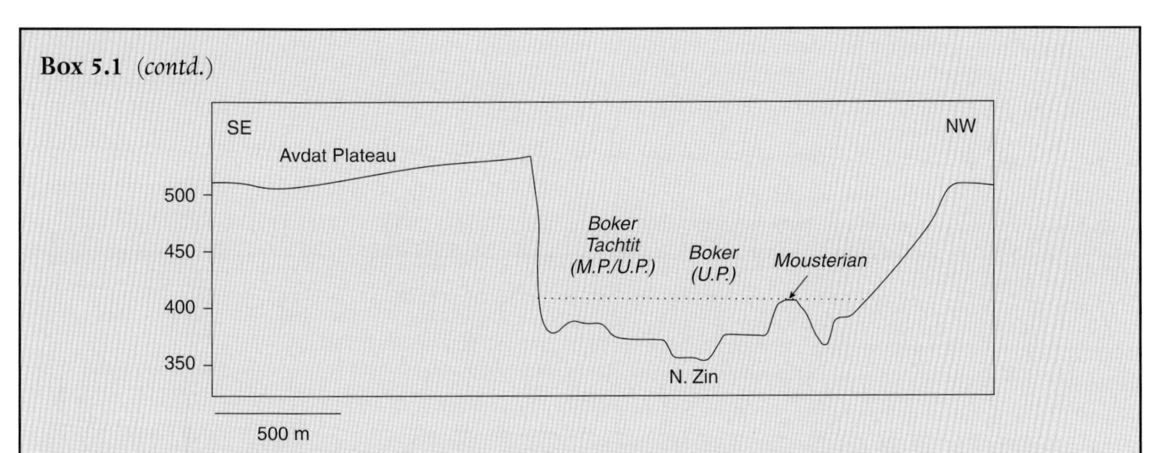

FIGURE 5.19 Generalized cross section through Nahal Zin showing the changes in elevations of landscape surfaces through time. The Mousterian surface is coeval with site D6 discussed above and can be traced through much of the Nahal Zin area. Its age is not definitively known, but recent estimates put it at about 200,000 years old, indicating that at this time, the valley was about 60 m higher than it is today. The stream subsequently incised its channel eroding into the underlying limestone, chalks, and marls of the local bedrock. This downcutting ceased with the onset of alluvial deposition associated with the Boker/Boker Tachtit terraces, which began about 45,000 years ago as based on the radiocarbon dates from Boker Tachtit (Marks, 1983). Alluvial build-up continued through the middle part of the Upper Palaeolithic about 25,000 years ago, as marked by the presence of the upper levels of Boker. Shortly after this time, the stream began to incise its channel, a phenomenon which continues to today. M.P. = Middle Palaeolithic; U.P. = Upper Palaeolithic.

AD 1,250 (e.g. ballcourts being replaced by platform mounds; the reduction in number of settlements but increase in size), and the collapse of the Hohokam in AD 1,450 and the abandonment of canals. A summary of thorough geoarchaeological examination of natural exposures and those from backhoe trenches is given in Figure 5.21. As can be seen, several aggradational phases could be recognized, accompanied by downcutting and infilling. Waters (2000) noted that during the period AD 1,050 to 1,150, the Gila River channel incised and widened (Unit IV), which appears to have coincided with a stage of high magnitude flooding in this area. The braided channel cut into the floodplain that had been previously stable for close to 1,000 years; such conditions made it difficult to organize transport of water over the highly porous stream gravels as channel location shifted continuously: "Any temporary diversion dams constructed to get water into headgates [of the canals] would have been vulnerable to being washed out with such channel environment" (Waters, 2000: 552–553). Waters links these hydraulic issues to changes in labor organization, canal engineering, and increased and more diverse food production. Furthermore, he notes that after this time, the channel appears to have stabilized with the deposition of silty clay and the formation of a paleosol, which was accompanied by probable farming and the establishment of villages on this stable surface. Little or no geomorphic change appears to have taken place at the end of Hohokam occupation, and Waters concludes that ". . . the reorganization of the Hohokam at the end of the Classic must have been the result of other factors" (Waters, 2000: 554).

In the Central Plains (Kansas and Nebraska in the United States), extensive work by Mandel has been instrumental in reconstructing the geomorphic history of this large region

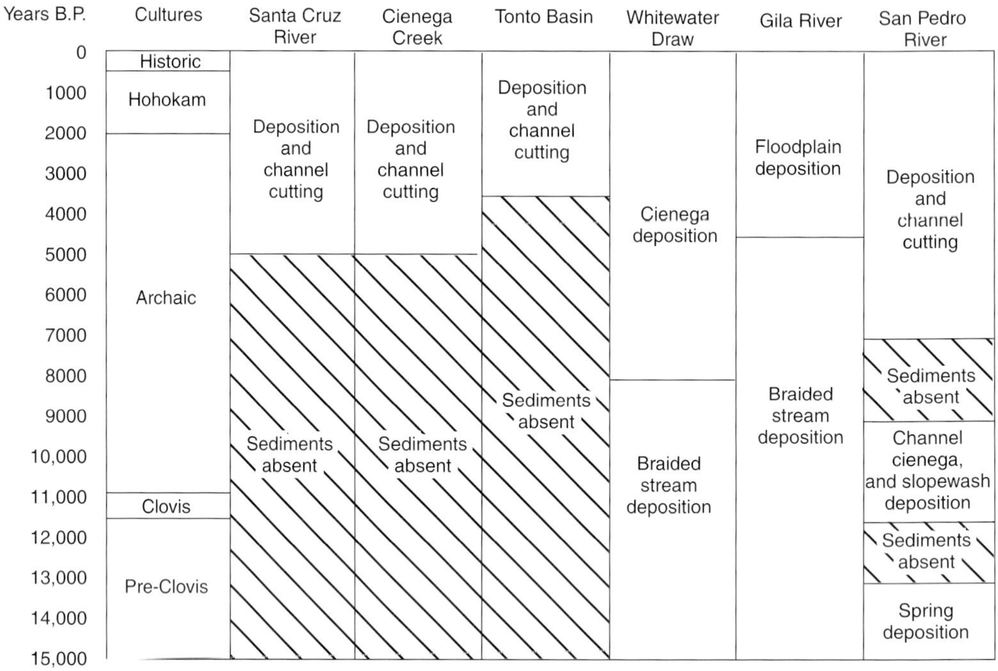

Years B.P.	Cultures	Santa Cruz River	Cienega Creek	Tonto Basin	Whitewater Draw	Gila River	San Pedro River
0 1000 2000 3000 4000 5000 6000 7000 8000 9000 10,000 11,000 12,000 13,000 14,000 15,000	Historic Hohokam Archaic Clovis Pre-Clovis	Deposition and channel cutting Sediments absent	Deposition and channel cutting Sediments absent	Deposition and channel cutting Sediments absent	Cienega deposition Braided stream deposition	Floodplain deposition Braided stream deposition	Deposition and channel cutting Sediments absent Channel cienega, and slopewash deposition Sediments absent Spring deposition

FIGURE 5.20 Alluvial sequences from several drainages in southwestern Arizona. These sequences show different events (e.g. erosion, deposition in streams, and cienega-marshy depressious) taking place in different valleys at different times. Erosion and lack of sediments in some valleys explains the differential distribution of Clovis sites in the area (modified from Waters, 2000, figure 3).

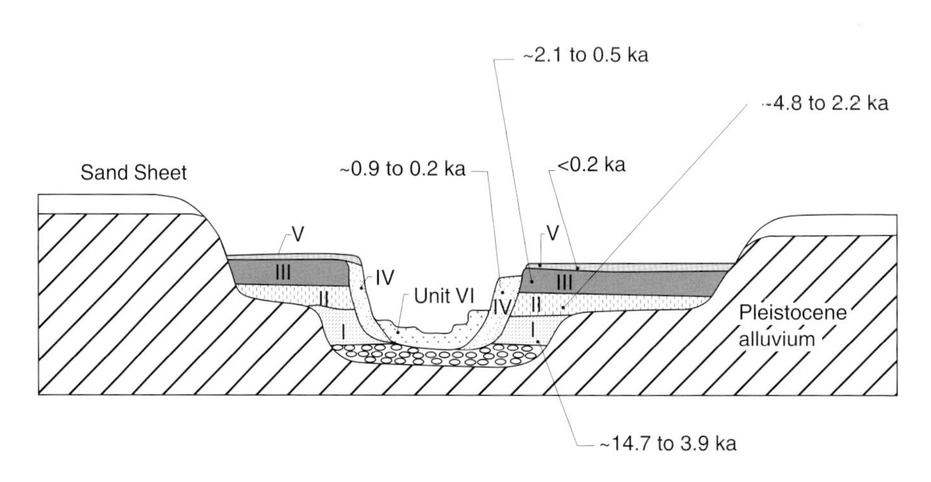

FIGURE 5.21 Composite cross-section from Middle Gila River, Arizona, illustrating terraces, stratigraphic units and associated ^{14}C dates (Modified from Waters, 2000, figure 4).

and showing how this knowledge can be used to understand the presence and temporal distribution of Paleoindian and Archaic sites in the area (Mandel, 1992, 1995, 2000). He used data from his own field work and those from the literature (e.g. unpublished dissertations, CRM reports) to document Holocene alluvial stratigraphy within parts of the Platte, Missouri, Kansas, and Arkansas River drainages. As in the southwest, perceptible gaps in the archaeological record can result from burial of the archaeological material, and erosion and removal of deposits containing cultural materials.

His results showed significant differences in the presence of deposits according to stream size, in that large and small streams exhibited different age distributions of the sediments (Fig. 5.22). He noted, for example, that large streams contained early, middle, and late Holocene alluvium, whereas small streams exhibited only late Holocene deposits; early and middle Holocene deposits can be found within alluvial fans where small streams enter major valleys. Furthermore, the overall paucity of radiocarbon dates (usually determined on soil organic matter) between 7 and 6 ka (roughly the time of the Altithermal in this area) is ascribed to the dynamic nature of valley sedimentation, which prevented the formation of soils. It appears that erosion of the uplands and much of the early Holocene alluvium in small streams was removed and redeposited in the alluvial fans through the Middle Holocene (ca 8 to 4 ka). These patterns can be documented elsewhere in the Midwest, and when combined with palaeoclimatic data (pollen, palaeobotanical evidence, computer models of climate), the inference is that they result from climatic change over the entire region. Specifically, they point to warmer and drier conditions associated with zonal (i.e. west to east) atmospheric circulation. In the later Holocene, in contrast, flow became more mixed with a shift toward more meridional (south to north) flow, and greater influence of moisture coming from the Gulf of Mexico. This shift corresponds to increased precipitation, and a concomitant change in amount and type of vegetation, such as short-grass prairie to mixed and tall-grass prairies, and increased forest cover. In turn, this led to ". . . reduced erosion rates on hillslopes, which in turn would have reduced mean annual sediment concentrations in small streams" (Mandel, 1995: 60) and greater storage of alluvium.

The importance of such detailed work over a large region is to demonstrate that the lack of Archaic sites in the interval 8 to 4 ka is a function of geomorphic controls rather than prehistoric lifeways. Such geomorphic processes either removed sites or buried them. Mandel points out that prior work relied on shallow testing of the landscape (e.g. shovel testing or surface survey) to locate sites. It is clear that such strategies will lead to a disproportionate amount of younger sites which occur at the surface, and will ignore the potential for locating earlier archaeological material. Furthermore, he concludes, that a basin-wide approach is needed to reveal the geomorphic and archaeological history, including the major gaps.

In the Southern Plains, alluvial sequences and terrace formation have been documented along many rivers in Texas and Oklahoma (Ferring, 1990; Ferring, 1992, 2001; Nordt, 1995, 2004; Waters and Nordt, 1995), revealing deposits that span what appear to be the bulk of human settlement in this part of North America. The Aubrey site in north Texas, for example, contains among the earliest dated Clovis remains in North America (Ferring, 1995, 2001; Humphrey and Ferring, 1994). It consists of a variety of sediments that include a basal channel lag/bar, overlain by lacustrine/spring deposits and flood plain silts, capped by soil formation.

A very similar setting is found at the Wilson–Leonard site, north of Austin (see Box 2.1). Here, along a lower terrace of Brushy Creek, is a ca 5 m thick sequence spans Paleoindian through Late Prehistoric archaeological remains, roughly 11 ka through 4 ka (uncalibrated) (Bousman *et al.*, 2002; Goldberg

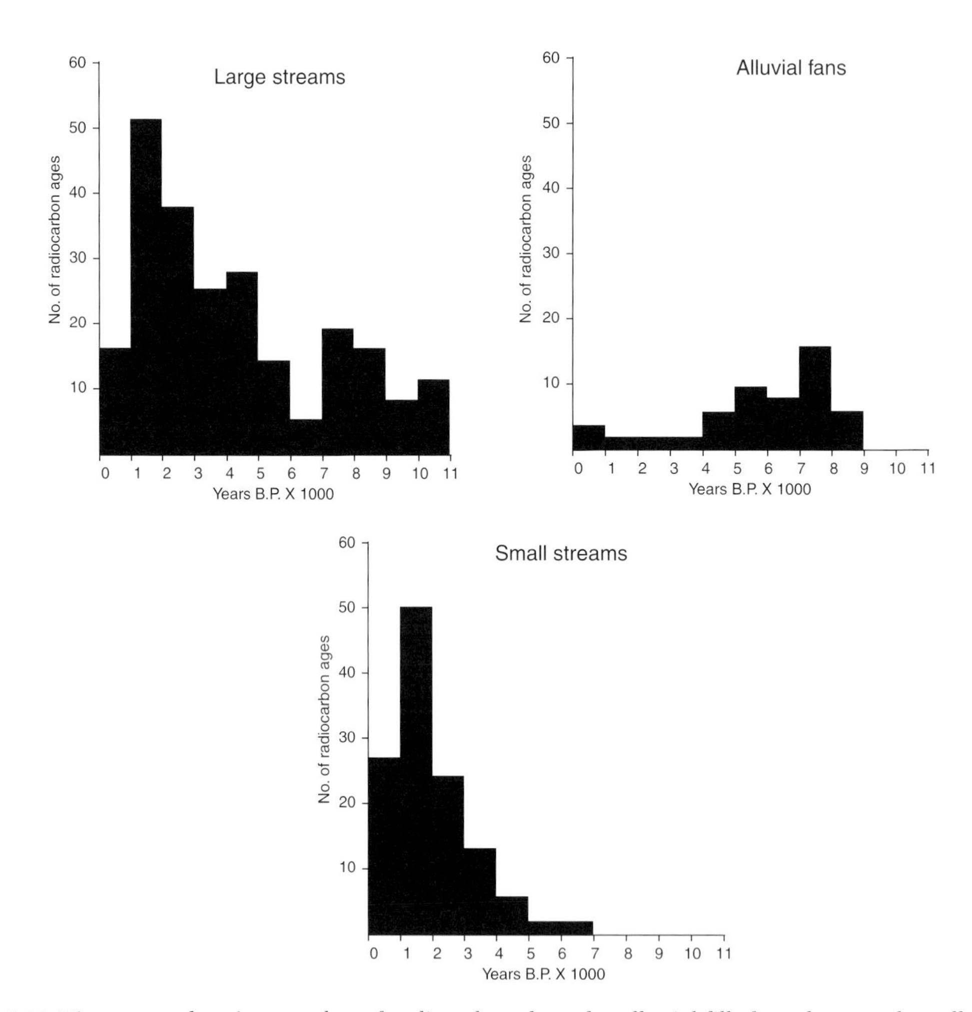

FIGURE 5.22 Histograms showing number of radiocarbon dates for alluvial fills from large and small streams, and alluvial fans in the Central Plains (modified from Mandel, 1995, figue 8).

and Holliday, 1998). The deposits consist of alluvial gravel that grades upward and laterally into organic-rich cienega deposits and flood plain silts, capped by the formation of a weak soil. These in turn, are covered by colluvial silt and fine gravel that become increasingly anthropogenic toward the top of the sequence, where numerous burnt rock ovens occur in association with Early Archaic hunters and gatherers. The sequence points to an overall decrease of alluvial activity during the late Quaternary with a drier spell occurring at the beginning of the Holocene. This is also confirmed from palaeobotanical and molluskan data (Bousman *et al.*, 2002). In any case, the site was occupied on and off for 7,000 years of fluvial and colluvial deposition.

5.5.2.2 Sites in the Old World

In the Old World, terrace sequences are abundant in the southern and eastern half of the Mediterranean Basin, where numerous prehistoric and archaeological sites have been well documented in fluvial and fluvio-lacustrine deposits (Goldberg, 1986, 1987, 1994; Hassan, 1995; Macumber *et al.*, 1991; Rosen, 1986; Schuldenrein and Clark, 1994, 2001; Zilberman, 1993) (see Fluvial Box). Among the earliest and best preserved ones found with prehistoric materials in clear context date to Late Middle Pleistocene or Late Pleistocene times and are associated with Middle Palaeolithic sites and artifact scatters (Fig. 5.7b). The earliest part of the sequence begins with the accumulation of well-rounded gravels linked to Middle Palaeolithic artifacts that locally occur within or on top of them. These gravels occur within well-defined terrace morphologies (e.g. Fig. 5.4b) or as remnants within or on the side of valleys. Unfortunately, chronometric dating of these early terraces has been elusive, and in need of clarification. The presence of fossil springs and relatively high amounts of arboreal pollen in some of these sites (D-35 in the Central Negev) suggested markedly wetter climates at this time (Goldberg, 1977; Horowitz, 1979).

Most of these gravels were eroded with up to 20 m of vertical downcutting during the later part of the Middle Palaeolithic, roughly between 70 and 45 ka, although the date of the onset of incision is approximate. In turn, incision was followed by renewed and widespread aggradation of massive fine-grained silts and clays that began at the very final stages of the Middle Palaeolithic and beginning of the Upper Palaeolithic, roughly 45 ka. These deposits can be found over the entire area, from southern Sinai (Gladfelter, 1992; Phillips, 1988), the Negev, and into Jordan. Their lithologies and bedding characteristics indicate that they represent massive inputs of originally aeolian derived sediments that were washed out of the atmosphere and concomitantly eroded off the slopes and introduced into the fluvial system. Damp and swampy (*paludal*) conditions existed in a climate that was overall wetter than the present; the reasonable possibility of summer rainfall (at present only 50–200 mm of precipitation falls in the winter) have been suggested (Sneh, 1983).

Upper Palaeolithic aggradation ceased about 22 ka and in most areas was followed by incision until about 15–14 ka, although of a more modest scale than the previous downcutting cycle. The end of the Late Pleistocene (ca 15–14 to 10 ka) is overall marked by renewed sedimentation but on a smaller scale. In many places, the deposits are locally reworked alluvium and colluvium from the previous depositional cycle but locally, high water tables attest to overall wetter climates for the Epi-Palaeolithic. In fact, Epi-Palaeolithic sites are widespread over the entire zone covering much of the Sinai and the Negev (Goring-Morris, 1987).

The early Holocene is represented by marked incision: in Nahal Besor ~20 m of incision took place within <8,000 years, and was interrupted by silty terrace deposits that interfinger with Chalcolithic sites and cultural deposits (Goldberg, 1987; Goldberg and Rosen, 1987; Shiqmim) dated to about 7–5 ka. It is not a coincidence that the stream was aggrading at this time of occupation of this relatively large village in the desert (Levy, 1995), and it is likely that the climate was wetter than at present.

Much of the later part of the Holocene is characterized by downcutting and erosion exception for a period of extensive alluviation of ca 1 m thick silts with *in situ* Byzantine remains throughout the Negev and East Sinai. Many of these deposits appear to have been cultivated by the Byzantines.

Causes of aggradation and incision in this region appear to be clearly driven by climate. This is shown by the fact that similar depositional and erosional features occur over a large area (Sinai, the Negev, and West Jordan) in which streams drain both to the Dead Sea (then about −175 to −200 m below sea level, versus ca −405 m today) and to the Mediterranean. This latter aspect rules out base level changes as an influence. Similarly, a marked period of incision along Nahal Besor, a major stream that drains much of the western Negev, took place at the very beginning of the Holocene, just when worldwide sea levels were rising. Furthermore, it is interesting to note that excavation and survey data suggest that for many locations, particularly for the Upper Palaeolithic and Epi-Palaeolithic there is a general trend toward increased number of sites-including surface sites-not associated with fluvial deposits during periods of sedimentation. The suggestion again, is that in most general terms, these aggradational phases point to wetter climates in what is now a hyper arid to arid terrain.

Finally, we can note that extensive fluvial studies have been carried out in Asia. For example, in the Upper Indus region of Pakistan, recent geoarchaeological work by Schuldenrein *et al.* (2004) has shown that alluviation slowed and pedogenesis ensued during the Early Holocene; it appears that at this time lateral accretion deposition gave way to entrenchment in these larger floodplains. Such changes appear to be climatically induced and related to changes in intensity of the southwest monsoon.

5.6 Lakes

5.6.1 Introduction

Lacustrine environments and associated wetlands are linked to fluvial terrains by the fact that many are fed by streams. In the chain of lakes that dotted the East African and Levantine Rift systems, they served as attractive locations to early hominins coming from Africa, up through to the Middle East. They continued to be a focus well into the Holocene, where for example, numerous habitations built out into lakes in Switzerland, and along the margins of the Great Lakes in the United States. In this section we will briefly examine some of the principal aspects of lacustrine systems and how humans interacted with them.

5.6.2 Characteristics of lakes

Lakes are closed bodies of standing water that vary considerably in size, both during past decades and centuries, but also over the course of the Quaternary. The basins in which they are formed have numerous origins, including rift valleys, volcanic and meteorite craters, glacial depressions left by decaying ice (kettles) or retreating ice (moraines), alluvial floodplains (oxbows and avulsed channels), or karstic depressions (sinkholes). Nevertheless, lakes are short-lived features on the earth's surface, and are subject to desiccation or infilling.

Generally, lakes are categorized as either open lakes or closed lakes. Open lakes (*exorheic*) are those which have an outlet, and consequently remain fresh, without concentrations of salts. They also tend to be stable and have shorelines (Boggs, 2001) with only some slight, short-range fluctuations in lake level.

Closed lakes (*endorheic*), on the other hand, have no outflow and dissolved solutes are concentrated; in arid and semiarid areas, evapotranspiration generally exceeds any inputs from rivers or springs. In these same areas, lakes can be only temporary features on the landscape (*ephemeral lakes*), where their basins can be filled only for short periods of time, much of the time the lake bottom is dry. In more extreme areas of the landscape, lake bottoms can be seen but are essentially dry for most of the time (*playa lakes*), although in more humid episodes during the Quaternary, they were perennial, or at least seasonal features on the landscape (cf. Fig. 15.5). Closed lakes are unstable and subjected to large inter- and intra-annual fluctuations in volume and position of the shoreline. Because of this sensitivity to external inputs of water and sediment, and associated changes in water chemistry that influences the biota (e.g. pollen, diatoms), lakes tend to provide valuable paleoclimatic records, either preserved within the lake (*lacustrine*) deposits themselves (Begin *et al.*, 1980) or in the record of shoreline changes (Enzel *et al.*, 2003; Klein, 1986).

Lacustrine sediments are varied and several facies can be recognized. Knowledge of this can aid in understanding the location and function of sites in now exhumed, fossil lacustrine deposits. Facies differences are tied to the effects of sedimentary input, depth of water, and the chemistry of the water; many of these factors can also be reflected in the biota within the lake. Clastic sediments are transported into the lake from streams, and much of the coarser load is dropped there along the margins (see below). Away from the lake margins, the finer material is carried in suspension by currents, with coarser, silty-sized material being transported closer to the shore, and finer material farther, where it eventually settles to the bottom. Winds promote surface turbulence and currents, which helps keep finer material suspended. At the same time, wind-induced waves and currents may also redistribute coarser materials around the coastal margins (Nichols, 1999).

The deep water facies then consists of silts and clays, and organic matter. In lakes, in colder climates such as Norway and Sweden, alternations of mineral and organic layers form on a seasonal basis, with coarse material accumulating during periods of snow melt in spring and summer, whereas the finer, organic fraction builds up in winter when the lakes are frozen. The production of these *varves* is widespread in Late Weichselian lake sediments (Ringberg and Erlstrom, 1999), and are instrumental in dating glacio-lacustrine deposits there (Ringberg, 1994). In addition, the organic fraction can be dated by radiocarbon. Finally, we note the abundance of climatic reconstructions made from pollen, diatoms, and other microbiological records that have been extracted from cores taken from lake bottoms. These are valuable archives of information that indirectly place archaeology in an accurate chronological and palaeoecological setting.

In arid and semiarid areas, evapotranspiration is strong, and these lakes and playas contain abundant precipitated salts, with calcium, sodium, potassium, and magnesium serving as the principle cations (Table 5.3); carbonates, sulfates, chlorides, borates, and nitrates representing the major anions. Different minerals have different solubilities: so calcite, for example, is less soluble than gypsum or halite and would precipitate first; gypsum is next, whereas halite would precipitate last. In arid soils, spatial partitioning can be seen of these minerals, with calcite occurring closest to the surface and halite at increased depth.

Of more direct interest to archaeologists, however, is the fact that former human occupation took place along or close to lake margins (Feibel, 2001). This setting can include not only the lakeshore itself but also the streams and

Table 5.3 Major evaporate minerals (modified from Reineck and Singh, 1980, table 19)

Mineral	Composition
Carbonates	
Calcite	$CaCO_3$
Aragonite	$CaCO_3$
Dolomite	$CaMg(CO_3)_2$
Magnesite	$MgCO_3$
Natron	$Na_2CO_3\ 10H_2O$
Trona	$Na_2CO_3 \cdot NaHCO_3 \cdot 2H_2O$
Sulphates	
Anhydrite	$CaSO_4$
Gypsum	$CaSO_4 \cdot 2H_2O$
Glauberite	$CaSO_4 \cdot Na_2SO_4$
Epsomite	$MgSO_4 \cdot 7H_2O$
Thenardite	Na_2SO_4
Chlorides	
Halite	NaCl
Sylvite	KCl
Borates	
Borax	$Na_2B_4O_7 \cdot 10H_2O$
Nitrates	
Soda Niter	$NaNO_3$
Niter	KNO_3

wetland areas that adjoin them. Thus stratigraphic sequences in these fluvio-lacustrine settings may include a variety of facies and lithologies that result from contributions from both neighboring streams and the lake. The latter can be caused by rising or lowering of the lake level, which in turn can be caused by tectonic movements within the basin or drainages, or climate changes. Consequently, deposits and archaeological sites in lake-margin settings may be influenced by a variety of depositional and postdepositional factors. These may include:

Changes in lake level may be expressed in different ways. A lowering of lake level exposes any sediment that was previously deposited at or below water level. This fosters pedogenesis and allows for any deposits or archaeological materials previously emplaced to be covered by alluvium, or be eroded if channels migrate (either by meandering, avulsion, or braiding), over the occupation area. In such situations, we may see stratigraphically a shift from finer beach sands, for example, to well-rounded, imbricated channel gravels. Sand dunes, if present, may also encroach upon sites. On the other hand, with rises in lake level, shorelines may transgress over occupation surfaces and deposits. Depending upon the rate of lake level rise, sites can be quickly inundated and preserved with fine-grained deeper water lacustrine deposits, perhaps richer in organic matter. In contrast, with slow rise in lake level and high energy beach settings, previous deposits and artifacts can be subjected to erosion, transport, or destruction by wave activity. The latter is particularly true in the case of major storm events.

Rises in lake level also are associated with higher groundwater tables, possibly transforming previously dry areas into swampy, less habitable ones. At the same time, diagenesis becomes more prominent, and depending on the specific environment, can be expressed by increased swampy (reedy) vegetation and rooting, cementation by carbonates, or the formation of iron pans (see Chaper 3).

5.6.3 Geoarchaeological examples

Archaeological sites in lacustrine and fluvio-lacustrine settings can be found throughout the world, particularly in the mid-latitudes and tropics where some of the earliest human sites can be found. The East African Rift (Kenya and Tanzania) and Jordan Valley, for example are associated with prehistoric sites for over the past 1–2 million years (Bar-Yosef and Tchernov, 1972; Feibel, 2001; Mallol, 2004; Potts *et al.*, 1999). The Lower Palaeolithic site of 'Ubeidiya is situated on the west bank of the Jordan River in the Central Jordan Rift Valley of Israel, ca

FIGURE 5.23 Field photograph of tilted sediments from 'Ubeidiya Formation, Jordan Valley, Israel. These chalky lacustrine marls are stained with goethite and overlain on the right by gravelly clays of alluvial origin (see also living floor I26 in figure 1.3c).

3.5 km south of the Sea of Galilee (Lake Kinneret). It dates to about 1.4 million years and is the oldest early Acheulean site outside of Africa (Bar-Yosef, 1994).

The stratigraphy of the site is rather complex (Picard and Baida, 1966; Tchernov, 1987). In brief, the lithic materials are found within the 'Ubeidiya Formation, fluvio-lacustrine deposits composed of shoreline and fluvial conglomerates interfingering with freshwater lacustrine clays and marls; incipient soil formation punctuate these deposits (Mallol, 2004), which were folded and faulted during the Middle Pleistocene. Occupations took place within this fluctuating shoreline/fluvial environment (Fig. 5.23).

Although the general scenario outlined above has been known for close to 40 years, the detailed aspects of the location of individual occupations have been made clear only recently by the microstratigraphic study of the deposits using micromorphology (Mallol, 2004) (Fig. 5.24). Results for the Layer I26 complex, for example, one of the most striking units in the field (Fig. 1.3c, Chapter 1), show that it represents mudflats and beaches that were subjected to periodic desiccation, allowing for

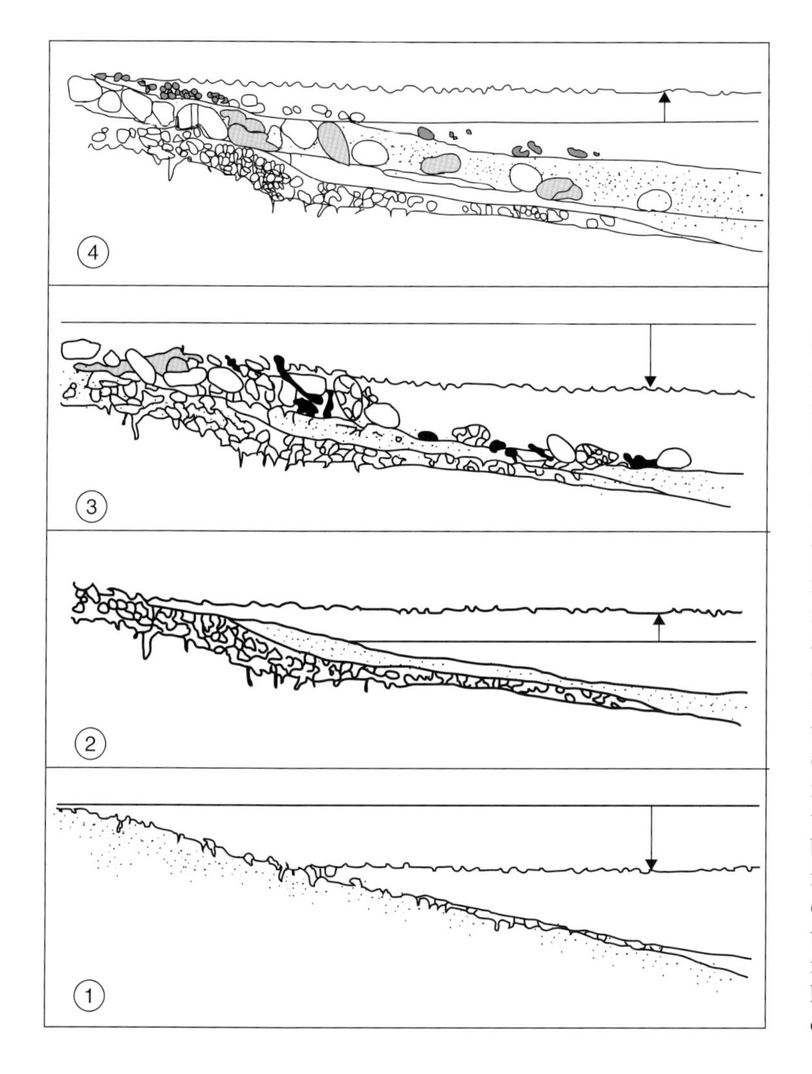

FIGURE 5.24 Reconstruction based on micromorphology of the sedimentary environment associated with layer I26 at 'Ubeidiya. (1) Low lake level exposing mudflat which is subjected to desiccation, cementation by carbonate, and break up by wave action; (2) rise in level of the lake, resulting in reworking of carbonate gravel and other surface clasts and covering by lacustrine silts; (3) drop in lake level with weathering and slight reworking of shoreline deposits, including remains of abandoned occupation that were situated on a pebbly beach; (4) rise in lake level which reworked and redeposited small clasts (as well as lithics) on top of the original occupation; this wave reworked material grades laterally lakeward into lacustrine deposits.(From Mallol, 2004).

the formation of gypsum and other evaporites and prehistoric occupation. These margin deposits were also subjected to wave action which reworked some of the geological and archaeological materials. These results enabled Mallol (2004) to follow lateral facies changes within lithostratigraphic units and make detailed correlations not previously apparent.

The Jordan Valley also contains lacustrine relicts of Late Pleistocene and Holocene age, many of which articulate with archaeological sites. The Epi-Palaeolithic site of Ohalo II (ca 19.5 ka BP), for example, is remarkable for its preservation of architectural and botanic remains, in large part due to submersion by the Late Pleistocene Lisan Lake (see Fig. 2.9d), and one of its present relicts, the Sea of Galilee (Tsatskin and Nadel, 2003). Archaeological material is found resting upon the Lisan lacustrine deposits, which through time became increasingly shallow as shown by the inputs of basalt derived near shore sandy deposits. Water level fluctuated, resulting in periodic submergence of the sediments, accompanied by truncation; a greater overall exposure is found in higher, more shoreward parts of the site, which shows bioturbation and incipient soil formation. Further south in the Dead Sea basin, archaeological and geological evidence, with archaeological structures of known age showing impacts of shorelines much above those of present levels can be found (Enzel *et al.*, 2003; Klein, 1986).

In Australia, the earliest sites on the continent are found in *lunettes*, formed on the lee side of lake basins (see fig. 6.3). At Lake Mungo, for example, a human burial occurs within a complex sequence of lake and aeolian deposits separated by distinct paleosols (Bowler *et al.*, 2003). Fluctuations in lake levels occurred between 50 and 40 ka which coincided with human occupation.

In a somewhat comparable setting and latitude in the New World, many Paleoindian sites occur in association with playa, lacustrine, and palustrine deposits (e.g. diatomites) on the Great Plains (Holliday, 1997). The two principal types of deposits exposed at the surface are

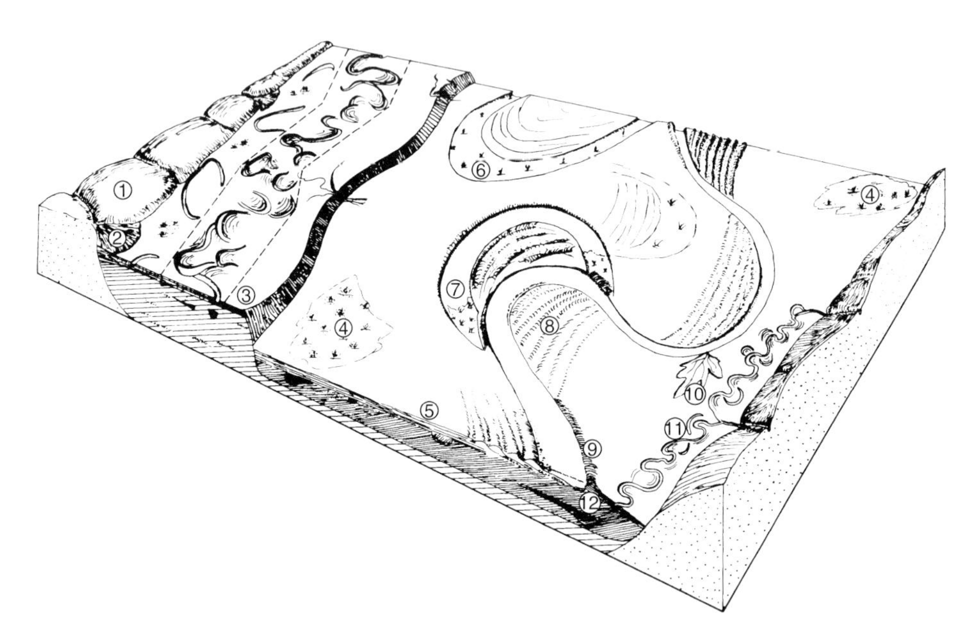

FIGURE 5.25 Schematic summary of the types of archaeological context that can be found in a meandering river system (modified From Gladfelter, 2001, figure 4.2).

TABLE 5.4 Associations of archaeological material with alluvial contexts (numbers are keyed to fig. 15.25) (partly modified from Gladfelter, 2001, table 4.1)

Geomorphic aspect	Possible archaeological associations
1) Dissected upland expressed as a bluff above the valley	Sites of any age possible. Soils may be polygenetic and palimpsests of several occupations are possible, making it difficult to distinguish individual occupations
2) Alluvial fan	Sites of any age that post date
3) Alluvial terrace with older, inactive meanders. The meander belt is shown by dashed lines. A paleosol is developed on the surface of the terrace	Sites within the meander belt cannot be older than the meandering activity, and sites inside of a particular meander loop cannot be older than the meander loop. Buried sites that survived past meandering activity will be old
4) Backswamp along lower, interior setting on the floodplain and flood basin	Surface and near-surface sites may be preserved and the oldest sites on the floodplain are likely to found in these settings
5) Contemporary floodplain, with convex shape and meander scars of a former river channel. Some features outlined in greater detail below (e.g. 8)	Location and age of sites is a function of the degree of geomorphic activity. For active meanders, older surface sites may be rare; age of buried sites conditioned by rate of vertical accretion. Within a meander loop (e.g. 8) surface sites become younger toward the outside of the loop
6) Infilled scar of meander cut-off and degraded point bar deposits	Sites predate location of present channel. Deposits within scar are likely to be organic rich
7) Oxbow lake in meander cut-off plugged at both ends with clay	Sites associated with lake can be any age that postdate its formation. Sites can be buried in levee deposits or on surface of point bars
8) Point bars	Sites on point bars or in swales between them. No site can be older than the expansion of the meander loop
9) Natural levee on outside of meander loop	Sites within levee deposits of active channel are penecontemporaneous with active channel; burial at great depth is not likely
10) Crevasse splay formed where natural levee is breeched	Brief occupation likely while splay is active, but more attractive when it becomes inactive
11) Yazoo-type river that flows in lower, interior margin of the floodplain near the base of the bluff. Source of water comes from bluff with much lower discharges than in major river	Sites older than meandering of Yazoo stream may survive beyond its meander belt in vertical accretion deposits of major stream

dark grey, slightly calcareous mud and a light grey, loamy, highly calcareous deposit. These types of deposits appear to represent wetter ("pluvial") conditions during which the basins were perennially filled or nearly so.

5.7 Conclusions

This chapter touched on only some of the major aspects of the fluvial and lacustrine environments, and the geoarchaeological issues that might be associated with them. Some of these features are summarized in Figure 5.25 and Table 5.4, which portrays some of the archaeological contexts that can be found in the meandering river. While specifically aimed at the meandering river context, it illustrates some of the major questions concerning fluvial geoarchaeology. These include: (1) site visibility – whether archaeological material is not present on the surface because of erosion or burial (the value of examining a large region both on the surface and with depth was demonstrated by Mandel's work), (2) dynamics of the landscape – how active, for example, is lateral planation and accretion, and will sites only a few hundred years old be visible on the surface due to this "etch-a-sketch" erasure of flood plain features? Clearly, older landscapes on higher terraces are liable to have thick, polygenetic soil formation, which may be associated with multiple, likely mixed, occupations from different time periods [e.g. Fig. 5. 25 (1)]. Thus, when working within the geoarchaeological contexts of alluvial deposits, one must observe closely the deposits and facies on the surface and at depth, and obtain proper chronometric control. The latter can be done preferably by the indirect dating of their contents (^{14}C dating of organic matter or other archaeological materials); recent advances in OSL dating to date the deposits directly look promising.

Some of the same issues can be found in lacustrine environments, particularly in those positions in the landscape that are marginal to the lacustrine and fluvial environments, such as at 'Ubeidiya, where streams flowed into the lake. Site or artifact integrity can be either destroyed by rising or lowering lake level by the higher energies associated with waves along beaches or fluvial channels. In the case of Ohalo II, on the other hand, conservation of structures and deposits is in part due to inundation of the site covered with low energy fine-grained deposits. Preservation of the materials in such instances is crucial for establishing subsistence strategies practiced by the occupants (Tsatskin and Nadel, 2003; Weiss *et al.*, 2004).

6

Aeolian settings and geoarchaeological environments

6.1 Introduction

Normally, when one thinks of the aeolian setting, notions of sand dunes and vast sand seas come to mind. Indeed, windblown sand covers about 25% of the earth's surface (Cooke *et al.*, 1993) (Fig. 6.1). Yet, aeolian activity is not just confined to blowing sand around: it is responsible for eroding and depositing vast quantities of silt- and clay-size dust; the Sahara alone appears to produce between 60 and 200 million tons of dust per year (Cooke *et al.*, 1993). Similarly, aeolian processes are responsible for producing markedly eroded landscapes composed not only of softer aeolian, fluvial, and lacustrine sediments (Brookes, 2001), but also of harder bedrock. From the geoarchaeological point of view, this erosion can bring about the complete removal or dispersal of the interstitial sediments and archaeological remains, resulting in the conflation and superposition of archaeological finds from what were once different layers and instances of occupation.

The aeolian environment and associated deposits constitute a significant geoarchaeological resource, as many sites that play an important part in human history and evolution are found in such deposits. Such sites occur not only in low latitude arid areas with drifting sand dunes, but also in the higher latitudes of Western, Central, and Eastern Europe, and Central and northeast Asia. Here, finer grained aeolian dust accumulations (loess) derived from or associated with glacial activity, blanketed many of these areas when glaciation was more active and extensive, and winds were stronger; loess is also found in low latitude deserts. Set into these loess deposits are numerous Palaeolithic sites, and because of the lower energies associated with dust accumulation the degree of preservation of the archaeological record and its contextual framework is generally high, especially in comparison to higher energy fluvial deposits. In this chapter we discuss some of the basic elements of aeolian settings, their landscapes and deposits, and relate them to past occupations and environmental changes.

6.2 Sandy aeolian terrains

Sandy terrains are now generally found in the arid parts of the globe (Fig. 6.1), where aridity has a number of causes. These include (1) subsidence of air associated with subtropical high pressure belts (most of the world's deserts are found here, such as the Sahara),

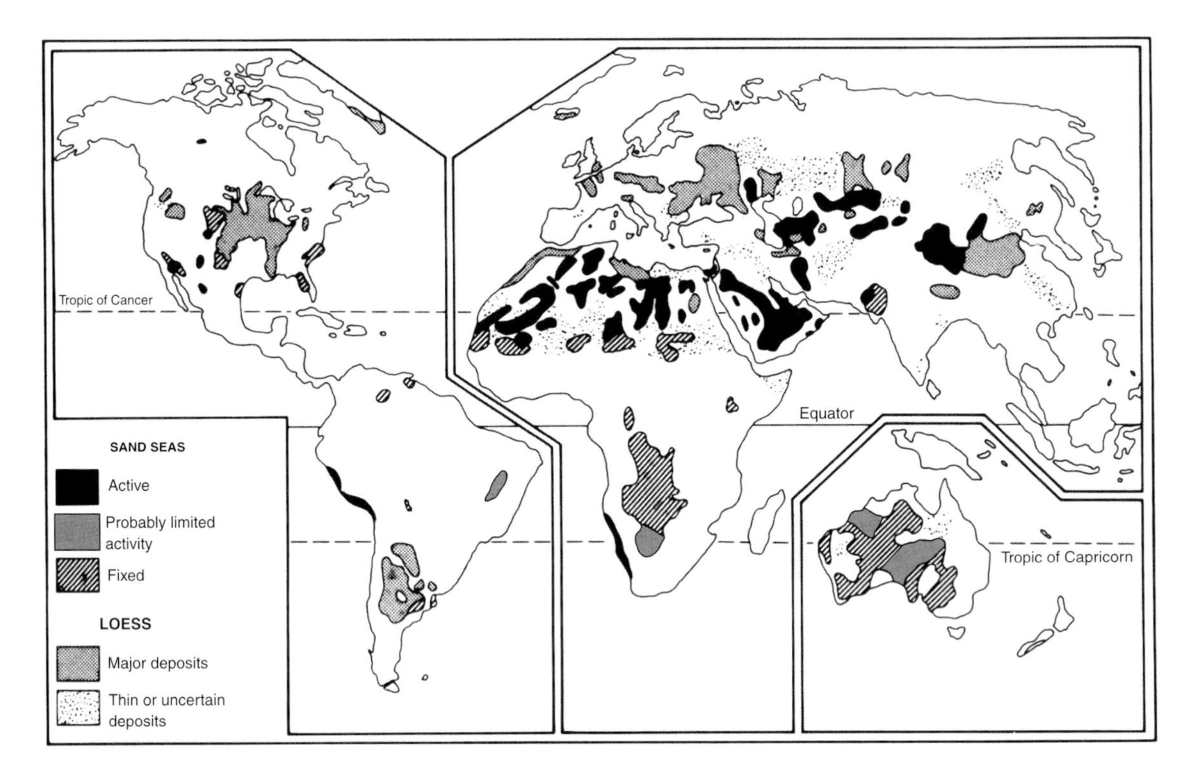

FIGURE 6.1 Distribution of sandy and silty aeolian deposits over the globe (modified from Thomas, 1989, figure 11.1).

(2) continentality and their positioning in the center of large continental land masses (e.g. deserts of China), (3) orographic factors related to uplifted terrains (e.g. Californian deserts, Dead Sea Rift) (Cooke *et al.*, 1993). On the other hand geoarchaeological (Haynes, 1982; McClure, 1976) and pollen evidence (Ritchie and Haynes, 1987) show that environmental conditions have changed dramatically in the past, permitting earlier habitation in what are now among the driest parts on earth.

Sand moves when the critical threshold friction velocity of the wind is exceeded, and the entrainment of grains is generally related to particle size and wind velocity, although other factors such as surface roughness, soil moisture, salt precipitation, and surface crusting play a part (Ritter *et al.*, 2002). Bagnold (1941) showed graphically the relationship between grain diameter and critical threshold. As such, the most easily erodible size is about 0.1 mm microns (Fig. 6.2). For grains less than this size, greater wind velocity is needed to entrain particles because of cohesion effects and low surface roughness. On the other hand, once grains are set in motion they impact stationary grains whose entrainment velocities ("impact threshold" in Fig. 6.2) are lower than the fluid threshold for equivalent sized grains.

Once grains begin to move, they are transported by different processes (Fig. 6.3). Creep generally involves coarser sand (ca >150 μm) and movement by rolling at the near surface (Cooke *et al.*, 1993). Reptation involves finer sand than creep, whereby "the low-hopping grains dislodged by the descending high-energy particles" (Cooke *et al.*, 1993: 254). With saltation, although similar to reptation, grains are ejected into the air and pick up additional momentum from the wind; on landing

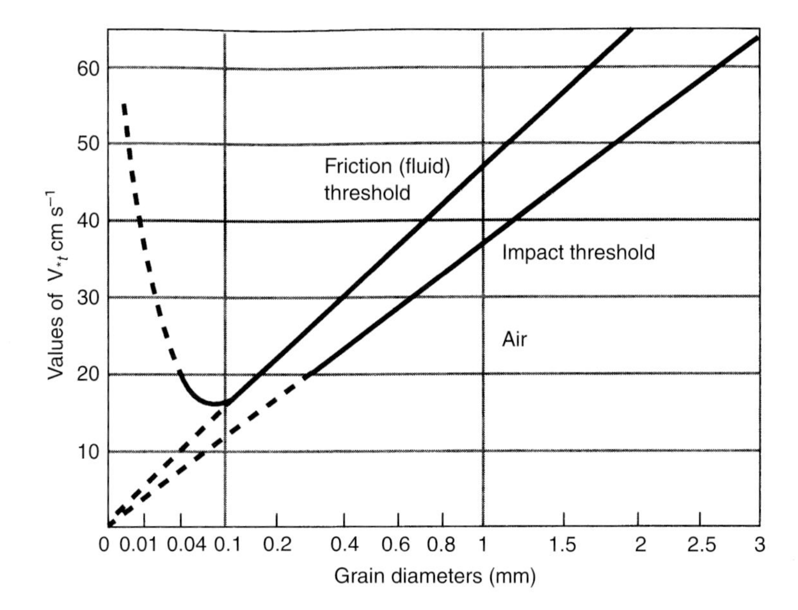

FIGURE 6.2 Diagram showing the relationship of particle size in millimeter to threshold velocity as determined by Bagnold (1941). (modified from Ritter *et al.*, 2002; Figure 8.7).

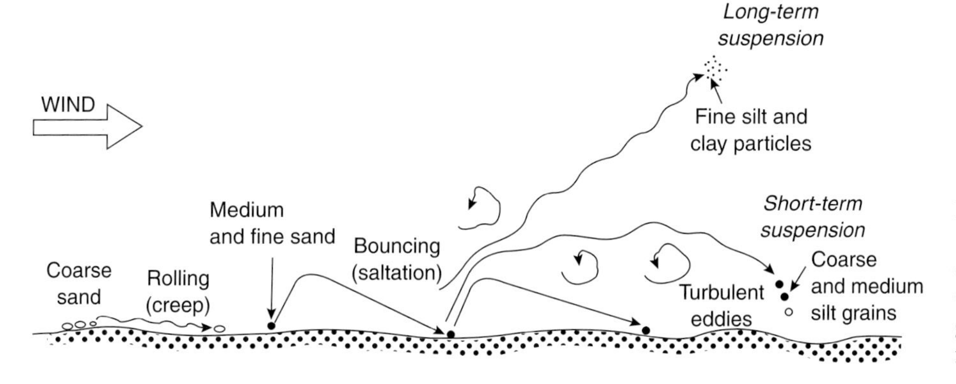

FIGURE 6.3 Schematic view of aeolian transport of sedimentay grains (modified from Pye, 1995).

they have enough energy to result in the bouncing up of several grains on the surface on impact. The finest, silt sizes (<62 μm) are carried in suspension with turbulent motion (Cooke *et al.*, 1993) and can end up being embedded in soils or as large accumulations of loess (see below); they are capable of being transported over great distances (e.g. from the Sahara to northern Europe, and westward to the Caribbean; see Pye, 1987).

6.2.1 Aeolian erosion

Erosion by wind is expressed at both the smaller, site scale and larger, regional scale. Both scales affect the geoarchaeological record. At a small scale, wind erosion by abrasion can result in the grinding away of rock fragments, including lithic implements. Abrasion, usually by sand-sized particles, can result in the formation of ventifacts, typified by facetted and pitted rocks and boulders, or smoothed and polished surfaces with elongated grooves (Fig. 6.4a,b). Measurement of the alignment of the latter can be used to infer the direction of the strongest or most prevailing wind direction (Ritter *et al.*, 2002).

At a larger scale erosion may sculpt entire exposures of bedrock resulting in the formation of yardangs. These are elongated hills which parallel the prevailing wind direction

FIGURE 6.4 Erosional effects of the wind. (a) Ventifacts from the western desert of Egypt, exposed on the surface and covered with a thin veneer of sand; knife handle is 12 cm long; (b) wind-polished limestone showing elongated grooves that parallel the strongest or prevalent wind directions; in the background on the left is a small yardang; (c) large yardang in western desert of Egypt, north of the Kharga Oasis and not far from the location in (b).

(Fig. 6.4c,d). Typically, the windward side is blunter than the leeward end, which tapers downwind.

From a more geoarchaeological perspective wind erosion may play an important role in the integrity of archaeological assemblages, although not necessarily. The principal risk is the effect of deflation in which fine grained material is blown away leaving a lag deposit of heavier, stony objects (see Box 6.1). Consequently, artifacts originally contained in different deposits from successive occupations can be found together within the same "assemblage" after the finer interstitial material has been removed. Deflation surfaces are common, and are particularly important in Old World settings, where time depth provides the opportunity for repeated and long time deflation. An abundance of sites in Egypt (and Sinai), Israel, and Jordan have been surveyed and excavated over the past several years, and many show signs of deflation (Goring-Morris, 1987; Wendorf and Schild, 1980).

Evaluating the effects of deflation is important in order to properly interpret the archaeological record. In cases where distinct lithic tool types

Box 6.1: Aeolian features in desert environments[1]

The Eastern Saharan and Arabian Deserts run from Sinai, Egypt, across the Negev in southern Israel into Jordan and the Arabian Peninsula of southwest Asia (Fig. 6.5). This terrain is among the most arid parts of the world, with precipitation being <100 mm per year and averaging <5 mm per year in the Arabian Peninsula. Although prehistoric sites have been recorded from the entire region (Phillips, 1988; Wendorf *et al.*, 1994; Wilkinson, 2003), eastern Sinai in Egypt and the western Negev Desert of Israel, have received intense attention revealing numerous archaeological and prehistoric sites in aeolian terrains (Goring-Morris, 1993; Goring-Morris and Belfer-Cohen, 1997). These include both sandy dune and loess deposits, as well as fluvially and colluvially reworked sands and silts.

During the Late Pleistocene (within the past 50,000 years), a good part of the area received large inputs of aeolian dust. This accumulated either as primary dust fall on upland surfaces (Bruins and Yaalon, 1979; Rognon *et al.*, 1987; Wieder and Gvirtzman, 1999) (Fig. 6.6). In addition, striking amounts of loess was penecontemporaneously reworked along slopes and into most drainages from Wadi el-Arish in Sinai to Nahal Besor in Israel, and all drainages in between (Goldberg, 1986; Zilberman, 1993) (Fig. 6.7). These alluvial deposits are actually paludal (swamp-like) and have been interpreted to indicate during the past ca 40,000 years or so, more gentle and sustained rainfall patterns, even throughout the year (Horowitz, 1979; Sneh, 1983).

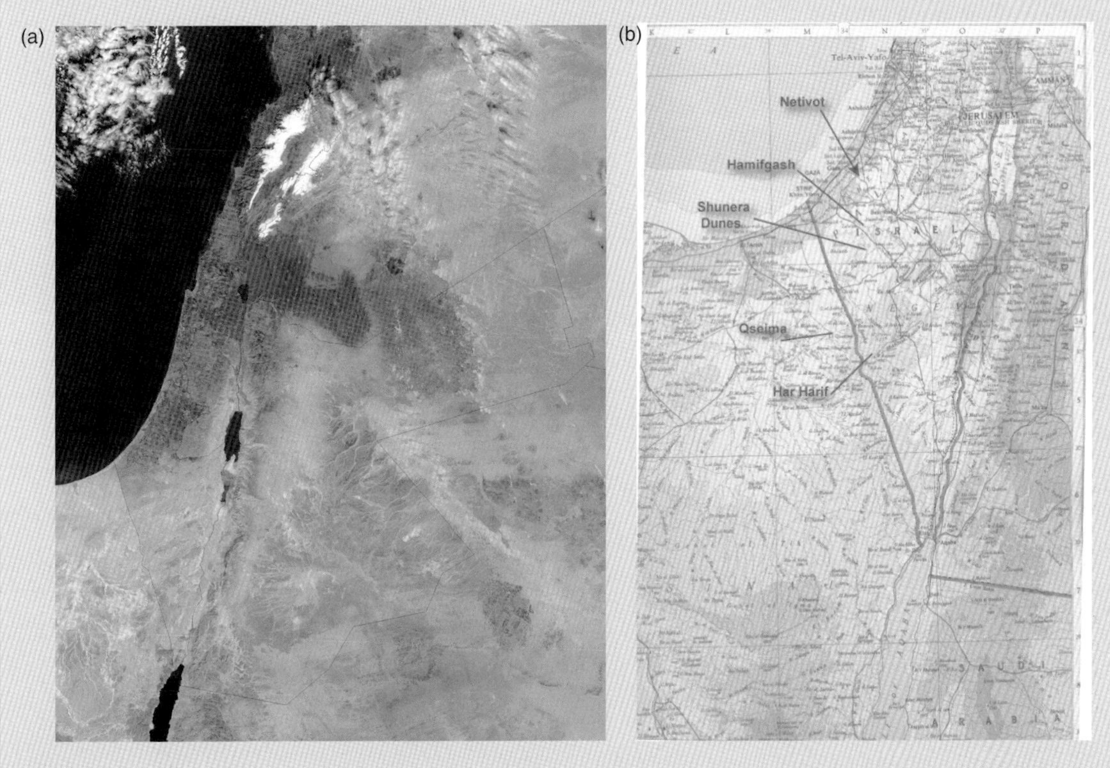

FIGURE 6.5 Sites and aeolian deposits in the Sinai and Negev Deserts. (a) Satellite photograph of southeast Mediterranean area showing the largely unvegetated nature of the region except for a stretch along the northern Levantine coastline of Israel and Lebanon; (b) map showing distribution of aeolian sand and silt deposits and localities mentioned in the text. *(cont.)*

Box 6.1 *(contd.)*

Such conditions were obviously attractive to prehistoric settlers, and we see many Upper Palaeolithic and Epi-Palaeolithic sites intercalated within these deposits from this entire area at this time.

In many place in northern Sinai and the western Negev, the loesses have been fluvially reworked with silts filling major drainages (Goldberg, 1994; Goring-Morris, 1987; Zilberman, 1993). In the later part of the Pleistocene (starting at about 25,000 BP) we see the relative reduction in dust accumulation and its fluvially reworked counterpart, and a concomitant increase in sand. The latter, early on, is expressed as sandy additions to the alluvium. Incision into alluvium becomes widespread, and is also somewhat coeval with the deposition of well-defined aeolian sand deposits, either as localized sand accumulations (Fig. 6.8) or as more massive linear dune features. The latter are strikingly visible in the field and on aerial photographs today (Fig. 6.9) from which it is obvious that these sand dune bodies and morphologies are plastered onto the existing Upper Palaeolithic fluvial terrain, which existed during wetter times.

Moreover, we see that these lithological and geomorphological changes took place across the landscape in a time-transgressive fashion: in northern Sinai (e.g. Qseima area, Qadesh

(a)

(b)

FIGURE 6.6 Loess deposits. (a) Section of loess and paleosols in an upland position of the Netivot area of the western Negev, Israel (cf. Fig. 6.5). The bulk of the profile is composed of calcareous clayey silty loess which has a relatively high proportion of finer silt and clayey washout dust (see text). It is interspersed with a number of calcic paleosols which point to intermittent periods of leaching and wetter climates. Only a few scatters of nondistinctive artifacts, however, were found within these deposits. Meter and hammer are for scale. (b) Holocene loess deposits also occur in the region but they are much rarer. Shown here is a thin (ca 10 cm) accumulation of loess on the high plateau (ca 1000 m elevation) of Har Harif (cf. Fig. 6.5b). The loess is below the feet of prehistorian Nigel Goring-Morris and overlies a concentration of Chalcolithic (ca 6,000–7,000 BP uncalibrated) flint artifacts, which can be seen here in the foreground eroding out from underneath the loess cover. Unlike in glacial terrains, loess in this part of the world is commonly associated with wetter conditions, in which aeolian dust is washed out of the atmosphere during rainfall events; these conditions also promote a macro- or micro-vegetative cover (Danin and Ganor, 1991) that aids in trapping the loess. Evidence from a variety of sources points to a noticeably wetter environment during the Chalcolithic in this part of the world (Goldberg and Rosen, 1987; Goodfriend, 1991). *(cont.)*

Box 6.1 *(contd.)*

(a)

(b)

FIGURE 6.7 a,b Two views of sediments in the Hamifgash area, the confluence of Nahals Beer Sheva and Besor, both major drainages in the Western Negev of Israel (cf. figure 6.5). In Figure 6.7a light colored silty and sandy alluvium are visible. Persons in the center are standing next to the excavation of a late Middle Palaeolithic site, Farah II. This is found at the base of a ca 18 m of fluvially reworked loessial and sandy deposits, consisting of cetimeter-thick sandy and silty beds as shown in Figure 6.7b). These sediments represent paludal/fluvial deposition that might have occurred throughout the year, in contrast to winter only rainfall occurring today. Within these overlying deposits, many Upper Palaeolithic and Epi-Palaeolithic sites can be found here and elsewhere along this broad coastal plain, from the western Negev, southwest into the northern part of Sinai. They testify to the attraction of these water-rich environments during the wetter conditions of the late Pleistocene (Goring-Morris, 1987).

(cont.)

Box 6.1 *(contd.)*

Figure 6.8 The hammer here is resting on the remains of cross-bedded sandy dune deposits (center) in the area of Qseima near Qadesh Barnea in eastern Sinai. In this area, these sands date to about 19,000 years ago and represent an incursion of sand that transgressed across northern Sinai and into the western Negev at this time. Glistening against the reflection of the sun and resting on these sands are deflated scatters of Epi-Palaeolithic (ca 13,000–14,000 BP) lithics of a large, partially eroded site. Note that in eroded terrains such as these where deflation is prominent, many of the lithic assemblages are likely to be mixed and it is not possible to discern individual occupation events.

Figure 6.9 Aerial photograph of the Hamifgash area (+) in center (cf. figure 6.5b). The generally smooth-textured area just to the south is silty alluvium that accumulated from roughly about 40/50,000 to 10,000 years ago. The linear features in the lower left-hand corner of the photograph are dunes, which rest upon these earlier fluvial deposits. Width of photo is ca 20 km.

(cont.)

Box 6.1 *(contd.)*

Barnea; Figs 6.5b and 6.8), the first sand incursions occur about 20 ka, while in the Hamifgash region in the western Negev, increased sand is noted above an early Epi-Palaeolithic site (Hamifgash IV; ca 16,000 BP); here, a veneer of sand covers Neolithic remains. In the nearby Shunera Dunes area just to the south (Fig. 6.9), Terminal Upper Palaeolithic material dated to just before 16,000 BP is well contextualized within dune sand (Goring-Morris and Goldberg, 1990). Thus we see an overall aridification taking place at this interval (roughly 25–15 ka), culminating just after the late glacial maximum.

[1] NB: All radiocarbon dates are uncalibrated.

Figure 6.10 (a) Dunes and sites along the border between Sinai and the Negev in the area of Nizzana and Shunera (see Fig. 6.10b). Note the linear texture of the topography in the upper half of the illustration which reflects the late Pleistocene linear dunes that cover the area. Illustration courtesy, N. Goring-Morris from Davidson and Goring-Morris, 2003. *(cont.)*

Box 6.1 *(contd.)*

FIGURE **6.10** (*Cont.*) (b) Aerial photo of part of Shunera area shaded in Fig. 6.10a above, showing linear dune in upper half and incised channel in lower half. Exposed next to the incised channel are late Quaternary fluvial silts (7–8 m thick) marked by an upward increase in sand and containing Upper Palaeolithic artifacts. These were subsequently eroded and followed by the formation of a calcic paleosol that is associated with the an eroded late Upper Palaeolithic site (Shunera XV; ca 18,000 BP) and a Kebaran site (Shunera XVII; ca 17,000 to 15,000 BP) that are found within thin sandy veneers that cover the terrace developed on the Upper Palaeolithic silty alluvium (Goring-Morris and Goldberg, 1990). Dune formation had already begun by ca 16,000 BP as demonstrated by the presence of the Terminal Upper Palaeolithic site of Shunera XVI, which is directly associated with the accumulation of these sands. Photo courtesy: N. Goring-Morris. (c) The figure in the foreground is standing on an eroded bedrock slope that descends to the north, into the canyon shown in Figure 6.10b. In the background is the southernmost linear dune depicted in Figure 6.10b. (d) Field view of excavation of Shunera VII, an Early Natufian site eroding out of sands in the foreground. In the background, the southernmost dune shown in Fig 6.10b is associated with numerous prehistoric sites that range from Geometric Kebaran (ca 14,500 BP) through Harifian (ca 10,700 to 10,000 BP); a deflational surface that postdates the Harifian contains Chalcolithic or Early Bronze Age pottery (ca 6,000 to 4,500 BP) (Goring-Morris and Goldberg, 1990). Photo courtesy: N. Goring-Morris.

are found together (e.g. Middle Palaeolithic Levallois core mixed with Epi-Palaeolithic bladelets), the effects of deflation are easily recognized. In cases, however, where nondiagnostic lithics occur, it is more difficult to detect mixing of occupations and assemblages. Fortunately, in most desert terrains, rocks exposed on the desert surface form desert varnish, which is a shiny, dark reddish or blackish coating about 100 μm thick. It consists of varying proportions of iron and manganese (which influence the color) and various clay minerals; it is thought to form from physicochemical and biogeochemical processes (Dorn, 1991; Dorn and Oberlander, 1981; Watson, 1989a). Although the chronometric dating of desert crusts has proven to be problematic (Bierman and Gillespie, 1994; Dorn, 1983), qualitative evaluation of mixing of archaeological surface finds in deflated terrains may be undertaken by examining different degrees of development on lithics (cf. Helms *et al.*, 2003). In addition, a lack of small lithic debris or a high proportion of cores and other heavy, large elements would suggest some removal of fine material by deflation or runoff; context is the best means to evaluate which of these alternatives is most reasonable.

In spite of the problems of mixing assemblages, deflation is a significant process in the past, as well as in the present. The present-day formation of deflation basins, or blowouts, is helpful in exposing stratigraphic sequences of deposits that normally would be concealed beneath the surface, thus providing windows into former landscapes, deposits, and sites. In antiquity, the formation of blowouts results in depressions in which water collects, even if seasonally or during wetter climates. They thus serve as the loci for floral, faunal, and human activity, and ultimately as traps for depositions and archaeological remains. At the Casper site in Wyoming, for example, an elongated depression (ca 93 × 22 × 2 m) formed about 10,000 radiocarbon years ago, and was filled in with lenticular accumulation (Unit B) of well rounded sand with frosted and pitted surfaces that exhibited some low angle cross-bedding (see Chapter 1; and below). Within the deposits were the remains of 74 *Bison antiquus* representing a "communally-operated bison trapping, killing, and butchering station which can be associated with the Hell Gap cultural complex" (Frison, 1974). These deposits and setting were interpreted to be a deflation hollow, similar to ones forming today in the area (Albanese, 1974), and were protected from subsequent deflation by the overlying lacustrine deposits of Unit C, which filled in the original blowout depression.

Similar aeolian situations are well documented in north-eastern Sinai (Egypt) and the Negev (Goring-Morris, 1987; Goring-Morris and Goldberg, 1990; Wendorf *et al.*, 1993, 1994). In the Nabta Playa area of southern Egypt, Holocene deflation of over 6 m has exposed lacustrine deposits which overlie Terminal Palaeolithic artifacts that rest on a cemented, fossilized dune. Not far away, in the Bir Sahara/Bir Tarfawi area, deflation is also responsible for excavating more than 8 m of sediment below the plateau level; this was later filled with aeolian sand with "numerous but heavily sandblasted Mousterian artifacts . . . " (Wendorf and Schild, 1980: 25).

6.2.2 Sand-size aeolian deposits

Coarse grained aeolian deposits are usually sand size and larger (>62 μm), and on the whole consist of quartz, which is most resistant. Depending on locale and age, other components are feldspar and *heavy minerals* (e.g. hornblende, zircon, garnet, magnetite, hematite; see Chapter 16), although these are subject to weathering especially if the material has been recycled during a long geological history. In the case of dunes (see below), wind is selective in the size of the material it transports: either finer sizes are removed (*winnowed*), or coarser sizes remain as lag deposits upwind. Consequently, aeolian sands tend to be well sorted (see Chapter 1) and increasingly negatively skewed (proportionately more finer material and less coarser material) (Watson, 1989b); however, much variation can be seen within the same deposit (Ahlbrandt and Fryberger, 1982). In addition, quartz sand grains tend to be subrounded and larger sizes tend to be increasingly better rounded, although shape is commonly a function of the nature of the material at the source; finer sand and silt sizes (<100 μm) tend to fracture during production and transport and consequently are angular to subangular (Watson, 1989b).

6.2.2.1 Bedforms

Sandy deposits accumulate in a variety of forms that are a function of source and supply of sand, wind strength and direction, nature of substrate, and seasonal changes (Reineck and Singh, 1980). These bedforms occur at various scales ranging from ripples on the centimeter/decimeter scale up to large sand seas (*draas*), which cover large areas of the landscape, on the order of kilometers.

Wind ripples range in size from 0.5 to 1.0 cm in height with wave lengths from 1 to 25 cm, features which are controlled by wind velocity and particle size. Granulometrically, they are composed of fine to coarse sand (ca 62 to 700 μm) (Boggs, 2001). In section, they are asymmetrical, with the windward side exhibiting slopes of about 10°, whereas the lee side has slopes of about 30° (Ritter *et al.*, 2002; Thomas, 1989). They are formed either aerodynamically as a result of shear stress of the wind on the sandy substrate, or as a result of bombardment by grains moving by saltation or creep (impact ripples). In the latter case, which is more common, grains moving by saltation, creep, and reptation, climb up the windward slope of the ripple; finer grains are removed leaving behind relatively coarser sand on the surface of the ripple. The coarse sand moves along to the crest, where it ultimately rolls down the windward side of the ripple (Fig. 6.11).

Through time, the ripple moves downwind, similar to dunes (see below) and as a result, the internal organization of a ripple resembles those of dunes with foreset bedding (Livingstone and Warren, 1996). Fine lithic debris can be moved along ripple surfaces, whereas coarser artifacts (e.g. cores) can be left immobilized with drifting material moving past it as the ripple migrates downwind.

6.2.2.2 Dune forms

Sand dunes and associated dune forms are more significant to geoarchaeology from a landscape and evolutionary perspective (Fig. 6.12). On this scale, their presence/absence and form may be significant of long-term environmental changes, such as climatic amelioration or desiccation. Furthermore, the movement of sand may result in the erosion or burial of archaeological sites, thus removing them from the stratigraphic record or preserving them for later excavation.

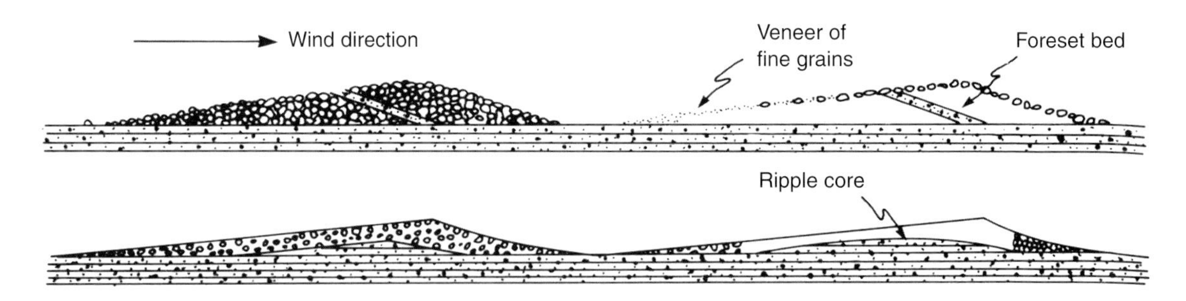

FIGURE 6.11 Internal organization and grain sizes of wind ripples showing coarse veneer and dipping foreset beds (modified from Cooke *et al.*, 1993, figure 19.6).

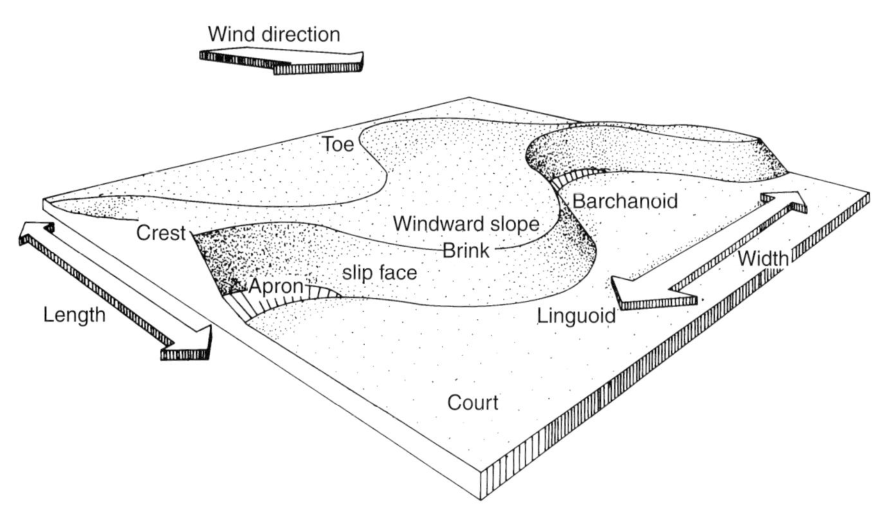

FIGURE 6.12 Nomenclature used to describe different parts of a dune (modified from Cooke *et al.*, 1993, figure 23.1).

TABLE 6.1 Classification of aeolian dunes by Livingstone and Warren (1996), which is essentially based on morphology

Dunes					
Free				Anchored	
Transverse	**Linear**	**Star**	**(Sheets)**	**Vegetation**	**Topography**
Transverse	Sandridge	Star	Zibar	Nebkha	Echo
Barchan	Seif	Network	Streaks	Parabolic	Climbing
Dome				Coastal	Cliff-top
Reversing				Blowout	Falling
					Lee
					Lunette

Dunes are the products of deflation and deposition of sand size material, which is sculpted into a variety of forms that are controlled by factors such as wind direction and speed, sand supply, vegetation, and grain size (Ritter *et al.*, 2002; Thomas, 1989). They differ from ripples, which are small scale bedforms and which form at lower flow velocities (Boggs, 2001). Numerous classifications of dunes exist that reflect shape, origin, wind characteristics, surface conditions, and whether they are mobile ("free") or fixed. One such classification by Livingstone and Warren (1996) is given in Table 6.1. A similar classification makes use of form and also characteristics, such as wind regime and a number of slip faces, (Ritter *et al.*, 2002) (Table 6.2 and Fig. 6.13a).

Transverse dunes are those whose slip faces are oriented in the same direction in response to a generally unimodal wind regime. They tend to be relatively broad (low length : width ratios) and are asymmetrical along the axis of sand movement, where sand is transported in a direction perpendicular to the orientation of the dune crest (Fig. 6.13a,b). Forms range from elongated, continuous ridges to discontinuous, individual dunes (barchans). These latter types, with crescentic shape and horns pointing downwind, form in areas of restricted sand supply. They typically have tabular to planar cross-beds in the center of the dune at about 34° (Fig. 6.14). Coalesced barchans result in barchanoid dunes and form with greater sand supply. Heights of transverse dunes range from as little to 0.5 m, up to >100 m (Livingstone and Warren, 1996).

Linear ridge dunes embody sand transport along the length of the crest of the dune, with slip faces found on both sides of the dune (Fig. 6.13b). They are relatively narrow and long (see Box 6.1), are symmetrical in section through the crest of the axis, and may reach up to tens of kilometers long (Cooke *et al.*, 1993). Normally, only one face is active, and activity can vary from daily to seasonally (Livingstone and Warren, 1996). Internal dune structure reflects this with sets of steeply dipping laminae that form within a given period and set up from others by marked bounding surfaces (Figs 6.13b and 6.15). A variant of the linear form is the *seif* dune, which has sharp, sinuous crests and can have lengths up to hundreds of kilometers (Ritter *et al.*, 2002). They are widespread in the Namib Desert, the eastern Sahara, and in Arabia, and the Simpson Desert of Australia, where, in the last case, dune crest are strikingly parallel or subparallel and commonly join with a characteristic Y-junction (Thomas, 1989). Lateral spacing between dune crests is quite regular, and can vary between 0.5 and 3+ km (Cooke *et al.*, 1993). Their origin is complex but it is generally accepted that they form in a generally bidirectional wind regime.

Reversing dunes are a form of dune networks in which the wind regimes are completely opposite and the ridges are asymmetrical (Boggs, 2001). As described by Cooke *et al.* (1993) for Oman, larger dunes are formed during the summer when the southwest monsoon is active. In winter, on the other hand, the wind direction is reversed, with smaller dunes being formed.

Star dunes appear to have a chaotic set of slip faces, with many pointing in several directions; they tend to have a somewhat radiating pattern

TABLE 6.2 Classification and illustration of major aeolian dune types (modified from Reineck and Singh, 1980; Ritter *et al.*, 2002; Thomas, 1989), based on morphology, number of slip faces, and wind regime

Form/dune	Number of slip faces	Major control on form	Formative wind regime
Transverse dunes			
Barchan dune	1	Wind regime and sand supply	Transverse, unidirectional
Barchanoid ridge	1	Wind regime and sand supply	Transverse, unidirectional
Transverse ridge	1	Wind regime and sand supply	More directional variability than for barchans
Linear Dunes			
Linear ridge/seif	1–2	Wind regime and sand supply	Bidirectional/wide unidirectional
Reversing dune	2	Wind regime and sand supply	Opposing bimodal
Star dune	3+	Wind regime and sand supply	Complex, multidirectional
Zibar	0	Coarse sand	Various
Dome dune	0	Coarse sand	Various
Blowout	0	Disrupted vegetation cover	Various
Parabolic dune	1	Disrupted vegetation cover	Transverse, unidirectional

with ridges emanating from a central higher area. Near this peak area, slopes range from 15° to 30° but these are lower away from the central area. Strata within star dunes consist "mostly of low- to moderate-angle ripple strata, with many gently sloping truncations surfaces" (Cooke *et al.*, 1993: 401). Star dunes occur in sand seas, where the source of sand is abundant, such as the Grand Erg Oriental in North Africa (e.g. http://daac.gsfc.nasa.gov/DAAC_DOCS/geomorphology/GEO_8/GEO_PLATE_E-6.HTML) (Livingstone and Warren, 1996). They appear to form in areas with multidirectional wind regimes.

Zibar are extensive dunes that have no slip faces and are characterized by their composition of coarse sand and hard surfaces. They tend to have low relief and are straight to parabolic in form (Cooke *et al.*, 1993). Many are found to occur upwind from sand seas which have finer grain sizes than the zibar, suggesting that the finer sand has been winnowed from the latter (Livingstone and Warren, 1996).

Dome Dunes have no slip face and are composed of fine grained, unimodal sand, with the majority being about 1 to 2 m high (Cooke *et al.*, 1993). Morphologically, they may be indistinct

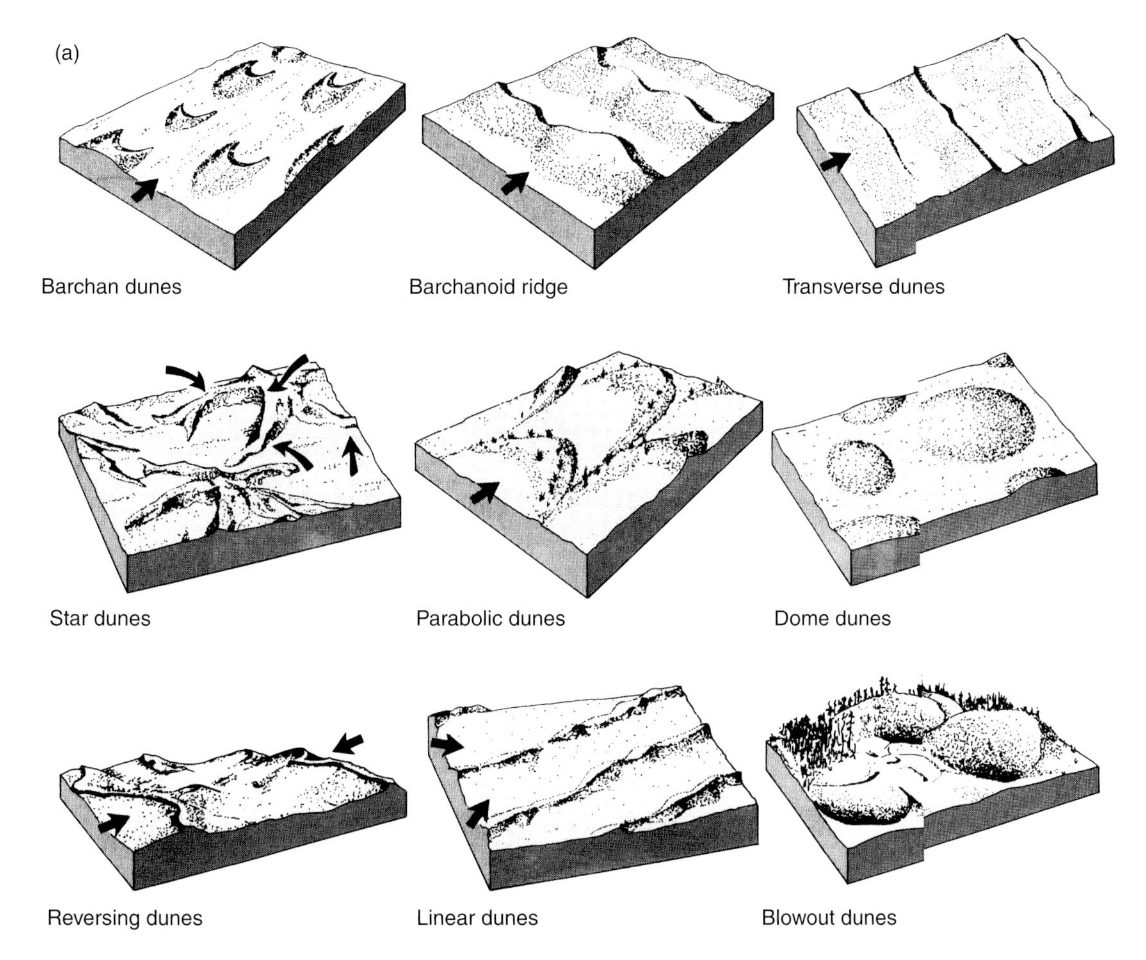

FIGURE 6.13 (a) Major dune types with arrows representing the principal wind directions (modified from Ritter *et al.*, 2002 after McKee, 1979). (b) Stratification associated with different dune types (modified from Brookfield, 1984, figure 7).

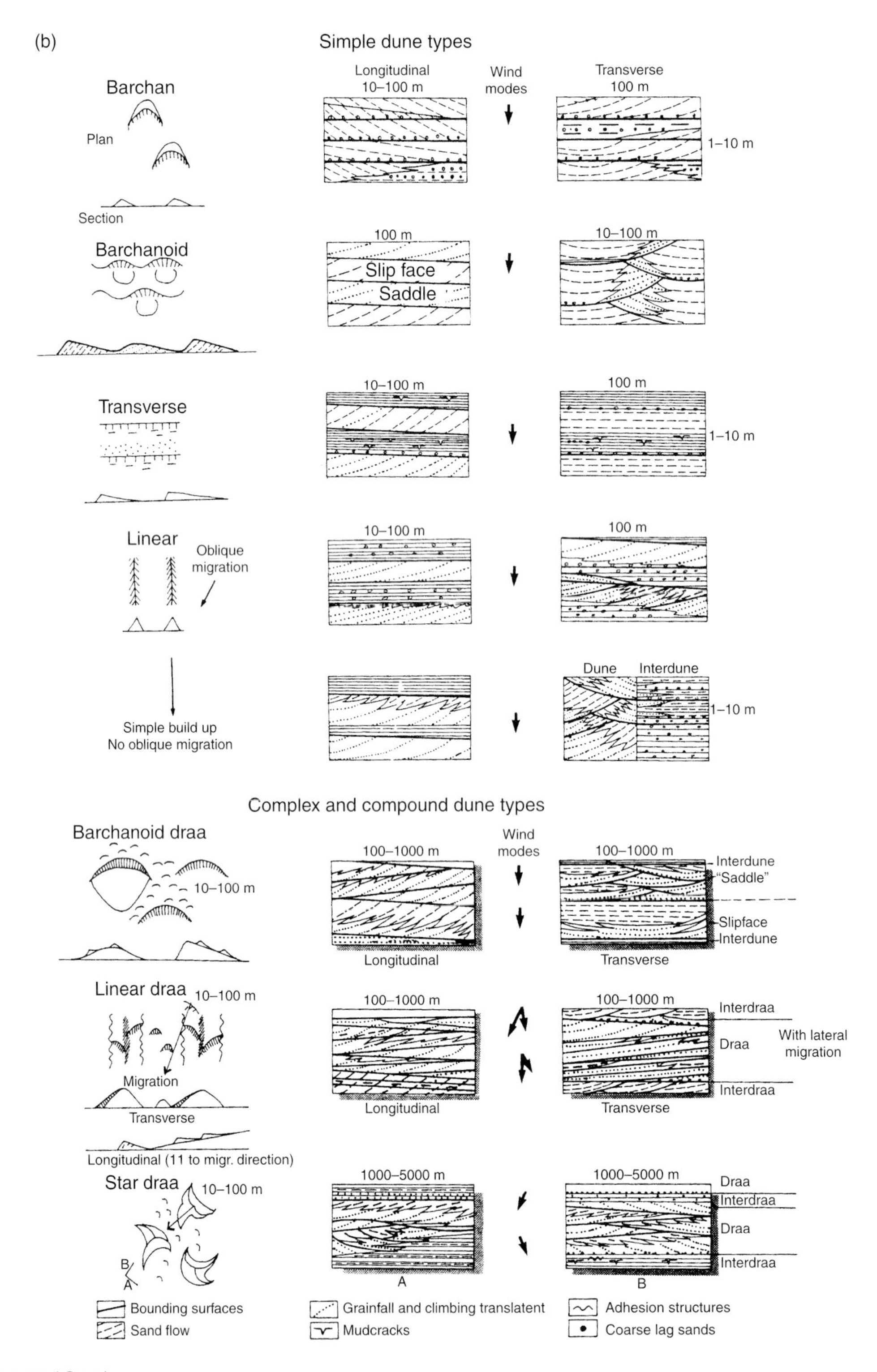

FIGURE 6.13 (*Cont.*)

(a)

(b)

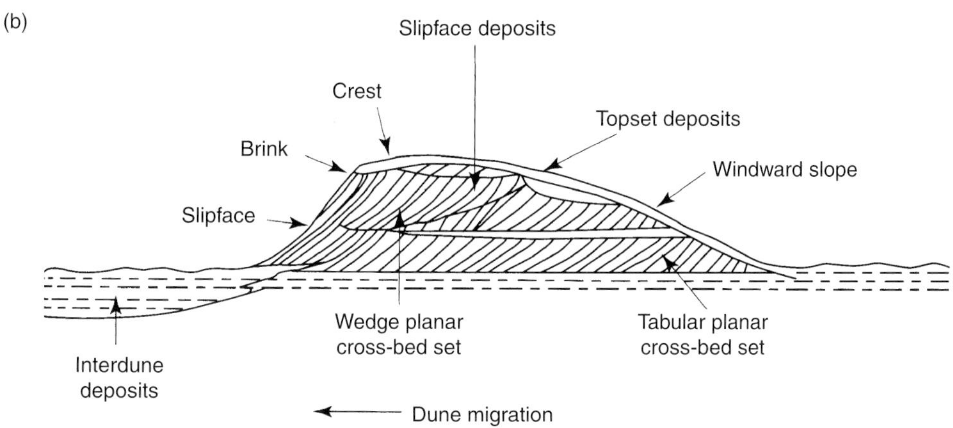

FIGURE 6.14 Barchan dune. (a) Photograph of barchan dune in Sinai, Egypt. Note the smaller scale ripples on the surface of the dune. The double arrows in the middle ground point to fluvial terrace consisting of reworked sandy deposits that are capped with a rubefied calcic horizon overlain by Epi-Palaeolithic (Kebaran; ca 13 ka) implements from sites Lagama 1 C and 1 F (Goldberg, 1977); slightly younger Harifian (ca 10.5 ka) material is scattered on the terrace riser between the terrace tread and the barchan; (b) schematic view through a barchanoid dune ridge showing types of bedding and relation to lateral, interdune deposits which are finer grained (modified from Boggs, 2001, figure 9.21).

from zibar described above (Livingstone and Warren, 1996).

Sandsheets are areas that have very little relief and subtle dune morphologies. Many serve as substrates over which other types of dunes drift (Cooke *et al.*, 1993). The largest sandsheet occurs along the border between Egypt and the Sudan, which has been the focus of prehistoric research for several decades by F. Wendorf, SMU, Dallas (see above). In this area, the sand sheet is capped by a soil developed in Neolithic times, clearly attesting to wetter conditions at this time. Ripples, consisting of coarse sand, are the predominant bedform, with very low angles of dip (Cooke *et al.*, 1993).

Anchored Dunes are those which are produced by topographic barriers or by fixed vegetation. In both cases these obstacles affect the flow conditions. For obstacles with slopes <30° sand is usually blown up and over the obstacle; in the case of *climbing dunes*, for example, sand is trapped on the upward slope, usually 30° to 50° (Livingstone and Warren, 1996). On the lee side of wide obstacles, calm air leads to the accumulation of sand along the lee slope of the obstacle (*falling dunes*) (Livingstone and Warren, 1996); with narrow obstacles, *lee dunes* are produced.

Plants serve as anchors to dunes and even today are commonly used to stabilize moving sand in cases of coastal erosion. Trapping vegetation can be an isolated plant to more extensive vegetated vary from areas such as those along coasts (Fig. 15.9b). Small clumps of vegetation will tend to produce individual mounds, whereas plants that spread laterally, will create more undulating topography (Livingstone and Warren, 1996). Stratification in vegetated dunes consists of steeper angle sets, and lower angle truncation surfaces; most beds dip at about 12° (Cooke *et al.*, 1993).

Parabolic dunes are U- or V-shaped in plan view, and the open end points into the wind. They are associated with vegetated areas at the margins of deserts, such as the Thar Desert in India. They appear to develop from *blowouts* whereby material is eroded from the underlying sediment, while at the same time the flanks are stabilized by vegetation (Cooke *et al.*, 1993). However, many instances of parabolic dunes show that they transgress along the same course, resulting in a nested pattern.

Coastal dunes (see also below; Chapter 7) are found not only along vegetated coasts, such as Cape Cod, United States, but also in areas flanking hyperarid desertic terrains, such as in the Negev/Sinai. In the latter case, coastal sand is transported northward along the coast by longshore drift from the Nile. This littoral transport, coupled with low relief and relatively shallow shelf and constant winds, results in the sand being blown up off the beach and accumulating on coastal cliffs or behind them. Many Holocene archaeological sites occur just beneath these coastal sands, while resting on Pleistocene aeolianites that also contain archaeological sites (Bakler, 1989; Frechen *et al.*, 2004; Gvirtzman and Wieder, 2001; Neev *et al.*, 1987; Tsoar and Goodfriend, 1994).

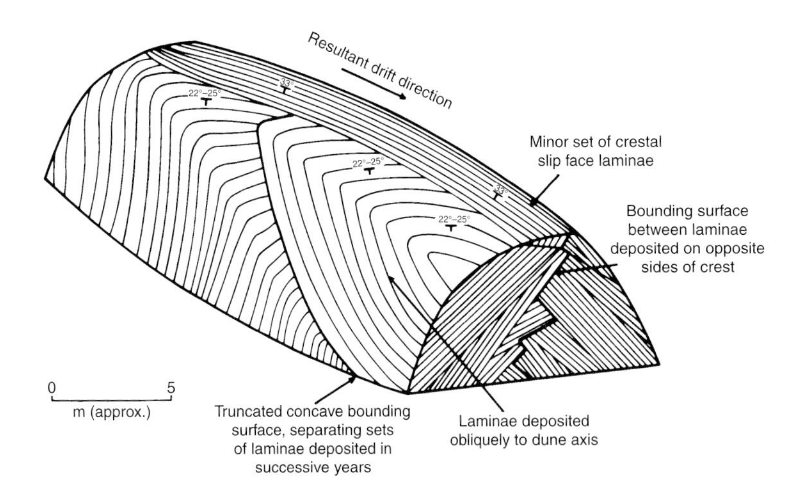

FIGURE 6.15 Internal structure of linear dune showing sets of laminae demarcated by bounding surfaces (from Thomas, 1989 after Tsoar, 1982).

Interdunal areas Areas between dunes, so-called interdunal areas, are confined by dune bodies (Boggs, 2001: 290). They may be deflationary as in the case of blowouts (see above) in which case little sediment accumulates. In cases where sedimentation occurs, the deposits show low angle sedimentary structures or are massive as a result of heavy bioturbation. Both wet and dry interdune areas are found. In dry interdune areas, the deposits tend to be coarser, poorly sorted, and poorly laminated. In wetter interdune areas, which are commonly moist throughout the year, the sediments are finer grained silts and clays, with freshwater mollusks, ostracods, and diatoms (Boggs, 2001). Where strong evaporation occurs or in areas undergoing desiccation, evaporates (carbonates, halite, and gypsum) precipitate, forming *sebkhas*. As pointed out above, wet interdune areas were attractive to game and humans in the past.

6.3 Examples of sites in dune contexts

Many interesting and significant archaeological sites are found in dune and associated arid land contexts, both in the New World, but especially in the Old World, in the desert zones of the Sahara. These geoarchaeological examples present stratigraphic sequences that reflect upon both local and regional environmental changes, which ultimately are linked with former climates.

In Wyoming, for example, Quaternary dune fields are characterized by thin, localized sandy deposits such as the Ferris Dune field where stabilized and parabolic dunes occur (see also comments above about the Casper site) (Albanese and Frison, 1995). During the interval of ca 7,660 to 6,460 BP, 9 m of sand accumulated during a marked drought. Overlying this are 8 m of aeolian sand containing lenses of interdunal pond deposits (ca 2,155 BP), attesting to more moderate conditions. Nearby, a 4 m sequence of sands associated with archaeological remains show at least two buried, truncated calcic horizons, attesting to stabilization and climatic change. In the Killpecker Dune Field, dome, transverse, barchan, and parabolic dunes occur. An iron-stained basal sand resting on Pleistocene deposits is overlain by sand with interdunal pond deposits at the top containing Cody Complex cultural material at the Finley site (ca 9,000 BP); the upper unit is 3,000 years younger at its base, and much younger at the toe (ca 755 BP) (Albanese and Frison, 1995).

Older and more sweeping geoarchaeological sequences occur in North Africa and southwest Asia, in the areas of Egypt, Sinai, and Israel. In the Bir Tarfawi and Bir Sahara areas in southern Egypt, lacustrine and aeolian deposits testify to alternating and marked differences in geological and regional environments and climates (Wendorf and Schild, 1980; Wendorf *et al.*, 1993). At site BT-14, for example, the stratigraphic sequence begins with topset and foreset beds of a large dune whose eroded surface displays wind blasted Aterian artifacts (≈Middle Palaeolithic). In turn, these sands were deflated and the resulting depression filled with carbonate and organic rich lacustrine silts (ca 44,190 BP) (Wendorf and Schild, 1980) that grade laterally into sandy beach deposits with numerous Aterian artifacts; a wetter climate is clearly indicated. These deposits were subjected to late Pleistocene–Early Holocene deflation, and eroded surfaces are dotted by recent dunes that tend to fill depressions (Wendorf and Schild, 1980: 38).

Similar types of events are well documented in the northeast, in Sinai, Egypt and the Negev, Israel, although the record of aeolian and fluvial activity is more pronounced (see Box 6.1 and Box 5.1). Gebel Maghara in northern Sinai is a ca 40 km long breached fold structure that contains aeolian and fluvially reworked sand and lacustrine deposits, with rubefied and calcified paleosols and associated

Upper Palaeolithic and Epi-Palaeolithic sites (Bar-Yosef and Phillips, 1977; Goldberg, 1977). In the Wadi Mushabi valley, the base of the Later Quaternary sequence is represented by massive, hard/compacted sand which cover the center and southwest part of the basin. The surface is locally rubefied (reddened) and is covered with ventifacted stones; carbonate nodules can be seen eroding out of these sands. Due to poor exposures, no artifacts were observed in the sands, yet they are dotted with Epi-Palaeolithic (Geometric Kebaran and Mushabian, ca 13,000–14,000 BP) sites. Elsewhere in the area, similar types of massive sands contain Upper Palaeolithic sites dated to about 31,000–34,000 years ago (Bar-Yosef and Phillips, 1977). Overlying the basal deposits are thin, patchy veneers or more massive sands, some of which are linear dunes. These sands are of various ages: the Epi-Palaeolithic site of Mushabi V, for example, occurs within a loose, massive, coarse sand that rests upon the soil of the basal sand; the material is *in situ* as demonstrated by the presence of three intact hearths. On the other hand, the slightly younger, Late Epi-Palaeolithic site of Mushabi IV is found in a more compact and poorly sorted sand. Both sites indicate a minor movement of sand that rests upon a stabilized, pedogenic surface that predates these later dunes by several thousand years. Moreover, the soil developed on the basal sand points to a wetter regime in which clay translocation and carbonate nodule formation could develop (Goldberg, 1977). Correlative wet intervals can be seen over much of the Sinai and Negev at this time (Belfer-Cohen and Goldberg, 1982; Goldberg, 1986).

Eastward into the Qadesh Barnea area of easternmost Sinai, and across into the Negev, aeolian deposits play an important part in documenting the geological and environmental history of this area during the late Pleistocene (Upper Palaeolithic, Epi-Palaeolithic), and Early Holocene (Neolithic) (Fig. 6.16). At present this area is covered with a system of southwest–northeast or west–east trending linear dunes that cover an area of about 10,000 km^2 (Goring-Morris and Goldberg, 1990). Intensive and extensive prehistoric research in this area has provided high resolution stratigraphic sequences, which owing to the associated presence of datable materials (e.g. charcoal, shell, carbonate nodules) has resulted in a rather clear picture of the landscape and the people who occupied it (Goring-Morris, 1987). We note for example, that dune encroachment began prior to the Late Glacial Maximum, reaching north central and north eastern Sinai about 25,000 years ago from the west. It steadily progressed eastward into the Negev, arriving in the Nahal Sekher (see Box 6.1) region about 5,000 years later.

Such progressive landscape and climatic aridification led to the blockage of many wadis (ephemeral streams) and the formation of localized playa/sebkha-type environments. This situation is in contrast to the preceding interval encompassing late Middle Palaeolithic and Early Upper Palaeolithic sites. During this interval of ca 50,000–40,000 to 25,000 BP wetter climates prevailed, as expressed by massive fluvial accumulations of silts and reworked sands within most of the drainages of the area; the dunes can be clearly seen in the field and in aerial photographs to rest unconformably on the previously formed fluvial terrain. Furthermore, *in situ* sites demonstrate that dunes were also locally mobilized in Holocene times, during the Neolithic, Chalcolithic, and Byzantine Periods (Goring-Morris and Goldberg, 1990). Finally, it should be noted, that although luminescence dating of such deposits is now more accurate than even a decade ago, the presence of datable materials from prehistoric sites serves as an independent dating source. Their excavation has the potential to provide higher resolution stratigraphic information than that obtained from examining the aeolian deposits alone.

Coastal sands and caves. Coastal cave sites are remarkable in their ability to trap aeolian sediments from the outside (see chapter on caves).

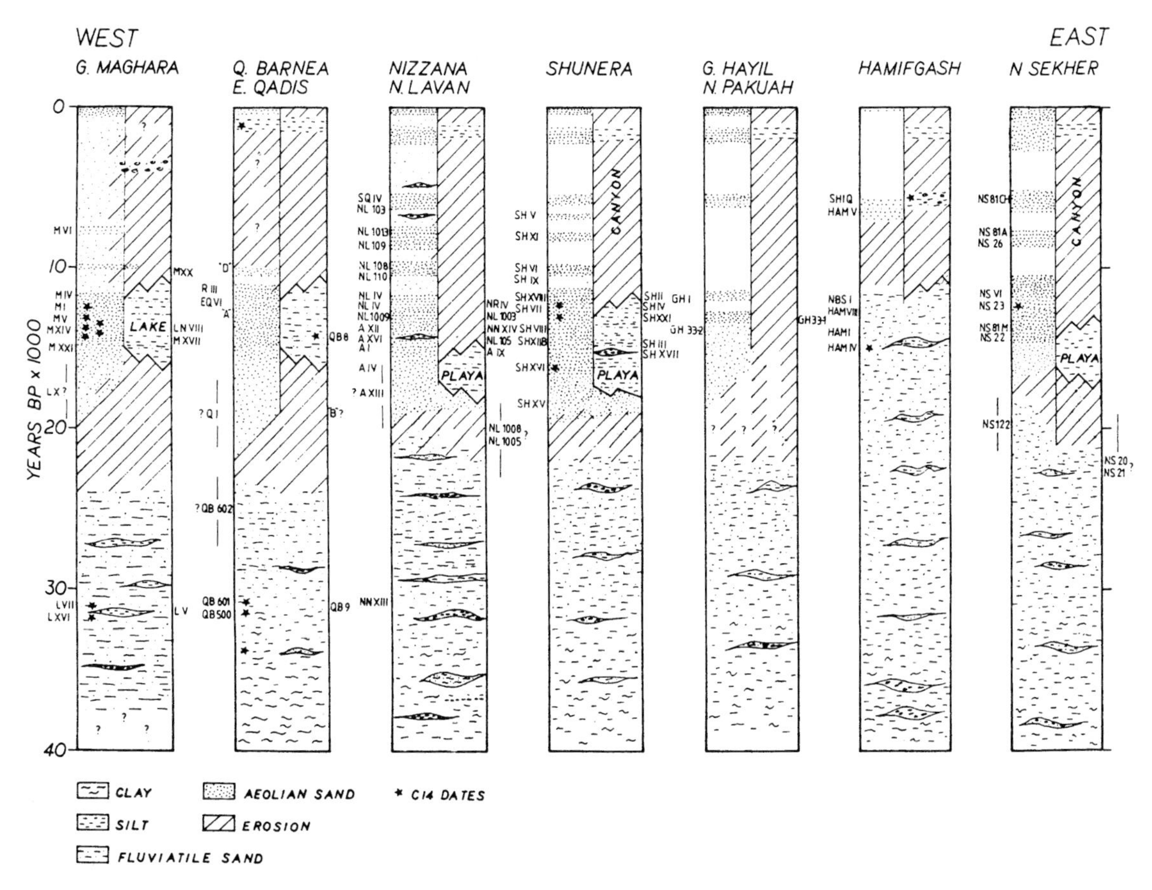

FIGURE 6.16 Geomorphic evolution of Northern Sinai and the Western Negev, showing the time transgressive nature of Late Pleistocene dunes from west to east. Indicated on the sides of the panels are the names of prehistoric sites; radiocarbon dates are also shown. (For details, see Goldberg, 1984, Goldberg, 1986; Goring-Morris, 1987; Zilberman, 1993) (modified from Goring-Morris and Goldberg, 1990, figure 4).

Many of the cave sites along the Mediterranean and shores of Africa are noted for the proximity of their aeolian inputs to the sea. Tabun Cave in Israel, for example, contains over 15 m of sandy and silty aeolian sediments associated with Lower and Middle Palaeolithic artifacts (Garrod and Bate, 1937; Goldberg, 1973; Jelinek *et al.*, 1973). In early investigations its infilling was framed within a "short" chronology (Farrand, 1979) and tied sea level fluctuations that date to only the last interglacial and younger periods. Recent ages furnished by TL and ESR techniques (Mercier *et al.*, 1995, 2000; Rink *et al.*, 2003), point to much greater ages that span over ca 200,000 years of accumulation which appear to provide too much time for the rather orderly infilling of the cave as shown by its sedimentology (Goldberg, 1973). It is now clear that the depositional history associated with such a long period of time is not evident and in serious need of clarification.

On the other side of the Mediterranean, sandy deposits are also a major constituent of the infillings in Gorham's and Vanguard Caves in Gibraltar (see Fig. 7.4). Together, these sites have thick deposits with Middle and Upper Palaeolithic occupations, and the hope of recent excavations was that they would contribute to our understanding of the extinction of Neandertals in Iberia (Barton *et al.*, 1999; Goldberg and Macphail, 2000; Macphail and Goldberg, 2000). In this regard too, the dating of the deposits is crucial to unraveling their geological histories. Efforts, however, have proven unsatisfactory, both from the technical point of view of securing accurate TL dates, but also from the standpoint of site formation. The key strata that span the Middle Palaeolithic and Upper Palaeolithic units are bioturbated and slumped, rendering any radiocarbon dates from these deposits suspect (cf. Pettitt, 1997). Nevertheless, it seems that the bulk of the massive sandy deposits at both sites accumulated within a period of <100,000 years (ca OIS 5, 4, 3); the Upper Palaeolithic layers at Gorham's Cave, for example, are predominantly cultural and not aeolian.

Similar types of sandy aeolian deposits are found in many coastal sites from South Africa. These range from sandy intercalations with cultural deposits during the Middle Stone Age (MSA) to more massive, typically sterile dune sands between the MSA and Late Stone Age (LSA). An excellent example is Die Kelders Cave, situated about 100 km east of Cape Town. Here detailed granulometric and micromorphological analysis was key to revealing the nature of site formation processes at the site, including its aeolian history (Goldberg, 2000; Marean *et al.*, 2000; Tankard and Schweitzer, 1974, 1976). Current research in the Mossel Bay area is aimed at documenting the type and age of sandy incursions in coastal prehistoric caves and settings during the Quaternary (Marean *et al.*, 2004) parallel to other efforts to resolve stratigraphic and dating of the aeolianites (Bateman *et al.*, 2004).

6.4 Bioturbation in sandy terrains

It is important to mention a recurrent theme that arises with archaeological sites in sandy sediments: does a buried artifact assemblage result from aeolian sedimentation, colluviation, or bioturbation (Frederick *et al.*, 2002; Leigh, 2001). Deciphering the effects of these processes is crucial if we are to interpret the archaeological record correctly.

In spite of difficulties, appraising such problems can be effectively accomplished in the field through detailed documentation of the site's geomorphic setting and microstratigraphy, and in the laboratory by investigating bulk properties such as granulometry, heavy mineral content, magnetic susceptibility, and phosphate content. Micromorphology is very effective in evaluating the effects of bioturbation, as microscopic signatures are often quite distinctive. A lack of correspondence of different measurements (e.g. radiocarbon dates, artifact abundance, phosphate content), with depth, for example, suggests that bioturbation may indeed be responsible for displacing artifacts and burying them; similarly the co-occurrence of lithic tools not normally associated (e.g. Middle Palaeolithic and Upper Palaeolithic artifacts) might point to bioturbation (Frederick *et al.*, 2002; also Stiner *et al.*, 2001).

6.5 Fine grained aeolian deposits

Fine grained aeolian deposits, including primary wind blown loess and its derivatives (e.g. colluvial loess) cover about 5% of the earth's land surface, with the thickest and most widespread deposits being in China, Central Asia, Ukraine, Central and Western Europe, the Great Plains of North America, and Argentina (Fig. 6.17). Most geoarchaeological related contexts are found in the Old World, due to the relatively late arrival of humans in North and South America, although some sites are associated with Holocene loess (Thorson and Hamilton, 1977). Overall, in the New World, loess may comprise a sediment source that ultimately may be part of the sedimentary record of archaeological sites,

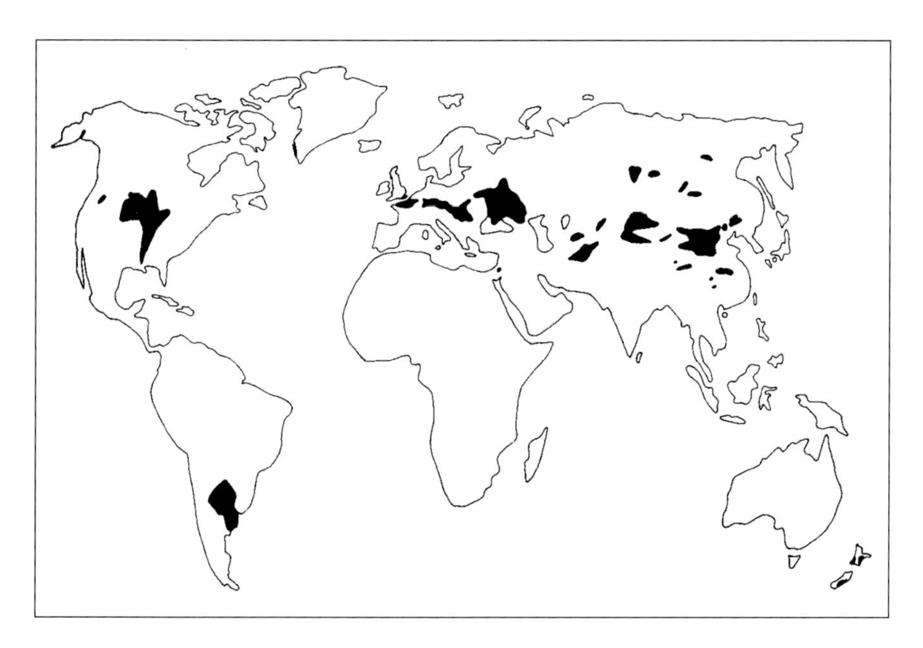

FIGURE 6.17 Distribution of major loess deposits in the world (modified from Pye, 1987, figure 9.1).

for example, Dust Cave, Alabama (Goldberg and Sherwood, 1994; Sherwood, 2001), or the substrate upon which occupation takes place. In addition, some soils from the southern Plains, in the Edwards Plateau for example, contain embedded aeolian silt (Rabenhorst *et al.*, 1984) derived from the High Plains (see Chapter 5); similar aeolian embedding of silts in soils occurs in the Mediterranean as well (Yaalon and Ganor, 1975).

The geoarchaeological importance of loessial deposits is that they are widespread, have accumulated over long time periods, and contain abundant soils of varying types that have palaeoenvironmental significance which can be retrieved relatively easily. Furthermore, important prehistoric sites, such as Willendorf in Austria (Haesaerts and Teyssandier, 2003) are found within loess deposits and associated soils. Thus, not only do these loesses and soils constitute one of the best and most continuous continental records of deposition, but also they supply high quality, detailed paleoenvironmental information, before, during, and after prehistoric occupation. Such information is crucial in reconstructing prehistoric lifeways and resolving thorny issues, such as the arrival of the earliest modern Humans in Europe (Conard and Bolus, 2003).

Loess can be characterized as "a terrestrial (i.e. subaerial) windblown silt deposit consisting chiefly of quartz, feldspar, mica, clay minerals, and carbonate grains in varying proportions" (Pye, 1987: 199). It can also contain phytoliths, heavy minerals, volcanic ash, and salts. Loess is unstratified and typically consists of silt-size grains (ca 20–40 μm) and is positively skewed. It may also contain clay- (<2 μm) and sand-(>63 μm) size material of varying proportions, and is commonly, but not always, calcareous (calcite and dolomite). It is porous, tends to keep vertical faces when exposed, and is easily erodible (Pye, 1987).

The origin of loess particles has not been clearly determined, and it appears that there are several mechanisms that can produce silt-sized grains that can be made available for wind transport (see Pye, 1987 for details). Some of the processes include (1) glacial grinding and frost-related comminution; (2) liberation of silt sized material from original rocks, (3) aeolian abrasion, (4) salt weathering, (5) chemical weathering, (6) clay pellet aggregation, and (7) biological processes (Pye, 1995). Glacial grinding, for example, is clearly pertinent to silt and loesses are produced proximal to glaciated terrains, whereas aeolian abrasion and salt weathering are more closely linked to the formation of silt size particles in desertic areas. As in the case of western and central Asia, for example, weathering of rocks takes place in the high mountains, and silt is transported and deposited onto the lowlands as alluvial deposits. Here, it is eroded and transported in the atmosphere, eventually deposited and accumulated as loess. Typical areas that can supply dust for transport and eventually end up as loess deposits include: (1) areas proximal to glaciers: glacial outwash and braided fluvio-glacial channels, (2) dry lake and stream beds, sebkhas, lunettes, alluvial fans, and existing loess deposits that are being deflated due to lack of vegetative stabilization, and (3) alluvial floodplains.

Aeolian transport of dust occurs when the wind exceeds fluid threshold (Bagnold 1941; Pye, 1987). As pointed out above, greater velocities are needed to move both coarser and finer grains, as the latter have internal cohesive forces that require greater velocities to overcome this cohesion; finer material also tends to have higher moisture content. Similarly, the binding effects of salts, clays, and microbiological crusts tend to inhibit deflation of finer particles (Pye, 1987). In fact algae and mosses serve as dust traps in arid areas. Surface crusting – whether biological (Danin and Ganor, 1991) or physical, produced by slaking – inhibits dust entrainment. Yet, once the crust is broken (e.g. by fluvial activity, ploughing, construction, trampling, vehicle movement) deflation is greatly facilitated.

For a typical windstorm event the smallest particles (<10 μm) with low settling velocities, are carried into the atmosphere. The coarser silt and fine sand fractions have higher settling velocities and are transported for shorter distances and periods of time; coarse and medium silt grains can be transported in the order of tens to hundreds of kilometers (Pye, 1995, figure 10; Fig. 6.18).

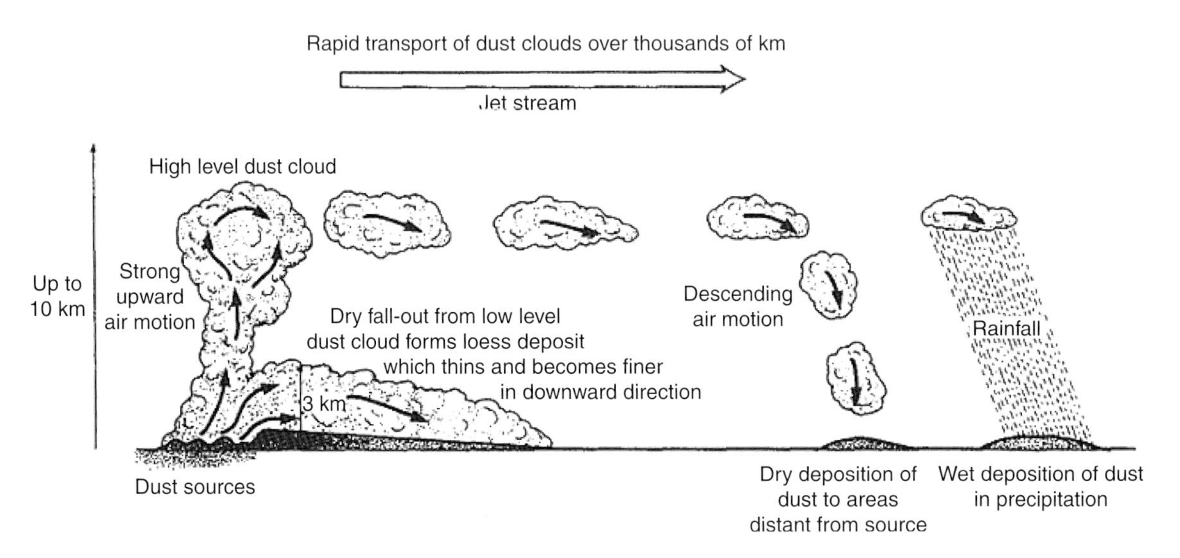

FIGURE 6.18 Generalized model showing methods of transport and deposition of aeolian dust (modified from Pye, 1995, figure 10).

Dust can be transported in both the lower and upper levels of the atmosphere, which is a function of local versus regional wind systems. In general, trade winds, which do not extend high into the atmosphere transport coarser material over shorter distances where it is deposited as dry fallout (Fig. 6.18). On the other hand, thunderstorms and strong advection transport large amounts of finer dust into the atmosphere where it can be transported for long distances, up to thousands of kilometers (Goudie and Middleton, 2001; Sarnthein and Koopman, 1980).

Deposition of dust is accomplished by a number of ways depending on local and atmospheric conditions. These comprise (1) gravitational settling of individual grains, or aggregates of grains (bound by electrostatic charges or moisture), (2) downward turbulent diffusion, (3) downward movement of air containing dust, and (4) particles being washed out of the atmosphere by precipitation. Under dry accumulation, deposition takes place with reduced wind velocity or with an increase in the surface roughness. However, if such material falls on a bare, unvegetated, or smooth surface, it is not likely to be trapped permanently but will be subsequently eroded and redeposited. On the other hand, dust is likely to remain if it falls on a moist surface (e.g. lake, playa, sebkha, interdunal areas) or one with vegetative cover (including both larger plants such as trees and cryptogams such as algae and mosses) (Danin and Ganor, 1991); on rough surfaces, such as alluvial fans, dust can settle between boulders (Pye, 1995).

Such dry loess accumulations can be relatively coarse, with grain size and thickness decreasing away from the source. This type of loess (Fig. 6.19) is typical of dust derived from outwash plains or from large streams, such as the Mississippi River. In desert areas, dust may be laterally contiguous with sand dunes blown from the sedimentary source (fluvial source). In this case, sand accumulates adjacent to the source and the amount of finer material increases downwind and away from the source (Fig. 6.19). This situation is evident in Israel where sandy deposits in the arid south grade into finer, siltier, and clayier loess and reworked loess as one proceeds away from larger wadis (ephemeral streams) (Yaalon and Dan, 1974). In this latter case, fluctuations in climate can shift the lithological and depositional boundaries and thus the stratigraphic loess records in these localities can be sensitive palaeoclimatic indicators (Bruins and Yaalon, 1979; Goodfriend, 1990).

In both of the above cases, finer-sized dust remains in suspension and is only deposited when it is washed out of the atmosphere through precipitation (Fig. 6.12). This type of accumulation is widespread, and is well documented in Europe, where dust of both Sahelian and Saharan origin can be found accumulating on surfaces (Littmann, 1991). In the Mediterrancan region, *terra rosa* soils have high quartz contents, which are clearly not derived from the underlying limestones that have little insoluble residues (Rapp and Nihlén, 1991). In addition, many vertisols in Israel owe their properties to the imbedding of smectitic clays of aeolian origin (Yaalon and Dan, 1974; Yaalon and Kalmar, 1978), whose shrink–swell properties and mixing, can have disastrous effects on the integrity of archaeological assemblages. In Hawaii, studies by Jackson *et al.* (1971) revealed the presence of quartz in soils developed on basaltic substrates.

Finally, topographical factors may lead to dust accumulation, such as that on the windward side of obstacles where a reduction of wind velocity may take place as well as an increase in trapping vegetation (Fig. 6.19) (Rognon *et al.*, 1987). In the Negev Desert, Gerson and Amit (1987) found significant accumulations (0.5–0.8 m) of aeolian silt within roofless Middle Bronze Age I buildings which took about 1,000 to 2,000 years to accumulate. These values are much higher

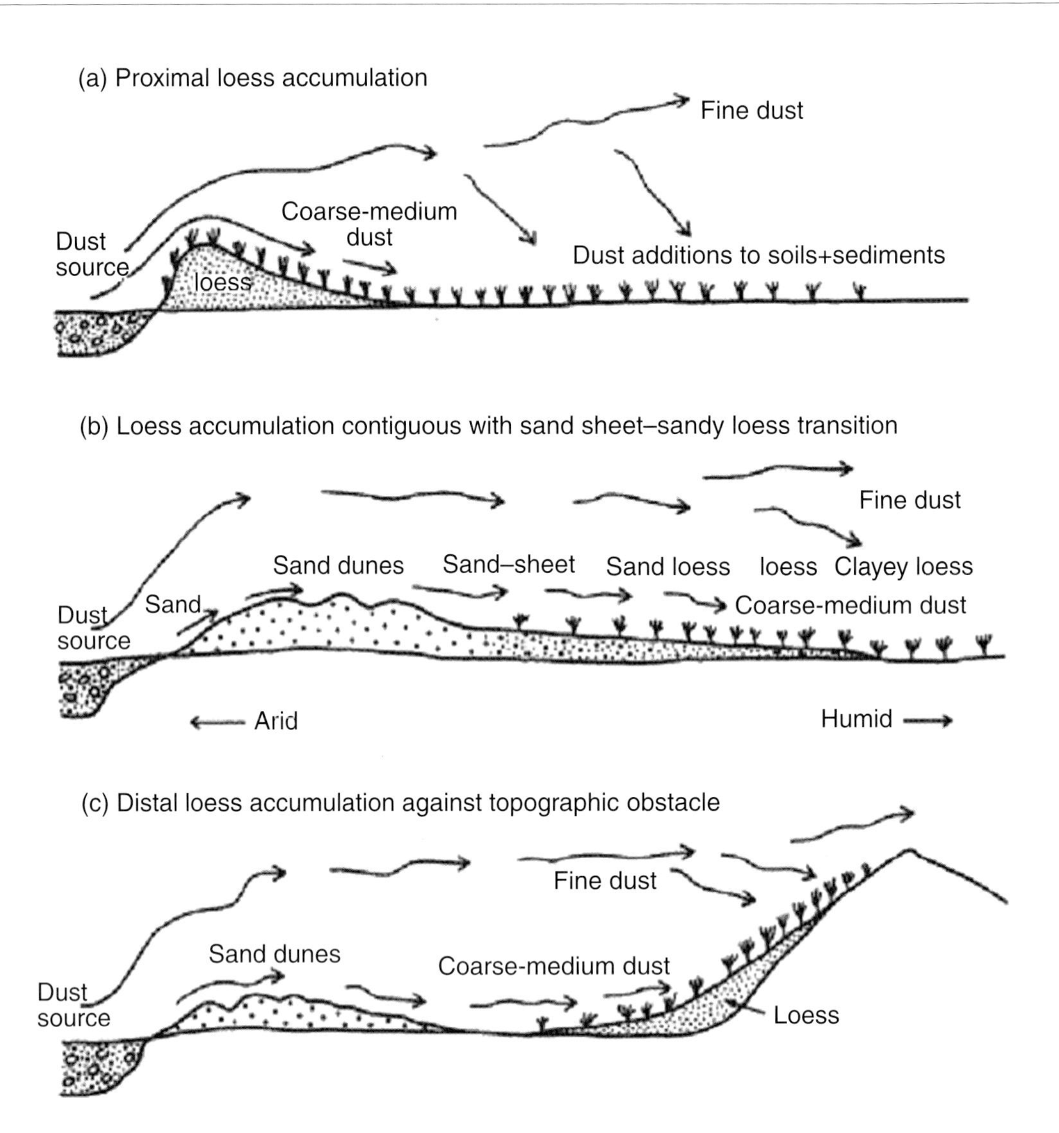

FIGURE 6.19 Mechanisms of dust entrapment and accumulation (modified from Pye, 1995, figure 11).

than present-day rates of accumulation in the Negev, which are influenced by human activities there (Yaalon and Ganor, 1975).

In sum, four phases are associated with the formation of loess: (1) formation of predominantly silt-sized grains, (2) aeolian transport under conditions of strong, turbulent winds, (3) deposition, and (4) post- and syn-depositional alterations (Pye, 1987). In Europe, during glacial times, dust was deflated from outwash belts and river bars and accumulated in the large band situated between the Alpine glaciers and those from the Scandinavian ice sheets. Dust accumulation took place during principally cold, stadial phases, and was commonly syn-depositionally modified by cold climate processes such as solifluction, cryoturbation, and ice wedging. During milder Interstadial events, modification of the accumulated deposits took place, principally represented by various types of pedogenesis, depending on climatic conditions.

Although loess is strictly speaking an aeolian deposit, other silty deposits are either derived from, or simply associated with loess accumulations. These are secondarily reworked loess deposits, and naturally have similar overall textural characteristics, although they tend to be more poorly sorted and occur in aggregate form; some are gravelly. These "secondary" deposits have a variety of names, depending on the language and culture. In English, "alluvial loess," "colluvial loess" are apparent. In Germany, where loess deposits are well expressed, the following distinctions are made: (1) "Fließlöß" (literally "running loess") in which loess is redeposited by sheetflow, resulting in bedded silty deposits, (2) "Schwemmlöß" is equivalent to alluvial loess, occurs in fluvial channels, and may contain boulder beds, plant remnants, and snail shells (Schreiner, 1997). In a comprehensive summary of loess deposits in Central Europe, and their correlation with the deep sea stratigraphy, two different types of loessial sediment, "pellet sands" and "marker" that are found in depressions in the Czech Republic and that have environmental significance are described (Kukla, 1977). The former are composed of sand-size aggregates of silt and clay that are produced during torrential rains following a long dry season; the latter take the form of thin bands of silt resulting from large scale dust storms. Both are associated with snail faunas that need warm and dry summers. A furthermore demonstrates the correlation of loesses and their paleoclimatic sequences (as based on the microstratigraphy and geomorphology, paleosols, malacology, and palaeomagnetic data) with the deep-sea record in terms of its continuity and recording of individual climatically forced events.

Similarly, loess deposition was not continuous, and during frequent breaks of deposition, different types of paleosols formed on these stabilized surfaces (Fedoroff and Goldberg, 1982; Haesaerts *et al.*, 1999, 2003). The differences in these soils are dramatic, and testify to evolving paleoclimatic conditions over large areas and within a specific locality. In Belgium, for example, (Haesaerts *et al.*, 1999) used macro-, meso-, and micromorphological observations to document 14 phases of pedosedimentary development. Early pedogene-sis is characterized by Alfisols with textural B horizon and clay translocation. Subsequent soils were developed on loess, colluvial loess, and soliflucted materials. These soil types include among others, (1) a Greyzem (US: Mollisol) deep, dark, organic-rich soils, with deep leaching of carbonates formed "in a continental climate with cold snowy winters and warm summers" (Haesaerts *et al.*, 1999: 17), (2) a humiferous, chernozem-type soil formed in humid conditions "with a steppe or forest-steppe vegetation," (3) a subarctic humiferous soil, in which oxidation-reduction and polygonal patterning and platy soil structure attest to seasonally humid conditions and cold conditions.

Loess deposits that cover parts of Brittany, Normandy, and Belgium originate from the English Channel which was repeatedly exposed as sea level fluctuated in response to glacial growth. In Central Europe (e.g. Austria, Czech Republic, and Hungary) where loess deposits are up to tens of meters thick, aeolian dust was derived from the outwash plains of the Alpine foreland, as well as the Danube Basin.

Loess deposits in northwest Europe are generally thin and less expressed on the landscape, lacking the vast loess covers seen in Central and Eastern Europe and further eastward. Yet, notable loess exposures associated with prehistoric remains do exist. The legendary prehistorian François Bordes was a geologist by training, and his doctoral dissertation was concerned with the lithic industries found in the Paris Basin northwest of Paris (Bordes, 1954). Early on, he defined a *loess cycle* that started with a band of pebbles at the base accompanied by solifluction; both represent the onset of a cold deterioration of the climate. As the climate continues to deteriorate, loess is deposited under cold and dry conditions. At the end of the loess cycle, climatic amelioration associated with an interstadial or interglacial climatic warming, results in a cessation of loess deposition and the formation of various types of paleosols described above. Although he clearly documented the association of lithic artifacts with the different loesses and paleosols (e.g. Acheulean industries were found in the "Older Loess"), we now know that the dating, and stratigraphic and palaeoclimatic sequencing of these deposits was not correct. Nevertheless, his early work showed the presence of lithic industries within the loess sequences.

In the United Kingdom, loess occurs mostly as patchy, eroded deposits (brickearth) associated with the last glaciation (Weichselian). While not extensive, numerous studies have been conducted on the palaeoclimatic significance of the paleosols associated with them (Kemp, 1985, 1986; Kemp *et al.*, 1994; Weir *et al.*, 1971).

Extensive and thick loess deposits and paleosols are found in the arid and semiarid Central Asian areas of Tajikistan and Uzbekistan, where together, they have a thickness of up to 200 m and range from Upper Pliocene to Upper Pleistocene date back to over 2 million years (Dodonov, 1995; Dodonov and Baiguzina, 1995) (Fig. 6.20). They have been studied in detail, including their palaeomagnetism, macro- and snail fauna, palynology, and pedology. Sections from the area all contain paleosols – grouped into pedocomplexes – that can be correlated; the Brunhes/Matuyama boundary is systematically located between paleosols 9 and 10, and the Laschamp excursion occurs between paleosols 1 and 2 (Ranov, 1995). The soils in the lower part of the sequence consist of red-brown soils, whereas those in the upper part are brown soils and are thicker, due to penecontemporaneous loess accumulation during pedogenesis (Dodonov, 1995).

Pollen analysis at the Lakhuti section, for example, has shown a systematic palaeoclimatic change from hot and occasionally moist conditions in the Early Pleistocene to systematically cooler and more arid climates in the Late Pleistocene. At this same profile, several important Palaeolithic localities have been found within the loesses in association with paleosols that are up to 50 to 60 m below the surface, and which predate the Brunhes–Matuyama boundary. At the Karatau site (ca 200,000 BP) (Dodonov, 1995) choppers, "Clactonian" flakes, and unifaces occur within the sixth pedocomplex. At site of Lakhuti (ca 130,000 BP) Mousterian elements with rare Levallois technology are found within the sixth and fifth pedocomplexes (Fig. 6.21). Upper Palaeolithic material is normally found within the third to first pedocomplexes.

So far, it appears that these finds are always located within the lower part of the paleosols and not within the loesses (Dodonov and Baiguzina, 1995; Ranov, 1995). Thus it seems that, overall, the soils are formed during warmer and wetter interglacial episodes. Although TL dating is problematic, and some correlations are made on the basis of relative positions of the soils within the sequence (Ranov, 1995). More detailed work is needed to establish direct links between paleoenvironments, site occupations, and their contents, a task that is hampered by the small size of the lithic assemblages, and a lack of ". . .'normal' archaeological layers, such as kitchen debris

FIGURE 6.20 Typically thick accumulation of loess from Uzbekistan. This profile is over 100 m thick and exhibits numerous paleosols as shown by the darker bands running through the lighter aeolian deposits; a Mousterian occupation occurs about one-third of the way down from the top.

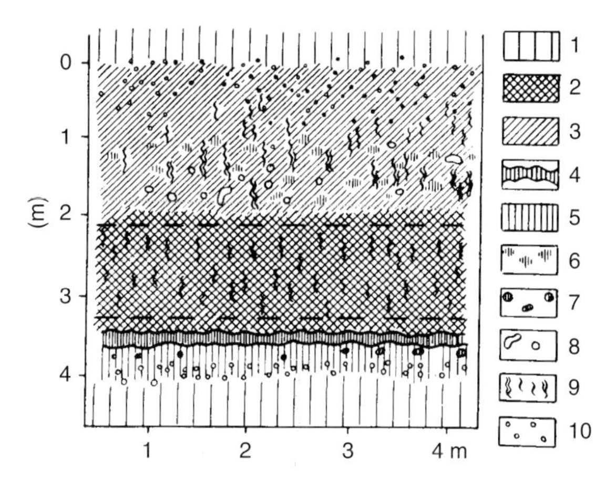

FIGURE 6.21 Schematic profile of fifth pedocomplex at the Palaeolithic Lakhuti section. 1) Loess; 2) brown paleosol; 3) cinnamon paleosol; 4) carbonate crust; 5) illuvial carbonate horizon; 6) carbonate concentrations; 7) carbonate concretions; 8) krotovina; 9) phytogenic pores and traces of roots; 10) traces of invertebrate fauna. Dashed lines delineate the level of Palaeolithic finds. (modified from Dodonov, 1995, figure 5).

where animal bones have been the meals of prehistoric people, stone structures, and . . . a complete absence of hearths" (Ranov, 1995: 735).

Nevertheless, the loess deposits and soils are strong paleoclimatic indicators: part of Central Asia, loess formation correlates with increase of the Siberian–Mongolian high-pressure system and dilution of monsoon activity, and to the contrary the soil processes correspond to high intensity of "summer" monsoon activity and a weaker influence of the Siberian–Mongolian high–pressure centre. In the western part of Central Asia the role of western air-mass transfer carrying the moisture from the North Atlantic and Mediterranean is very significant. In this area of Central Asia loess forming epochs are related to reduction of western airmass transfer and strengthening of the Siberian–Mongolian anticyclone with simultaneous development of the polar air-mass, while the soil processes are connected with strong activity of western air-mass transfer and a weak influence of the Siberian-Mongolian high–pressure system. (Dodonov and Baiguzina, 1995: 719)

The younger loesses of Western and Central Europe, datable by ^{14}C of charcoal and bones, mostly found in association with archaeological sites, has been documented with extraordinary clarity by P. Haesaerts and coworkers. Their work should be taken as a geoarchaeological yardstick for meticulous field observations, sample collection, and cleaning of samples for

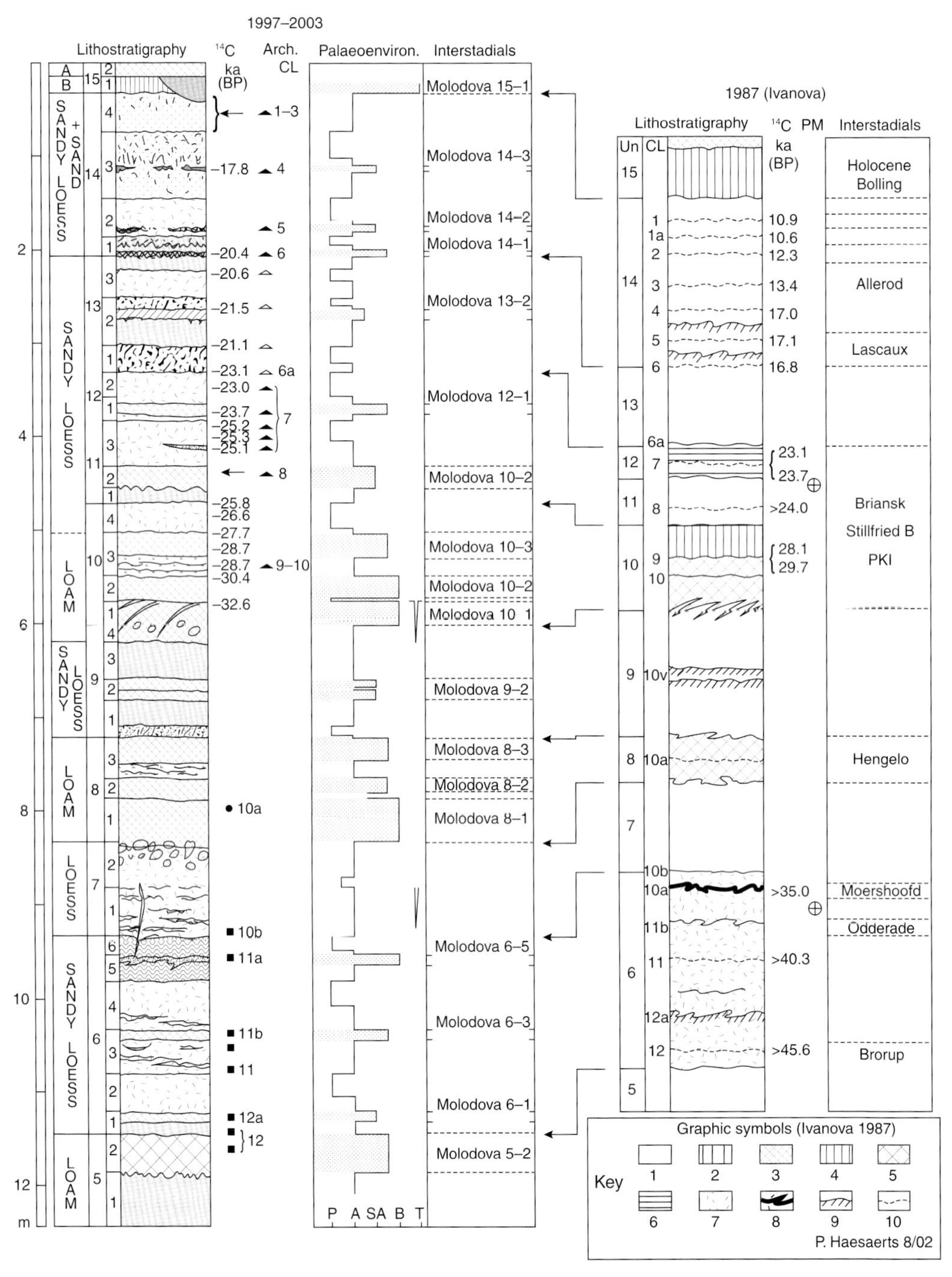

FIGURE 6.22 Simplified stratigraphic drawing of loess section from Moldova V (Ukraine) showing various lithostratigraphic units, soil horizons, and soil features (e.g. gleying, banded B horizon, biogaleries). The illustration provides the essential depositional and postdepositional elements of the profile, along with environmental data and environmental correlations. These are all readily understandable just through observation. 1) loess 2) B2t horizon (Holocene); 3) Al horizon (Holocene); 4) humiferous horizon; 5) fossil soil; 6) gley with; 7) ocher-grey loess; 8) soot layer; 9) gley 10) cultural layer (modified from Haesaerts *et al.*, 2003).

radiocarbon dating. In addition, the stratigraphic drawings (see Fig. 6.22 for a simplified view) are outstanding: they document the sedimentological and stra-tigraphic information and portray it in a way that concisely and efficiently conveys the entire geological (depositional and postdepositional) story at a glance (Haesaerts and Teyssandier, 2003; Haesaerts *et al.*, 1999, 2003).

In the East Carpathian region for example, they were able to reconstruct a precise pedosedimentary and paleoclimatic sequence (spanning ca 33,000–10,000 BP) that incorporates Middle Palaeolithic (Moldova V, Ukraine) through Late Palaeolithic (Epi-Gravettian; Cosautsi, Moldova) sites (Haesaerts *et al.*, 2003). Such sequences not only depict the nuances of rapid climate change during this time period but also indicate that the area was more or less continually occupied, although the types of sites changed as did the environments: Aurignacian occupations, for example, consist of workshops, whereas Gravettian occupations (ca 23,000–26,000 BP) are distributed over larger areas and appear to be associated with long-term seasonal encampments. During a colder and then drier time slice of 23,000– 20,000 BP, occupation is scarcer and represented by small concentrations that seem to represent short-term occupation. Epi-Gravettian (20,000–17,200 BP) settlement represents a return to long-term seasonal occupation in a "contrasted and relatively moist climatic context" whereas few remains are visible for the very cold conditions between 17,200 and 15,000 BP.

In most of North America, extensive loess deposits occur over much of the mid-continent region in the area of the Mississippi Basin, with patches of loess being found scattered locally over the eastern part of the United States. Unfortunately, most of the loess is essentially Pleistocene in age – although Holocene loess is recognized (Mason and Kuzila, 2000) – and thus predates the arrival of most humans into North America; it therefore has only minor geoarchaeological relevance. In Alaska, however, close to a major source of human migration from Asia, Thorson and Hamilton (1977) excavated the stratified Dry Creek site in the central part of the state. This site occurs within a sequence of loesses and interstratified sand units and paleosols that overly Pleistocene glacial outwash. The three oldest soils are immature tundra soils (Cryepts) with organic A horizons on mineralogenic B horizons; they do not form in the region today. Archaeological components consisting of artifacts, vertebrate faunal remains, and pebbles displaced by humans are found in four distinct layers within the loesses. The lower three are similar to material in upper Palaeolithic sites in northeast Siberia, but here date to about 11,000–8,500 BP (uncalibrated) and are associated with conditions of the late glacial. The youngest archaeological occupation which was less intense than the earlier ones, is associated with a forest soil and was occupied between ca 4,700 and 3,400 BP(uncal).

6.6 Conclusions

Aeolian deposits, whether in sandy or silty terrains, constitute a remarkable geoarchaeological resource. On one hand, loess sites in Central Europe and Western Asia, for example, contain some of the most significant sites for documenting important archaeological questions, such as the replacement of Neanderthals by anatomically modern humans. Because of fine grained sedimentation, coupled with various types of soil formation, they represent faithful indicators of past environments and archives of human activities. On the other hand, sandy deposits, particularly those in coastal settings, for example, in the Mediterranean and South Africa, are notable for their connection to eustatic (worldwide) sea level changes. In these sites (e.g. Gorham's Cave, Die Kelders, Blombos) the sandy intercalations form the substrate for human occupations and also seal the deposits, thus permitting high resolution temporal images of living surfaces, and windows overlooking specific human activities.

7

Coasts

7.1 Introduction

In this chapter we discuss coasts in terms of their cultural importance. Coastal morphology, sea level dynamics, and sediment types are also explained before examples of the utilization of coastal resources by humans are described.

An understanding of coasts is important in archaeology because they were often preferentially occupied by humans, and for the reason that their resources were utilized by human societies. It is therefore important to be able to identify the different types of settings and deposits formed in coastal areas in order to understand past coastal morphologies. For example, the 500,000 year old early hominid and coastal site of Boxgrove, West Sussex, United Kingdom, is now located along a former coastline (palaeocoastline) that is 46 m above present sea level, and some 11 km inland from it (Roberts and Parfitt, 1999) (Box 7.1). The Roman coastal port and fort of Pevensey is now nearly 2 km from the sea, but was at the head of a peninsula adjacent to a 10 km deep inlet – now infilled with (pelo-calcareous alluvial gley) soils formed in marine alluvium (Jarvis *et al.*, 1983). On the other hand, as sea levels rose during the Holocene many coastal sites were drowned. Such rising sea levels not only preserved some unique and important sites (such as the submerged Neolithic village of Atlit-Yam off the coast of Israel) (Galili and Nir, 1993) and organic remains, such as boats, but they have also buried evidence for former shoreline occupations associated with the colonization of the New World (Fladmark, 1982). Equally, coastal resources, which include foodstuffs, fodder plants (reeds and other plants for matting and baskets), seaweed (for manure) (Conry, 1971), and salt (briquetage), are often recorded in near-coast terrestrial sites. It can be noted that, because of glacio-isostatic rebound of the FenoScandinavian region over the last 13,000 years, coastline settlements are presently found up to 100 km inland (Linderholm, 2003). In some regions, coasts, and their sediments and populations have also been subjected to the threat of tsunami ("tidal" waves induced by shifts in the seabed by seismic/earthquake activity) and major environmental alternations such as El Niño (Nunn, 2000). Underwater archaeology, emerging features, and palaeofeatures cannot, however, be simply termed "time capsules" because they were inundated, by the sea, as a wide variety of processes can affect underwater sites (Stewart, 1999). Nevertheless undisturbed drowned stratified sites have been found in some quiet "back-waters" of the Baltic inundated during the Holocene (Skaarup and Grøn, 2004).

7.2 Palaeo sea shores and palaeo coastal deposits

Coastal sediments can best be described according to depositional energy and degree of salinity.

Box 7.1 Boxgrove (United Kingdom) – the marine and salt marsh sequence

The 500,000 year old early hominid and coastal site of Boxgrove, West Sussex, United Kingdom, is now located along a former coastline (palaeocoastline; Fig. 7.1a) that is 46 m above present sea level, and some 11 km inland (Roberts and Parfitt, 1999). Here, flint freshly weathered from the chalk cliffs and beach cobbles provided both cortical nodules for hammerstones and raw material for Lower Palaeolithic Acheulian hand axe production. Three interglacial marine cycles were identified at Boxgrove based upon beach sand, beach gravel, and cliff fall deposits, the last cycle being contemporary with early hominid occupation (Colcutt, 1999). In the fine sand-size (63–125 μm) beach sands (Unit III), iron stained root traces of terrestrial plants reflected periods of marine regression. The overlying lagoonal sediments (Slindon Silt; Unit IVb) are composed of both thick (10 mm) and thin (0.5 mm) sandy laminae that are interbedded with very thinly laminated (0.15–0.30 mm) mainly coarse silt (31–63 μm) and calcareous clay (Figs 7.1b,c). This bedding shows very little bioturbation, the reasons for which are still being studied. More important, however, is the evidence of in situ butchered large animal remains (e.g. horse) and hand axe manufacturing debris within these mudflat sediments (Figs 7.1d–e). The undisturbed nature of these artifacts within intact laminated sediments implies that butchery and artifact manufacture could have taken place over a very short space of time, possibly as little as a 4–12 h period of low tide when the mudflat was exposed. Fine sediments deposited during the ensuing high tide then sealed the scatter. In such ancient sediments as at Boxgrove, minerals framboids of black pyrite become oxidized into reddish iron oxide pseudomorphs of pyrite. Organic laminae have similarly been transformed into "ironpans."

The mudflat deposits marked the final marine regression at Boxgrove and a developing freshwater and terrestrial environment. Subaerial weathering included sediment ripening, the homogenization of sedimentary bedding, the removal of carbonates, structural formation, and activity by small mammals and soil invertebrates and this produced an Entisol, whose origin was very difficult to prove both in the field and in thin section because of post-depositional geogenic processes (Macphail, 1999). On the other hand, the important presence of shrews, moles, and voles (*Microtus*), the last aiding the biostratigraphic dating of the site, demonstrated that a terrestrial soil (Unit IVc), had formed. This soil formation was succeeded by a rise in ground water, peat formation, and accompanying terrestrial drainage, an environment where both amphibians and fish are recorded. This environmental change caused the Boxgrove palaeosol to become totally transformed, through Na^+ induced slaking and loss of iron through reduction (see Box 7.2). Iron is now only preserved as infrequent root mark pseudomorphs reflecting the presence of salt marsh plants. Most of the iron, in fact, became concentrated into a form of bog-iron, the multilayered salt marsh peat being transformed into a 20 mm thick iron and manganese pan (Unit Va). Most of the hominid activity was focused around a freshwater pond, where a tibia and two teeth were recovered (Roberts *et al.*, 1994).

(cont.)

Box 7.1 *(cont.)*

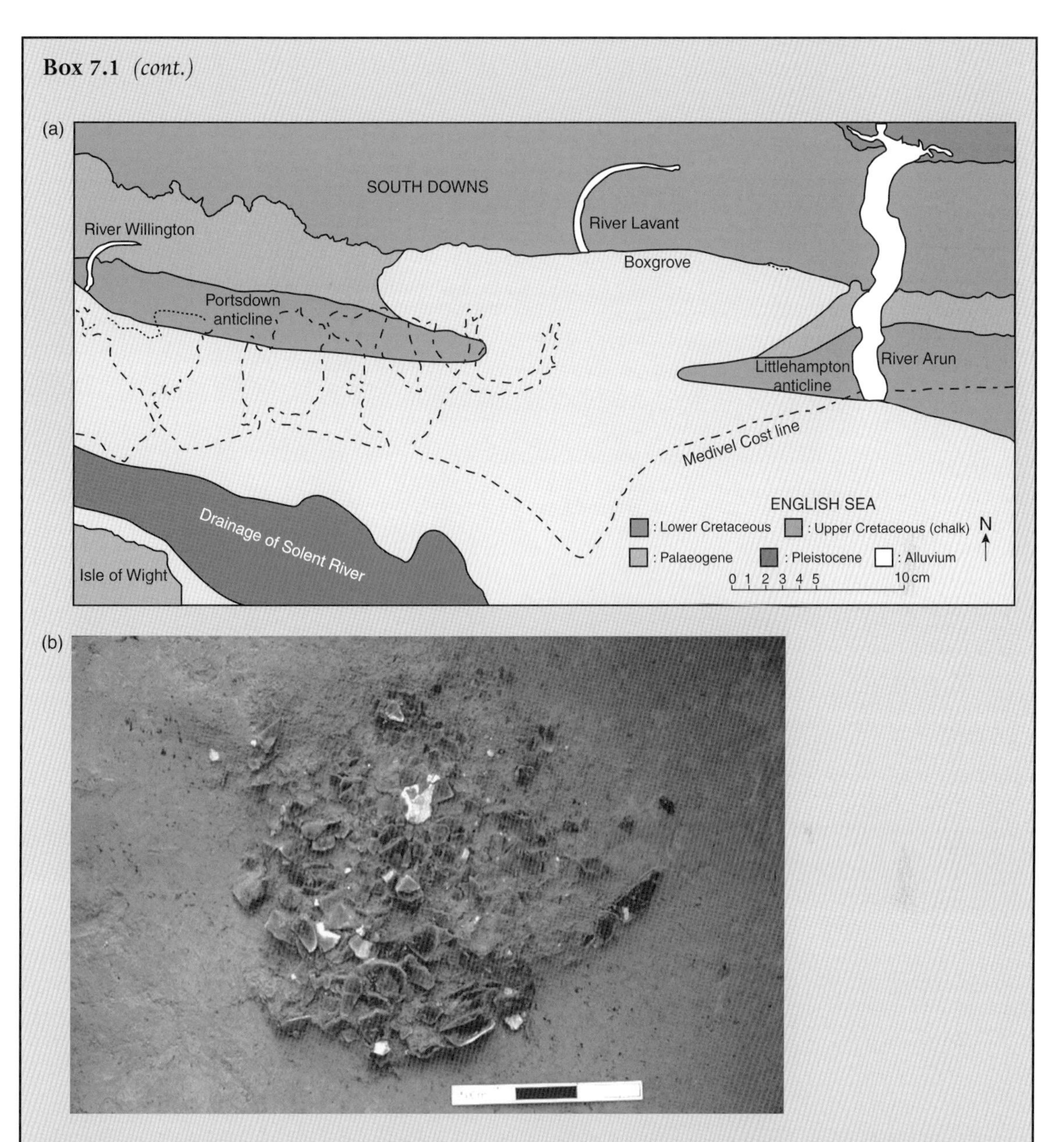

FIGURE 7.1 (a) Reconstruction of Middle Pleistocene paleo-coastline at Boxgrove (United Kingdom), with kind permission of Mark Roberts (Roberts *et al.*, forthcoming); (b) *in situ* flint scatter Unit IVb, mudflat and lagoonal deposits; (c) Unit III – fine beach sands with intercalated silty clay layers resulting from marine regression (plane polarized light [PPL], frame ~9 mm); (d) flint flake from the above scatter; flint is embedded in Unit IVb intercalated silts and chalky muds indicative of tidal sedimentation sealing the flint scatter (PPL, frame length ~5.5 mm); (e) As in Figure 7.1.2c, but under crossed polarized light [XPL] showing high interference colors of quartz silt and medium interference colors of chalky (calcitic) mud.

(cont.)

Box 7.1 *(cont.)*

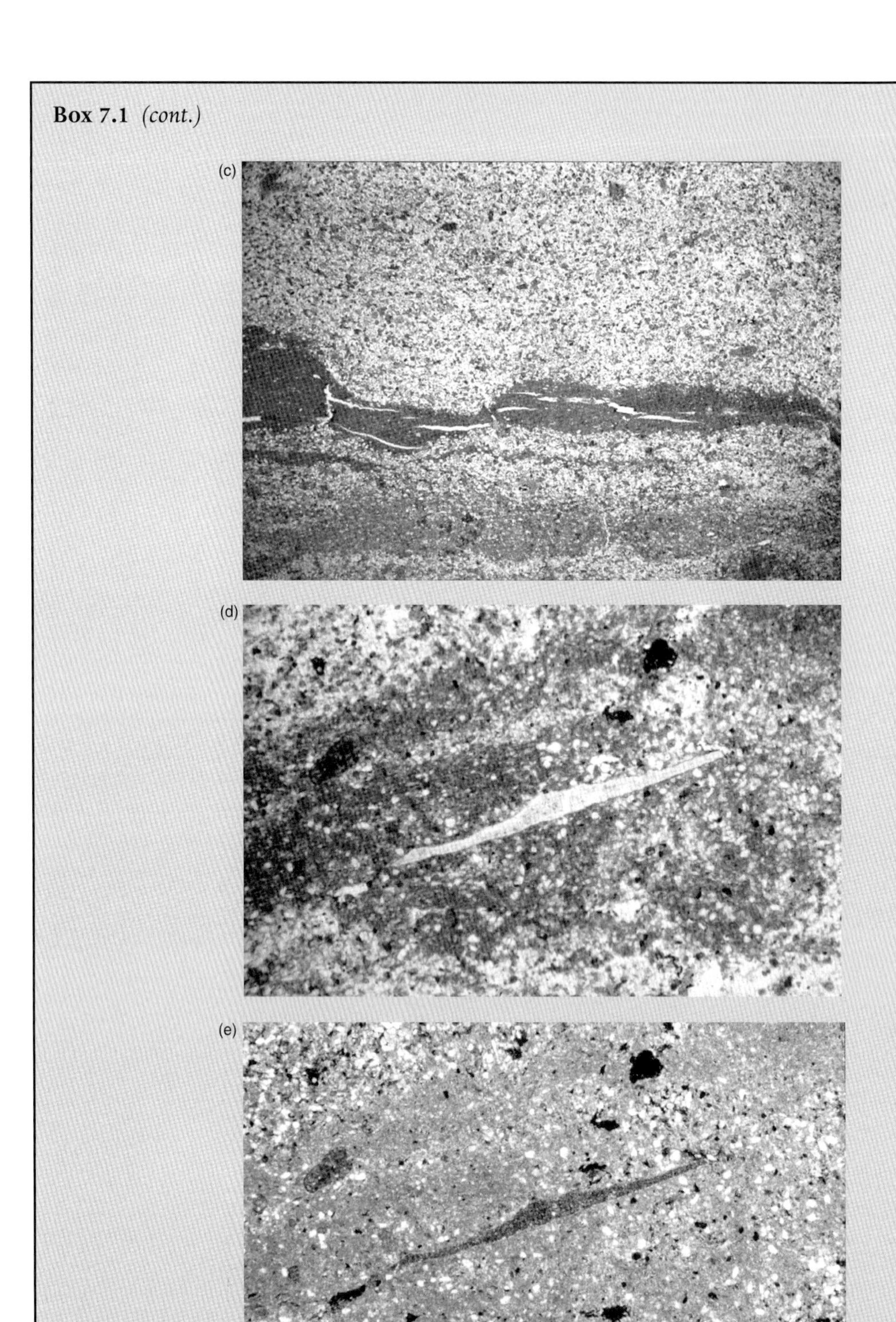

FIGURE 7.1 *(Contd.)*

For instance, the highest energy is recorded at the beach through wave action, whereas deltas and tidal estuaries are less saline than totally marine environments. Depositional energy and character are recorded, for example, in grain size and bedding type (see Chapter 1). Salinity is mainly studied through analysis of faunas, because sodium (Na^+) salts (halite) are extremely labile (mobile) materials, and rapidly leached from ancient sediments. Fauna are very sensitive to both salinity and sedimentary environment, and can include fish and mollusk, with microfossils such as diatoms, Foraminifera, and Ostracods, producing very sensitive datasets. Further, the remains of large marine animals, which range from turtles and seals to whales, can be found in coastal occupation sites; deposits in Norse Orkney, for example, linked coastal farming with the fishing industry (Linderholm, 2003; Simpson *et al.*, 2000, 2005). These have the potential of providing important cultural and environmental information.

Former shorelines are commonly preserved in Quaternary settings. They can be documented in a number of ways that include raised beaches, erosional notches (knickpoints) and platforms, as well as beach sands, gravels, and remains of marine fossils. Studies of raised beaches – especially those associated with older prehistoric remains – were the main method of investigating eustatic (worldwide) palaeo-sea levels, before developments in bathymetry (Zeuner, 1959). The rise of chronometric methods increased the spatio-temporal resolution of these changes. Investigations of coastal shell middens have shown that humans exploited various coastal niches. For example, gastropods (e.g. winkles and limpets) were collected along the high tide limit of rocky shores, whereas bivalves (e.g. clams) were gathered from lower energy environments of sandy beaches and mud flats, examples of which can be found in South Africa, California, and Gibraltar (Fa and Sheader, 2000). When such studies are undertaken, the changing morphology of coasts through time has to be considered. The rias (drowned valleys) of southwest England and Brittany formed as sea levels rose during the Early Holocene, with local affects sometimes being exacerbated by local tectonic activity. The coastal plain that had been present along the Italian Riviera, and exploited by Upper Palaeolithic and Neolithic populations, now no longer exists: the cave of Arene Candide that had been previously occupied by these populations, now lies 300 m directly above the Mediterranean Sea (Maggi, 1997). Similarly, land bridge avenues of migration of southeast Asia and the Alaskan arctic became fragmented into islands.

7.2.3 Coastal environments

Coasts can be divided into a series of environments that are important to the human use of shorelines and their immediate hinterland (Fig. 7.3). These are:

- the high energy cliff/beach zone
- the coastal dune area
- low energy estuarine mudflat and lagoonal environments and
- saltmarsh, mangrove, and other swampland.

7.2.3.1 The high energy cliff/beach zone: sediments and sedimentary structures

An idealized coastal sequence could comprise a gentle (ca 10°) seaward sloping wave-cut platform formed out of the local bedrock (Fig. 7.3). The high water mark is indicated by a sea cliff, similarly formed out of the local geology, and the presence of mollusks such as limpets. Marine deposits grade from:

- At and near the base of the cliff, meter-size beach boulders derived from the cliff
- Beach "gravels" composed of cobbles (>60 mm), pebbles (4–60 mm) and gravel

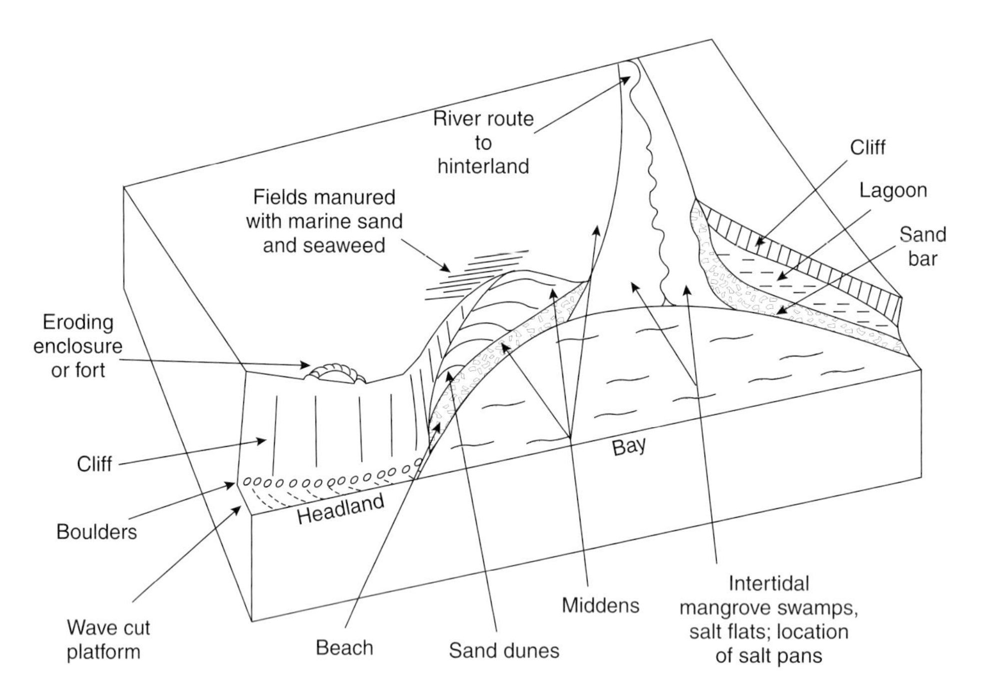

FIGURE 7.2 Diagram showing possible coastal zones – high energy cliff/beach zone, coastal dune area, low energy estuarine mudflat/saltflat, mangrove, lagoonal and saltmarsh areas, and the wide variety of possible archaeological features.

sensu stricto (2–4 mm), stretching some 50 m from the base of the cliff and

- Beach sands (50–2,000 μm) that can extend hundreds of meters seawards.

Coarse beach deposits are derived from the reworking of fluvial gravels, conglomerates, tills or from bedrock and materials eroded from coastal cliffs. The composition of ancient beach gravels (e.g. cobbles) at Boxgrove, United Kingdom, has been studied, in order to trace their origins (provenance) that include autocthonous nearby cliff fall versus local long shore movement, or distant deposition by glacial ice (Roberts and Parfitt, 1999). In addition, the shape characteristics (angularity/roundness; see Chapter 1) of beach gravels (e.g. subangular modal class) are generally different from those of coarse fluvial deposits that often include larger proportions of well-rounded clasts. As a caveat, however, it must be mentioned that some ancient beaches can contain large numbers of extremely well-rounded material, testifying to considerable high energy working (Bridgland, 1999). In any case, it should be remembered that although it has been claimed that beach pebbles may tend to become flattened in the swash zone (zone of back-and-forth movement of breaking waves), particle shape is strongly influenced by the lithological characteristics of the original material (Boggs, 2001; Pettijohn, 1975).

Beach sands that vary in grain size from 100 to 500 μm can display excellent sorting and are generally negatively skewed (skewness of >1); in the offshore direction where water deepens, sediments become finer grained and less well sorted (Reineck and Singh, 1986: 137, table 9). Individual sand grains also have distinctive surface features when observed under the SEM (e.g. Bull and Goldberg, 1985; see Chapter 1). A variety of bedding types may be observed within beach sands. These could include planar and cross-bedding, which may help to identify upper and lower beach facies, and direction of wave energy (Reineck and Singh, 1986). Such beach sands, when viewed in thin section have a massive structure, and a porosity consisting of packing voids (Fig. 7.1c). Beach sands are sometimes thrown up by storm events, and beach boulders may form anomalous inclusions through the impact of tsunamis in tectonically active areas, such as the Caribbean and Japan (Okinawa) (Pepe Orthiz, personal communication; Akira Matsui, personal communication). As a caveat, however, tropical storms (hurricanes and typhoons) can also readily shift and deposit coarse materials (Meredith Hardy, personal communication).

7.2.3.1.1 Fauna and resources

Some beach sands typically contain large numbers of faunal remains, termed bioclasts. When formed of mollusk (bivalve and gastropoda), and foraminifera shells, and ostracod valves that are all composed of calcite and aragonite, these sands are calcareous, and called bioclastic sand. Such fossils provide environmental information (e.g. sea temperature, energy, salinity) and give proxy dating, based upon biostratigraphy (e.g. microscopic nanofossils such as coccoliths), for the Quaternary. The analysis of coastal middens (see below), other open-air sites and caves shows that, although foodstuffs like limpets (Patella sp.) are less often eaten nowadays (Western Isles of Scotland, southern Italy), they were commonly consumed in earlier times, such as on the Neolithic Mediterranean coast (e.g. Arene Candide). Other edible shellfish of this environment include the gastropods whelks (e.g. Nucella sp.) and winkles (*Littorina* sp.), and the bivalve mussel (*Mytilus* sp.). Beaches are also the habitat of numerous types of edible bivalves, such as cockles (e.g. *Cerastoderma* sp.), razor shells (*Ensis* sp.), and clams. Shellfish were also used (Fenton, 1992) for fishing bait. Marine *Argopecten* and *Chione* shells at St. Elijo Lagoon, California provide much of the dating evidence for the ca 4,000–7,000 year old middens there (Byrd forthcoming). It should be noted that concentrations of mollusk shells are not always middens. Channels sometimes contain abundant shells, but these are death assemblages concentrated by flowing water, with shells sometimes displaying a preferred orientation. Equally, sand worms and sea birds are known to selectively concentrate shells, the last termed ornithogenetic shell concentrations (Balaam *et al.*, 1987 Reineck and Singh, 1986: 155;). Rocky shores and beaches are also foci of sea bird activity that may provide eggs, meat, and oil. The last is also obtainable from cetaceans (seals), and the exploitation of turtles would also yield meat and eggs. Cliff-dwelling birds would include raptors and scavengers, and bird bones concentrated. The last in their pellets should not be confused with that of birds consumed by humans.

7.2.3.1.2 Postdepositional changes

Relict beach deposits may be cemented to varying degrees. Such cementation can range from small amounts of microcrystalline secondary calcium carbonate (both calcite and aragonite) that coats packing voids or forms (micrite) bridges between particles to completely cemented sandstones (beachrock) (Box 7.1). The latter may form rather quickly, being penecontemporaneous with the accumulation of the sand. Where sediments are stained by secondary iron and manganese, this may highlight both bedding and bioturbation features produced by burrowing mollusks or root traces developed during ephemeral exposure to subaerial weathering (Colcutt, 1999).

7.2.3.2 Coastal dunes

On beaches, the sand located above sea level is exposed to wind action, and on stable shores, wide sand dune belts can form, and these are 4.5–9 km wide along Pacific coast from Oregon to California, United States. Coastal sand dunes also characterize Les Landes, the coastal area northwest of Bordeaux in France where large tracts of coniferous woodland have been inundated (FAO map soil type Calcaric Regosols on recent dune sand) (Reineck and Singh, 1986: 350). Coastal sand dunes can also accumulate against cliffs. One notable example is at Arene Candide, Liguria, Italy, where the cave was hidden by a sand dune, now quarried away, which was nearly 300 m high (Maggi, 1997; Figure 10.1a). Caves at Gibraltar are similarly partially infilled with dune sand (Stringer *et al.*, 2000).

Sand dunes normally show cross-bedding and are characterized by exceedingly well-rounded sand grains compared to beach sands and offshore beach deposits (e.g. roundness value of 0.35–0.4), and are only a little less rounded than terrestrial sand dunes (see Chapter 1). At Vanguard Cave, sands dip (slope) from the shore into the cave and are indicative of a coast characterized by sand dunes when sea level was lower between 25,000 and 100,000 years ago. Here sands are interdigitated with silts originating from phreatic cave activity. Mousterian combustion zones are found on both silty and sandy substrates, and at one combustion zone, sand dune burial was so rapid that the ash layer was perfectly preserved, despite wind activity commonly reworking ash crystals at such cave sites through deflation (Macphail and Goldberg, 2000). At the coastal site of Brean Down, Somerset, United Kingdom, both Quaternary and Holocene dune sands that had accumulated against a limestone promontory, were studied by Ian Cornwall and Richard Macphail (1990). Sand dune formation seems to have restarted during the Beaker period (ca 4,000 BP), burying a Beaker ploughsoil and infilling some possible ard marks in one location and inundating a probable grassland surface in another. Ensuing Bronze Age settlements occur between sand blow phases, which seem to mark marine regression affecting the local estuary of the

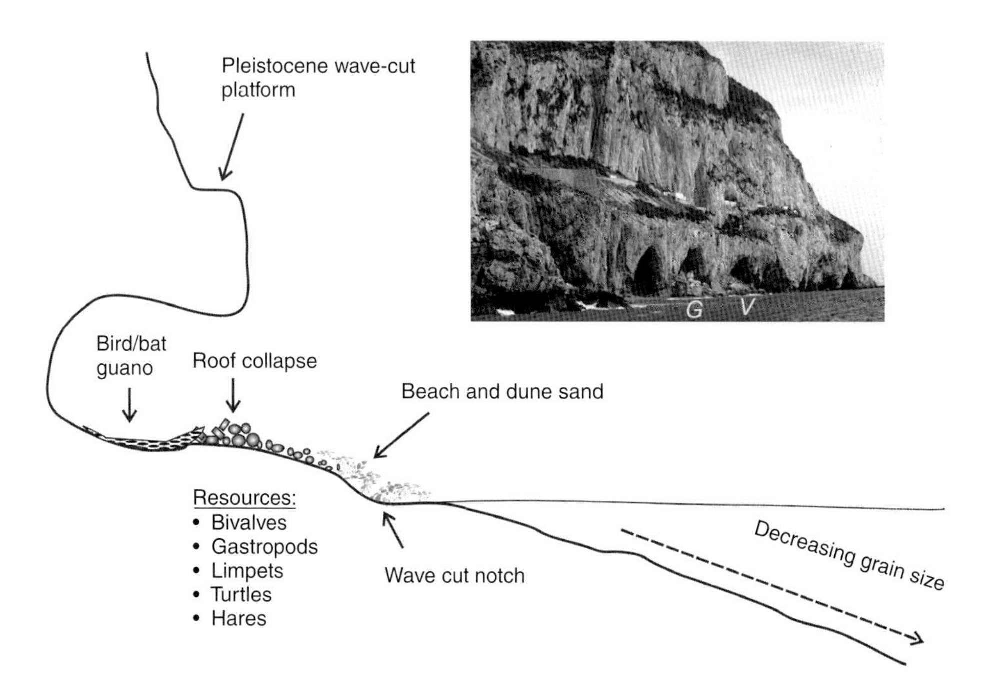

FIGURE 7.3 Section across cliff/beach zone showing sediment lithology, features, and resources; Gibraltar Caves – Middle and Upper Palaeolithic Gorham's Cave (G) and Middle Palaeolithic Vanguard Cave (V).

River Severn (Bell, 1990). *In situ* combustion zones are characterized by stratified sequence of wood charcoal and wood ash crystals, whereas locally reworked combustion zones are composed of blown sand, ash crystals, and charcoal fragments.

Thus dune sand deposits, although primarily composed of resistant minerals such as quartz, can include local material such as ash, and sand-size bone and mineralized coprolites – as at the Gibraltar caves. It is also noticeable that if charcoal is present it is also rounded, but often with a larger grain size compared to the sand matrix; wind being able to transport larger pieces of charcoal compared to sand because of its lower specific gravity. Chemical studies of cave deposits at Gorham's and Vanguard Caves show that loss on ignition (organic matter estimate) and phosphate content of blown sand can be enhanced by organic matter (e.g. 0.1–2.5% LOI) and by fine fragments of bone and guano or coprolite (410–2300 ppm P), but still contain less organic matter and very much less phosphate compared to some guano layers (3.1–5.9% LOI; 18,500–72,100 ppm P). Commonly dune sands include fine shell fragments (e.g. mollusks and foraminifera) and other marine carbonate, typically giving them an alkaline pH in an unweathered state (e.g. max. pH 8.7 at Vanguard Cave).

7.2.3.2.1 Postdepositional changes and soil formation in coastal dune sands

Like beach sands, dune sands can become cemented by iron or calcium carbonate according to geogenic conditions, for example by groundwater. Sand dunes may be stabilized by vegetation, which is often nowadays purposely carried out using marram grass (*Ammophila arenaria*), blown sand losing velocity in grass and becoming deposited (Fig. 7.4). Fungal hyphae and some lichens are also able to begin the stabilization process of sand dunes, and surface living soil invertebrates begin to inhabit the litter layer; the 1.5–3.5 cm thick humic AC horizon can develop over a period of 13 years. This has been shown by analogous studies of truncated acid sands (Mücher, 1997). Calcareous sands are more fertile and likely to undergo a more rapid soil development, but this is accompanied by decalcification. Various workers agree to a rate of 0.03–0.04% removal of calcium carbonate

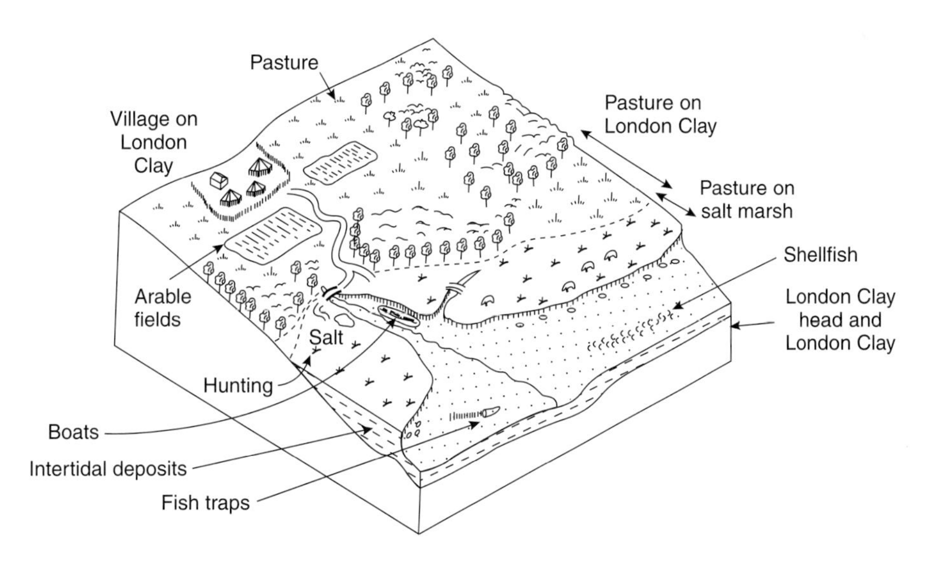

FIGURE 7.4 Estuarine environment and landuse – Later Bronze and Earlier Iron Age Essex Coast (Wilkinson and Murphy, 1995, figure 132); a boat and human and animal footprints were found in the intertidal deposits at Goldcliff, Gwent, Wales, United Kingdom (Bell *et al.*, 2000).

per year, but decalcification has been shown to be favored by low temperature and high percolation rates, as for example, during the early Post Glacial period in present-day temperate regions, whereas later Holocene warmer temperatures caused this rate of dissolution to decrease (Catt, 1986; van der Meer, 1982). In areas such as Romney Marsh, Kent, United Kingdom, where there is a prograding (expanding) shoreline, typical sand-pararendzinas (e.g. Typic Udipsamments, Calcaric Arenosols) occur on recent sand dunes with a sparse cover of Marram grass, whereas on the older dunes inland, weathering has induced the formation of sand-rankers (Quartzipsamment; Arenosol) and these have an acid flora that includes gorse (*Ulex europaeus*)(Avery, 1990; Jarvis *et al.*, 1984). In the latter soils the calcium carbonate content was reduced to 0–0.4%, with the original alkaline pHs being lowered to pH 3.9–5.8, possibly since medieval times.

7.2.3.3 Low energy estuarine mudflat and lagoonal environments

Fine sediments composed of silt and clay are deposited in protected coastal environments where there are offshore bars (e.g. lagoons), and in and near estuaries (Figs 7.4 and 7.5). Such environments have proven to be rich in archaeological remains because incremental Holocene sea level rise buried sites under marine and intertidal muds (Box 7.2). In addition localized marine regression (retreat) and transgression (advance) led to changes in morphology of coastlines that were often utilized by humans. Modern examples of such coastlines (Reineck and Singh, 1986) show a series of sand-flat and mud-flat deposits, with, at the top of the shoreward sequence, salt marsh deposits (see below), and these were variously utilized as a resource base from prehistory to the modern period (Bell *et al.*, 2000; Wilkinson and Murphy, 1986).

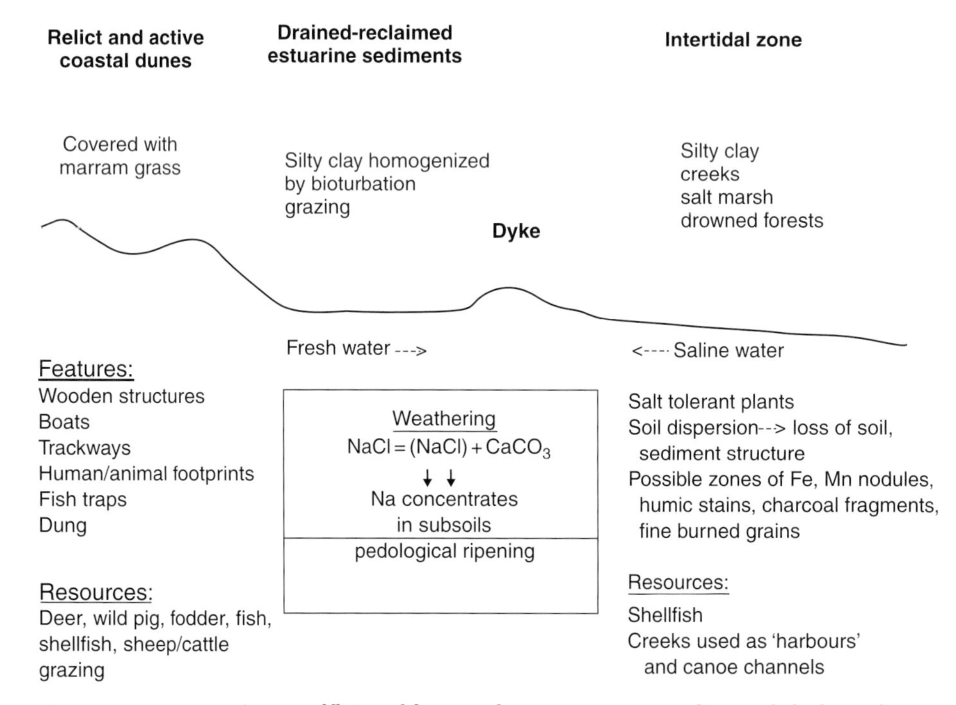

FIGURE 7.5 Low energy estuarine mudflat and lagoonal environments: sediment lithology, features, and resources.

7.2.3.3.1 Intertidal sediments and sedimentary structures

Horizontal, thinly laminated fine deposits that result from tidal deposition, become dominant in contrast to the massive and sometimes cross-bedded sands of beach environments. Lagoonal and mudflat deposits can include coarse (fine sand) as well as fine (silt and clay) laminae, with some laminae rich in detrital organic matter. An example of a marine beach – intertidal mudflat/lagoonal succession is given from the Lower Palaeolithic site of Boxgrove, United Kingdom (Box 7.1). Such lagoonal/estuarine sediments have been found sealing old terrestrial landscapes, where they reflect sea level rise during the Early Holocene. Coastal peats, the tree-stump remains of forests, Mesolithic middens, hearths and Neolithic to Iron Age occupations, have all been found along the present intertidal zone of southern England. Extremely well-preserved Meosolithic forests are reported from the Danish Baltic (Skaarup and Grøn, 2004). Holocene inundation and sea level rise is similarly recorded by drill cores from the Yangtze delta, in China, where it is argued that such environmental stress around 4,000 years ago caused a major cultural discontinuity (Stanley *et al.*, 1999).

One of the first geoarchaeological and environmental studies of a coastal midden was carried out at Westward Ho!, Devon, United Kingdom, where the late Mesolithic site (ca 6,000 BP) at −2.20 m OD only became exposed when storm scouring of the modern sandy beach coincided with spring tides, roughly once every one or two years (Balaam *et al.*, 1987) (see Fig. 7.6). The site is located near the confluence of the rivers Taw and Torridge. Here, marine "blue clay" (Munsell color 5Y5/1) up to 0.9 m thick occurs over an eroded terrestrial substrate composed of fine clay loamy Pleistocene (Ipswichian?) alluvium (29% clay, 34% silt, 38% sand) that is characterized by polygonal patterning (periglacial patterned ground). In contrast, the overlying "blue clay," a fine silty clay loam (~32% clay, ~52% silt, ~16% sand), is massive and strongly influenced by pyrite (3.2%; point counting), which produces pHs as low as pH 2.6 (cf. acid sulphate soils). Only 1.4% void space was recorded, along with very small amounts of detrital amorphous organic fragments (2.8%; 0.49% organic C). Archaeomagnetic dating of this "blue clay" gave it a date in the region of ca 8400–7800 BP. The site was then affected by marine "regression," but possibly this was only a local phenomenon, and as suggested by Kidson's work in the Bristol Channel, influenced by such factors as changes to the sedimentation rate and/or coastal geomorphology (Kidson, 1982). At any rate the "blue clay" was occupied by late Mesolithic humans (see below), and ensuing freshwater/terrestrial conditions led to the formation of an "Atlantic" woodland peat, characterized by the pollen of lime, oak, elm, ash, hazel, and willow trees and an associated woodland insect and land snail fauna (Balaam *et al.*, 1987). An upper "blue clay" (36% clay, 42% silt, 22% sand) sealed this peat after ca 5,000 BP. Fish, crustacea, and mollusks such as clams and oysters are typical resources of the intertidal zone.

7.2.3.3.2 Postdepositional changes to intertidal sediments

Massive silty clay deposits and laminated silts are susceptible to bioturbation by plants and animals, or to disruption by mud crack formation and the development of crystals of salt (halite) and gypsum along with other secondary minerals such as trona (sodium bicarbonate $Na_3(HCO_3)(CO_3) \cdot 2 \cdot H_2O$), siderite ($FeCO_3$), and pyrite ($FeS_2$) (Kooistra, 1978; Miedema *et al.*, 1974; Reineck and Singh, 1986). All these are typically found in such environments. Much archaeological work has been carried out on the Thames foreshore in London (Sidell, 2003). Indeed, the mortar floor and its "foundations" of slag and charcoal at the Shakespearean Rose Theatre rest on fine intertidal deposits that contain organic fragments and secondary pyrite and gypsum (Macphail and Goldberg, 1995). When the

Tower of London Moat deposits were investigated and interpreted it had to be remembered that there were both inputs from the London Clay substrate, which had been excavated for Henry III's constructions, and the River Thames itself (Keevill, 2004). The London Clay supplied relict (Tertiary Period) pyrite whereas the Thames estuarine sediments themselves contained contemporary sulphate that also produced secondary gypsum crystals and pyrite (Macphail and Crowther, 2004). The heavy metal content, however, was mainly associated with Medieval and post Medieval contamination associated with cess and industrial activity. At Westward Ho! and on the Essex coast the presence of monocotyledonous plants are also recorded. The many creeks and first order channels that often drain such environments produce gullies, leading to the development of cut-and-fill features within the sedimentary sequence. These features may contain death assemblages and also be the locations where organic artifacts such as fish traps and even boats are preserved; the footprints of Mesolithic people have also been found (Bell et al., 2000).

When terrestrial soils are affected by saline water and estuarine sediments, some marked changes occur. The primary transformation is the slaking of fine soil, which is produced by Na^+ ions causing the collapse of soil structures and the dispersal of fine soil down profile (Macphail, 1994). Arti- and ecofact distributions are unaffected by this pedological process, however, and remain as well-sealed *in situ* material (Box 7.2). A model of the effects of drainage on lagoonal deposits have been monitored from Dutch polders, where the tree growth, invertebrate, and small mammal activity produced "zoologically ripened" Entisols 6 to 12 cm in thickness over a 10 year period (Bal, 1982).

7.2.3.4 Salt marsh, mangrove, and other swamplands

Along low-lying coastlines, saltmarsh occurs above the tidal mudflats of western Europe (Fig. 7.2). This is the equivalent to areas of mangrove swamp in the tropics and subtropics. Swamplands with mangrove or the swamp or bald Cypress (*Taxodium distichum*), as found for instance, along the margins of the coastal plain in the southeast United States, were occupied by hunter–gatherers, including more recently the Seminole. Such environments are also found in Mesoamerica and are the location of Mayan "raised fields" (Jacob, 1995). Salt marshes have also provided important grazing land.

7.2.3.4.1 Nature of deposits

The deposits of the salt marsh and other swamplands are very similar to those of mudflat and lagoonal environments, but have been variously affected by subaerial weathering, surface and channel water flow, and biological activity. Different marine animals exploit sand and clay (mud) flat niches, and most salt marshes display a succession of less salt-tolerant plants landward. This is because the landscape receives freshwater from the landward side, but is influenced daily by saline water, that at maximum high tide may well flood the whole area. This explains the practice of building sea defenses (e.g. dykes) along the coast of the North Sea in Europe at least from the Medieval period onward (Figs 7.4–7.5).

Commonly relict intertidal deposits retain some sedimentary elements, such as laminae, but because of biological activity, some or total homogenization takes place. Subaerial weathering often removes some of the salt (NaCl) and calcium carbonate ($CaCO_3$) from the upper ripened horizon of the soil/sediment (Fig. 7.5). This produces calcareous and noncalcareous pelo-alluvial gley soils (Vertic Fluvaquent; Eutric Fluvisol; Avery, 1990). These soils have a saturation extract conductivity ($S\,m^{-1}$) increasing from 0.16 (0–5 cm) to 0.47 (98–120 cm), and show for example an exchangeable %Na increase from 7.4–10.6% (0–15 cm) to 22.7–30.2% (70–85 cm) (Hazelden *et al.*, 1987; Jarvis *et al.*, 1984). These workers also found that microrelief has an

Box 7.2 Drowned coasts of Essex and the River Severn, United Kingdom

Recent coastal erosion and industrial development projects in the United Kingdom have been surprisingly effective in uncovering artifacts, features, and land surfaces that became drowned as sea levels rose during the Holocene, a well-publicized example being "Sea Henge," a circle of well-preserved large wooden posts at Holme-Next-The-Sea on the Norfolk coast. In fact, submerged wetland landscapes have been surveyed in much detail in the United Kingdom, and include the Humber Wetlands, the Lincolnshire and Cambridgeshire Fenlands, the drowned coastlines of Essex and Thames estuary, the Somerset Levels, the Severn estuary (e.g. Goldcliff on the Welsh side), and North West Wetlands (French, 2003; Wilkinson and Murphy, 1986, 1995). The first evidence of this phenomenon, in the form of submerged forests, was noted in the eighteenth century, although historically recorded from Wales in ad 1170 by Giraldus Cambrensis (Bell *et al.*, 2000: 3). Major coastal survey in Essex (the Hullbridge Survey) began in the 1980s (Wilkinson and Murphy, 1986). Much of the Baltic and the North Sea between England and Europe was dry land during the Early Holocene, but as the ice sheets melted, a sea level that was some 45 m below present mean sea level (9,500–9,000 BP) began to rise rapidly at around 2m a century (Devoy, 1982; Wilkinson and Murphy, 1995)(Ole Grøn, personal communication). Estimates of the rising sea level at the River Severn site of Goldcliff, Gwent, have been calculated at this time at 16 mm per year (i.e. 1.6 m/century), diminishing to 8.5 mm per year by ca 6,600 BP. The earliest excavated occupations that have been found date to the Later Mesolithic, and include a hearth and footprints. An important Mesolithic midden was also found at Westward Ho!, Devon, approximately 100 km to the southwest of Goldcliff (Balaam *et al.*, 1987). Whereas the Neolithic (The "Stumble" and other River Blackwater sites) and Bronze Age are well represented on the Essex Coast, the next major period of occupation at Goldcliff is the Iron Age (Bell *et al.*, 2000). Typically, at sites in both Essex and Gwent, occupations mainly occur in loamy "soil" substrates formed in sand and clay-rich Pleistocene Head deposits. They are sealed by the lowermost silt-rich clay deposits that record estuarine inundation of the sites.

At Goldcliff, the Mesolithic population occupied the site during a period of marine regression (from around 5,600 cal. bc), and probably hunted wild pig, deer, waterfowl, and trapped fish (Bell *et al.*, 2000). The geoarchaeological study of the bone and charcoal-rich Mesolithic hearth site included a loss-on-ignition (LOI), phosphate and magnetic susceptibility (χ) survey (Crowther, 2000) and soil micromorphology (Macphail and Cruise, 2000). Although most phosphate concentrations were low (P, <0.400 mg g^{-1}), around background levels, there were localized patches of high concentrations (P, >2.00 mg g^{-1}; max. 3.46 mg g^{-1}); and there is a highly significant statistical correlation between P and LOI for the occupation as a whole. On the other hand, the magnetic susceptibility values (χ and χ_{max}) (see Chapter 16) are low, and there is no correspondence with the concentrations of charcoal that were believed, at the time, to represent *in situ* burning. The main cause of these results from magnetic susceptibility could either be the poor correlation of the χ survey and the mapped charcoal scatter, or simply the marked postdepositional effect of gleying and associated low concentrations of iron (Crowther, 2000). The soil micromorphology investigations here and in

(cont.)

Box 7.2 *(cont.)*

Essex may also help understand these findings. The Mesolithic hearth at Goldcliff is represented by compact dark reddish brown soil that includes charcoal and burned bone, and although biological working fragmented this hearth/old land surface, its burned state allowed it to resist many of the effects induced by marine inundation. On the other hand, charcoal can also be found both embedded across the boundary between the old ground surface and the estuarine clay, and at the base of this layer. This phenomenon, which was also recorded at The Stumble in Essex, demonstrates the lateral mobility of charcoal when sites become inundated. The present distribution of charcoal at such sites does not therefore necessarily correspond to its original pattern.

Once-terrestrial soils are normally completely transformed by increasing site wetness and marine inundation. This phenomenon is recorded both in the geological record (Retallack, 2001), and locally in the Fenlands and Essex coastline of the United Kingdom (French, 2003). At a series of Neolithic coastal sites in Essex, both inundation (Fig. 2.7) and regression are recorded. For example, at Purfleet on the River Thames (London), an estuarine (diatoms) clay loam accumulated that related to the Thames II transgression of ca 6,500–5,400 BP, and a Neolithic woodland soil developed in this during the Tilbury III regression (ca 4,930–3,850 BP)(Wilkinson and Murphy, 1995). The last was evidenced by a small number of finds, including polished axes, a woodland land snail fauna that had developed over some centuries, and a microfabric relict of a possible woodland topsoil. Biological activity removed any evidence of original sedimentation. Subsequently (after 3,910 BP) the site became wetter and was sealed by a 0.77 m thick freshwater wood peat containing ash, alder, and yew trees. The buried soil microfabric of organo-mineral excrements, thin (<1 mm) and very broad (3–5 mm) burrows, and woody roots, is presently partially preserved by iron stained sodium carbonate and micritic calcium carbonate (14.0% carbonate) and typical minerals of estuarine deposits, such as pyrite are present (Macphail, 1994). On the "estuaries" of the rivers Blackwater and Crouch presumed argillic brown earth soils formed (in Head of London Clay origin; Fig. 7.4) under an oak and hazel dominated woodland (pollen evidence) (Wilkinson and Murphy, 1995). This woodland was probably influenced by local clearance around occupations such as at Neolithic site of The Stumble, where flints, pottery, burned flint, and charcoal of cereals and wild plant food were found. The overlying massive estuarine clay loam that accumulated after 3,850 BP, exhibits coarse vertical plant root channels, and contain few horizontally oriented planar voids relict of decayed detrital plant matter. It was found that at the five sites investigated, the underlying 100 mm thick "topsoil" is composed of very pale massive silt loam (11% clay, 76% silt, 12–13% sand) with a very low porosity (5–10% voids). In marked contrast the underlying "subsoil" is a silty clay loam and clay loam (e.g. 21–34% clay, 27–69% silt, 10–39% sand), with very abundant finely dusty, and moderately to poorly oriented clay void coatings and infills. It is quite clear that these textural pedofeatures owe nothing to an earlier and assumed terrestrial woodland soil here, but rather totally reflect a transformation induced by saline water inundation. Here, Na^+ induced complete slaking (soil dispersion) of the Neolithic topsoil, as in saline soils generally (see Chapter 3). The current presence of pyrite

(cont.)

Box 7.2 *(cont.)*

framboids and gypsum also mirror this transformation into estuarine soils (Saline and Acid-sulphate alluvial gley soils; Wallasea 1 soil association) (Avery, 1990; Jarvis *et al.*, 1984). This has meant that most of the clay has been dispersed down-profile, whereas archaeological materials such as pottery and coarse charcoal were unaffected and remain at or near the surface of the once-biologically worked Neolithic soil. Finally, at one location, now sealed by peat, inundation led to "waves" of humic matter being washed down-profile into the newly formed massive "topsoil." These formed pseudo-laminae that could be mistaken for relic sedimentary bedding (Lewis *et al.*, 1992).

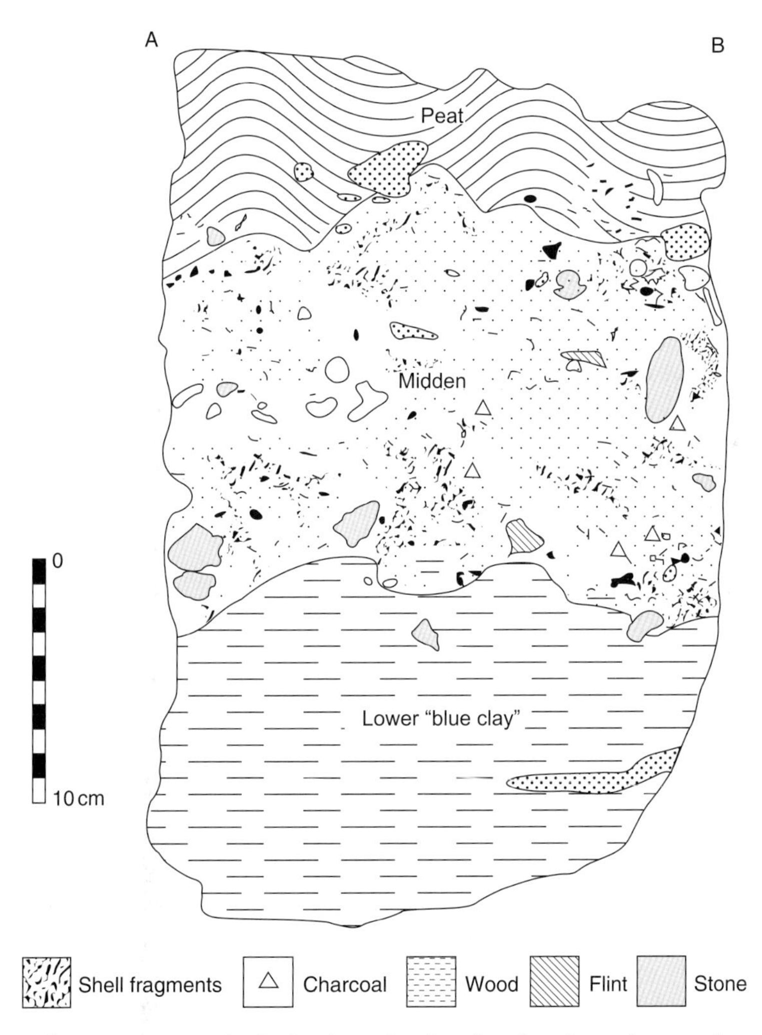

FIGURE 7.6. Westward Ho!, Devon, United Kingdom. Section drawing through estuarine sediments and Mesolithic midden (modified from Balaam *et al.*, 1987).

affect on Na content of topsoils, with pools and creek sites containing more Na (12–13 % exchangeable Na) compared to neutral ground and levees (9% exchangeable Na). They also calculated a dispersion ratio when investigating structural stability of the soils, based upon percentage exchangeable Na, organic C and $CaCO_3$, with organic matter and calcium carbonate promoting better structured well-flocculated soils and Na promoting dispersion (see Chapter 3; Box 7.2). As noted above, when Early Holocene soils became inundated they became deflocculated ("dispersed"). Near-shore terrestrial soils may also be influenced by Na from sea spray, leading to accelerated down-profile clay translocation (V. Holiday, personal communication).

Bronze and Iron Age salt marsh deposits have been studied from the Severn Estuary (between South Wales and England, near Bristol). It was found that magnetic susceptibility measurements reflect only amounts of burned soil present because the iron minerals that contribute to "topsoil" magnetic susceptibility enhancement have been transformed, or the iron forming them removed, by gleying (Crowther, 2000, 2003). A common soil micromorphological phenomenon is the presence of intercalatious and other textural features that infill relict void space between now-poorly preserved peds. This is again the result of deflocculation of ripened "soil" surfaces. Voids tend to be either vesicles (complete water saturated collapse) or closed vughs (massive inwash of fines). Also channels and burrows can become infilled with silty estuarine material. Organic matter from the ripening "humic" Ah horizon, because of postdepositional oxidation–reduction is often only represented by iron and manganese staining, although this sometimes may be as pseudomorphs. The effect of inundation and renewed sedimentation is less clear when salt marsh deposits are flooded as compared to terrestrial soils (Box 7.2), because the ripened sediment is very similar to the new deposit, and deflocculation takes place masking the boundary. Often an old land surface may only be vaguely identified in the field through a concentrated horizontal zone of iron and manganese nodules (Allen *et al.*, 2002). It then takes soil micromorphology to recognize a buried layer more precisely.

7.2.3.5 Intertidal, salt marsh, and swampland resources

Many kinds of fish, shellfish, amphibia (turtles), and game were exploited from the intertidal zone. Late Mesolithic occupations in the United Kingdom show that deer and wild pig were hunted. In later prehistory to the modern period, an additional important land use was grazing (Wilkinson and Murphy, 1995: 220) (Fig. 7.4). It is likely that both sheep and cattle were grazed on salt marshes, with the exploitation of cattle that were stabled on the salt marsh itself, being indicated by cattle bones, cattle lice, and cattle foot prints at Goldcliff (Bell *et al.*, 2000). The intertidal deposits studied at this site showed that creeks (palaeochannels; see Chapter 5) were used as tracks, as further suggested by the presence of fragments of dung. Hoof prints show slow inwash of muddy sediments, and palaeochannel infills demonstrate how it is likely that cattle were involved in eroding the contemporary peat surface. The salt marsh became dry enough at times for heather to grow on the ancient sphagnum peat surface, and it could be inferred from the presence of *in situ* charred heather roots that the pasture was managed by fire. Such locations are also essential for communications by coastwise traffic, and boats were used to move cattle and sheep.

7.2.3.6 Middens

Anthropogenic deposits are found in coastal sites throughout the world, reflecting the many activities associated with the sea for tens of thousands of years. Among the earlier occurrences of fish and shellfish exploitation are Palaeolithic cave sites in the Mediterranean

FIGURE 7.7. Estuarine inundation silts burying prehistoric old ground surface (Kubiena box sample) at the Blackwater estuary, Essex, United Kingdom.

and Middle Stone Age coastal sites in South Africa. Gorham's and Vanguard Caves, Gibraltar contain numerous shellfish remains in spite of the fact that the sea was over 4 km away. A similar setting and type of exploitation was noted for Die Kelders Cave (South Africa) where numerous shell fish remains survived extensive diagenesis after having been taken from the sea, which was likely tens of kilometers distant at that time (Goldberg, 2000).

Middens form an archetypal coastal occupation deposit that is found worldwide. The northwestern coast of North America contains numerous shell middens that have been painstakingly investigated by Stein and her team (Stein, 1992). Their work sets the geoarchaeological standard for shell midden study. More recently soil micromorphological studies of Californian shell middens have shown the importance of identifying the effects of bioturbation and weathering on these deposits (Goldberg and Macphail, forthcoming). Food waste such as shell and bone, hearth debris (charcoal and ash), and contemporary environmental conditions (e.g. recognized from pedofeature analysis), can all be differentiated from "simple" postdepositional events. For example, scavengers comminute bone and leave coprolitic material.

Similarly in the 1980s at Westward Ho!, detailed analysis of a midden were carried out (Balaam *et al.*, 1987). Some characteristics can be noted. The midden is both more poorly sorted and coarse (14% clay, 37% silt, 48% sand) compared to the underlying more clay and silt-rich "blue clay" (see above), although this mineral material is of alluvial/fen carr origin (see Fig. 7.6). In addition it is more organic (9.2% Organic C) and contains, 2.7% rock fragments and 18.7% shell inclusions (point counting). Despite being located in wetland (land snails present), this midden was sufficiently bioturbated so as not to reveal any archaeological stratigraphy. On the other hand, fine mineral and organic laminae were recognized under the microscope showing how the midden built up alongside alluvial and peat forming sedimentation.

A major activity recorded in the "red mounds" of Eastern England is salt making, in which the red color comes from rubefied materials such as briquetage (Wilkinson and Murphy, 1995). At another salt making site, Brean Down, Somerset, Bronze Age huts were located in a coastal environment dominated by blown, medium size sand, and this ubiquitous material is found intercalated between hearth layers. Round house floors, however, were constructed of calcareous estuarine silts, that were humic and contained marine diatoms (Bell, 1990). Similarly, the early medieval site of Yarmouth, Norfolk (United Kingdom) includes structures where ground raising and floors have been made out of estuarine silts, and occupation deposits within these structures contain fish bone; Yarmouth developed into an important Medieval port. In Orkney, other Northern Isles, and Scandinavia, midden deposits contain fish bone and phosphatic fish bone residues that reflect the industrial scale of fishing in medieval times (Simpson *et al.*, 2000). Seaweed has been a traditional ingredient for manuring to aid the formation of "cultosols" such as plaggen soils, for example, in Ireland (Conry, 1971). The possibility that seaweed was utilized as a manure in inland Sussex, United Kingdom, during the Bronze Age, was suggested by Bell (1981) from the presence of small marine shells (Fig. 7.2). Similarly, links have been made between humans and the sea from the identification of chemical elements and isotopes ($^{12}C/^{13}C$) in human bones that indicate a diet rich in marine resources (Reitz and Wing, 1999: 247 ff.).

7.3 Conclusions

The influence of coasts on humans can therefore be detected down to the elemental level. Many sites from coastal areas have been preserved, and it has been important to understand the processes that form coastal sediments in order that such brief occupation episodes as that recorded in Unit IVb at Boxgrove are not misinterpreted as terrestrial land surfaces. Equally, the correct identification of estuarine-sealed terrestrial landscapes in the present-day intertidal zone has been significant. Lastly, the exploitation of marine materials inland, such as salt, fish, and seaweed, has been recognized. Finally, working along shorelines has its own peculiar dangers, for example, from rock fall to rapidly rising tidewater (Bell *et al.*, 2000).

8

Caves and rockshelters

8.1 Introduction

One of the most fascinating of geoarchaeological environments, and one that has attracted the attention of professional and amateur archaeologists and the average person on the street is that of caves and rockshelters. Much of prehistory – particularly in western Europe – began in caves and rockshelters, whose form and contents can be readily seen in the landscape. Thus, because caves provide a "protected" environment in landscapes, which through time have often become lost through erosion, we see that an apparent disproportionate amount of information about human evolution, art, lithic traditions, and faunal exploitation are evidenced from famous cave sites. Some examples are, Zhoukoudian (China), numerous sites in France (too numerous to list, but Lascaux, Combe Grenal, Pech de l'Azé, le Moustier, La Micoque, La Quina, Le Ferrassie, Arcy-sur-Cure are prominent) and the Middle East, including Tabun, Amud, Kebara, Hayonim in Israel, Shanidar in Kurdistan, and Yabrud in Syria (Table 8.1).

Other than being protected and thus attractive places to use and live in, caves and rockshelters constitute intriguing depositional environments, especially from the geoarchaeological point of view. Typically, materials that accumulate in them, both geogenic and anthropogenic, are inclined to stay there. These accumulations, however, are commonly modified by physical and chemical syn-depositional (contemporaneous) and postdepositional processes, that are both of human and "natural" origin. Thus, unlike open-air sites that are typically subjected to higher energies (e.g. from streams), which can result in complete or partial removal of evidence of former occupations, caves commonly tender the opportunity to form and preserve thicker, more widespread, and richer archaeological sediments. The commonly anthropogenic nature of cave sediments (e.g. ashes, hearths, organic-rich remains and guano; cf. Chapter 10), therefore can provide refined insights into past human activities and behavior. This chapter examines cave and rockshelter formation, their sediments, human activities recorded in them, and how their sediments have been used to reconstruct past environments.

8.2 Formation of caves and rockshelters

We distinguish here rockshelters from caves, since their differences in form also condition the types of depositional and postdepositional processes that are recorded. Rockshelters are typically wider than they are deep and have an overhang. They can develop in limestone, as in southwestern France (Laville *et al.*, 1980), or in clastic rocks as in much of the southeastern

Table 8.1 Important caves and rockshelters discussed in the book: their location, ages, and significance

Site	Location	Time period	Significance	Selected references
Zhoukoudian, Locality 1	China	LP	Thought to have among the earliest uses of controlled fire	Goldberg *et al.*, 2001; Jia, 1999; Jia and Huang, 1990; Liu, 1985, 1987; Qian *et al.*, 1982; Teilhard and Pei, 1934; Teilhard, 1941; Weiner *et al.*, 1998, 2000; Xu *et al.*, 1996
Lascaux	France	UP	Noted for its spectacular wall paintings	Knecht *et al.*, 1993; Leroi-Gourhan, 1987; Leroi-Gourhan and Allain, 1979
Combe Grenal	France	MP	Thick sequence of deposits; the virtual type section of the Middle Palaeolithic in southwestern France	Beyries, 1988; Bordes, 1972; Bordes *et al.*, 1966; Chase, 1986
Pech de l'Azé	France	MP	Three collapsed caves/rock shelters in the same area of the Dordogne with abundant Middle Palaeolithic lithics, fauna, and geological studies	Bordes, 1954, 1955, 1975; Goldberg, 1979a; Laville *et al.*, 1980; Schwarcz and Blackwell, 1983
Le Moustier	France	MP	The type site for the Mousterian and a classic rockshelter filled with *éboulis* and lithics	Peyrony, 1930; Valladas *et al.*, 1986
La Micoque	France	MP	The type site of the Micoquian Industry	
Arcy-sur-Cure	France	MP; UP	Middle/Upper Palaeolithic Transition	Bailloud, 1953; Girard, 1978; Girard *et al.*, 1990; Leroi-Gourhan, 1961, 1988
Gorham's Cave	Gibraltar	MP; UP	A thick sandy sequence with sandy, anthropogenic, and biogenic layers, with localized diagenesis	Barton *et al.*, 1999; Goldberg, 2001a; Goldberg and Macphail, 2000; Macphail and Goldberg, 2000; Waechter, 1951; Zeuner, 1953
Vanguard Cave	Gibraltar	MP	A large heap of sand with localized finer silty sands containing small hearths and occupations	Barton *et al.*, 1999; Goldberg, 2001a; Goldberg and Macphail, 2000; Macphail and Goldberg, 2000; Waechter, 1951; Zeuner, 1953
Tabun Cave	Israel	LP; MP	A thick sequence of sandy, silty sand and anthropogenic deposits that have been heavily modified by diagenesis. The Neanderthals remain unscathed	Albert *et al.*, 1999; Bull and Goldberg, 1985; Goldberg, 1973; Goldberg and Bull, 1982; Goldberg and Nathan, 1975; Jelinek, 1982; Jelinek *et al.*, 1973; Mercier *et al.*, 1995; Tsatskin *et al.*, 1992

Kebara Cave	Israel	MP; UP; EP; NAT	Recently excavated site with great detail concerning Neanderthal burial, anthropogenic deposits (many combustion features), mineralogy, and diagenesis and lithics	Albert *et al*., 2000; Arensberg, 1985; Bar-Yosef, 1991; Bar-Yosef and Sillen, 1994; Bar-Yosef *et al*., 1992; Baruch *et al*., 1992; Goldberg and Bar-Yosef, 1998; Goldberg and Laville, 1988; Laville and Goldberg, 1989; Porat *et al*., 1994; Schick and Stekelis, 1977; Schiegl *et al*., 1994, 1996; Stekelis, 1954; Turville-Petre, 1932; Valladas *et al*., 1987; Weiner *et al*., 1993,1995
Qafzeh Cave	Israel	MP; UP	Anatomically modern humans that predate Neanderthals from Kebara in stony, and anthropogenic diagenetically altered deposits	Aitken and Valladas, 1992; Rabinovich and Tchernov, 1995; Schwarcz *et al*., 1988; Valladas *et al*., 1988; Vandermeersch, 1972
Amud Cave	Israel	MP	Neanderthal finds and phytoliths in an ashy matrix	Chinzei, 1970; Chinzei and Kiso, 1970; Hovers *et al*., 1995; Madella *et al*., 2002; Suzuki and Takai, 1970
Hayonim Cave	Israel	LP; MP; UP; EP; NA	A cave with a deep cultural sequence, anthropogenic deposits, and marked diagenesis	Goldberg, 1979b; Weiner *et al*., 1995, 2002
Arene Candide Cave	Italy	UP; NEO	A site showing clear stabling activities during the Neolithic	Courty *et al*., 1991; Macphail *et al*., 1994, 1997
Die Kelders Cave	South Africa	MSA; LSA	Interbedded sand and anthropogenic units with many ashy combustion features	Goldberg, 2000; Hahn, 1995; Marean *et al*., 2000; Schweitzer, 1979
Klasies River Mouth	South Africa	MSA	Site noted for its succession of hearths and last interglacial date	Deacon *et al*., 1986; Singer and Wymer, 1982
Blombos Cave	South Africa	MSA; LSA	A remarkable cave noted for its early use of ocher, bone points, and lithic industry	Henshilwood *et al*., 2001
Hohle Fels Cave	Germany	MP; UP	Site with deposits that span the MP/UP boundary and anthropogenic burned deposits; early mobile art	Schiegl *et al*., 2003
Geißenklösterle Cave	Germany	MP; UP	Site with deposits that span the MP/UP; early ivory figures	Conard and Bolus, 2003; Conard *et al*., 2004; Dippon, 2003; Hahn, 1988
Yabrud Cave	Syria	LP; MP	Thick LP, MP sequence with possible climatic significance	Farrand, 1970, 1979; Goldberg, 1969; Rust, 1950
Douara Cave	Syria	MP	Well-studied cave with many interdisciplinary studies	Akazawa, 1976, 1979; Endo, 1973, 1978a,b; Hirai, 1987; Koizumi, 1978; Matsatuni, 1973; Saito and Tiba, 1978; Uesugi, 1973

Table 8.1 (*Cont.*)

Westbury cave	LP		Biostratigraphic record of pleisticene climatic fluctuations, including units apparently contemporary with coastal Boxgrove (see Box 7.1)	Andrews *et al.*, 1999, Macphail and Goldberg, 1999
Gough's Cave	UK	UP	Controversial site relating to cannibalism in the Upper Palaeolithic	Churchill, 2000; Currant, 1991; Macphail and Goldberg, 2003; Parkin *et al.*, 1986; Stringer, 2000
Meadowcroft Rockshelter	USA	Paleoindian Archaic	Noted for its controversy concerning early radiocarbon dates and the arrival of humans in North America	Adovasio and others, 1984; Beynon, 1981; Beynon and Donahue, 1982; Donahue and Adovasio, 1990; Goldberg and Arpin, 2001
Dust Cave	USA	Paleoindian Archaic	A remarkable sequence of geogenic and anthropogenic deposits, including Paleoindian prepared surfaces	Driskell, 1994; Gardner, 1994; Goldberg and Sherwood, 1994; Goldman-Finn, 1994; Hogue, 1994; Homsey, 2003; Meeks, 1994; Sherwood, 2001; Sherwood and Goldberg, 2001; Walker, 1997; Walker *et al.*, 2001

LP = Lower Palaeolithic; MP = Middle Palaeolithic; UP = Upper Palaeolithic; EP = Epi-Palaeolithic; NAT = Natufian; NEO = Neolithic; B = Bronze Age; I = Iron Age; LSA = Later Stone Age; MSA = Middle Stone Age

United States (Donahue and Adovasio, 1990). In many places such as southwestern France, they are produced by differential weathering of the bedrock, in which weaker (generally fine-grained sediments, e.g. shales) are weathered more intensely than harder strata (Fig. 8.1). In the fluvial environment, they can be produced by streams undercutting cliff edges, and in coastal areas, wave-cut niches are also well known. In either case, unlike caves (see below) rockshelters

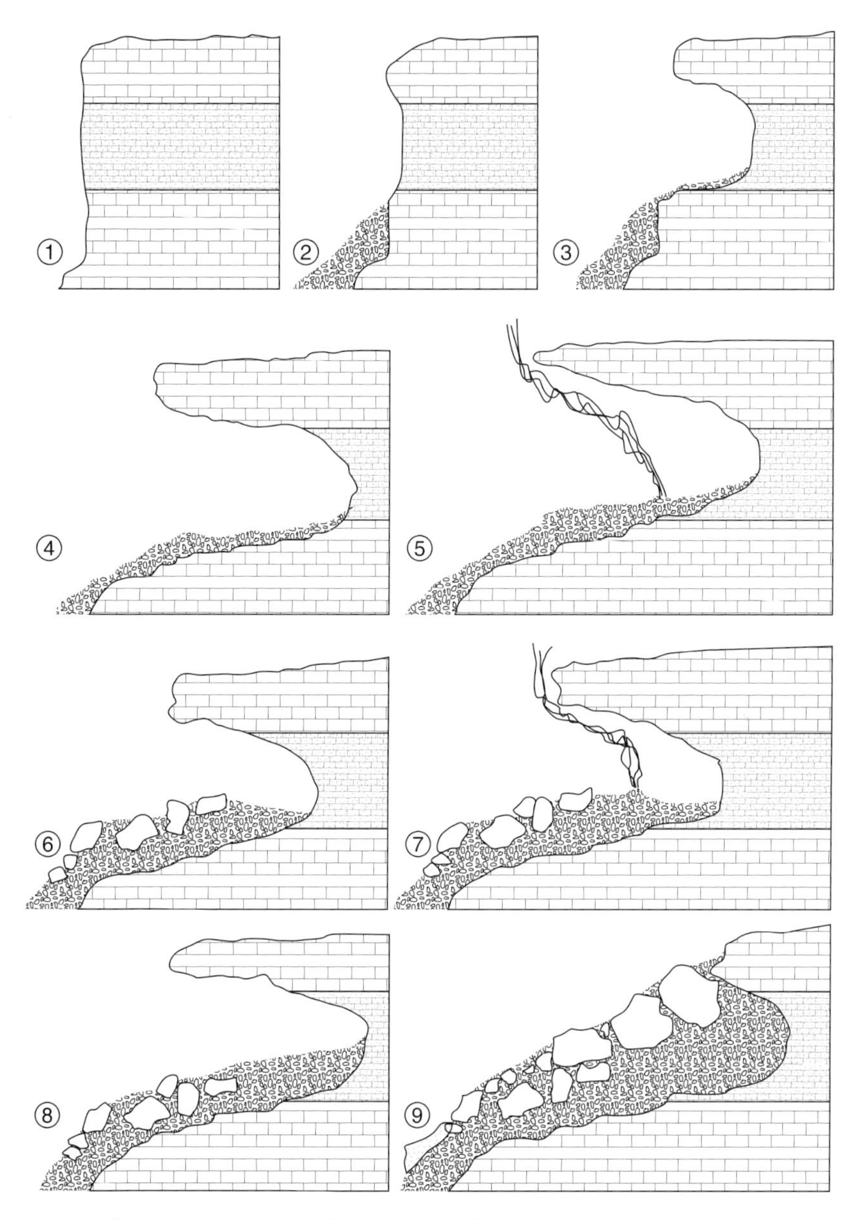

FIGURE 8.1 Depiction of evolution of a rockshelter using those in southwestern France as a general model. The initial cavity is formed by differential weathering of the limestone, which contains more easily weatherable strata. With time, the cavity deepens as rocks fall from the roof and walls, ultimately leading to collapse of the roof, either in stages or completely. Human occupation can occur throughout this development (modified and simplified from figure 3.3 in Bordes, 1972).

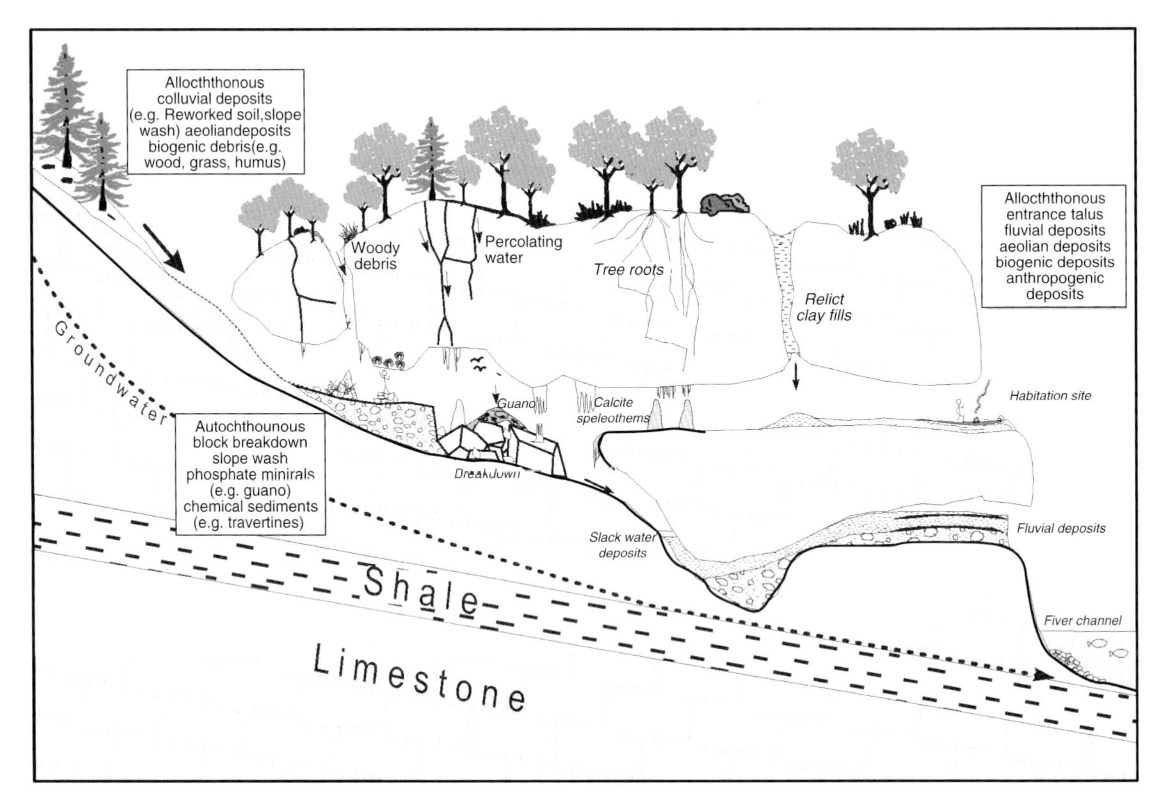

FIGURE 8.2 Schematic representation of the types of processes operating in a karstic cave environment. Indicated are physical and chemical inputs and outputs, as well as transformations, such as precipitation and dissolution (modified from Gillieson, 1996, figure 1.1).

are more subaerially exposed and subjected to atmospheric weathering processes, such as freeze-thaw, solifluction, leaching, and cementation; these processes are more typical of pedological postdepositional modifications found in open-air sites. They resemble the passive karst settings discussed in Woodward and Goldberg (2001), in which original karst processes, such as solution/dissolution, dripping vadose water, and so on, are reduced.

Caves, on the other hand, have much larger-scale features, and their networks can extend for tens or hundreds of kilometers and can be more than a kilometer below the earth's surface (Gillieson, 1996) (Fig. 8.2). Although most human activity is concentrated at and toward the entrances, it is quite clear that rituals were carried out very deep within some caves (Sherwood and Goldberg, 2001b). The overwhelming majority of caves are formed by the dissolution of carbonate rocks (limestone and dolomite), although many are found in friable sandstones, and some are developed in soluble material such as halite salt (Frumkin, 1998; Frumkin *et al.*, 1994). Their atmospheres tend to be more humid than those found in rockshelters, a factor that favors biological activity and various forms of chemical and biological weathering (Woodward and Goldberg, 2001).

8.3 Cave deposits and processes

Deposits from both inhabited and uninhabited caves from throughout the world have been investigated for decades. Geological deposits include those derived from both inside (autochthonous or endogenous) and outside (allochthonous or exogenous) the cave (Tables 8.2 and 8.3).

TABLE 8.2 Types of cave sediments (modified from Sherwood and Goldberg, 2001)

Sediment type	Autochthonous (endogenous)	Allochthonous (exogenous)
Clastic sediments		
Weathering detritus	•	
Block breakdown (*éboulis*)	•	
Grain breakdown by dissolution, abrasion of rock walls by humans[a]	•	
Entrance talus	•	•
Infiltrates (drip)	•	•
Fluvial deposits		•
Glacial deposits		•
Aeolian deposits		•
Biogenic debris		
bird and bat guano	•	○
gastroliths	•	○
carnivore coprolites and bone	•	•
wood, grass	○	•
humus	○	•
Anthropogenic deposits	○	•
Micro-artifacts (bone, shell, lithics, etc)		
transported soil/sediment		
wood/charcoal		
ash		
Chemical sediments		
Travertines	•	
Evaporites	•	
Guano, phosphate, and nitrate minerals	•	
Resistates	•	
Ice	•	

Locale: Likelihood • = high; ○ = low
[a]Hughes and Lampert, 1977

8.3.1 Autochthonous deposits

Gravity accumulations produce a common type of deposit, called *éboulis* after the French word for rubble. *Éboulis* consists of large rocks detached from the walls and roof, or individual grains liberated by the disintegration of the bedrock by dissolution (Donahue and Adovasio, 1990; Goldberg and Arpin, 1999). In Europe, the definitive process of detachment is caused by freezing and thawing of water that has percolated into cracks and fissures in the bedrock (cryoclastic deposition; see "mechanical weathering" in soils, Section 3.2.1) (Bonifay, 1956; Laville *et al.*, 1980; Schmid, 1963). Loosening of bedrock blocks by dissolution also occurs, but it is notably less prominent than the results of freeze-thaw. Accumulation

TABLE 8.3 A sampling of the more important syn- and postdepositional processes and agents acting in prehistoric caves

Physical
Penecontemporaneous reworking by wind, water, and humans
Slumping and faulting
Trampling by humans and other animals
Import and accumulation of bones and organic matter by animals
Burrowing and digging by humans and other animals
Vertical translocation of particles by percolating water
Dumping of material, midden formation
Hearth cleaning and rake-out
Chemical
Calcification (speleothems, tabular travertine, tufa, cemented "cave breccia"
Decalcification
Phosphate mineralization and transformation, and associated bone dissolution
Formation of other authigenic minerals: gypsum, quartz, opal, hematite

of sand abraded from cave walls by inhabitants of sandstone shelters has also been documented in Australia (Hughes and Lampert, 1977).

Éboulis is common in many of the deposits in caves and rockshelters mentioned above from western Europe and have been studied in considerable detail (Fig. 8.3). Laville *et al.* (1980), for example, employed a variety of techniques and parameters, such as grain size and shape (angularity and roundness) as indicators of weathering and frost-cracking, acidity (pH), and calcium carbonate content as indicators of weathering and soil formation. The results of these studies were originally thought to provide paleoclimatic sequences that could be correlated throughout the Dordogne region of southwestern France. Layers with abundant coarse *éboulis*, for example, were considered to represent relatively cold climates, whereas red layers within the *éboulis* were thought to be

FIGURE 8.3. Photo of *éboulis* from the Middle Palaeolithic site of Pech de l'Azé IV, France, recently excavated by H. Dibble and S. McPherron. The smaller blocks of *éboulis* represent a relatively steady rain of roof fall, whereas the large blocks at the very bottom and in the upper part, represent large-scale roof collapse that took place at more punctuated intervals. The distance between the vertical lines is ca 1 m.

paleosols that developed under more temperate conditions. Paleoclimatic sequences among different caves served as bases for correlation of the deposits and their lithic industries. This strategy not only supplied background paleoenvironmental information to the archaeological context, but also served as a means of indirectly dating the caves and their artifact sequences, as dating techniques other than radiocarbon were not widespread at the time.

We now know that these pioneering efforts were a bit overzealous, as climatic factors responsible for the production of *éboulis* are much more complex than originally envisioned (Bertran, 1994; Lautridou and Ozouf, 1982). In addition, TL dating of burned flints demonstrated that the inferred climate-correlations were largely inaccurate by several millennia (Valladas *et al.*, 1986). Similarly, micromorphological work showed that reddish stratigraphic units are not soils but deposits of reworked Tertiary weathering products derived from the plateaux above the caves and rockshelters (Goldberg, 1979a); soils in the strict sense, as opposed to exposed occupation or weathering surfaces, do not form in caves (see below). Lastly, Campy and Chaline (1993) demonstrated that many temporal gaps exist in the depositional sequences of many of these caves, thus making correlations on the basis of matching of apparent climatic cycles tenuous at best. Nevertheless, the formation of *éboulis* is likely tied to colder conditions, although it is difficult to state more than that.

8.3.2 Allochthonous deposits

Materials are also transported into caves and rockshelters from outside. These items can include reworked soils that emanate from cracks in the roof, runoff (Goldberg and Bar-Yosef, 1998; Macphail and Goldberg, 2000), fluvial deposits (Sherwood and Goldberg, 2001a), and aeolian inputs. The latter are particularly prominent in prehistoric sites along littoral parts of the Mediterranean climatic zone. Massive accumulations of sand – along with interspersed occupations – can be seen in Gorham's and Vanguard Caves, Gibraltar (Macphail and Goldberg, 2000), Tabun Cave in Israel (Jelinek *et al.*, 1973), and in many sites along the South African coast, such as Blombos and Die Kelders Caves (Henshilwood *et al.*, 2001; Marean *et al.*, 2000); in these sites, the influx of sand is related to glacio-eustatic sea level changes and as such they are useful in the reconstruction and correlation of regional landscapes. Finer grained silts can also be blown into caves in the form of atmospheric dust produced by dust storms; silty deposits at the Middle Palaeolithic/Upper Palaeolithic sites of Hohle Fels and Geißenklösterle in Swabia, Germany, are of such a loessial origin (Hahn, 1988).

8.3.3 Geochemical activity

Geochemical activity is common in caves, but less so in rockshelters. It usually takes the form of accumulation and oxidation of organic matter, and the precipitation and dissolution of various minerals in the sediments, including bones. The most common process is precipitation of calcium carbonate, which occurs in various forms that include stalactites, stalagmites, and other ornamental forms (Gillieson, 1996; Hill and Forti, 1997). Massive, bedded accumulations of calcium carbonate are generally known as flowstone or travertine and can cover or intercalate with archaeological deposits (Fig. 8.4). In suitable instances, where levels of contamination by detrital components (e.g. silt and clay) are low, these carbonate accumulations can be dated with Uranium-series methods (Rink, 2001). At the Palaeolithic site of El Castillo in Spain, for example, Bischoff *et al.* (1992) dated a flowstone separating Acheulean and Mousterian layers to 89+11/−10 ka BP, a date which turns out, "calls into question the temporal significance of these archaeological designations" (p. 49).

Carbonate-charged water dripping on detrital sediment can lead to the total impregnation

FIGURE 8.4 Photo of entrance to Gorham's Cave, Gibraltar. Shown behind the figures in the foreground are interdigitated soft and calcite-cemented sand and clay, capped with a stalactite. The overall resistance of this mass of sediment is caused by the continued although variable dripping of carbonate-charged water: in wetter intervals – or when the rate of detrital sediment was lower – carbonate cementation extended further away from the pillar as can be seen by the wedge-shaped nature of the cemented layers.

and cementation of the underlying deposits, resulting in hard, concreted masses, commonly called "breccia" or "cave breccia"; such deposits are widespread in the Mediterranean region, such as Italy. Animal and human bones, seeds, and other organic materials tend to be well preserved in these breccias, and thus they serve as valuable stores of environmental and cultural information.

Dissolution of calcium carbonate can result from a number of factors. Among the most prominent is the acid conditions produced by guano, the decay of organic matter, and carbonic acid derived from carbon dioxide dissolved in water as it passes through the soil (see also "weathering" in Chapter 3). Associated with this carbonate dissolution, particularly in caves of the Mediterranean region, is the transformation and formation of a number of phosphate minerals (Table 8.3) (Karkanas *et al.*, 1999; Weiner *et al.*, 2002). The geochemistry of these interactions is quite complex although recent research on modern deposits has begun to reveal some of the major processes and conditions operating within the caves. Shahack-Gross *et al.* (2004) for example, showed the rapidity of

TABLE 8.4 The authigenic minerals associated with cave sediment diagenesis (modified from Karkanas *et al.*, 2002)

Mineral	Formula
Brushite	$CaHPO_4 \cdot 2H_2O$
Calcite	$CaCO_3$
Hydroxyapatite	$Ca_{10}(PO_4)_6(OH)_2$
Fluor-carbonated apatite (Francolite)	$Ca_{10}(PO_4)_5(CO)_3F_3$
Dahllite (carbonated apatite)	$Ca_{10}(PO_4,CO_3)_6(OH)_2$
Ca, Al-phosphate amorphous	Ca, Al, P-phase, non-stoichiometric
Crandallite	$CaAl_3(OH)_6[PO_3(O_{1/2}(OH)_{1/2})_2]$
Montgomeryite	$Ca_4Mg(H_2O)12[Al_4(OH)_4(PO_4)_6]$
Taranakite	$H_6K_3Al_5(PO_4)_8 \cdot 18H_2O$ or $K_x[Al_{2-y}(H_3)_y](OH)_2$ $[Al_xP_{4-x-z}(H_3)_zO_{10}]$
Leucophosphite	$K_2[Fe_4^{3+}(OH)_2(H_2O)_2(PO_4)_4]$ $2H_2O$
Nitratite	$NaNO_3$

formation of certain minerals under acidic conditions – within decades. As organic matter degrades, the amount and availability of Al, K, and Fe increase, whereas nitrogen and sulfur decrease. Minerals, such as brushite, taranakite, nitratite, and gypsum (Table 8.4) form first, although they are transient. With time, as organic matter continues to decrease but still under acid conditions, clay minerals start to degrade, releasing Al and Fe, and possible formation of authigenic opal. Other phosphatic minerals (e.g. dahllite, crandallite, montgomeryite, taranakite, and leucophosphite – Table 8.4) also form under complex diagenetic pathways (Karkanas *et al.*, 2000).

Implications of these diagenetic changes are many. The most apparent is the localized to total dissolution of bone in many cave deposits, particularly those in Mediterranean environments, such as Tabun, Hayonim, Kebara (Goldberg and Bar-Yosef, 1998; Weiner *et al.*, 2002). Thus, the distribution, presence, or absence of bones, can be mostly a function of geochemical processes and not to human activities. Using a number of laboratory experiments, it was shown that bone apatite (dahllite) is preserved in pHs above 8.1; in neutral and alkaline conditions bone will be preserved but the apatite will be recrystallized.

More subtle, however, are the diagenetic effects in which radioactive elements, such as potassium, can be concentrated. In turn, these localized concentrations can affect the dosimetry measurements needed to obtain chronometric dates using TL and ESR techniques (Weiner *et al.*, 2002) (see Chapter 16), thus yielding incorrect dates if they are not taken into account. Finally, results of mineralogical studies of diagenetically altered cave deposits suggest that the occurrence of the more insoluble minerals, such as leucophosphite, can be used to infer the presence of former fireplaces in cases where their actual presence is no longer visible (e.g. Schiegl *et al.*, 1996).

8.3.4 Biogenic deposits

Biogenic inputs in caves are signified by bat and bird guano, as well as the remains of denning activities by carnivores and omnivores. Hyenas, for example, can produce noteworthy accumulations of bones, coprolites (generally apatite), and organic matter (Goldberg, 2001b; Horwitz and Goldberg, 1989). Deposits linked to cave bear occupations have also been inferred from phosphate-rich sediments associated with both the bones of their prey and the bears themselves (Andrews *et al.*, 1999; Schiegl *et al.*, 2003). In Middle Eastern prehistoric caves, pigeons generate millimeter-size rounded stone gastroliths ("crop stones"), indirectly indicating a significant but now invisible source of guano (Goldberg and Bar-Yosef, 1998; Shahack-Gross *et al.*, 2004). The identification of the presence of such gravel-size gastroliths at Arene Candide

Box 8.1 Kebara Cave, Israel

The site of Kebara Cave, about 30 km south of Haifa, Israel, is developed within a Cretaceous limestone reef on Mt. Carmel (Fig. 8.5) and is one of the best studied prehistoric cave sites in the Old World. It represents an excellent example of a modern, multidisciplinary excavation in which a team from a variety of scientific backgrounds (e.g. prehistory, geoarchaeology, geophysics, geochemistry, zooarchaeology, palaeobotany), and countries (e.g. Israel, France, United States, Spain, United Kingdom) focus their talents to elucidate a wide range of archaeological and environmental questions and issues (Bar-Yosef *et al.*, 1992). Such problems include dating the deposits and associated lithic, faunal, and human remains, evaluating past environments within and outside of the cave, reconstructing of human behavior and evolution, and the use of space in the cave during the Middle Palaeolithic and Upper Palaeolithic (Fig. 8.6a,b). From the geoarchaeological point of view, Kebara has a rich history and exhibits sediments of geological and human origin that have been modified by geogenic and anthropogenic agents. Only a few of these aspects are presented here.

One of the most striking facets of the deposits at Kebara is the abundance of ashy combustion features that provided new insights into Neanderthal activities associated with fire (Fig. 8.7) (Meignen *et al.*, 1989, 2001). Among others, these features include well-defined combustion structures, with shapes ranging from lenticular to tabular (Box Fig. 8.7a), and massive bedded ash dumps (Box Fig. 8.7b). Thus, specifically different types of activities associated with the Neanderthal occupants are evident by examining the sediments themselves, such as *in situ* burning events and

FIGURE 8.5 Kebara Cave as seen from a high bluff on Mt. Carmel looking to the southeast. The cave is formed within a Cretaceous limestone reef during pre-Pleistocene times. At the mouth of the cave is a small rise that marks the accumulation of soil material that was washed down from the slopes above the cliff face.

(cont.)

Box 8.1 *(cont.)*

the dumped ashes. The differentiation of these deposits and their interpretation constitute the fabric of the site's stratigraphy. These accumulations and associated activities reflect upon the length of occupation episodes and the localized use of space in the site.

In addition to anthropogenic deposits which constitute the majority of the Middle Palaeolithic deposits at the site, geogenic sediments also occur, and are particularly marked in the Upper Palaeolithic deposits. Much of the sediment is ashy and organic rich, with some additions of aeolian quartz sand and silt; these deposits were emplaced

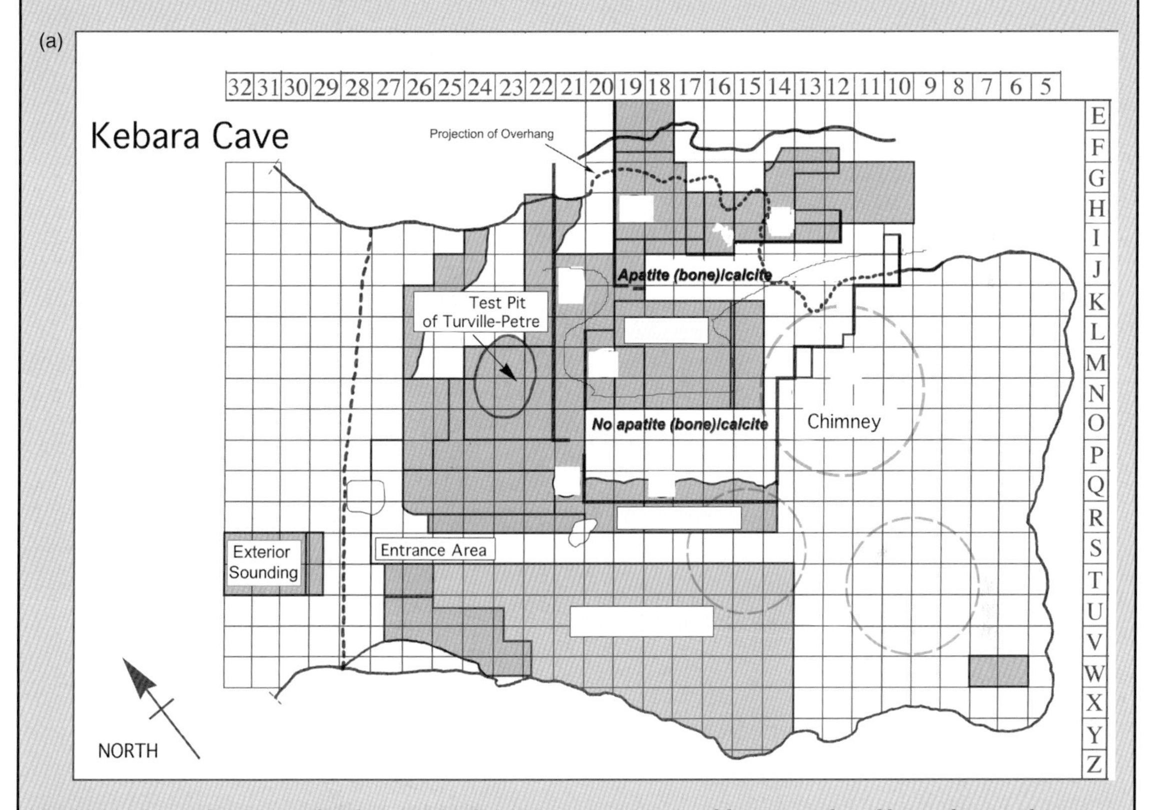

FIGURE 8.6 (a) Plan view of Kebara Cave showing excavations and location of profiles and areas shown in photos. The curved line shows the boundary between layers that contain apatite (bone) and calcite (to north) from those in which bone has not been preserved but dissolved as a result of diagenesis. The recognition of the causes of such distribution patterns places constraints on how we interpret the archaeological record. (b) Cross section of Kebara Cave. The archaeological remains within the layers are as follows: A: Historical; B: Natufian; C: Kebaran; D: Aurignacian; E: Early Upper Palaeolithic (Units I–III, IV); F: Mousterian. Note the build-up of roof fall and other sediment at the entrance, beneath the brow of the cave; this accumulation helps create a primary, depositional relief, with a natural slope toward the back of the cave (right). This slope was enhanced by a major subsidence episode that took place at the end of the accumulation of the Middle Palaeolithic deposits, prior to Upper Palaeolithic occupation (Laville and Goldberg, 1989) (modified from Bar-Yosef *et al.*, 1996).

(cont.)

Box 8.1 *(cont.)*

by runoff, which emanated from the entrance of the cave beneath the drip line and flowed toward the rear of the cave. This deposition was promoted by the relief from the front to the back of the cave, which resulted from two factors. The first is that reworked soil materials derived from the slopes above the cave built up as a mound beneath the brow of the cave (cf. Fig. 8.5 and 8.6b) which raised the height of the entrance section. In addition, at the end of the Middle Palaeolithic and beginning of the Upper Palaeolithic, the deposits in the direction of the rear of the cave subsided on the order of several meters (Laville and Goldberg, 1989); this was a result of reorganization of sedimentary material that probably plugged the chambers within the subsurface karstic cave network, and similar types of subsidence can be seen today in areas of active karst such as Florida. In any case, with this substantial relief from front to back, deposits from the entrance were easily transported toward the back of the cave during heavy rainfall events, one of which we witnessed in October of 1984. Interestingly, when viewed in thin section, these well

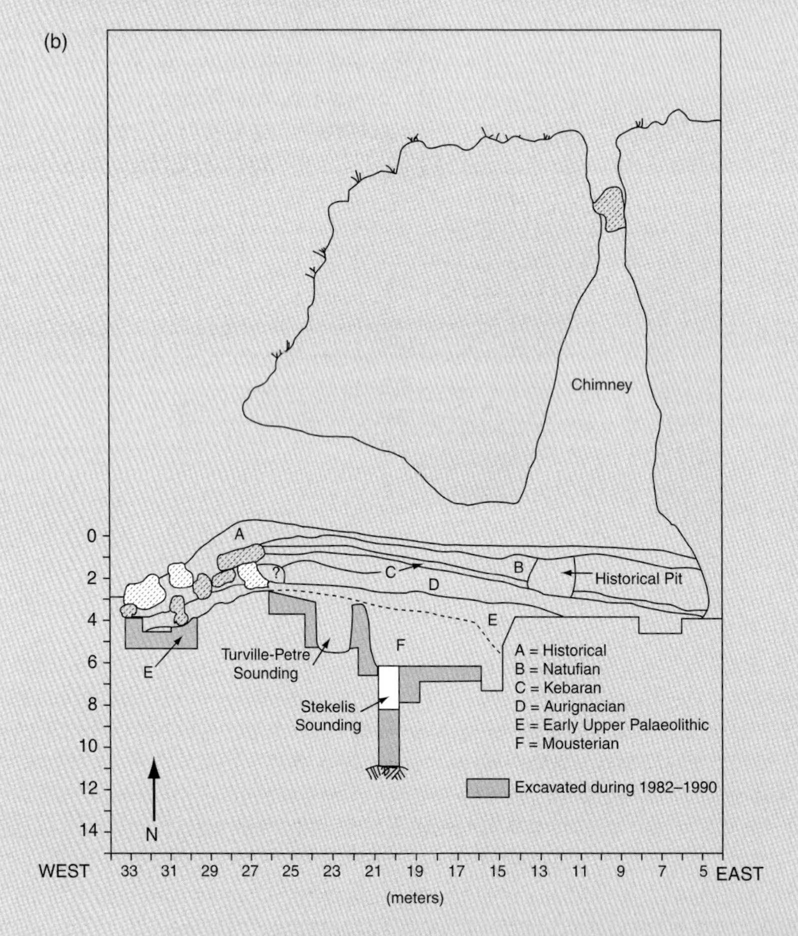

FIGURE 8.6 *(Cont.)*

(cont.)

Box 8.1 *(cont.)*

(a)

(b)

FIGURE 8.7 Features associated with burning at Kebara. (a) The southwestern corner of the Kebara excavations (Sq. O22) showing the abundance of combustion features marked by the black (charcoal-rich) and white (ashy) lenses. These features represent both burning events that took place within hollowed depressions (lenticular shapes), as well as fires made on flat surfaces (tabular/linear shapes). The patchiness or absence of distinct burned layers in the upper left-hand part of the photograph is due to bioturbation by small and mid-size rodents; the large elliptical hole in the middle could be from a porcupine. Bars on scale are 10 cm. (b) A thick accumulation of ashes against a bedrock wall in the northern part of the site (near Sq. I13). Thin section analysis revealed that this is an accumulation of finely bedded ashes mixed with sand/granule size bone and aggregates of terra rosa. It provides insights into Neanderthal behavior by signifying that this mass of deposit is a result of ash dumping. Scale is 20 cm long.

(cont.)

Box 8.1 *(cont.)*

(a)

(b)

Figure 8.8 Two views of the Upper Palaeolithic deposits at Kebara. (a) Shown here are the well-bedded sediments on the southern section (Q15–20 in) that dip to the rear of the cave (left) as a result of slumping of the interior sediments. Some burrowing is evident, as shown by the circular feature in the center right of the photograph. The hearth at the very left can be seen in the east profile. (b) East profile (along Squares N, O, P, Q 14 in Box Plate 8.1a). The deposits in this section are somewhat more weathered and burrowed than those in (a), but the bedding is still evident. Diatoms were recovered from the darker layers next to the dolomite rock in the center of the photograph. This boulder is interesting in that not only is it a piece of roof fall, but it exhibits a ca. 10-cm thick crust composed of a number of phosphate minerals. These phosphates vary from carbonate apatite adjacent to the unaltered rock in the center, to increasingly more insoluble phosphate minerals such as leucophosphite at the very outer rim (Weiner *et al.*, 1995).

(cont.)

Box 8.1 *(cont.)*

bedded Upper Palaeolithic deposits (Fig. 8.8) are composed not only of terra rosa reworked from soil above and outside the cave, but also fragments of Middle Palaeolithic combustion features, such as charcoal and ashes, that were being eroded from near the cave entrance and transported toward the rear. These thick, well bedded Upper Palaeolithic deposits indicate prolonged periods of deposition by runoff, and furthermore suggest wetter climatic conditions during this time, roughly ca. 43 to 30 ka (Bar-Yosef *et al.*, 1996). This last inference was supported by the presence of diatoms in some of the Upper Palaeolithic layers whose analysis indicated the presence of a continually wet substrate (B. Winsborough, personal communication).

Diagenetic alterations are widespread at Kebara and they have serious implications for interpreting the site's history and the use of the cave. The most striking feature is the lack of bones in more than half the excavated area of the cave (Fig. 8.5). The immediate question that comes to mind, 'is this lack of bones the result of primary factors, such as lack of discard or removal/cleaning by humans, or some other taphonomic factor?' The answer is revealed by detailed mineralogical analysis of the deposits (Weiner *et al.*, 1993), which indicates that the absence stems from the dissolution of bone by acids derived from organic matter (guano, human refuse). Such information clearly helps to demonstrate that bone distribution in many parts of the cave is due to postdepositional alteration and not to human activity. In turn, this knowledge influences the way we look at the zooarchaeological record and how it can be interpreted at the site. The second implication is that in many locations, not only have the bones disappeared, but calcite (both as *éboulis* and as ashes from fires) has been dissolved or replaced by phosphate minerals that range from apatite to leucophosphite and taranakite (see text). These diagenetic alterations are responsible for concentrating radiogenic elements that in turn affect the dosimetry calculations, which constitute an essential part of thermoluminescence and electron spin resonance dating techniques. They are prominent techniques in dating deposits beyond the range of radiocarbon.

was important; otherwise some other, and bizarre explanation, for example, a physico-chemical/sedimentary depositional process (e.g. fluvial activity) would have had to be envisaged (Macphail *et al.*, 1994). Both anthropogenic and natural accumulations of plant remains (e.g. grass, wood, roots, seeds) can originate from vegetation growing at the cave entrance, which has been washed or blown into the cave; they can also be brought into the cave for use as bedding, fuel, or shelter, or even as nesting material (Wattez *et al.*, 1990). Quantitative phytolith studies of cave sediments from Israel have been instrumental in evaluating these biogenic/anthropogenic inputs (e.g. Albert *et al.*, 1999, 2000).

8.3.5 Anthropogenic occupation

Anthropogenic deposits are common in caves, and in the past these have been largely overlooked. People track in materials from outside, either on their feet or as material attached to hides or carcasses, although such processes are difficult to document precisely. More prominent are fireplaces, hearths, and other features associated with fire. Although isolated hearths are not uncommon, striking combustion features are found in Palaeolithic caves from the Middle East, for example, Tabun, Kebara, Hayonim (Goldberg and Bar-Yosef, 1998; Meignen *et al.*, 1989) and Middle Stone Age sites in South Africa such as Klasies River Mouth, Die Kelders, Blombos (Marean *et al.*, 2000; Singer and Wymer, 1982). Those at Klasies River Mouth, Die Kelders (Plate 8.1), and Kebara (see Box 8.1) are particularly noteworthy for their degree of preservation and abundance. At Kebara different types of combustion features could be recognized, including a massive accumulation of ashes that were dumped by the Neanderthal inhabitants (Meignen *et al.*, 2001). These combustion features provide high resolution windows into human activities and technologies related to fire, including the type of combustible, the duration and intensity of burning, cleaning, trampling, and dumping.

On a similar note, caves figure into the first uses of controlled fire. These include Swartkrans in South Africa and Zhoukoudian, Locality 1 in China. Recent geoarchaeological research was directed to evaluating the nature of the latter, which roughly range in age from 250 to 600 ka. Micromorphological and FTIR analyses of the often-quoted "hearths" in Layer 10, are in fact poorly sorted interbedded accumulations of sand, silt, and clay-sized mineral and organic matter (Plate 8.2a, b) (Goldberg *et al.*, 2001). Such fine-grained, laminated deposits more reasonably accumulated in standing water depressions, which were likely associated with the presence of the Zhoukou River, which at that time occasionally flowed into the cavern.

Caves could also be sites of inhumations. Some of the best preserved remains of Neanderthals have been found in caves, for example, at Gibraltar (Stringer *et al.*, 2000), and Gough's Cave, Cheddar Gorge, United Kingdom is famous (infamous) for its collection of supposedly cannibalized skeletal remains (Stringer, 2000). Both Final Palaeolithic and Neolithic "cemeteries" have been found at Arene Candide, Liguria, Italy, with the latter including up to 40 burials, some in stone cists and some associated with the use of ocher (Maggi, 1997). Moreover, this cave is the archetypal example of Holocene anthropogenic deposits, which are typical of the Mediterranean region, and which display either a homogeneous or black and white "layer cake" appearance (Macphail *et al.*, 1997). These layers are several meters thick and are best preserved from the Early Neolithic (ca 5,500 BC) to the Late Neolithic/Chassey (ca 4,000 BC). The homogeneous deposits represent a dominance of domestic occupation, which is reflected in the large amounts of associated eco- and artifacts. On the other hand, the layered sediments are often the in situ remains of burned stabling floors, reflecting periods when the cave was seasonally occupied by pastoralists who used tree leaf hay to fodder their animals (Courty *et al.*, 1989, 1994). This model of sedimentation has been found to hold true in France, Switzerland, and across northern Italy (Binder *et al.*, 1993; Boschian and Montagnari-Kokelji, 2000).

8.4 Environmental reconstruction

Climate and environmental changes can be recognized in cave sediments, although it requires careful investigation to extract them from the sediments. As discussed above, the preserved coarse fraction of *éboulis* initially played a major role in climatic interpretation (Laville, 1964). More recently however, the finer fraction of the sediments, as revealed by soil micromorphology in particular, disclosed some interesting palaeoclimatic information. Micromorphological analysis of sediments at Hohle Fels Cave, southwestern Germany, for example, showed a marked increase in cryoturbation during the Late Pleistocene in deposits that spanned the Gravettian through Magdalenian periods (Goldberg *et al.*, 2003; Macphail *et al.*, 1994;

Schiegl *et al.*, 2003); the Mousterian hearths at Grotte XVI, France are strikingly deformed by cryoturbation (Karkanas *et al.*, 2002). In the western Mediterranean area, Courty and Vallverdú (2001) analyzed three caves and rockshelters, and asserted that warm conditions are associated with dripping water and biogenic precipitation of carbonates. This finding is similar to that from the Earliest Holocene deposits at Gough's Cave, Cheddar, United Kingdom (Macphail and Goldberg, 2003). On the other hand, cooler conditions are associated with rapid deposition of exogenous sediments or the accumulation of debris associated with freezing activity. They noted, however, that the role of these detailed climatic inferences in human behaviors or activities remains to be demonstrated. More easily, variations in sedimentation and associated syn-depositional processes, have occasionally been well related to Pleistocene climatic fluctuations as observed from the associated faunal evidence (Andrews *et al.*, 1999).

8.5 Conclusions

Caves and rockshelters are special geoarchaeological situations, as they can yield faithfully recorded documents of both environmental change as well as past human activities. This statement is particularly true for prehistoric time periods, such as the Lower and Middle Palaeolithic, where anthropogenic deposits – and their inferred human activities and behaviors – are overall much better preserved in caves than in their open-air counterparts. The detailed study of cave deposits by micromorphology and other micro-techniques has revealed many new insights into interpreting the geological and archaeological records for both prehistoric and later periods. Issues of bone preservation and associated diagenesis, sfor example, not only have ramifications about interpreting the archaeological record, but also in evaluating dates that are derived from objects and the deposits that enclose them.

Part II

Nontraditional geoarchaeological approaches

Introduction to Part II

Geoarchaeology has traditionally been practiced at the landscape/geomorphic and site level, concentrating on site chronologies and regional correlations (Haynes, 1990); less emphasis was placed on significance of individual strata at the microscale level. Early geoarchaeological research in the United States, particularly for Early Paleoindian sites was often concerned with the geological setting of the sites, stratigraphy, and Quaternary landscape changes. These approaches still tend to dominate geoarchaeological research there, as much less geoarchaeology is concerned with detailed study of individual deposits, particularly anthropogenic ones. In Europe, landscape-level geoarchaeology was also prominent until the past few decades with a strong emphasis on Quaternary palaeoenvironments and climatic change. Bordes (1954), and Zeuner (1959) for example, are towering examples of this approach. More recently, concomitant with increased importance of salvage archaeology, attention has expanded to include the developing field of anthropogenic influences on the landscape. Furthermore, the deposits that people left behind have become a focus of attention, and this has led to the use of new research techniques and approaches to infer the activities and site formation processes that produced them. These issues are considered in this section, which focuses on human occupations and their deposits.

9

Human impact on landscape: forest clearance, soil modifications, and cultivation

9.1 Introduction

The issue of human impact on the landscape is an important one, both from archaeological/historical and present-day perspectives. What was the effect on the vegetated landscape for example, by Neolithic people in the Near East who manufactured burned lime to produce plaster floors (Garfinkel, 1987; Rollefson, 1990) (see Chapters 10 and 13)? Similarly, was soil erosion in the Mediterranean and the formation of alluvial and colluvial fills due to human intervention or some natural cycle in the earth's geomorphic rhythm? Understanding such phenomena is just as significant for today, as we search globally for methods of sustainable agriculture and woodland resource management. Thus, the impact of human activity on the landscape in the past needs to be recognized. In this section, we explore some of the ways that past human activities can be detected, using soils and geomorphological records.

Past impacts can be assessed in a number of ways. These can range from the detection of, (1) the smallest influence that aids the recognition of an early human presence (e.g. forest clearance and associated minor soil disruption) to (2) major effects such as landscape and soil modifications. The latter could include the formation of overthickened cultosols such as plaggen soils, infilling of dry valleys with colluvium (see Chapter 4), and the construction of terraced and/or paddy field landscapes. As noted in Chapter 3, pedogenic processes are governed by the five soil forming factors, and the factor *organisms* can sometimes be overly influenced by humans. First, humans can directly affect the plant cover, through forest clearance, maintenance of grazing by fire and animal husbandry, and by the growing of "crops," which can include woodland exploitation through to arable farming. This chapter describes human impact from the least obvious affects on the soil, starting with forest and scrub clearance activities and how this can affect soil fertility, both positively and negatively.

9.2 Forest clearance and soil changes (amelioration, deterioration, and disturbance)

The terms "forest" and "woodland" are used differently around the globe. For example, forest is a legal term in England for land that was set aside for hunting that is not necessarily totally wooded, such as the Medieval "New Forest," Hampshire/Dorset. The terms virgin forest, old growth woodland, and primary woodland are discussed in Peterken (1996, 16 ff.). For our purposes human populations first cleared "virgin forests" and "primary woodland," but there has also been clearance of secondary woodland and scrub, and in Europe especially, woodlands have been managed in various ways (coppicing for charcoal, tree-shredding for leaf hay fodder), all of which may have impacted on soils.

The major implied effect of humans on the edaphic (plant and soil) environment in British prehistory was recognized by Dimbleby (1962), a pioneer in environmental archaeology working from the 1950s. Through pollen analysis of barrow-buried soils he showed that woodland had been replaced by heath plants, which have shallow rooting systems compared to broadleaved trees. Thus, heath plants do not aid in soil nutrient recycling and moreover, produce chemicals (e.g. tannins) that slow down organic matter breakdown and mobilize sesquioxides (see Chapter 3). A French contemporary of Dimbleby, P. Duchaufour categorized human effects in both soil degradation and amelioration (Duchaufour, 1982; 137–141). Using present-day edaphic studies, buried soils, pollen, and ^{14}C dating, these workers were able to recognize one of the first links between woodland clearance activity, the maintenance of open areas by fire and grazing, and the development of leached soils (e.g. albic horizons, Spodosols) formed on poor geological substrates. Duchaufour also identified the connection between montane terracing, and the expansion of alpine grass (hay) meadows at the expense of conifer (e.g. larch) forest, and the replacement of podzols by brown soils with a mull horizon, a form of soil amelioration (Courty *et al.*, 1989: 315; Duchaufour, 1982: 140). Their pioneering insights are an encouragement to present-day workers who are discovering anomalous soil characteristics that are difficult to explain simply through natural pedogenic processes and that may well be human induced.

In the case of buried soils, the most important first step is to be able to differentiate ancient features from those caused by the process of burial, as studied by experiments (Chapter 12), and by other postdepositional effects, for example bioturbation (Johnson, 2002). One way to achieve this is through soil micromorphology, and is done by ascertaining the "hierarchy of the features," that is, by identifying the temporal ordering of soil formation and pedofeatures according to their geometrical arrangement in the thin section and by following the law of superimposition (Courty *et al.*, 1989) (see Chapter 16). It is well established for example, that in some podzols, in the lower and least leached part of the spodic (Bh/Bs) horizon, sesquioxidic coatings can be found superimposed on clay coatings. This may provide a proxy history of increasing soil acidity, with primary clay translocation (under probable broadleaved woodland) being followed by podzolization that perhaps developed under an ensuing ericaceous species-dominated vegetation. At some sites, the last followed clearance and was maintained by grazing and burning (Dimbleby, 1962, 1985). This occurred on poor sandy substrates across Europe, and although mainly dating to the Bronze Age, lowland and upland podzolization can be traced to both Mesolithic and Neolithic sites through to Iron Age and Medieval periods, the last usually due to large-scale monastic sheep grazing (Romans and Robertson, 1975; Simmons, 1975).

The onset of podzolization was therefore asynchronous across Europe because of these human impacts, and thus the presence of podzols should never be employed as a single "marker horizon" or be used to provide some kind of proxy dating – as was attempted in the past (Macphail, 1986; Macphail *et al.*, 1987). It has also been demonstrateol, that European Holocene podzols can equally form *naturally* under broadleaved woodland, by late prehistoric times (e.g. Bronze Age/Iron Age), and as also shown by the study of modern podzols (Dimbleby and Gill, 1955; Mackney, 1961; Macphail, 1992a).

Throughout the present-day boreal region of North America, northern Asia, and Scandinavia, podzols occur naturally under conifers. In North Sweden these podzols can be anomalously shallow, being only 30 cm thick at some Iron Age (here dating to the first millennium AD) and Viking sites. These "postglacial" soils have not undergone any kind of erosion, however, but have only undergone 2,000 years of development (instead of 10,000) because they were only recently exposed by isostatic uplift (Linderholm, 2003). Here then, is an example of an unexpected environmental situation specific to a particular regional and cultural human setting that would need to be understood in order interpret human activities for any given locality correctly.

Woodland clearance has been inferred from finding wood charcoal in buried topsoil horizons, and possible "slash and burn" agriculture has been identified in Scotland (Romans and Roberston, 1975). This has seemingly been replicated by Swedish experimental "slash and burn" cultivated podzols where again coarse wood charcoal was concentrated in the uppermost few centimeters of the soil (Macphail, 1998) (see below). Where treethrow subsoil holes contain charcoal of mainly one tree species, it has been suggested that the fallen tree that formed this feature, was burned *in situ* – especially when burned soil is also present (Barclay *et al.*, 2003; Macphail and Goldberg, 1990; Mark Robinson, personal communication). Such findings are not unequivocal evidence of purposeful clearance, however, because natural blowdowns occur during storms, and such opening up of the woodland landscape by human groups may only have been opportunistic. The presence of artifacts in treethrow holes is also not conclusive evidence of large-scale forest clearance, because again the use of such hollows could be opportunistic, and artifacts in the soil could also have fallen into the hollow as it became infilled by extant soils (Crombé, 1993; Newell, 1980).

Treethrow is a common natural occurrence (Stephens, 1956), and treethrow formations can give clues to the nature of past woodland (Langohr, 1993) (see Figs 9.1–9.7 and Plate 9.1).

FIGURE 9.1 Treethrow: modern example from the "hurricane" of 1987 that produced "blowdowns" of mature beech trees that were shallow rooted in thin decalcified drift (Clay-with-Flints) over chalk (Chiltern Hills, Hertfordshire, United Kingdom).

FIGURE 9.2 Treethrow subsoil hollow (section and part plan) formed by a "blowdown" during the Atlantic Period, ca 5,000 years ago as indicated by mollusks in the subsoil fill; Balksbury Camp, Hampshire, United Kingdom (Allen, 1995).

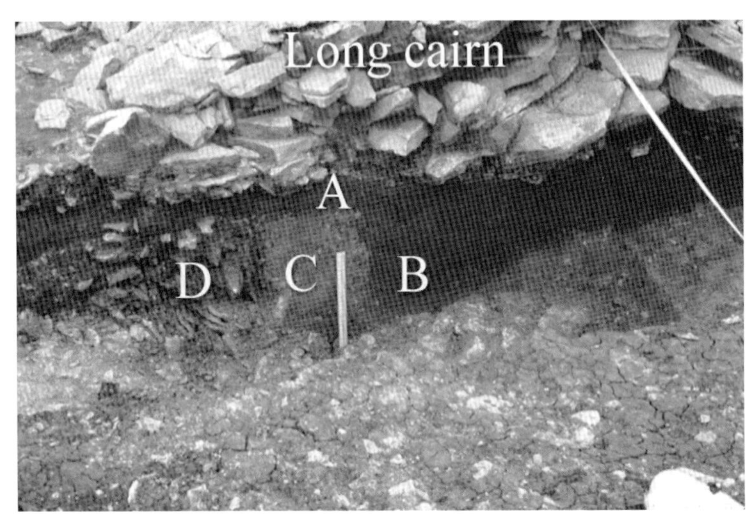

FIGURE 9.3 A section through Hazleton long cairn and its buried soil, showing the Neolithic old ground surface (A), the remains of an earlier "Atlantic Period" – (mollusks) treethrow subsoil hollow (B), up-throw of marl (C), and deeper rock (D) from the Jurassic Oolitic limestone geology.

FIGURE 9.4 The charcoal-rich early Neolithic old land surface (OLS) at Hazleton long cairn, termed the "midden area." This contained flint artifacts, pottery, domestic animal bones, and charred wheat and hazel nuts (Saville, 1990).

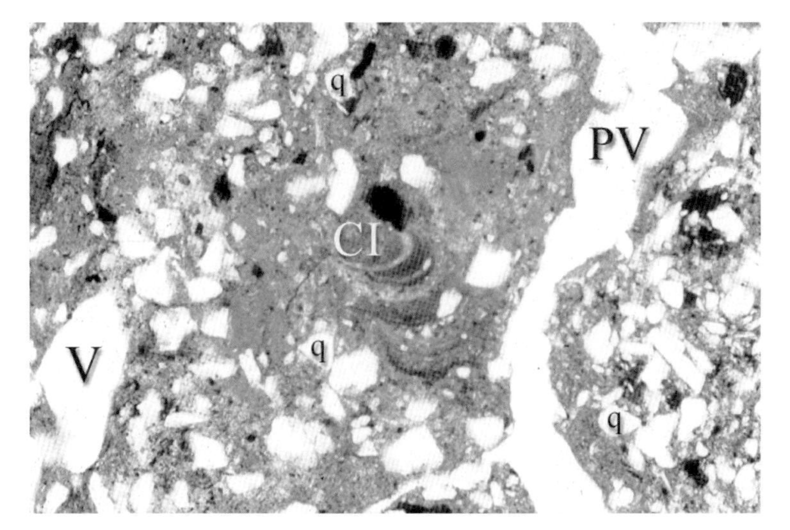

FIGURE 9.5 Photomicrograph of the "Atlantic Period" subsoil at Hazleton (Fig. 9.3 Area B), showing: quartz sand (q) in the clay loam soil; an open vugh (V); crescent infills (CI) of dusty clay; and dusty clay coated planar void (PV). These infills and coatings resulted from treethrow soil disruption. Plane polarized light (PPL), frame is ~1.35 mm. (Note: this is technically a Bt horizon formed under woodland, but here most of the textural pedofeatures – clay coatings – were formed by treethrow not *lessivage*).

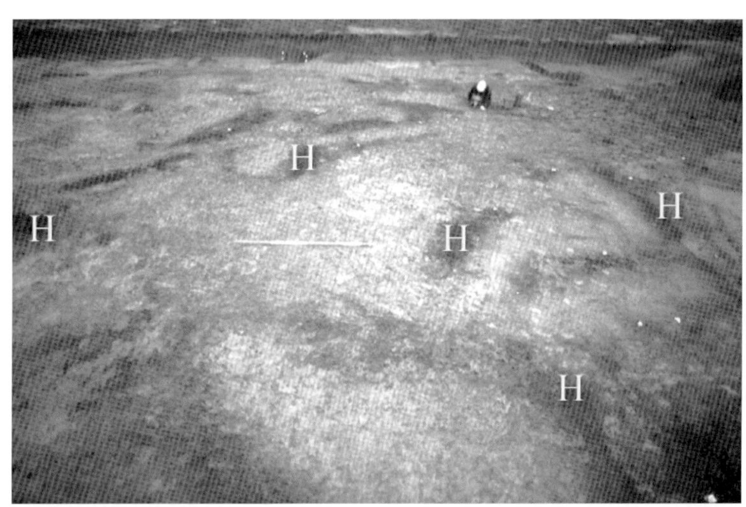

FIGURE 9.6 Prehistoric (alluvium buried) palaeo-landscape at Raunds, Nene valley, Northamptonshire, United Kingdom; a series of semi-intercutting treethrow holes are outlined by curved "humus" infilled boundaries (see Fig. 9.7). Most tree holes date from the Late Mesolithic/Early Neolithic to the Beaker/Early Bronze Age Period (Healy and Harding, forthcoming).

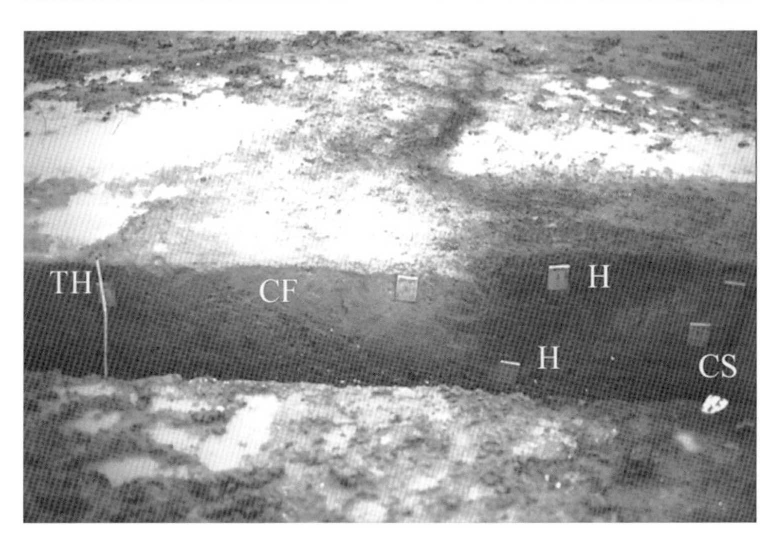

FIGURE 9.7 Raunds: section through 2.5–3.0 m wide treethrow hole (see Fig. 9.6) showing the wide humus (H) infill on the right, and thin humus (TH) on the left. The humus is now almost totally replaced by iron and manganese staining. The central fill (CF) is less humic and is characterized by dusty clay void infills (cf. Fig. 9.5). The control soil (CS) profile (i.e. the local natural *argillic* Bt subsoil) outside the tree hole is both poorly humic and contains few textural pedofeatures compared to the tree-hole (Macphail, forthcoming).

It is commonly thought that one method of forest clearance took the form of simply burning down trees. Deciduous trees do not readily burn unlike conifers, however, and other methods of tree removal must be envisaged. Recent studies have suggested ringbarking as a technique (where the bark is removed in a ring from around the base of the tree, for example, with a stone axe), and once dead, pulled down (Barclay *et al.*, 2003; Healy and Harding, forthcoming; Romans, personal communication). There is in addition, clear soil micromorphological and bulk data showing that trees, once pulled down (or blown over) were burned *in situ*; Stewart (1995: 37–38) has illustrated methods of using controlled burning for cutting down, and cutting through, cedar in northwestern America. The reader is referred to the many classic and modern studies on the character of ancient forests and woodland (Peterken, 1996; Tansley, 1939) and the effects of human impact upon them, for example, across Europe Berglund, 1986 Huntley, 1990 and the Near East (Zeist and Bottema, 1990). Soil evidence of forest clearance and woodland management is presented below.

An excellent example of early clearance and woodland management was identified for Megalithic Brittany, France. Gebhardt (1993) was able to integrate her soil and micropedological findings with pollen and charcoal data when identifying clearance, management by fire, and leaching of soils at classic Neolithic buried soil sites such as the famous location of the megalithic monument of Er Grah. Even on the acid granitic substrates of Brittany, there was an instance of calcareous ashes being preserved by rapid burial through monument construction. Similarly, the first effects of podzolization could tentatively be dated to Neolithic occupation of the same granite in Cornwall (southwest England) at the outer rampart of the Carn Brea "fort" (Macphail, 1990). Here, a highly leached topsoil (Ah and A2 horizon) had apparently developed during only 1–2 centuries of local land management that probably resembled that recorded by Gebhardt (1993). Only a weakly formed spodic horizon had begun to form at Carn Brea, however, and was superimposed on the acidic cambic Bw horizon that exhibited earthworm worked soil. In order to construct the rampart, this soil was probably cleared of shrubs, and burning had led to the present bAh/A2 (Albic) horizon becoming rich in fine charcoal and phytoliths. Trampling, probably by people as opposed to animals, had compacted the surface soil and formed dusty clay coatings in closed vughs, seemingly immediately ahead of rampart construction (Courty *et al.*, 1989; Macphail *et al.*, 1990). On the same Cornish granite, however, at the Bronze Age kerb cairn-buried site at Chysauster, the contemporary Cambisol had formed under open oak/hazel woodland with local (possibly on-site) cereal cultivation. Whereas the Carn Brea soil had been completely sealed, at Chysauster the "open" ca 0.50 m thick kerb cairn had allowed post Bronze Age podzolization to alter much of the buried soil even under a large kerb stone (Smith *et al.*, 1996). Similar studies have been undertaken in Scotland (Mercer and Midgley, 1997). The danger of mistaking postburial podzolization for ancient podzolization has been recognised for a long time, and the continuation of the chemical effects of podzolization was demonstrated at the 33-year-old Wareham Experimental Earthwork (see Chapter 12) (Fisher and Macphail, 1985; Macphail *et al.*, 2003; Runia, 1988).

Clearance and its effect on other soil types and environments have been less well studied, but the likely clues to clearance activity are the same (see below). In areas of tropical Oxisols the rainforest produces a highly nutrient (N–P-bases) rich humus, which can be utilized for a few years for shifting agriculture by slash and burn, a practice that can eventually lead to secondary savannah and the loss of this fertile humus layer (Duchaufour, 1982: 412–413).

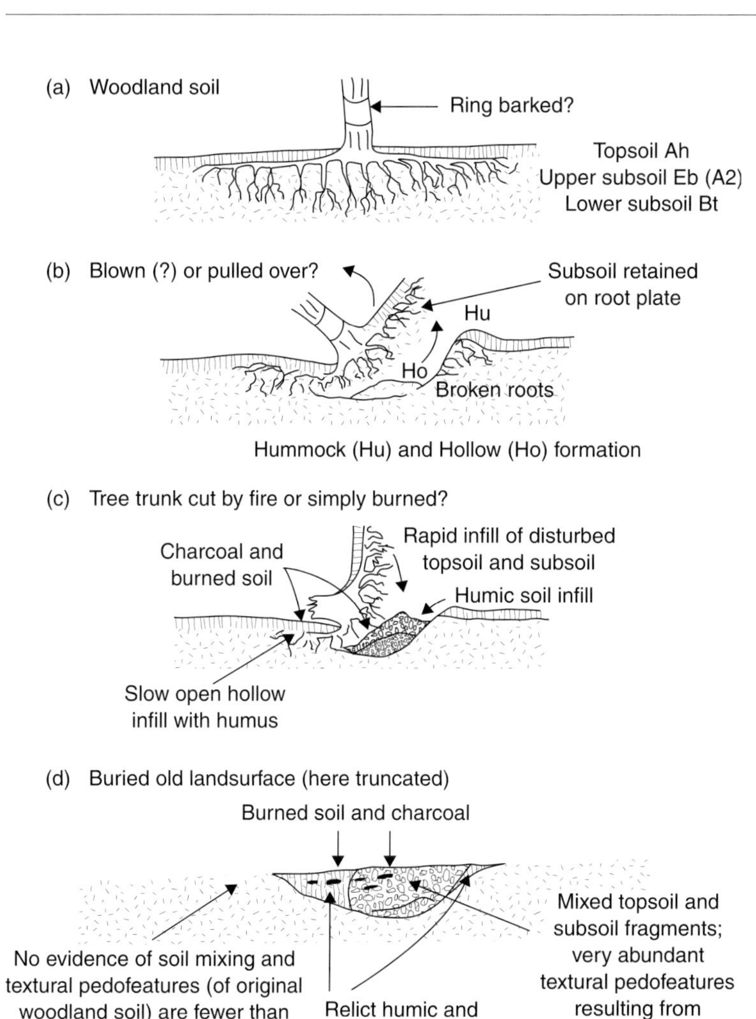

FIGURE 9.8 Model of treethrow field soil features; (a) broadleaved woodland soil where clay is translocated into the lower subsoil Bt horizon by *lessivage* (see figure 3.2b–d) – trees can be killed by ringbarking and are thus more easily pulled down and burned; (b) clearance may have taken place after trees were blown down or pulled down after ringbarking, and produced hummock (Hu) and hollow (Ho) landscapes; (c) *in situ* burning of the tree produced charcoal and burned soil (with an enhanced magnetic susceptibility), whereas mainly subsoil fell off the root plate, humic topsoil infilled beneath the tree trunk and around the edges of the treethrow pit; (d) treethrow disturbance produced a mixed soil infill and very abundant textural pedofeatures compared to undisturbed soils between treethrow features. (After Barclay *et al.*, 2003; Courty et al., 1989: 127; Langohr, 1991; Macphail and Goldberg, 1990; Macphail and Linderholm, 2004.)

9.3 Forest and woodland clearance features

Soil evidence identifying clearance of trees scrub woodland, and shallower rooting shrubs has been found at a number of archaeological sites. This evidence includes the identification of disrupted soils with possible inclusion of charcoal, phytoliths, and burned soil (Figs 9.1–9.10). These findings have been used to create some general models of vegetation clearance that can be utilized during soil investigations (Courty *et al.*, 1994, table 4; Macphail *et al.*, 1990, table 1; Macphail, 1992b, table 18.2). These models were also influenced by classic studies of American woodland soils and geology (Denny and Goodlett, 1956; Lutz and Griswold, 1939). In brief, shallower (0–20 cm) soil disturbance is caused if shallow rooting small plants are pulled or grubbed up, in comparison to deeper rooting trees and scrub (20/40–100 cm); some modern pit-and-mound (or hummock and hollow; Figs 9.7 and 9.8) topography reaches up to 4 m high (Peterken, 1996). In the case of shallower rooting shrubs,

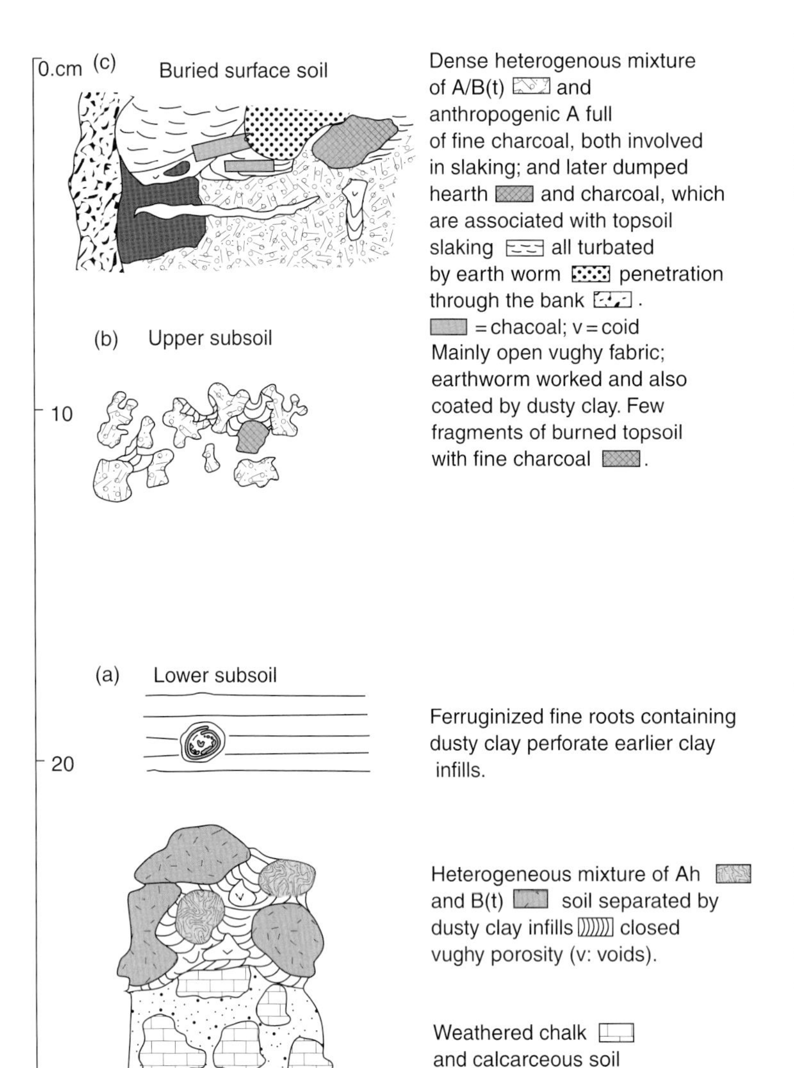
0.cm
10
20
30 cm
(c) Buried surface soil
Dense heterogenous mixture of A/B(t) and anthropogenic A full of fine charcoal, both involved in slaking; and later dumped hearth and charcoal, which are associated with topsoil slaking all turbated by earth worm penetration through the bank .
= chacoal; v = coid
(b) Upper subsoil
Mainly open vughy fabric; earthworm worked and also coated by dusty clay. Few fragments of burned topsoil with fine charcoal .
(a) Lower subsoil
Ferruginized fine roots containing dusty clay perforate earlier clay infills.
Heterogeneous mixture of Ah and B(t) soil separated by dusty clay infills closed vughy porosity (v: voids).
Weathered chalk and calcarceous soil

FIGURE 9.9 Model of soil formation resulting from Neolithic clearance at Maiden Castle, Dorset, United Kingdom and construction of enclosure bank – a soil micromorphological sequence (not to scale); (a) at 20–30 cm depth: clearance of woodland (land snail evidence) mixed decalcified Bt horizon soil (formed in non-calcareous drift) and chalk rock and chalky subsoil, with dusty clay infill features that also resulted from this disturbance becoming rooted by the revegetation of the area; (b) at around 10 cm depth: the once compact soil had been opened up and biologically homogenized by earthworm activity, but continuing surface disturbance (anthropogenic activity) led to voids becoming coated with dusty clay and mixing with burned charcoal-rich topsoils; (c) 0–5 cm: pre-enclosure activities and trampling had produced a puddled/slaked and mixed surface soil (probable bare muddy ground) containing burned soil and charcoal. Post bank-burial earthworm burrowing produced chalky soil infilled channels. (Macphail, 1991)

only A and A2 horizon material is likely to become turbuted, whereas if deeper rooting trees are cleared or pulled over, subsoil horizon fragments (e.g. Bt, Bw, C) become mixed with topsoil material (see Fig. 9.10; see also Box 3.1). Often this disruption is accompanied by loose soil being washed into fissures within the disturbed soil hollow, and orientation of the resulting textural pedofeatures is "right way up" (Figs 9.5 and 9.9). In contrast, fragments of relict mixed-in Bt horizon material may show intra-ped textural features that are not "right way up." These findings partly derive from the study of recently (World War II) cleared soil in the Italian Apennines, and numerous ancient examples of treethrow holes (Courty *et al.*, 1989: 286–290). Furthermore, natural treethrow holes dating to the Atlantic period in England (Balksbury Camp and Hazleton long barrow) have been differentiated from those which are

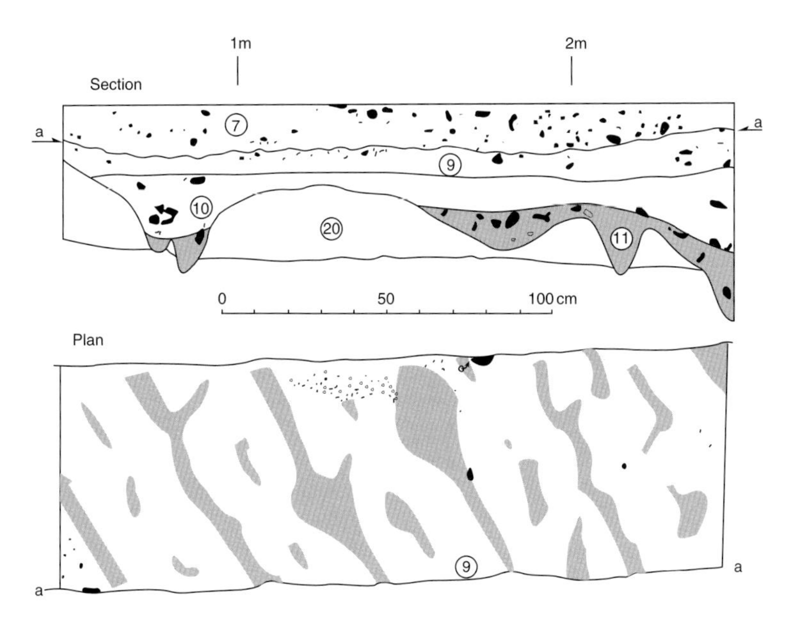

FIGURE 9.10 Beaker cultivation at Ashcombe Bottom: plan and section of possible ard marks at the Ashcombe Bottom site, Lewes, East Sussex, redrawn from Allen (1995: 80); a machine trench cut across the dry valley (Sussex Chalk Downs); layer numbers are as follows: 20 – valley bottom periglacial solifluction deposits; 11 – truncated early Holocene soil (argillic brown earth; Macphail, 1992a); 10 – Late Neolithic/ pre-Beaker colluvium; 9 – putative prehistoric old land surface formed of colluvium containing abraded Beaker pottery (consistent with ploughing) and featuring at its surface some 14 parallel grooves – possible ard marks (see Fig. 9.11); 7 – later prehistoric colluvium (see Chapter 4).

associated with human activity (e.g. Raunds). Pedological studies as well as personal observations (RIM) of pit-and-mound ground formed by past treethrow – including that caused by the hurricane that hit southern England in 1987 – show that treethrow holes receive (and sometimes hold) more water and humus than the surrounding soil (Peterken, 1996; Veneman *et al.*, 1984) (Figs 9.1–9.3; 9.6, 9.7). The remains of an Atlantic woodland molluskan fauna was recovered from within the subsoil hollow at Hazelton (Figs 9.3 and 9.4), despite decalcification of the (later) Neolithic ("top") soil, and at Balksbury Camp (Fig. 9.2) carnivorous mollusks were included in the species list (Allen, 1995; Saville, 1990). That subsoil hollows formed by trees become infilled with organic-rich soil has long been recognized in the archaeological record (Limbrey, 1975). In plan view (after the modern topsoil and overburden has been removed), these features, are 3–5 m across, and have a banana shaped dark infill on one side (Figs 9.6 and 9.7); The last is due to the concentration of relict organic matter here. This is sometimes strongly replaced by iron and manganese mottling where water tables have risen (as in the floodplains of the Upper Thames and Nene, United Kingdom, for example), but measurements of organic carbon and LOI can still be twice as much as those found in the rest of the soil fill or in the soil profile outside the hollow (see Table 16.4). At one site at Raunds, Northamptonshire, United Kingdom, where a fallen tree was burned in situ magnetic susceptibility enhancement was very high: with values ($\times 10^{-8}$ m^3 kg^{-1}) of 894 units (concentrated rubefied soil – burned "clay") and 96 units (fragmented rubefied soil) (Macphail and Goldberg, 1990). These high values of χ can be compared to other areas of the fill (8–10 units), including the dark iron and manganese mottled soil fill (22–44 units) and the overlying alluvium (11–26 units).

As discussed in Chapter 4, vegetation clearance activities on slopes can lead to erosion and colluviation, for example, as documented at

Brean Down, Somerset (Bell, 1990) (see Box 3.1). In the Italian Apennines, it is possible that mineralogenic infilling of low ground contributed to peat initiation (Cruise, 1990). A major impact on woodland and the landscape in general that is contemporary with colluviation and is dated to the Chalcolithic in Italy and the Beaker period in southern England (Macphail, 1992b) (see below).

9.4 Cultivation and manuring

Sites of ancient cultivation have been identified worldwide, from the Americas (Jacob, 1995; Sandor, 1992) and Europe (Gebhardt, 1990, 1992, 1995) to Asia (Matsui *et al.*, 1996; Miyaji, forthcoming; Sunaga *et al.*, 2003). The most obvious signs of cultivation are cultivation field features, such as ard marks, plough marks, and spade marks (Fig. 9.11; Box 9.1). These, however, must all be treated with caution, and not confused with drag lines, bases of wheel ruts, linear periglacial features, mole drains, and other modern features. Other larger field features (e.g. lynchets, cultivation ridges, and terraces) are also often a clear means to identify *in situ* areas of cultivation (Barker, 1985; Evans, 1999; Fowler and Evans, 1967; Sandor, 1992). The presence of cereal pollen, charred cereal grains, and soil micromorphological features, which may be indicative of cultivation, also need to be interpreted carefully. For example, cereal pollen is often strongly associated with areas of crop processing (Behre, 1981; cf. Bakels, 1988; Segerström, 1991). Equally, soil micromorphology data must be treated with caution, as a number of processes can produce features suggestive of cultivation, following the

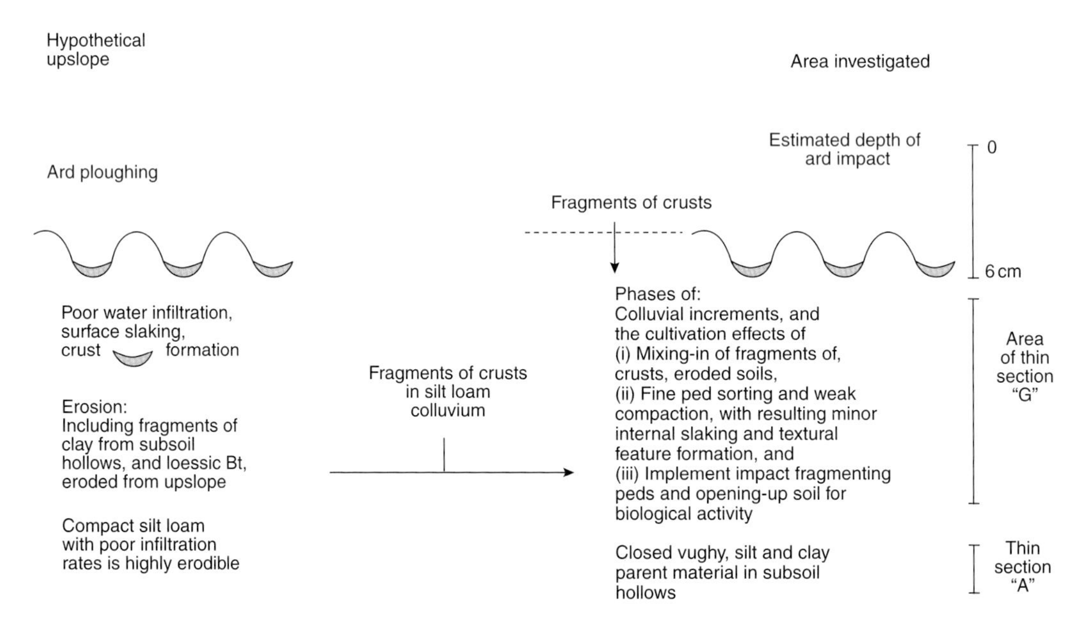

FIGURE 9.11 Beaker cultivation at Ashcombe Bottom: soil micromorphology and model of formation processes involved in developing a colluvium characterized by coeval biological activity and textural pedofeatures formation (from supposed ard impact) – including surface crusts (Macphail, 1992a; cf. Boiffin and Bresson, 1987; McIntyre, 1958a and 1958b).

law of equifinality (see Section 16.6). It is therefore crucial that all human and environmental factors are taken into consideration when employing soil evidence to locate cultivation (Carter and Davidson, 1998; Courty *et al.*, 1989; Macphail, 1998). Some generalizations can be listed as follows:

- *Period* (Neolithic through to Medieval and post Medieval): with markedly different impacts being envisaged, and manuring only being considered at Neolithic sites on a site-by-site basis whereas a heavy manuring regime is expected for many medieval and historic sites (Bakels, 1988; Simpson, 1997, 1998; van de Westeringh, 1988)
- *Culture* influences how cultivation is carried out and agriculture managed (Thomas, 1990): examples of different crops and cultures are: eastern Asian dry field and wet field rice cultivation (e.g. Yoyoi Period); Mesoamerican Maya lowland maize cultivation; semiarid cultivation in Middle and Near East Asia (prehistoric Upper Mesopotamia) and southwest United States (e.g. pre-Pueblo cultures); prehistoric cereal cultivation in western Europe (Engelmark, 1992; Matsui *et al.*, 1996; Miyaji, forthcoming; Sandor, 1992; Wilkinson, 1990, 1997)
- *Soil type and environment* will affect the way soils will respond to cultivation, and consequently the character of the archaeological soils we study now. Some examples of extreme variations of cultivated soils presented in this chapter and in the literature are: Spodosols *to* Rendolls, tropical *to* alpine soils; stable humic topsoils *to* truncated poorly stable subsoils; poorly humic soils with low biological activity *to* highly humic fertile and totally biologically worked soils (e.g. Gebhardt, 1990, 1992, 1993, 1995; Macphail, 1992, 1998; Macphail *et al.*, 1990)
- *Crop* old world cereals, maize, potatoes, rice, vegetables, and vines require different cultivation regimes, many of which will produce recognizable field, micromorphological, and chemical features
- *Cultivation regime* slash-and-burn, ridge and furrow, arable field, horticulture; wet or dry cultivation; terracing; high intensity infield and low intensity outfield manuring; "rural" manuring with animal dung and nightsoiling around populated areas – have been recognized in the archaeological soil record both as microscopic traces and through phosphate chemistry for example (Table 9.1),
- Cultivation tool (and method): hoe; shallow and deep (rip-) ards (Figs 9.1–9.11); mouldboard ploughs (Box 9.1) and swing ploughs; spades and digging sticks; impact of different animal traction; parallel or crisscross ploughing. All have had different impacts that range from local differences in chemistry and micromorphology to major erosion and colluviation (Gebhardt, 1992, 1995; Lewis, 1998; Macphail, 1992b; Mikkelsen and Langohr, 1996; Sunaga *et al.*, 2003; Van Vliet-Lanoë *et al.*, 1992, figure 9.4).

The only way to tackle such complicated issues is to utilize well-studied cultivation sites, as archaeological analogues of various ages and character, and to learn from experiments (see Chapter 12). An example of an archaeological analogue study is given in Box 9.1, for the site of Oakley, Suffolk, United Kingdom (see also Macphail *et al.*, 1990). Here, Late Roman/Saxon cultivation, as first indicated by field features (mouldboard ploughmarks) at field edge and within field locations, is examined in terms of its chemistry and micromorphology. An important issue here is the effect of cultivation and manuring on Spodosols, a practice that in The Netherlands eventually led to plaggen (very thick cultosol; Table 3.2) soil formation during medieval times (van de Westeringh, 1988). A simple finding at Oakley is the development of a fine tilth

Box 9.1 Cultivation at Late Roman/Saxon Oakley, Suffolk, United Kingdom

At the rural Roman site of Oakley (Suffolk), just outside and across the river Waveney from the small Roman town of Scole, Norfolk, a number of archaeological contexts were investigated. This included a late Roman/Saxon cross-ploughed field, where six thin section samples were taken to study the interior and field edge soils. Associated pollen samples were taken from two of the Kubiena boxes, before resin impregnation was carried out.

Modern soil cover

The soil history of the site is complicated and remains partly conjectural especially as regards the nature of the post-Roman fluvial inundation (Macphail and Linderholm, 1996). The dominant natural soils that appear from fieldwork to be of Roman age, are Aquods (typical gley podzols and humo-ferric gley-podzols; for example, Sollom 1 and Crannymoor soil associations respectively Hodge *et al.*, 1983). The same soils are found across the road at the Scole-Dickleborough site and would have been acid in antiquity (e.g. pH 4–5, cf. Avery, 1990: 266). The general area of Scole, however, has been broadly mapped as having a sandy brown soil cover Hodge *et al.*, 1983, but this is misleading. In addition, the site is bounded to the north by an arm of fen peat and river alluvium from the river Waveney (Mendham soil association Hodge *et al.*, 1983) and such sediments could have influenced the infill of the palaeochannel. Again, although not mapped, the site appears to also have received a later thin cover of probably eutrophic alluvium, inducing the anomalous present day pH 7.0–7.6 ($n = 4$) for the buried Spodosols here. Thus, the fen clay that buried the site may well have had pHs ranging from pH 6–8 (Avery 1990: 300 *et seq.*).

The buried Ap horizon has a little more organic matter (1.37–1.53% LOI) and 2% citric acid extractable P (194–228 ppm P) compared to the field's buried A2g horizon (0.49% LOI and 47.6 ppm P), and local Roman buried A2 horizons at Scole (0.44–0.59% LOI, 41.9–58.9 ppm P). The buried illuvial spodic B horizons can also be compared chemically (bBsg – 1.43% LOI, 213 ppm P; bBh 0.89% LOI, 120 ppm P). The very low magnetic susceptibility of the bAp ($\chi = 1.4-2.2 \times 10^{-8}$ Si kg^{-1}) can be compared to the extremely low readings for the bA2g ($\chi = 0.1-0.9 \times 10^{-8}$ Si kg^{-1}) and illuvial B horizons (bBsg: $\chi = 2.1 \times 10^{-8}$ Si kg^{-1}; bBh: $\chi = 24 \times 10^{-8}$ Si kg^{-1}).

Two areas of the ploughsoil were examined in thin section. Their subsoils showed two types of amorphous pedofeatures, (1) a spodic monomorphic sesquioxidic type of grain coating and (2) ferruginous nodular formations of hydromorphic (gley) origin (sample **3**). Another type of iron nodule in the ploughsoils is discussed below. A ploughmark from the interior of the field contains more frequent anthropogenic inclusions and features that are anomalous in a natural podzol, compared with the ploughmark by a field edge. Both areas contain burned flint, coarse wood charcoal and admixtures of (alluvial?) clay, with void coatings and earthworm burrows. Samples from the interior of the field (Fig. 9.12), however, have in addition, inclusions of brickearth (constructional material?), clay (raw material?) and anomalous calcareous materials such as a single biogenic calcite (earthworm) granule and an additional iron nodule type. The last contains rare calcite ash embedded with charcoal in this probable phosphatic amorphous iron nodule, as indicated by *in situ* vivianite (crystalline iron phosphate). It was also noted that this interior field ploughsoil also appears to contain more evidence of

(cont.)

Box 9.1 *(cont.)*

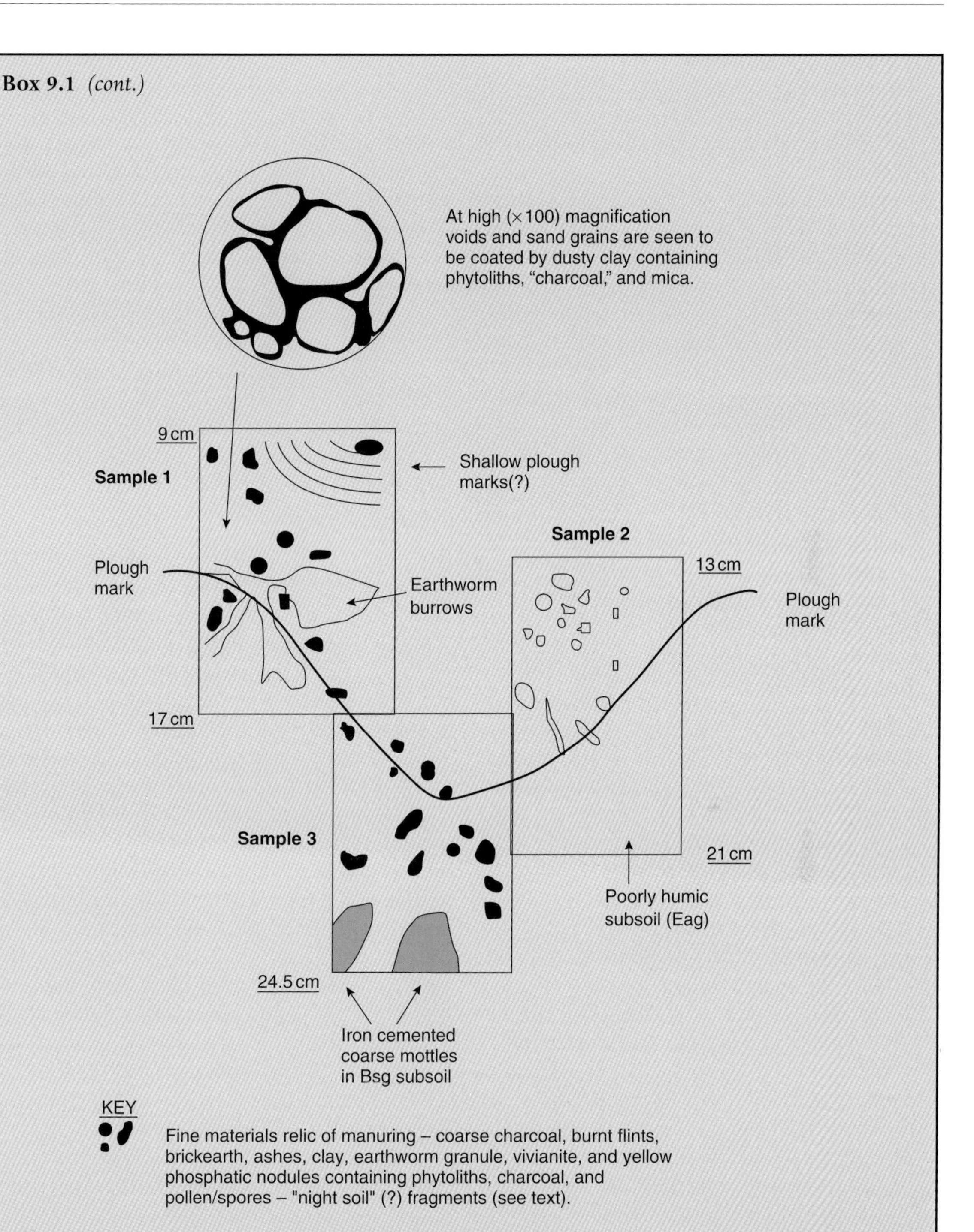
At high (×100) magnification voids and sand grains are seen to be coated by dusty clay containing phytoliths, "charcoal," and mica.
9 cm
Sample 1
Shallow plough marks(?)
Sample 2
13 cm
Plough mark
Earthworm burrows
Plough mark
17 cm
Sample 3
21 cm
Poorly humic subsoil (Eag)
24.5 cm
Iron cemented coarse mottles in Bsg subsoil
KEY
Fine materials relic of manuring – coarse charcoal, burnt flints, brickearth, ashes, clay, earthworm granule, vivianite, and yellow phosphatic nodules containing phytoliths, charcoal, and pollen/spores – "night soil" (?) fragments (see text).

FIGURE 9.12 Diagram showing three thin section samples (1–3) and their soil micromorphology across a late Roman/Saxon mouldboard plough mark at Oakley, Suffolk, United Kingdom (Macphail, 1998).

(cont.)

Box 9.1 *(cont.)*

earthworm burrowing compared to the field edge site.

At both locations anthropogenic inclusions are not presently restricted to the obvious mouldboard ploughmark (Fig. 9.12). The blackish humic soil, that displays very low interference colors (XPL) is rich in fine charred organic matter and includes phytoliths and mica, and this loosely infills packing voids between sand grains (enaulic related distribution). It can also form textural features (void coatings and infills). At the top of sample **1** curved laminar fills, some 1.5 cm deep and 3.5 cm wide, are present, possibly as humic sand infills of another shallower furrow. The laminae are picked out by thin concentrations of humic fine soil and iron staining, which may represent the effects of flooding that postdates the agricultural activity.

Evidence for the presence of ploughsoils:

1. macromorphological features indicative of deep (major field features) and shallow (surface furrow infill) ploughing (Barker, 1985; Fowler and Evans, 1967);
2. similarly, a pattern of anthropogenic inclusions in part respecting the morphology of the ploughmark suggesting manuring (cf. Wilkinson, 1990, 1997);
3. a general microfabric of grain coatings and loose infillings consistent with modern analogue studies of podzols transformed by ploughing, but with no evidence of the soils undergoing total slaking as would have been the case if ploughed when waterlogged (Jongerius, 1970, 1983).
4. dusty clay void coatings and infills resulting from the mechanical disturbance induced by ploughing (Jongerius; 1970 1983; see main text);
5. anomalous burrowing by earthworms and residual presence of calcareous materials in what was a naturally acidic and earthworm impoverished soil, as a response to the enhanced base levels that resulted from manuring (Courty *et al.*, 1989: 133; Wilkinson, 1990, 1997; Macphail, 1992a);
6. a specific chemical signature indicative of manuring, albeit low in total citric acid extractable P, but consistent with other cultivated podzols (Engelmark and Linderholm, 1996; Liversage *et al.*, 1987; Macphail, 1998; Pape, 1970);
7. a pollen spectra, albeit highly sparse, not inconsistent with the presence of cereals, grasses, and herbs, including weeds of fallow ground (Wiltshire, 1995).

This wide range of findings reflect probable arable activity accompanied by low intensity manuring at Oakley. In the Middle East, just the spread of residual artifacts around settlements has been utilized to imply manuring of arable fields (e.g. Wilkinson, 1990). At Oakley not only are highly residual materials such as burned flint, charcoal, and constructional waste (brickearth) present, but also there are clear examples of calcareous and phosphatic materials being dumped. A further speculation is that occasionally "nightsoil," which is regarded as typical Roman fertilizer (Dave Sankey, MoLAS, personal communication, 1996), can be identified. It is preserved in the form of phosphatic amorphous iron nodules, in which ashes and charcoal are embedded. (Similar "nightsoil" nodules of amorphous phosphate (P, Fe, and Ca) and neoformed vivianite, have been studied with microprobe from St Julien, Tours, France and characterize Late Roman and Early Medieval horticultural practices; Macphail and Crowther, 2005) (see also Table 9.2).

In summary, at Oakley nonintensive manuring (cf. plaggen soils; Pape, 1970) was presumably employed to raise the pH and general fertility of what must have been originally

(cont.)

Box 9.1 *(cont.)*

highly impoverished soils, the earthworm burrowing testifies to their success in improving the soils. Burrows and evidence of soil mixing by soil animals are present alongside dusty clay void coatings, thus apparently showing biological activity to be coeval with the formation of textural features in these cultivated soils. Such juxtaposition of features is typical of some ancient and experimentally cultivated soils (see main text).

composed of phytoliths, "clay," and fine organic matter that forms a loose fill between sand grains, and grain and voids coatings. Moreover, the presence of earthworm burrows (and a biogenic earthworm granule formed out of calcite) is totally anomalous in Spodosols, as is the presence of "exotic" inclusions. Current amounts of phosphate in the Ap horizons are not high, but around three times that in the bA2 horizon, and similar to amounts found in experimentally manured Spodosols in Sweden (see Chapter 12). It can also be noted that arable fields on the outskirts of Roman Canterbury, Kent, United Kingdom (now sealed by the late third century AD earth-based town ramparts), also received small amounts of town waste that included, "nightsoil," bone, and other coarse anthropogenic ("exotic"/allocthonous) inclusions (Table 9.1a). These comprised gravel-size flint, burned flint, wood charcoal, an example of articulated phytoliths (cereal processing waste?), and fragments of amorphous organic matter. Here, where the soils are naturally more fertile Alfisols, ploughsoils were found to have 1.4 to 1.9 times as much P as the underlying natural subsoil (Bt horizon formed in fine loam), and also show rather higher levels for magnetic susceptibility and heavy metal concentrations (Cu, Pb, Zn) (Macphail and Crowther, 2002).

Dark Age to Medieval cultivated and manured colluvial loessic soils at the Büraberg, North Hessen, Germany were found to contain 2.5 times as much phosphate and twice the magnetic susceptibility signal (χ_{max}; Section 16.4) as the substrate (Macphail and Crowther, 2003; Henning and Macphail, 2004). Similarly, Sunaga *et al.* (2003) found that volcanic mudflow buried Edo Period cultivated soils at two sites in central Japan that had a higher available phosphate content and bulk density, compared to buried uncultivated soils, which in contrast were characterized by more total carbon, total nitrogen, and "easily decomposable organic matter."

9.5 Landscape effects

Changes to the morphology of the landscape through cultivation activities, erosion, colluviation, and anthropogenic landscaping (e.g. terracing) are important to recognize. In Europe, large-scale studies of arable agriculture-induced colluviation have been conducted, for example, on the loess landscapes of France, Belgium, and Germany. In the Belgian Ardennes, thick colluvial deposits have been associated with medieval arable cultivation of buckwheat (*Fagopyrum esculentum*; Kwaad and Mücher, 1977, 1979). Van Vliet *et al.* (1992) suggested that there were, for western Europe as a whole, links between amounts of erosion and climatic impact (e.g. cold wet winters of the seventeenth-century Little Ice Age), deforestation (e.g. first clearances), agricultural practices (e.g. arable versus grazing; introduction of new technology such as swing plough, horse traction), demographic pressure (e.g. as lessened by the plague of the fourteenth century),

TABLE 9.1a Chemical and magnetic (χ) properties of samples analysed at Canterbury CW12; late third century AD Roman buried soils and town rampart (modified from Macphail and Crowther, 2002)

Context	Sample	LOI (%)	$CaCO_3$ (%)	pH (1:2,5, water)	Phosphate- P_i (mg g^{-1})	Phosphate- P_{oh} (mg g^{-1})	Phosphate- P (mg g^{-1})	Phosphate- P_i:P (%)	Phosphate- P_o:P (%)	χ (10^{-8} SI kg^{-1})	χ_{max} (10^{-8} SI kg^{-1})	χ_{conv} (%)	Pb (μg g^{-1})	Zn (μg g^{-1})	Cu (μg g^{-1})
South face															
Rampart	1a	1.34	0.224	7.9	0.752	0.267	1.02	73.8	26.2	26.1	1200	2.18	6.9	10.4	5.2
1212	1b	1.68	0.629	7.8	1.70	0.258	1.95	86.8	13.2	38.2	795	4.81	11.5	18.2	10.5
1212	1c	1.67	0.776	7.8	1.34	0.247	1.59	84.4	15.6	52.1	729	7.15	9.5	17.7	9.2
1212	1d	1.71	0.738	7.8	1.15	0.233	1.38	83.2	16.8	34.3	722	4.75	9.4	17.2	8.6
Natural	1e	1.27	0.738	7.9	0.744	0.218	0.962	77.3	22.7	20.0	795	2.52	7.5	9.5	4.5
East face															
1212	2a	1.76	0.751	8.0	0.971	0.298	1.27	76.5	23.5	47.0	801	5.87	9.8	16.1	8.4

Soils: Rampart = dump of loessic brickearth; 1212 = lightly manured (soil micromorphology) Ap brickearth topsoil; Natural = brickearth subsoil.

n. b: Pi = inorganic fraction: P_o = organic fractin; P = total Phosphorus.

TABLE 9.1b Chemical and magnetic susceptibility (χ) data from the site of the Dark Age Carolingian (eighth century) fortress of Büraburg, Nordhessen, Germany (Crowther in Henning and Macphail, 2004: 237)

Context/ sample no.	LOI (%)	Carbonate estimate[a] (%)	Phosphate- P_o(mg g^{-1})	Phosphate- P_i (mg g^{-1})	Phosphate- P (mg g^{-1})	Phosphate- P_o : P (%)	Phosphate- P_i : P (%)	χ (10^{-8} m^3 kg^{-1})	χ_{max} (10^{-8} m^3 kg^{-1})	χ_{conv} (%)
4/4	n.d.[b]	1	0.392	2.12	2.51	15.5	84.5	41.3	393	10.5
3/3	n.d.	2	0.358	2.25	2.61	13.8	86.2	41.2	417	9.88
3/3a	1.69	2	0.407	2.34	2.75	14.9	85.1	41.2	483	8.53
2/2	n.d.	0.1	0.295	1.20	1.50	20.0	80.0	44.9	346	13.0
1/1	n.d.	0.5	0.312	0.696	1.01	31.1	68.9	39.9	774	5.16

[a] Where values are recorded as 0.1%, no carbonate was detected.

[b] n.d. = not determined (insufficient sample).

Soils recovered in auger core: 4/4 = manured loessic ploughsoil colluvium; 3/3 and 3/3a = occupation debris soil with stones, mortar, charcoal, and possible burned soil, relict of razed eighth century AD Dark Age fortress; 2/2 = buried disturbed soil; 1/1 = weathered and stained Triassic Bunter Sandstone natural.

and economic pressures (e.g. expansion of arable agriculture during World War II in the United Kingdom). Similar trends can be identified in the United States where in the west, once-stable prairie soils were ploughed and became susceptible to both rainsplash erosion (see Chapter 4) and deflation ("Dust Bowl").

The use of terracing to mitigate soil loss and to create flat ground has already been described in Chapter 4, but terraces have also been employed to both manage water in arid areas (e.g. Mesomerica, the Mediterranean, and Middle East; French and Whitelaw, 1999; Sandor, 1990; van Andel *et al.*, 1990; Wagstaff, 1992), and to create "wet rice" paddy fields. In arid areas, soils used for agriculture need to be irrigated, and fields have associated channel systems. When dry soils are wetted through irrigation they may become water saturated and, in thin section, may exhibit a collapsed structure, which is manifested as a void type known as vesicles, where air bubbles were once trapped (Courty *et al.*, 1989: 136–137). When ancient soils are characterized by biological features (e.g. root channels and excrements of mesofauna) that indicate anomalous high levels of biological activity in "arid" soils, and relict vesicles are still visible in thin sections, these two types of pedo-features together may infer irrigated cultivation. In paddy soils the use of terraces to hold water can be suggested by microfeatures indicative of gleying (hydromorphism), such as iron and manganese nodules and soil impregnation (Bouma *et al.*, 1990; Matsui *et al.*, 1996). At Nara, Japan, a large thin section sample taken from the horizontal plane in an ancient paddy soil exhibited 2 mm wide hollow circles of iron and manganese impregnation grouped into five to seven of these circles. These have been interpreted as being relict of rice planting, when small bunches of rice were planted at a time (Matsui *et al.*, 1996). The adding of manure/"night soil" and the trampling of water-saturated soil by large animals (e.g. traction buffalo) and humans has produced distinctive dark, mottled, and massive structured soils. The effects of stock on Neolithic soils can be recorded both chemically (see Tables 16.4 and 16A. 5) and in thin section (see figure 16.5) (Macphail and Linderholm, 2004).

9.6 Conclusions

Human impact on the landscape has been almost ubiquitous across many parts of the planet. Sometimes this impact has been slight whereas in other areas it has altered the whole landscape. The role of the geoarchaeologist is to attempt to identify the part played by humans versus natural phenomena, such as climatic change. Workers in other environmental disciplines have the same task, and can provide vital complementary information. Often, sites are multi period, and the effects of each culture need to be differentiated in time and space, if possible. It is therefore necessary to be aware of the subtler influences of humans as well as major and more obvious impacts.

10

Occupation deposits I: concepts and aspects of cultural deposits

10.1 Introduction

One of the most important aspects of this book is its approach to archaeological sediments *sensu stricto*. This is a subject that is largely Passed over textbooks on geoarchaeology, and is actually explored in Courty *et al.* (1989) only within the topic of "anthropogenic features." It is only over the last 20 years that occupation deposits have begun to be studied in detail, with reporting of archaeological, ethnoarchaeological (Brochier *et al.*, 1992; Stiner *et al.*, 1995), and experimental studies increasing exponentially over the last ten years (see below). This research has demonstrated that such anthropogenic deposits *cannot* be studied by the standard methods and mind-frame used in traditional soil science and geology (traditional "geoarchaeology"). This is because such "traditional" analyses and approaches, no matter how detailed, time-consuming, or expensive, are unlikely to recover and reveal the full cultural and environmental potential of the deposits. Furthermore, this potential can often be found only through detailed microstratigraphic observations and analyses of microfacies *sensu stricto* (Courty, 2001) (also Chapter 2) using microscopic analysis, and bulk methods that are aimed to address the special nature of anthropogenic deposits (e.g. phosphate fractionation, magnetic susceptibility fractional conversion). Some archaeologists have also recognized the special character of occupation deposits: they are unique in their ability to reveal specific human activities such as *in situ* occupation, the accretion of floor deposits, or the dumping of hearth waste from elsewhere. This realization has led archaeologists to model such site formation processes accordingly (e.g. Schiffer, 1987) and has provided an essential link between theoretical archaeology and geoarchaeology.

Yet, as models are just a *starting point* in any study, they can be taken only so far and they must be adjusted or even rejected according to geoarchaeological findings. In any case, it is becoming increasingly evident that archaeological deposits are an integral part of the archaeological record and they have a story to tell about the human activities that occurred at any site (Goldberg and Arpin, 2003). To ignore them or study them in cursory fashion would be equivalent to looking only at the color of lithic artifacts, for example. The first part of this chapter examines the nature of occupation deposits, and the many possible ways that they may have formed. The ensuing Chapter 11 examines examples of occupation deposits from the Near East (tells), North America (mounds), and Europe (Roman, Saxon, and Medieval deposits).

10.2 Concepts and aspects of occupation deposits

We begin with a review of some of the concepts and aspects of archaeological deposits, and show how cultural archaeological questions can be addressed through geoarchaeological approaches. Later sections deal with archaeological deposits and their stratigraphic organization as expressions of *material culture, discard, use of space*, and other *formation processes* (see Table 10.1). We also discuss settlement–landscape interrelationships, the origin of occupation materials, and their depositional history, along with some examples of postdepositional transformations that affect occupation deposits.

Contextual analysis and depositional history. When Renfrew stated that "all problems in archaeology begin as a problem in geoarchaeology," he highlighted the fundamental importance of understanding the depositional context of all artifacts and palaeoecological remains (Renfrew, 1976). These remains have complex predepositional, depositional, and postdepositional histories (Clarke, 1973; Courty *et al.*, 1989). Detection and identification of these histories depend on methods and scale of excavation, sampling, and analysis, as well as research questions, concepts, and the perceived importance of each particular deposit and context (Hodder, 1999) (see Chapters 15–17).

Modeling site formation processes. Schiffer (1987) developed the concept of archaeological site formation processes, synthesizing and building on the work of other researchers (e.g., Butzer, 1982). He grouped these processes conceptually into natural formation processes (N-transforms) and cultural formation processes (C-transforms) to aid identification and discussion of the associated attributes. Many deposits are formed and altered by a complex interaction of these agencies and processes, as discussed throughout this book.

Traces of natural agencies and processes in archaeological sites can be identified by referring to a wide range of research in sedimentology/geology, soil science, and biological sciences, as well as geoarchaeological, experimental, and ethnoarchaeological research (see Part I, and Chapter 12). The nature and impact of these agencies and processes will vary according to sediment type, hydrogeology, topography, and climate (Chapter 3). More specific differences, however, are reflected in the nature of inhabited and constructed environments (e.g. roofed versus unroofed areas), variations in population density and intensity, and differences in the use of space. Two particular examples are illustrated by the formation processes responsible for *tells* and *mounds* (Chapter 11) and *dark earth* (Chapter 13). These materials and accumulations have been studied from both the aspects of (1) use of urban space, and (2) natural transformation of occupation deposits into "soil" (Goldberg, 1979; Macphail, 1994; Macphail *et al.*, 2003; Rosen, 1986). Additional research, however, is urgently needed to document the agents and means of their modification within a range of different built environments.

One starting place has been Butser Ancient Farm, Hampshire, which was founded by the late Peter Reynolds during the 1970s (see Chapter 12). Research activities at the Farm showed, for instance, how arable soils could be differentiated from pasture soils and how different use of space produces markedly different floor deposits. Moreover, the experiments provided some essential guidance on how such "deposits" could be recognized in the archaeological record. Other traces of cultural agencies and processes revealed by experimental and ethnoarchaeological approaches as well as empirical and interdisciplinary contextual analysis (Hodder, 1986; Matthews *et al.*, 1996) are also described in Chapter 12. Architectural structures and features are also a readily available source of materials that are subject to erosion, redeposition, and reuse (Chapter 13). The two main studies that have taken this into consideration are again dark earth and tells.

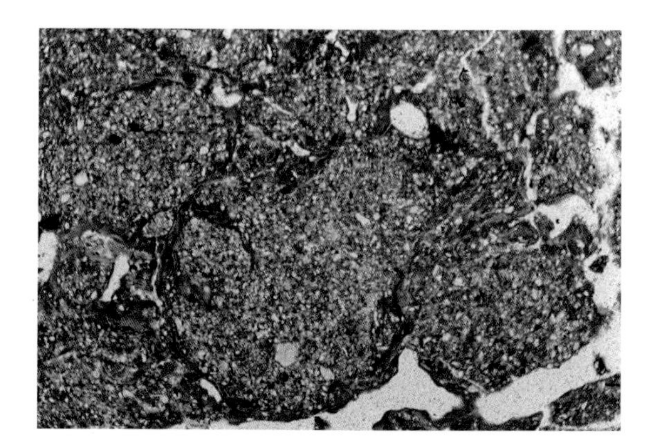

Plate 3.1 Boxgrove: Unit 11, Middle Silt Bed coarse and fine granular soil was affected by multiphase mechanical silt and clay inwash; the large amounts of textural pedofeatures in this layer may be contributing to the yellowish red field colors of this soil, rather than any "palaeoargillic" process, such as rubefication (reddening) of the clays that occurs in some true interglacials. Rather this microfabric records humid cold climate effects. PPL, width is ~3.35 mm.

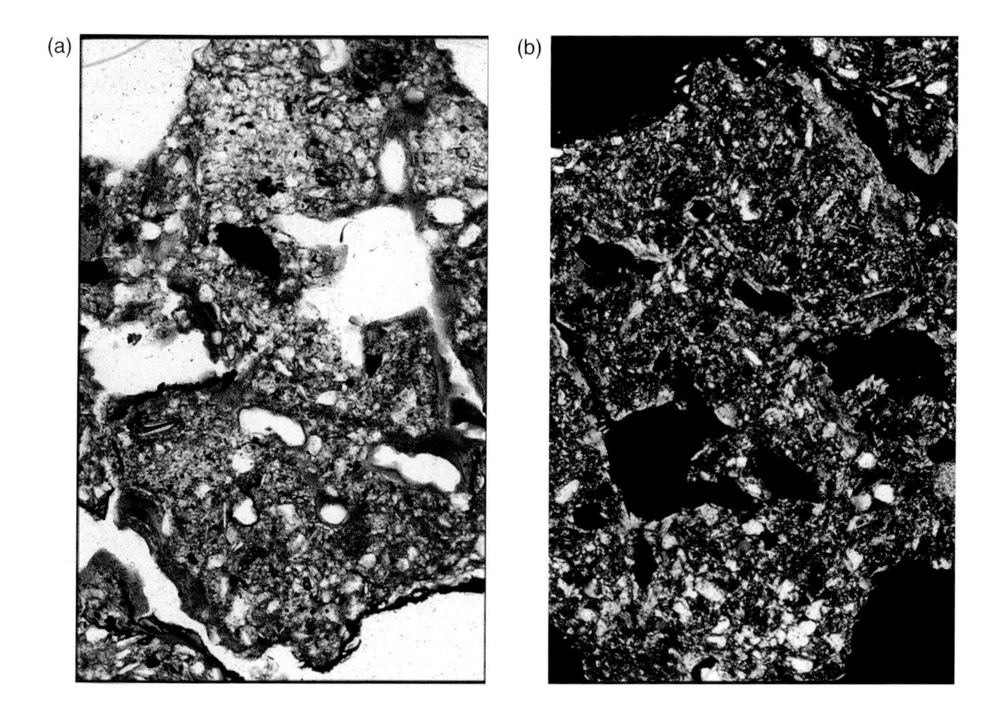

Plate 3.2 (a) Photomicrograph of diagnostic argillic horizon of an Alfisol. Here fluvial brickearth drift on the Chalk at Balksbury Iron Age Camp (Hillfort), Hampshire, UK (Ellis and Rawlings, 2001; Wainright and Davies, 1995), had developed into an Alfisol; note microlaminated clay coatings on ped surfaces and coating voids – although such coatings are often 'dated' to the Atlantic and Sub-Boreal Periods in Europe, it is quite likely that some of the clay translocation relates to human impact (See Chapter 9). PPL, frame length is ~5.5 mm;. (b) As Fig. 3.2c, but under crossed polarized light (XPL) – note high interference colours of oriented clay in the void coatings.

Plate 5.1 Detail of profile shown in Figure 5b showing the alternation of the sediments and lateral thickening and thinning of the tan loess and silt layers (Ls) and the grey clay (Gc). Note how the grey clays coalesce to the right but become separated by the sandy silts and sands to the left, both reflecting the different lithological source areas and intergrading of different microfacies.

PLATE 8.1 Photo of interbedded sand and combustion features at Die Kelders Cave, South Africa. Both the sand and some of the burned materials have been partially modified by wind and humans. The former rebedded the ashes and charcoal whereas the latter was responsible for displacing the combusted materials as part of activities such as rake out. The width of the profile is ca. 1.5 m.

(a)

(b)

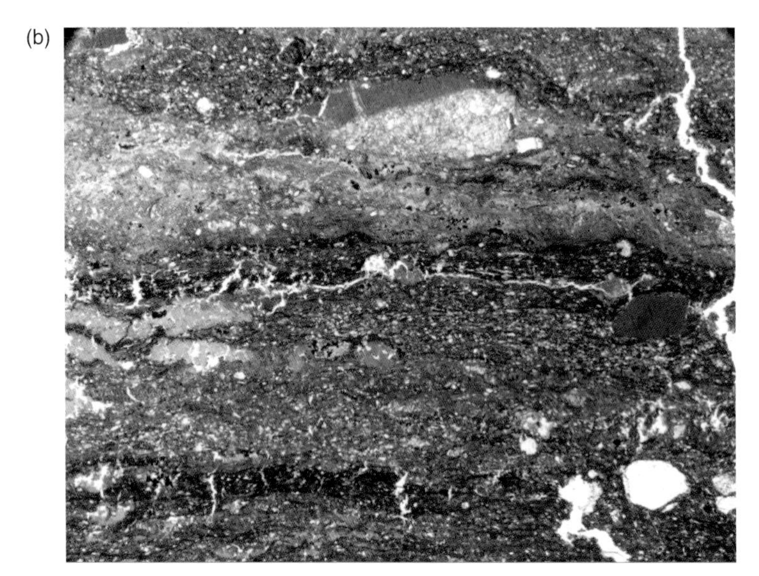

PLATE 8.2 (a) Putative hearth from Layer 10 in Zhoukoudian, Locality 1. The structured appearance was long thought to be a result of in situ burning activity associated with the controlled use of fire.
(b) Photomicrograph of thin section from middle of the deposit showing finely laminated mineral and organic material, only some of which is charred. These laminated sediments accumulated in standing water depressions within the cave (see Goldberg *et al.*, 2001 for details.). Plane-polarized light; width of field is ca. 6.4 mm.

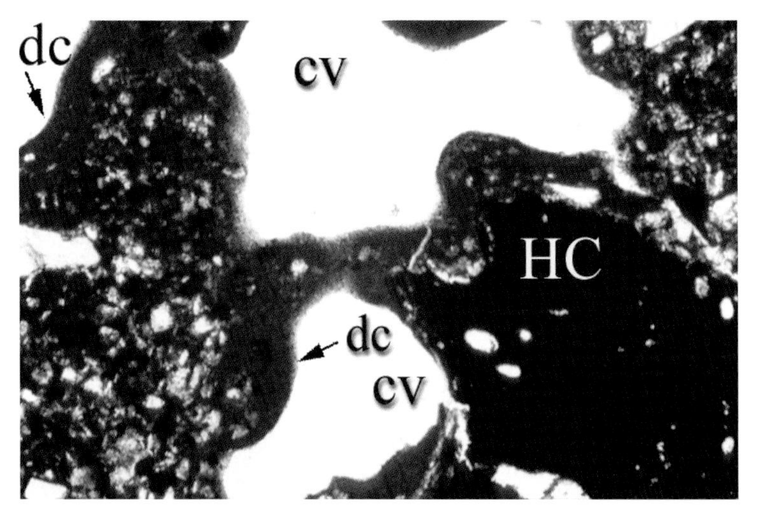

PLATE 9.1 Photomicrograph of Neolithic old land surface (Fig. 9.4 OLS) at the "midden area" at Hazleton, showing: very dusty soil containing microscopic and coarse (HC; hazelnut shell) charcoal, the occupation (cultivation, trampled) disturbed soil characterized by closed vughs (CV) formed by very dusty clay (DC) void coatings. PPL, frame is 1.35 mm.

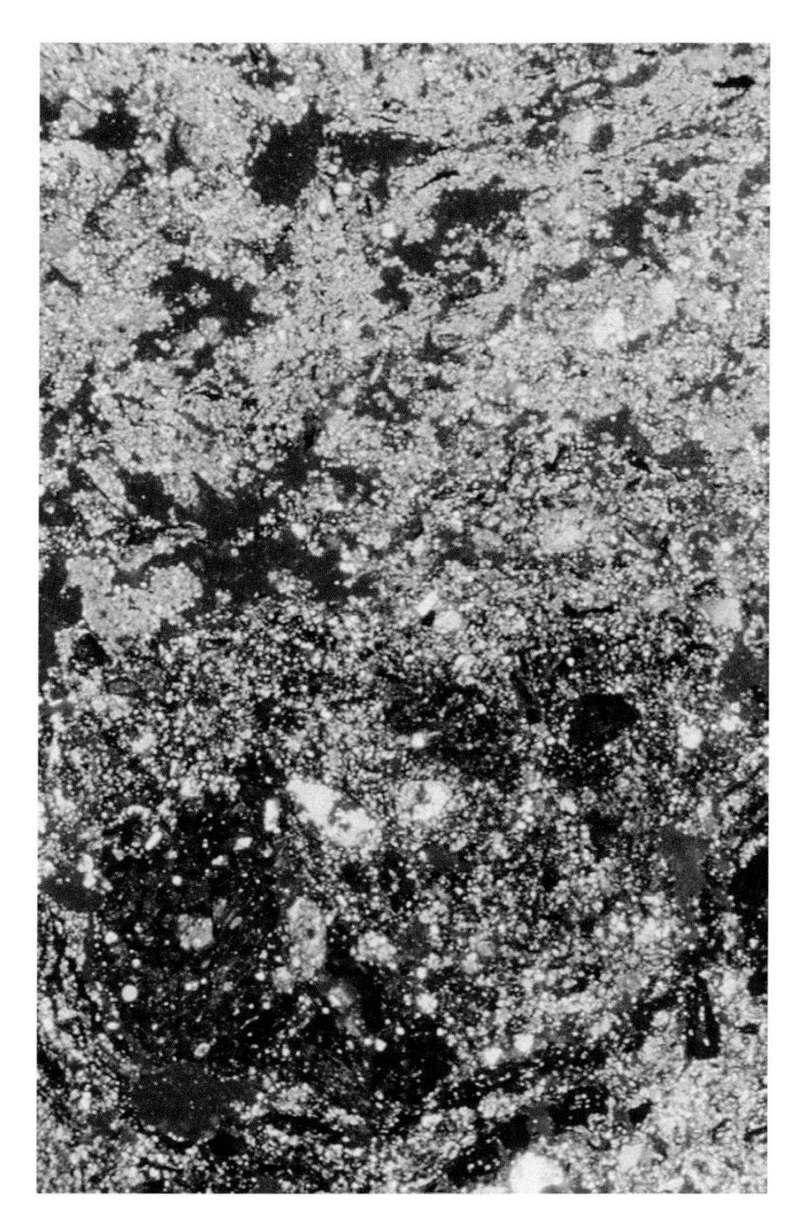

Plate 10.1 Neolithic pastoralism at Arene Candide cave, Liguria, Italy (see Fig. 10.1); *in situ* ashed stabling deposits; photomicrogragh of the junction between the phosphate- and dung-stained stabling floor 'crust', and the overlying ashed fodder and bedding layer; as Fig. 10.1c but under crossed polarized light (XPL), and showing birefringent ash that includes twig ash, faecal spherulites and calcium oxalate remains of tree leaves (Macphail *et al.*, 1997). Note rubefied stabling crust layer (frame width is ~5.5 mm).

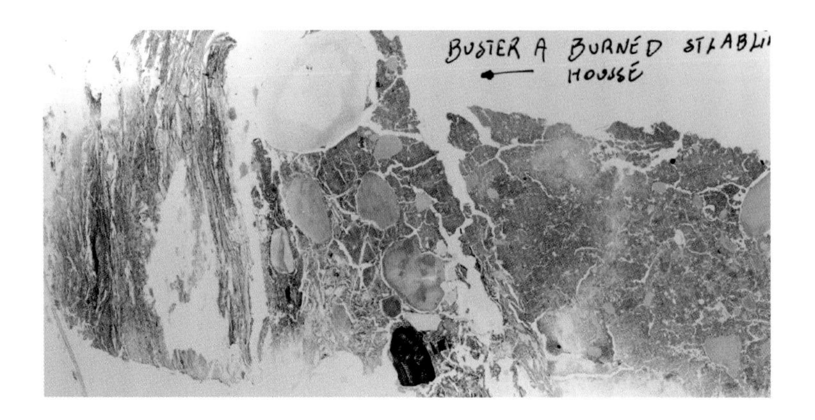

PLATE 12.1 Butser Ancient Farm, Hampshire, United Kingdom, 1990; 13 cm long thin section scan of the stabling floor and buried soil, from center of Moel-y-gar stabling roundhouse. 0–30 mm: compact organic stabling floor crust composed of horizontally layered plant fragments embedded in phosphate; 41% LOI; 6,000 ppm P. 30–90 mm: phosphate-stained stable floor with phosphatised chalk clasts and chalk soil, and included coarse wood charcoal; 32% LOI; 2840 ppm P. 90–130 mm: weakly phosphate-stained chalk soil with relict features of earthworm working; 23% LOI; 1460 ppm P.

PLATE 13.1 An example of plant-tempered burned daub from Ecsegfalva, Early Neolithic Hungary (Körös culture); daub contains alluvial clay clasts (C) and void pseudomorphs of plant (P) temper; note rubefied edge (RE). Length of thin section scan is ~55 mm.

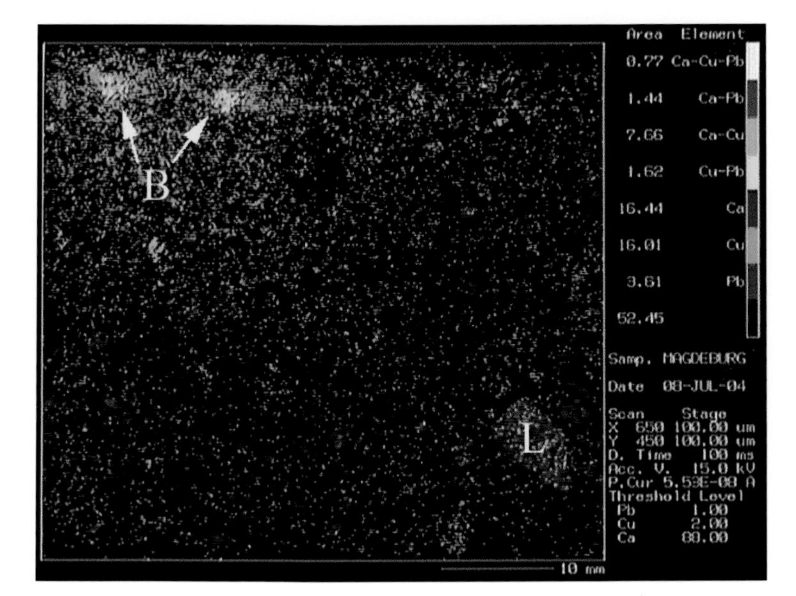

Plate 13.2 Twelth Century Magdeburg, microprobe maps of Cu (copper) showing location of bronze casting fragments – bronze bell manufacturing evidence; Bronze casting droplets (B) are a focus for Cu-Pb with Cu staining emanating out as corrosion products; there is also a scatter of calcitic (Ca) ash and limestone fragments (L). Scale bar =10 mm

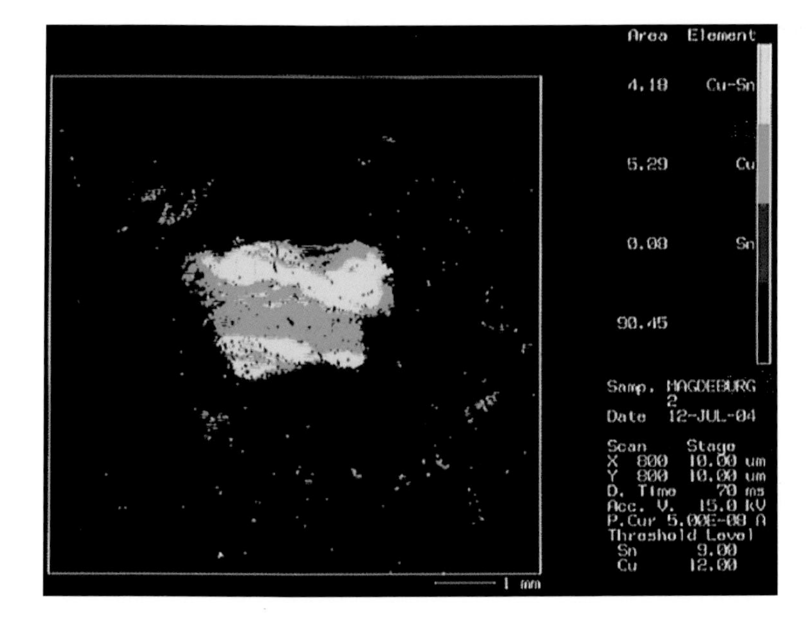

Plate 13.3 Detail of bronze casting fragment – microprobe map of Cu and Sn (tin). Scale bar =1 mm.

PLATE 15.1 Photograph of Southern end of wall profile 2 Hopeton Earthworks site, Ohio (above) and accompanying profile description (below). Note the soil nomenclature, type of information included in the description,and the inclusion of a archaeological "Stratigraphic Unit". [Photo and descrption finished by Rolfe Mandel see (Mandel *et al.*, 2003) for details].

TABLE 10.1 Micromorphological and macromorphological attributes from tell sites (Çatalhöyük, Turkey; Tel Brak, Abu Salabik, Iraq; Saar, Bharain); modified from Matthews *et al.*, (1997)

Context	Prepared or Unprepared Surfaces	Accumulated Deposits	Postdepositional modifications
Roofed Structures			
Food Preparation	Loam plastered surface	Discrete strong parallel oriented lenses of organo-mineral deposits with grindstone fragments; vegetal pseudomorphs and siliceous graminae plant fragments	Organic staining, bioturbation, salt formation immediately below surface
Food cooking adjacent to ovens and hearths	Plastered and compacted surfaces	Multiple layers of moderate parallel oriented loam with organo-mineral material, burnt fuel, oven fragments, bone, flint/obsidian flakes, charred grain, tubers, date palm phytoliths	Subhorizontal cracking in plaster floors, salts, bioturbation, organic staining
Storage in small rooms or bins	Clayey plaster with included organic matter, gypsum plaster	Charred cereal grains, either pure or mixed within building debris	Bioturbation
Reception/"clean activities"	Well prepared plasters often with finishing coats and impressions of mats/rugs	Thin lenses of charred and siliceous plant remains, sterile silty clay with strong parallel orientation	Organic staining and horizontal cracks
Ritual (associated with altars, sculptures, and wall paintings	Multiple plaster layers, often with fine coats, occasionally painted, impressions of mats/rugs	Burnt remains, waterlain crusts, red ochre remains	Organic staining, salts, and bioturbation
Probably Roofed			
Stables	Very few prepared surfaces, overall undulating	Interbedded lenses of fragmented dung pellets	Organic staining, salts, and bioturbation
Unroofed			
Domestic courtyards and streets	Few plasters, aggregate hard-core surfaces, unprepared surfaces, bitumen pathways	Layers or unoriented deposits with cultural references, reworked wind- and water-lain sedimentery clasts, uncompacted refuse deposits, undisturbed wind and waterlain deposits, dung	Salts, bioturbation, wind and water reworking
Civic, administrative, and ritual courtyards	Mudbrick foundations with baked brick, lime plaster, and plastered surfaces; few unprepared surfaces	Mineral-rich remains with some burnt and cultural refuse, thin layer of ash	Salts, bioturbation, wind and water reworking
Middens	Few prepared surfaces, mainly unprepared surfaces with different depositional episodes and in situ burning	unoriented massive deposits, some wind and water-laid deposits and in situ burning	Bioturbation, settling and compaction, organic staining

Many researchers are investigating traces of burning (e.g. combustion features – Meignen *et al.*, 2001), burned tree holes, trampling (e.g. floors versus open spaces, domestic versus stabling; Cammas, 1994, Matthews *et al.*, 1997; Macphail *et al.*, 2004) and weathering (e.g. experimental earthworks; Bell *et al.*, 1996; Macphail *et al.*, 2003). Nevertheless, more interdisciplinary research is needed to elucidate the processes involved in the preservation and alteration of different components and materials. Data on these subjects are particularly important, as understanding depositional history underlies all archaeological research, a wide range of cultural resource management decisions are dependent on evaluation of burial environment and postdepositional alterations (French, 2003; Macphail and Goldberg, 1995).

The significance of understanding site formation processes has been reemphasized by recent methodological and theoretical developments in archaeology. First, postprocessual archaeologists, in particular, are calling for greater transparency and reflexivity in archaeological excavation, sampling, interpretation, and presentation (Hodder, 1999; Lucas, 2001). Hence the emphasis in this book, on the presentation of primary field, bulk and micromorphological data, and explanations of the ways they can be used to make identifications and interpretations (Macphail and Cruise, 2001). A growing trend of publishing *everything* on CD-ROM or on the Internet is aiding transparency of reporting (Chapter 17). Secondly, many worldwide studies of sociocultural behavior, domestic and ritual practices, identity and agency, are searching for traces of uses and concepts of space in houses and settlements at increasingly higher scales of spatial and temporal resolution. These questions all require in-depth knowledge of site formation processes.

One of the chief broad areas of study is the analysis of occupation surfaces and the development of occupation sequences (the formation of occupation surfaces is covered in detail in Chapter 12). It should be noted, however, that a primary model was developed by Gé *et al.* (1993). But like all models, it is only a guide and not the *only* way occupation surfaces form or are preserved. We have encountered interpretational nonsense when this model has been applied blindly, for example, by inappropriately *fitting* layers into this stratigraphic model in order to *identify* "floors"; similarly the presence of "floors" was rejected because *all* the elements of this model were not present. Use of models needs to be based upon objective observation, and common sense should be the ultimate guideline.

Discard. Accumulations of occupation deposits and artifactual and palaeoecological remains have been classified along the following lines (Renfrew and Bahn, 2001; Schiffer, 1987):

- *Primary* residues or refuse deposited within an activity area (Fig. 10.1 and Plate 10.1) (See also Chapters 7 and 9).
- *Primary transposed* refuse deposited away from the original activity area, for example, in rubbish dumps or streets and lanes (see Fig. 11.5a–5b).
- *Secondary use-related* refuse, which has been removed from secondary refuse areas and reused (e.g. rubbish used to infill a building or manure a field; see Box Fig. 9.12).

In the last case, the nature of the finds in a fill may well suggest that "secondary use-related refuse" (sometimes termed "tertiary refuse") is present, but this does not always mean that the fill itself is "tertiary." An example of this is given from studies of Grubenhäuser (Box 11.1), where a "primary transposed" infill is composed of dismantled turf walls/roof, while the *finds* within the turf are in fact tertiary in nature.

Renfrew and Bahn (2001) suggest that activity areas and accumulations of deposits can be classified according to whether they occur in areas of resource procurement, manufacturing, use, discard, or reuse. Again, these inferences must be elucidated through detailed contextualization of

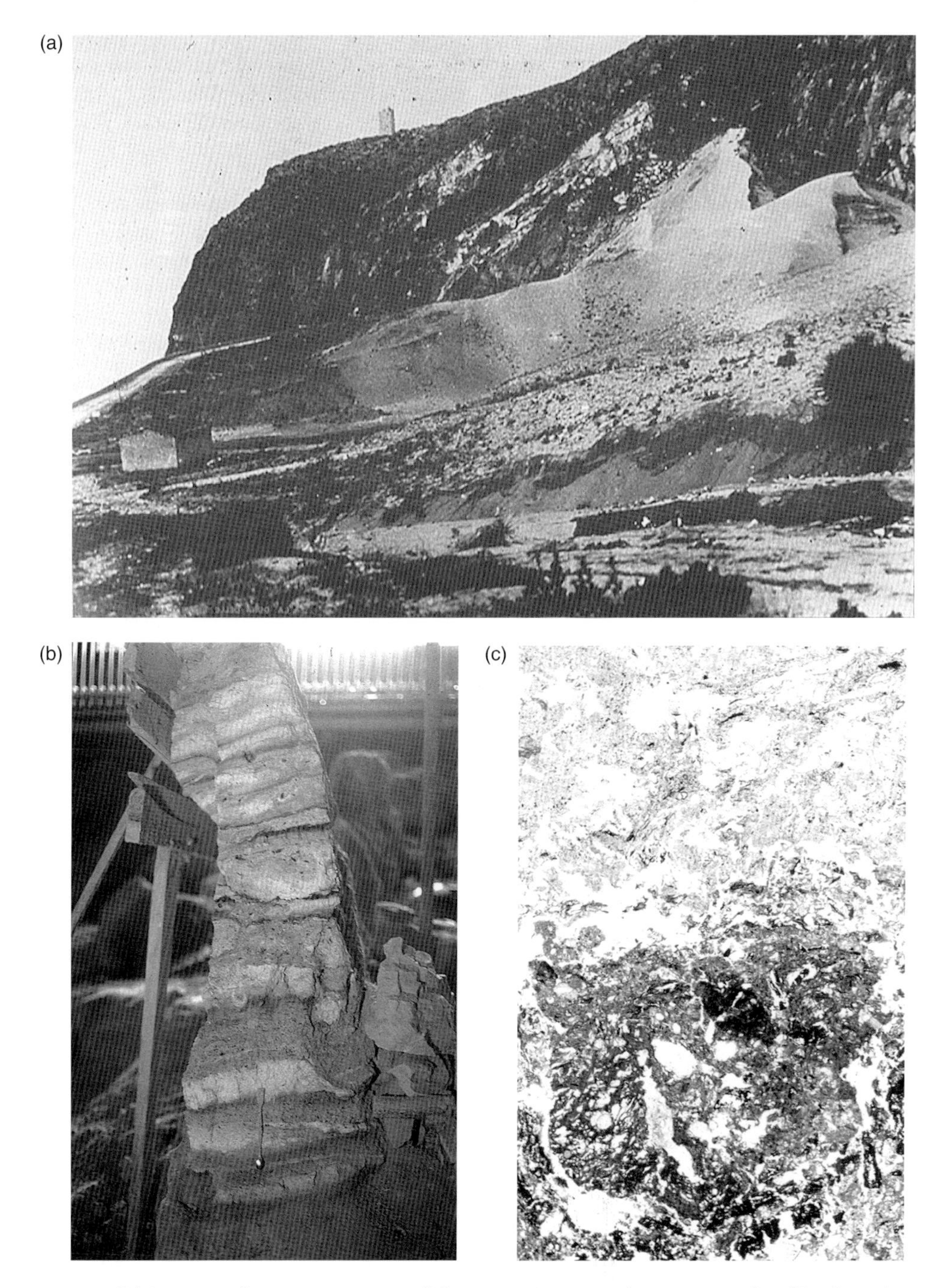

FIGURE 10.1 Neolithic pastoralism at Arene Candide cave, Liguria, Italy; an example of both "primary use related" (*in situ* stabling floor deposits), and "secondary transposed" refuse deposition (locally dumped stabling waste), best differentiated using soil micromorphology. (See Experimental data in Chapter 12). (a) Nineteenth-century view of the cliff and white sand dunes now quarried away. Note fort built against the *Saraceni* on top of the cliff. (b) Primary residues – rapidly accumulated Middle to Late Neolithic (Chassey) stabling layers (5,000 to 5,700 BP) composed of alternating brown ("stabling crust") and gray

the archaeological finds, as well as by using soil micromorphology. With such detailed observations, debates such as those between Binford (1981) and Schiffer (1987) will become irrelevant. A wide range of other contextual variations also affects the type, thickness, and frequency of deposits within archaeological sites, as discussed below.

10.3 Stratigraphic sequences as material culture; concepts and uses of space

There has been increasing awareness that "differential artifact densities cannot be read in a straightforward manner, but depend," at least in part, "upon rates of sedimentation. A social interpretation of sedimentation is just as necessary as a social view of the artifacts contained in the soil" (Gosden, 1994: 193), and consideration of natural agencies and microenvironmental conditions (Sherwood, 2001). Thus, when examining stratigraphy as part of the material culture, a number of caveats need to be kept in mind, as discussed below.

On examining urban deposits from a wide range of archaeological sites in different geographical regions, Butzer observed that the agents and processes in operation during phases of positive demographic expansion differ significantly from those in operation during negative demographic phases (i.e. site abandonment), when buildings were allowed to decay and debris built up in streets (Butzer, 1982: 90–1, Table 6.1). He also noted significantly, that:

> *The rate and type of build up differ on living floors, in streets and alleys, or in and around community structures, such as civic buildings, walls, terraces, and drainage systems. (Butzer and Freeman in Rosen 1986: xiii)*

Similar variations in sedimentary sequences within different rooms and areas of settlements have also been observed during ethnoarchaeological research. This research is illustrating the wealthy range of sociocultural and ecological decisions and contexts that influence the selection of materials for floors or surfaces, the nature of deposits that build up on those surfaces, and the type and extent of postdepositional alterations (Cammas, 1994; Cammas *et al.*, 1996b; Courty *et al.*, 1994; Matthews *et al.*, 1997, 2000) (Table 10.1). Modern experiments (Chapter 12), for example, have shown a clear difference between domestic and animal stabling space for the British "Iron Age" and indicated how these signatures may be preserved as both short-term and long-term features in the archaeological record (Macphail *et al.*, 2004; Reynolds, 1979). These approaches are engendering greater emphasis on sediments as material culture, structuring and structured by human behavior. They are part of a wider field of study, which is examining the symbolic and social aspects of architecture and uses of space (Parker Pearson and Richards, 1994). These microstratigraphic studies of uses of space are equally applicable in cave sites (Boschian and Montagnari-Kokelji,

FIGURE 10.1 (*Cont.*)
(ashed leaf hay fodder, dung, and wood) layers. The thickest brown layers however, are dumped ashed stabling waste. (c) Photomicrograph of junction between stable floor "crust" layer where bedding and fodder composed of leaf hay (e.g. *Quercus ilex* leaves and twigs), which has been compacted by animal trampling, and stained by phosphate-enriched dung and liquid waste. This material was burned after use, probably in springtime after over-wintering and the birth of lambs; the "wet" crust resisted total combustion compared to the gray ashed fodder and dung remains above (Maggi, 1997). Plane polarized light (PPL), frame height is ~5.5 mm.

2000; Cremaschi *et al.*, 1996; Macphail *et al.*, 1997; Macphail and Goldberg, 2000; Meignen *et al.*, 2001), and less permanent sites, such as modern pastoral tents (Goldberg and Whitbread, 1993). Often at hunter-gather sites, however, the best-preserved remains relate to combustion zones (Meignen *et al.*, 2001; Schiegl *et al.*, 2003; Wattez, 1992), rock ovens, and other "cooking" sites (Goldberg and Guy, 1996; Linderholm, 2003), and the burned soils of hearths, which tend to be stable and resist natural reworking.

Finds within activity areas and buildings are often few; they may represent discard/ abandonment activities and have been subject to disturbance (Schiffer, 1987). It is the smaller artifactual remains that McKellar and Schiffer suggest are "more likely to become primary refuse" even in areas that are periodically cleaned. The study of microstratigraphic sequences is enabling the recognition of traces of evidence from uses of space. Such traces are not merely derived from the time of abandonment, but developed throughout the life history of buildings, and their foundation and use. These factors vary according to the changing life history of the inhabitants and development of the settlement morphology. Numerous examples exist from Middle Eastern tells (Matthews, 1995; Matthews *et al.*, 1997), and from Early Medieval sites in Britain (Macphail *et al.*, "Guildhall of London"; Macphail and Linderholm, in press; Matthews, 1995; Matthews *et al.*, 1997; Tipper, Forthcoming) and Belgium (Gebhardt and Langohr, 1999; Langohr, 1991).

10.4 Time and scale

Although many archaeological questions relating to people, houses, and settlements are seeking resolution of daily, seasonal, and life cycle timescales, these have rarely been explored (Foxhill, 2000: 491). Many archaeological timescales are based on analysis of building phases and levels, but "the timescales underpinning maps and plans (based on excavation and survey data) are longer term and less finely tuned than the timescales of the social processes which created the data in the first place" (Foxhill, 2000: 496). Analysis of microstratigraphic sequences within buildings and settlements are providing an exciting new data source for examining these smaller timescales, which are more relevant to social and palaeoecological issues. Some of these are explored; below. The study of hearth structures, or tip lines in fills, for example, represent virtually instantaneous events in the archaeological record. They represent a contrast to an undifferentiated fill from a building or open-air space, such as a plaza. The investigation of timelines and the changing use of space through time and for different cultures is an ultimate goal and can be effectively studied using a microfacies approach (Courty, 2001; see below; Chapter 2). An example of how geoarchaeological studies contributed to the multiperiod site of Deansway, Worcester, is given below (see Figs 10.3 and 17.2) (Dalwood and Edwards, 2004).

One of the most fundamental associations of archaeological remains is the single depositional unit, at both macro and micro scales of analysis (Courty, 2001; Macphail, 2003a). Identification of a single depositional unit may vary according to scales of observation, individual perceptions, excavation strategies, and research questions (Stein and Linse, 1993). The significance of a single depositional unit has been widely discussed, and may represent a palimpsest of activities, depositional events, and alterations (Bordes, 1975). The correlation between depositional units, and natural events and human actions needs to be closely examined. Detection of individual daily, seasonal, or even annual activities and events at many archaeological sites may have been blurred or aggregated by trampling, sweeping, the use of mats or floor coverings, or natural events and postdepositional alterations. For example, it had first been assumed that the stabling events identified at Arene Candide were the product of single season activity, but it was only at the micro-scale that a single stabling episode could be differentiated from thin dumps of stabling refuse (e.g. Arene Candide, Fig. 10.1). Thus, correlations between stratigraphic sequences and time need to be closely investigated (Shanks and Tilley, 1987).

Some of these difficulties were recognized at Butser Ancient Farm, where deposits resulting from over a decade of stabling could not be divided into the known yearly stabling periods. Furthermore three different *stabling* microfacies types were recognized across the floor of the structure (Macphail and Goldberg, 1995). Although these different microfacies are essentially all contemporary and laterally connected, they could easily be misinterpreted as showing individual events of (1) a single (possibly cattle) stabling episode after a second stabling floor had been emplaced, (2) three separate stabling episodes by sheep and goats, and (3) postabandonment bioturbation (Fig. 10.2). This example illustrates the importance of taking lateral samples of activity surfaces and shows how fatal it could have been if this sequence had been interpreted (along with archaeological examples) without detailed discussions with colleagues who are experienced in the practical management of stock – namely C. Bakels (Institut vor Prae-en Protohistorie, The Netherlands – IPP), R. Engelmark (University of Umeå, Sweden), and P. Reynolds (Butser Ancient Farm, United Kingdom). It has to be understood that occupation deposits are far more difficult to interpret than geogenic deposits and natural soils. Whereas natural processes follow many known rules as the many textbooks on sedimentology and soil science show, human activities are idiosyncratic and do not appear to follow *laws*, although many have been sought (Courty *et al.*, 1989; Schiffer, 1987).

10.5 Settlement–landscape interrelationships

Study of deposits selected for architectural materials and surfaces can lead to consideration of a wide range of questions and an appreciation of the interconnections between people and sediments in settlements and landscapes.

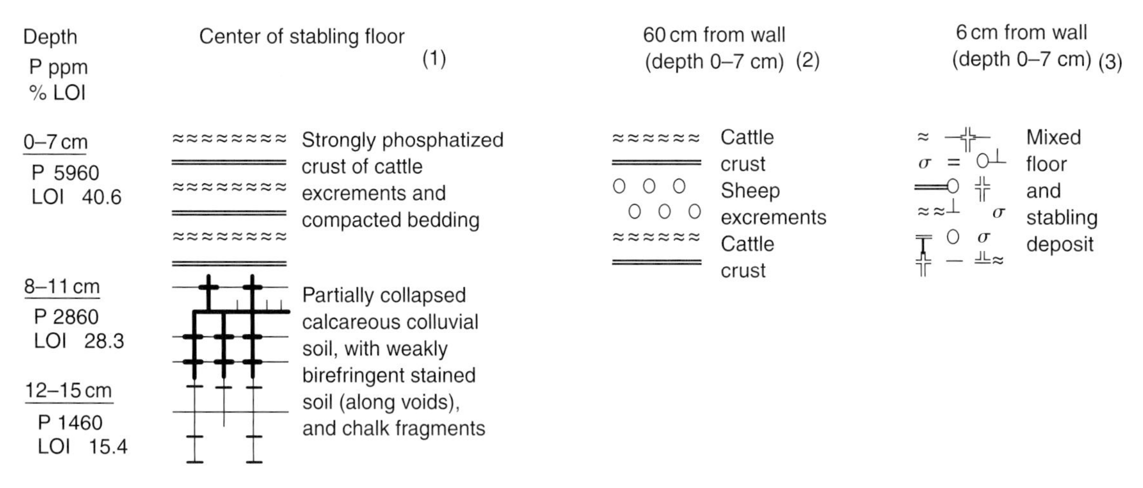

FIGURE 10.2 Diagrammatic representation of site formation process at the Moel-y-Gar stabling roundhouse, Butser Ancient Farm, showing different microfacies across the floor that "record" stabling between 1977 and 1990; phosphate (P), loss-on-ignition (LOI), and soil micromorphology (Macphail and Goldberg, 1995).

What was the source of these materials? Did they occur in natural or cultivated landscapes, private, public, or other categories? and Who had access to these materials? From how far and by what means were they brought to the site? Did their procurement correspond to other natural or human cycles and events, such as hunting or gathering expeditions? How, when, and by whom were these sediments manipulated and transformed into architectural materials and surfaces? Was the creation and application of these materials part of daily routines or rituals, or other events in the natural world and human life histories? Did the sediments have particular structural, chemical, symbolic, or aesthetic properties or meanings, either as finished materials and surfaces or as secondary infill?

The ethnoarchaeological research discussed above is illustrating ways in which these and other considerations may be part of human perception and treatment of sediments. In fact, it is probably easier to gain insights into the past from ethnoarchaeological studies in the Mediterranean region than perhaps from Roman and Early Medieval western Europe: in the latter case there are no *living* societies that replicate these lifestyles – although all preindustrial settlements can be perhaps modeled from unsophisticated towns in Africa (Galinié, University of Tours, France, personal communication).

On the other hand, the authors have noted that ethnoarchaeological experimental floors from domestic structures of early settler (United States), Jomon (Japan), Iron Age (United Kingdom) and Viking (Denmark) replications, all suffer from being too *clean* – for obvious modern health and safety reasons. For example, the characteristic highly phosphate contaminated floors from occupation sites where food and faecal remains can be abundant, cannot be compared chemically (e.g. P) to experimental floors (Macphail *et al.*, 2004). Alternatively, there can be good mirroring of measurements and patterns of magnetic susceptibility (see Fig. 12.3) (Allen, 1990).

There have always been links between geoarchaeology and petrographic analysis of ceramics. These include analyses to identify local sources of raw materials for pottery (Spataro, 2002), and to determine techniques of ceramic manufacture (Roux and Courty, 1998). Equally, there have been studies of adobe and daub-built structures (Whittle, 2000) (Chapter 13). Unfired clay and loam were identified within grubenhäuser fills at Anglo-Saxon West Heslerton, North Yorkshire (Fig. 11.8), and ascribed to pottery (Alan Vince, personal communication) and loom weight production, respectively (Macphail *et al.*, Forthcoming; Tipper, forthcoming). Clues to past industrial activities may be indicated by heavy metal analysis of bulk samples, although heavy metals can also be naturally concentrated in organic remains. Also, microscopic nonferrous metal slags and casting droplets (e.g. bronze), which are not picked up with a magnet, may be found and identified in thin and polished sections and through microprobe studies (see Chapter 13).

10.6 Origin and predepositional history of occupation deposits

Archaeological sediments within settlements often include a diverse range of components, of mineral, biological, and artifactual origin. Detection and identification of these elements depends on the type of analysis and scale of resolution during excavation, screening and flotation, macroscopic or microscopic observations, and organic or inorganic analyses (Barham, 1995) (Chapters 15 and 16). The environmental archaeology team at Umeå University, for example, screen for macro-remains before soil samples are sieved to 500 μm for use in chemical and magnetic susceptibility analysis. The identification of stone-size material is an important component of some studies of dark earth (Macphail, 1981; Sidell, 2000). The origin and predepositional histories of deposits can then be discerned by analysis of a range of attributes, including composition, diversity, abundance, size (fragmentation), shape (degree of angularity/roundedness), surface boundaries, and alterations of internal color and microstructure (resulting from weathering or burning, for example; Courty *et al.*, 1989; Pettijohn, 1975) (see Chapter 1). As many archaeologists are aware, small, rounded ("abraded") pottery fragments may be a result of regular plough damage.

Predepositional characteristics include traces and effects of differential burning, weathering, abrasion, and fragmentation. Effects of different burning conditions and temperatures have been studied (Courty *et al.*, 1989: 107) with regard to plant remains (Boardman and Jones, 1990; Wattez and Courty, 1987; Wattez *et al.*, 1990), bone and clays, and minerals (Rice, 1987; Schiffer *et al.*, 1994). Similarly, the effects of weathering have been studied in a range of subdisciplines, especially in relationship to burial (Bell *et al.*, 1996; Evans and Limbrey, 1974; Hedges and Millard, 1995; Weiner *et al.*, 1993). The movement of artifacts by trampling have also been investigated, sometimes in an attempt to replicate what has been found on-site during excavation (Barton, 1992; Gifford, 1985; Gifford-Gonzalez *et al.*, 1985; Villa and Courtin, 1983).

A range of human and natural phenomena can affect the character and concentration of artifactual components, including the nature of activities, individual discard practices and concepts of space, and micro-environment (Kelly and Wiltshire, 1996; Lawson *et al.*, 2000).

Artifacts can be concentrated through aeolian deflation, once-thick deposits becoming palimpsests. rapid sedimentation of fluvial or aeolian origin may lead to less concentrated numbers of artifacts. Open areas may receive mixtures of wind and water-sorted deposits, and anthropogenic materials – the latter perhaps from collapsed building debris, industrial activities, or from animal and human traffic (Cammas, 1994; Cammas *et al.*, 1996a; Macphail, 1994, 2003b; Matthews *et al.*, 1997). It should be possible to differentiate these from simple ground raising and constructional episodes (Fig. 10.4).

Deposits rich in plant and bone remains include items discarded next to ovens and combustion features (e.g. rake out; Meignen *et al.*, 2001), in middens, and occasionally in ritual areas where there has been burning. Deposits with heterogeneous components include burnt and unburnt aggregates in building infill and packing, and some midden deposits rich in diverse aggregates and plant remains. Concentrations of components from specific activities include lenses of flint or obsidian chips in knapping areas; food remains (e.g. egg shell, bone, and cereal waste); midden and waste from stock pounding (Figs 10.1, 10.2 and Table 10.1); and basalt grinding stones and fragments. The last case may only be significant with exotic materials, as for example, in Norman London where grindstones of German volcanic rocks were imported.

Plant remains within settlements represent one of the most abundant classes of remains from activities and constituents of deposits (Goldberg *et al.*, 1994; Matthews and Postgate, 1994, figures 15.8a,b). The range of plant remains identifiable in thin sections includes charred and siliceous remains (including phytoliths), desiccated and water-logged remains, pollen and spores, calcitic ashes, and impressions in surrounding sediments of plant remains that have since decayed (Courty *et al.*, 1989). Burnt and unburnt dung from omnivores, carnivores, and herbivores is also preserved and is identifiable in thin section, often with plant and bone remains of their diet (Horwitz and Goldberg, 1989; Macphail *et al.*, 1990, 1997; Matthews *et al.*, 1996). In some cases, dung of some herbivores has been inferred from the presence of calcitic *faecal* spherulites (ca 20 μm) (Brochier, 1983; Brochier *et al.*, 1992; Canti, 1997, 1998, 1999), However, as spherulites can form in numerous ways (bacterial activity; in kettles), and be ubiquitous in some calcitic occupation deposits, their presence does not always, *identify or locate* an *in situ* stable (Macphail *et al.*, 1997; Wattez, 1992; Wattez *et al.*, 1990). Neither does their absence negate an identification of a stock holding area, because the liquid waste produced by animals commonly leads to the leaching of labile calcium-rich minerals (e.g. monohydrocalcite, $CaCO_3 \cdot H_2O$; Brochier *et al.*, 1992; Heathcote, 2002; Macphail, 2000; Shahack-Gross *et al.*, 2003; 2004). Various calcium oxalate minerals can also be relict of plant materials on archaeological sites – root remains in pig coprolites, leaf remains in ashed leaf fodder.

10.7 Depositional history

Attributes that may furnish information on depositional agency and process and the nature of associated sociocultural and microenvironmental circumstances, include numerous lithological characteristics: deposit boundaries and thickness, particle size and sorting, component orientation and fabric, depositional relationship between fine and coarse materials, and pore pattern. Interpretations of depositional history can draw on comparisons to a large body of literature at both the macro and micro-scale in sedimentology, archaeology, and experimental and ethnoarchaeological research (Bell *et al.*, 1996; Goldberg and Whitbread, 1993; Macphail and Cruise, 2001; Metcalfe and Heath, 1990; Schiffer, 1987). Many of the interpretations also refer to basic soil and sediment micromorphology derived from empirical studies (e.g. Douglas, 1990; Reineck and Singh, 1986). Natural and cultural postdepositional agencies are discussed in Chapters 3 and 13. Natural depositional agencies are frequently modified within the habitation environment, and wind- or water-laid deposits can be reworked by trampling in occupied areas (Rentzel and Narten, 2000). Indicators of water-laid deposits incorporate areas of such conditions in the depositional record and include surface crust formation associated with rainsplash, and graded bedding in puddles and zones of sediment deposition (Courty *et al.*, 1989). (Some features relating to the exposure of bare ground to elements are also dealt with under the subject headings of *erosion* and *cultivation* in Chapters 4 and 9, respectively.) Well-sorted water-laid crusts may be identified in unroofed areas, street/lanes, and abandoned buildings or areas (Cammas, 1994; Cammas *et al.*, 1996b; Matthews *et al.*, 1997). Wind- and water-laid deposits are common in the Mousterian and Upper Palaeolithic contexts at Gorham's and Vanguard Caves (Macphail and Goldberg, 2000). (See chapters 5 and 7 environments).

Many accumulated deposits in occupation areas are generally poorly sorted with a range of aggregates and particle sizes. In general, some occupation sequences comprise thin lenses of deposits (<0.5–3 mm thick), with strong parallel orientation and distribution of components, suggesting periodic accumulation and compaction over time in some room deposits. In contrast, many depositional units in room infills are thick (>10 cm), and have unoriented and randomly distributed components, which indicate massive dumping. In addition occupation deposits can be moved and redeposited for an infinite number of reasons. Some examples are dumping of midden into abandoned house plots at Roman Southwark, backfilling graves, and establishing landscape gardens (Fulford and Wallace-Hadrill, 1995, 1996; Macphail, 2003c; Macphail *et al.*, 1995).

10.8 Postdepositional modifications

Any reading of the archaeological record must consider and identify the natural and cultural postdepositional agents and processes that have

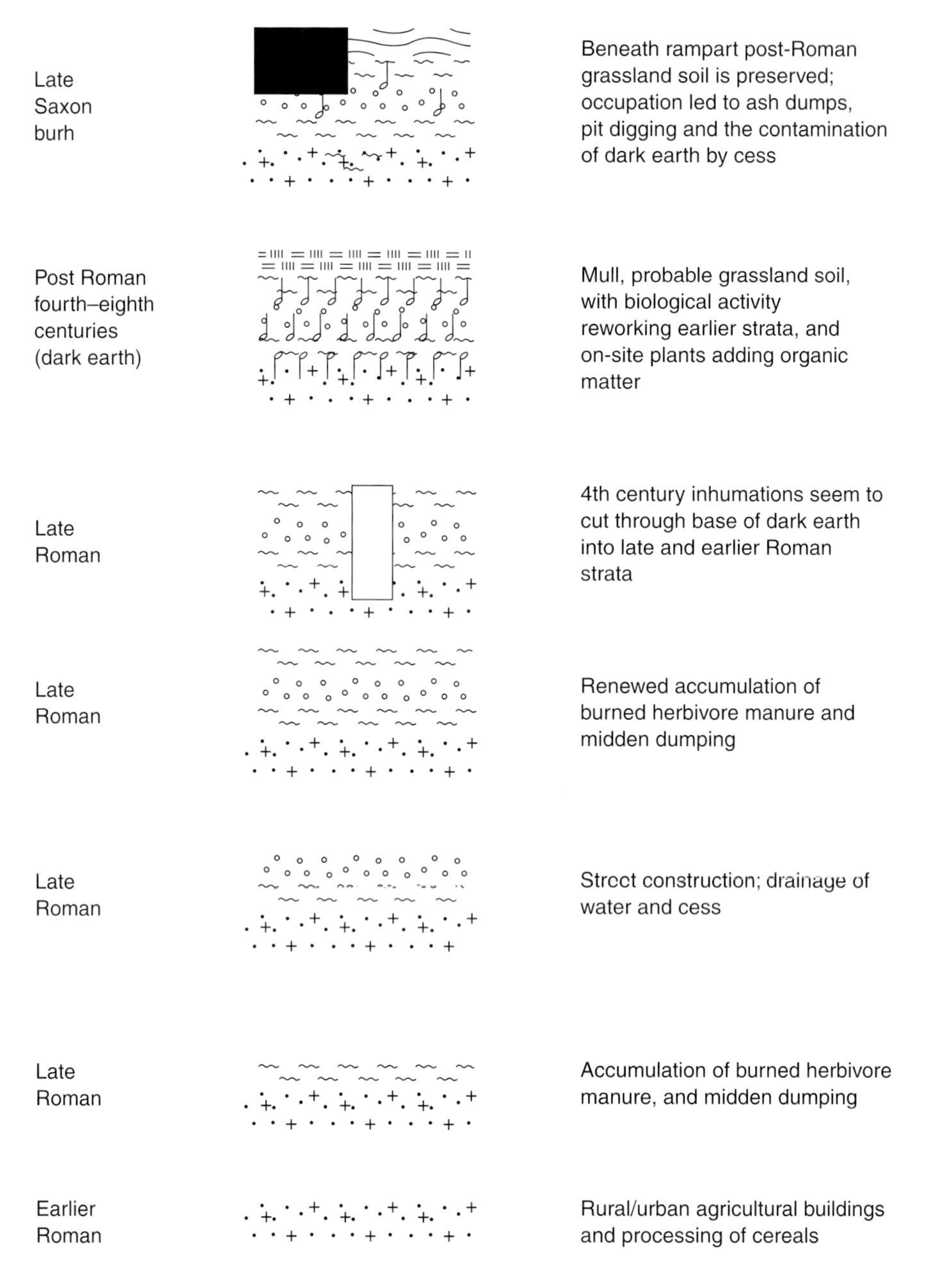
Late
Saxon
burh
Beneath rampart post-Roman grassland soil is preserved; occupation led to ash dumps, pit digging and the contamination of dark earth by cess
Post Roman
fourth–eighth
centuries
(dark earth)
Mull, probable grassland soil, with biological activity reworking earlier strata, and on-site plants adding organic matter
Late
Roman
4th century inhumations seem to cut through base of dark earth into late and earlier Roman strata
Late
Roman
Renewed accumulation of burned herbivore manure and midden dumping
Late
Roman
Street construction; drainage of water and cess
Late
Roman
Accumulation of burned herbivore manure, and midden dumping
Earlier
Roman
Rural/urban agricultural buildings and processing of cereals

FIGURE 10.3 Diagrammatic representation of how the palimpsest of Roman, Dark Age, and Saxon deposits across the Deansway site, Worcester illustrating the chronological land use reconstruction of the site (Macphail and Goldberg, 1995) (see Fig. 17.4).

transformed it. Analysis of these processes assessing the significance of the presence absence, and nature of certain components, and permit inference of palaeoenvironmental conditions (Fig. 10.4d). Good overviews of these are discussed in early publications (Butzer, 1982; Limbrey, 1975; Schiffer, 1987); a more recent, worldwide treatment can be

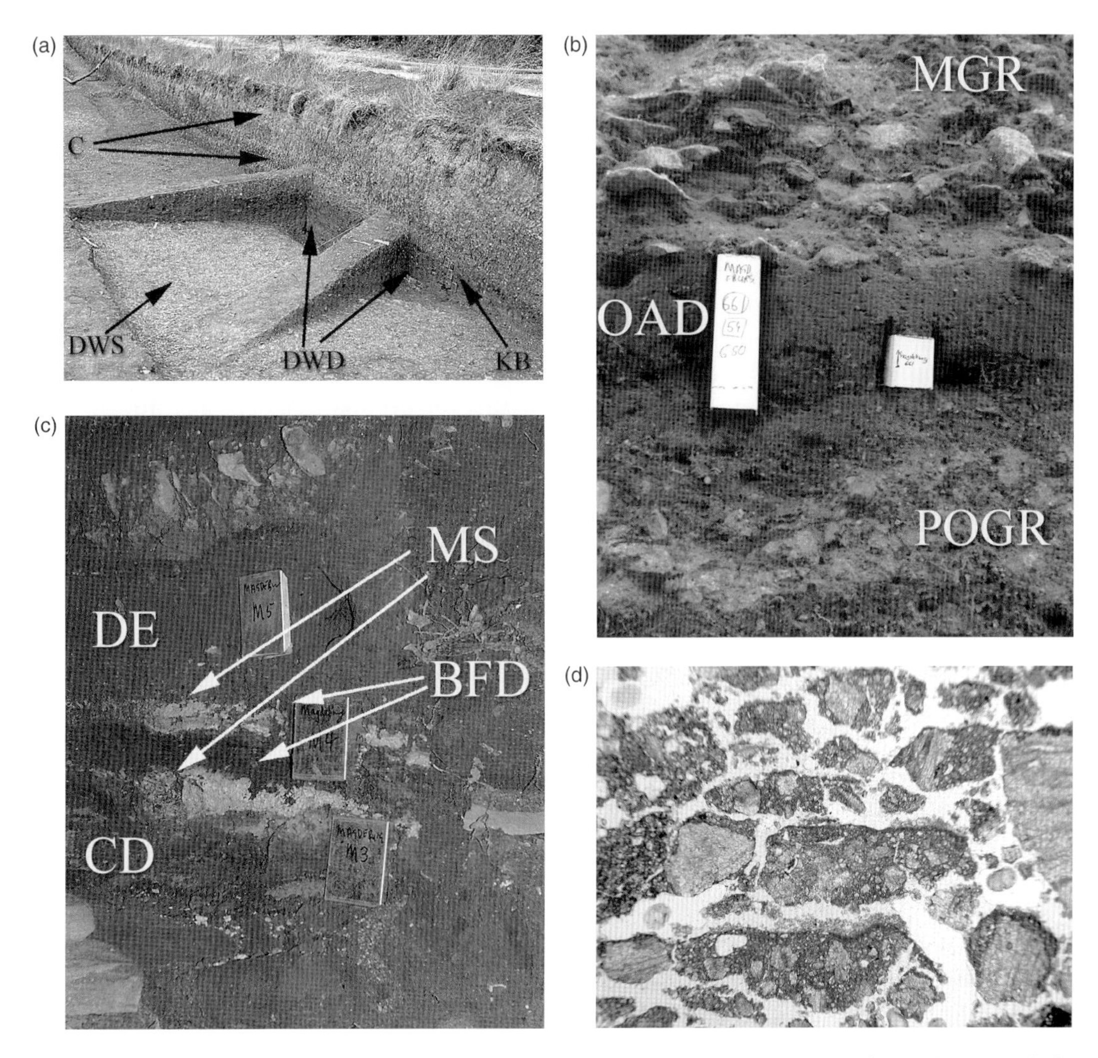

FIGURE 10.4 (a) Middle Saxon droveway at West Heslerton, North Yorkshire, United Kingdom, just below the scarp face of the Yorkshire Wolds (Chalk); note compacted chalky droveway surface (DWS), dark droveway deposits (DWD) – sampled by 5 Kubiena boxes (KB), and overlying Medieval and post-Medieval colluvium (C) indicative of severe ploughsoil erosion (Macphail *et al.*, forthcoming; Tipper, forthcoming). (b) Magdeburg Cathedral, Magdeburg, Germany, showing a depositional sequence of post-Ottonian (tenth century) Church ground-raising (POGR), twelfth–thirteenth century open air deposits (OAD) associated with bronze bell manufacture (see Figs 13.10 and 13.11) for the Magdeburg Cathedral (AD 1363), and overlying medieval ground raising (MGR) for the Magdeburg New Market area (Macphail and Crowther, in press). (c) Magdeburg tenth century Ottonian Church levels showing Kubiena boxes (from the base: M3, M4, M5) sampling constructional deposits (CD), mortar spreads (MS) or possible "floors" associated with church construction, beaten floor deposits (BFD; See Chapters 11 and 12) and dark earth (DE) that formed between the tenth and twelfth centuries (Macphail and Crowther, in press). (d) Magdeburg tenth century Ottonian levels; sample M4 (see Fig. 10.4c); photomicrograph of horizontally fissured ("frost" lenticular structure) mortar "floor" composed of limestone and a crushed limestone lime-based matrix. Plane polarized light, frame width is ~5.5 mm.

found in Boschian *et al.* (2003). Pedoturbation and pedological postdepositional affects are also discussed in Courty *et al.* (1989, chapter 8) and in Chapter 3. As an example, middens, because of their original organic character, are characteristically reworked by mesofauna, such as slugs, earthworms, other annelid worms, isopods, insects, and by small burrowing mammals. This reworking often occurs contemporaneously with midden formation. According to the cultural context, middens can be also worked by hyenas, *canids* (dogs, coyotes, wolves), and pigs, as demonstrated by the presence of their coprolites. This would help explain the common lack of reliable stratigraphy in some midden sites (Goldberg and Macphail, in press; see also Balaam *et al.*, 1987; Stein, 1992) (Chapter 7).

We have already noted that the best-preserved components and microstratigraphic sequences occur in anaerobic, (often waterlogged), and arid environments. Deposits in settlements and buildings that have been rapidly buried are also well preserved. These include rapid burial by volcanic ash, mudflows, or deliberate infilling in sites where houses or monumental buildings are constructed on top of earlier infilled buildings (see Chapter 11). Climatic, seasonal, and local conditions would have strongly influenced the nature and rates of decay, and between different activity areas (see Chapter 12). Urban morphology may well include industrial areas, middens, burials, pens, stables, and latrines, administrative and domestic buildings, open areas, and streets and lanes that were reorganized through the lifetime of the settlement.

10.9 Conclusions

This chapter has discussed the very special nature of occupation deposits, and how they cannot be studied simply by traditional geoarchaeological methods. It is only relatively recently that archaeologists in some parts of the world (particularly Europe) have begun to appreciate the fact that occupation deposits are an *essential* part of the material culture. By ignoring the value of these deposits archaeologists limit their ability to fully understand their sites (Goldberg and Arpin, 2003). The chapter has also provided an outline of the kinds of information that archaeological deposits potentially contain, reflecting, for example, the relevance to spatial distribution studies and the recovery of artifacts and ecofacts at sites. Some of the above models, concepts, and issues are elaborated further in later chapters on examples of occupation deposits from the Near East, North America, and Europe (Chapter 11), Experimental Archaeology (Chapter 12), and Human Materials (Chapter 13).

11

Occupation deposits II: examples from the Near East, North America, and Europe

11.1 Introduction

Occupation deposits *sensu lato* are ubiquitous to archaeological sites, for example the hunter-gatherer knapping scatter at Boxgrove (Box 7.1). This chapter, however, attempts to provide some examples of occupation deposits formed by complex societies from different regions and cultures around the world. First, we discuss tells and mounds, mainly from the Near East and North America, respectively. The second part of the chapter focuses upon urban archaeology as dealt with in Europe.

11.2 Tells

11.2.1 Introduction

Tells and mounds are the quintessential expression of occupation deposits and in situ habitation (Matthews, in press; Rosen, 1986; Jing *et al.*, 1995). Although tells are associated with Neolithic to Bronze Age Middle Eastern archaeology, large mound-like accumulations can be found worldwide. Mounds, for example, are common archaeological phenomena from central United States through to Central America. Cahokia, Illinois in the United States is a well-known mound of massive dimensions. Tells owe their shape and size to repeated phases of construction and occupation which through time have led to marked accumulation that commonly rise above the normal landscape. Thus tells have been identified through Turkey and the Levant. We will not be considering other massive constructions such as stone-built pyramids and mud brick built ziggurats, found in Egypt and Mesopotamia, which are further examples of monumental architecture. Other ad hoc, less obtrusive evidence of marked occupation in western Europe (e.g. midden deposits and anthrosols such as Dark Age dark earth and *terra preta*) are discussed in terms of their Medieval and Roman origins (see below and Chapter 13).

Archaeologically, tells and mounds are complex features that can represent a long-term focus for social, economic, political, and/or ritual activity. Thus, for complex societies they provide a deeply stratified record of social behavior, change, organization, and status that are often unmatched at other types of archaeological sites. As a result, their stratigraphic makeup is normally complex and difficult to decipher (Rosen, 1986). For this reason, there is a great need for geoarchaeological support to tease out the stratigraphic details in order to infer synchronic and diachronic activities.

In this section we discuss the most important geoarchaeological aspects of tells and mounds, such as material makeup, use of space and site formation processes, and the specific kinds of cultural and background environmental information that can be inferred from these studies. The latter include both local, regional landscape, and climatic histories. For simplicity, the more intensively investigated tells of the eastern Mediterranean region are discussed before New World settlement mounds. We discuss the major questions of:

- What are tells?
- How do they form?
- How do they decay/erode?
- What are the best ways to study tells?

11.2.2 Nature and geography of tells

A *tell* (the Arabic name for mound) is frequently used in modern names for archaeological sites. Other equivalents include Turkish *tepe* or *höyük*, and Iranian *tepe* (Fig. 11.1). Mounds are composed of cultural architecture stones and mud brick (Chapter 13) and associated debris. The last can be formed from leveled, collapsed, transported, and eroded complex features and materials which make up the tell settlement (see below). In the Old World the geographical distribution of distinctive settlement mounds extends from Hungary and Greece, to Turkey, The Levant, Mesopotamia, Egypt, North Africa, Iran, and Central Asia.

Tells vary in size from hundreds to thousands of square meters. The size of settlement alone, however, is not an indicator of the range of activities conducted at a site. Some smaller tell sites have proven not to be village settlements, but small ritual sites with a temple The largest settlement mounds in the Near East can exceed more than 1 km long and more than 40 m high; some of which were formed from the amalgamation of smaller mounds as at Tell Brak (Wilkinson *et al.*, 2003). Many sites expanded and contracted, or were even abandoned at different times. The density and proportion of domestic, ritual, administrative, and industrial areas on settlements mounds is also highly variable. Tells often represent whole cities and display all the intricacies of such complex settlements – public, private, and religious buildings, streets and open areas (cf. plazas), public utilities, craft areas, which are organized spatially within the settlement – evolved through time.

Although mounds in the Middle East are highly obtrusive, commonly rising tens of metres above a flat plain, chronologically comparable settlements in Europe have a different expression in the landscape. For example, the Potterne (Lawson, 2000) and Chisenbury (LBA/EAI) occupation deposits in Wiltshire, United Kingdom, are 1–2 m thick and extend over an area of 4–6 h. In India (Harappan culture) and China, there are important equivalents to tells. This style of settlement can be found to continue in Europe through to Roman and Medieval times, although technologically and culturally these differ in details of sophistication and type of organization and building materials (see below).

11.2.2.1 Architectural materials

A major component of tells are the architectural materials that express social, economic, cultural, and ritual activities (Matthews *et al.*, 1997; Parker Pearson and Richards, 1994). In essence, tells are built up from walls and collapsed or leveled remains of two types of architecture, composed of stone or mud brick. The building stones are either quarried materials from within or outside the site and are variously utilized and manipulated (cut, dressed). The other characteristic of tells is the occurrence of mud brick, free-formed or made using moulds, and generally intentionally unburnt. The mud brick is often composed of local soil or sediment (e.g. silty alluvium) tempered with plant fragments (e.g. straw and reeds; Figs 11.2 and 11.3; Courty *et al.*, 1989: 119), which can be assembled to form walls in a variety of ways (see Chapter 13). Walls can be constructed from individual bricks, mud-mortared bricks, mortared bricks, rammed earth, and clunch – hand puddle earth

(Norton, 1997). Other architectural materials include dressed-stone, fired mud bricks and tiles, fired-lime and gypsum plasters, bitumen, wooden timbers and reeds. The building components of tells are thus also good indicators of the range of accessible raw materials found in the local landscape (Goldberg, 1979a; Rosen, 1986). Mud brick walls and structures, which are not maintained and/or abandoned often decay, erode, can be redeposited as colluvium (Goldberg, 1979a), which can be difficult to differentiate from the original intact mud brick. Mud brick or rammed earth buildings are particularly susceptible to erosion by rainfall, rising groundwater, and reprecipitation of salts (Rosen, 1986: 11).

These basic components of tells typically get redistributed by various activities and mixed with other materials including organic matter, for example, beams, thatch, and are further mixed with cultural and natural materials in the following, for instance: floors, storerooms, streets, courtyards, stock management areas, and latrines. A summary of some of the common soil micromorphological attributes of tells investigated in Bahrain, Iraq, Syria, and Turkey are provided in Table 10.1, and is based on Matthews *et al.* (1997) (Tell Box 11.1).

11.3 Mounds

Mounded, massive anthropogenic accumulations are not merely confined to the Old World but are also widespread in North and

Box 11.1 Tells

Tells are prominent features of the Near Eastern landscape and Eastern Mediterranean landscape (Tell Fig. 11.1). Moreover, a lot of what we know from roughly the last 10,000 years of human history in the Levant, is revealed in their deposits. In fact, most of the settlement in the area that is now Israel and Palestine in pre-Hellenistic times (say, before the seventh century BCE) is found in tells (Mazar, 1990). In this area of the Levant,

FIGURE 11.1 The tell-like form of the Bronze Age site of Titris Höyük in southeastern Turkey is evident here. In fact, as is common in tells, the tell mound itself constitutes only one part of the occupied area; the flat spot in the foreground also contains evidence for human settlement in the lower city.

(cont.)

Box 11.1 *(cont.)*

most are between 3 and 8 ha, although Tel Hazor in northern Israel is more than 80 ha in size.

A major component of tells is mudbrick, which is commonly known by the more generic term, adobe. Raw material for mudbrick is usually quarried from exposures of silty and clayey alluvium (Bullard, 1970; Davidson, 1973; Goldberg, 1979a; Rosen, 1986), but where this is lacking at a site, other fine grained materials, such as loam and loamy sand can be used (Fig. 11.2). These finer grained components are commonly mixed together with straw for binding purposes. When these organic components eventually decay, they leave behind a characteristic form of elongated voids that are observable in the field, and more particularly in thin section (Tell Fig. 11.3).

FIGURE 11.2 Mudbrick wall from Bronze Age layer at entrance to Tel Akko, northern Israel. The mud bricks here are composed of reddish sandy loams, locally known as *hamra*, which is a type of paleosol developed on Pleistocene aeolianite ridges found along the coastal plain (Yaalon, 1967).

When mud brick features are abandoned or destroyed, the bricks begin to decay, with the primary weathering agent being water in the form of humidity, rainfall, or groundwater (Fig. 11.4) (Rosen, 1986: 10). In addition, repeated wetting and drying produces shrinking and swelling aiding breakdown. Crystallization of salts also have deleterious effects on brick stability.

Another common constituent of tells is plaster, which is made from lime or gypsum, the latter being a relatively late development. This material has a variety of uses that include plastered floors and walls, and as mortar used between bricks or to coat skulls and sculpture (Boulton, 1988; Goren *et al.*, 2001). One of the earliest uses of plaster comes from Pre-Pottery Neolithic B (PPNB) sites such as Yiftahel, Israel (Tell Fig. 11.5) and 'Ain Ghazal, Jordan. These and similar floors have featured in discussions about the amount of combustible needed in the production of the burned limestone to produce them, and the possible effects of deforestation in extracting wood for fuel (Garfinkel, 1987; Rollefson, 1990). Moreover, evidence of repeated plastering events reflect upon social stratification and access to labor and raw materials (e.g. Goren and Goldberg, 1991; Kingery *et al.*, 1988; Matthews *et al.*, 1996).

(cont.)

Box 11.1 *(cont.)*

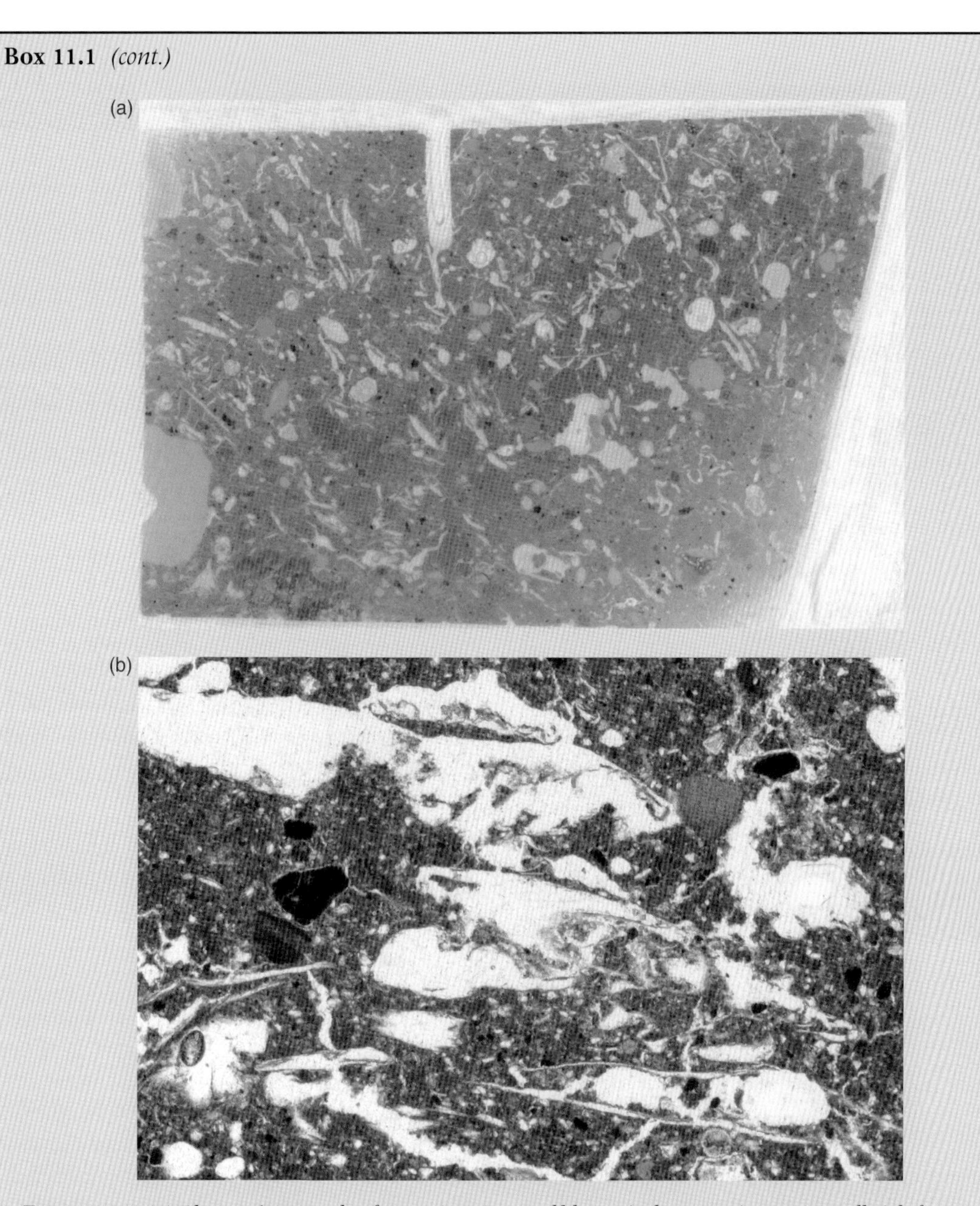

Figure 11.3 (a) Photomicrograph of Iron Age (ca twelfth to sixth centuries BCE) mudbrick from Tel Yoqneam, north-central Israel. The abundant porosity, embodied by elongated and circular elliptical voids, is produced by straw binder that has since decayed. Width of view is 54 mm. (b) Detailed, microscopic view of mudbrick showing the pseudomorphic voids of vegetation. Plane-polarized light (PPL); width of view is ca 4.7 mm.

(cont.)

Box 11.1 *(cont.)*

FIGURE 11.4 Face of the southern baulk of Square J/5 in the main hall (Locus 3146) of the Level VI temple (Late Bronze Age) in Area P from Tel Lachish, Central Israel (Goldberg, 1979). Unit 1: decayed mudbrick with chips of plaster and charcoal; Units 2 and 3: soft, crumbly mudbrick debris of slightly different color, with stone fragments and pebbles; Unit 4: finely laminated silt with rounded pebbles and rodent holes. Units 1–3 seem to be a result of simple collapse of brick wall(s), and not the result of human destruction by fire or other means as is not uncommonly postulated. Unit 4 material was deposited by runoff but it is not clear from where this material was washed.

FIGURE 11.5 Field photograph from PPNB site of Yiftahel, northern Israel showing one of several lime plastered floored surfaces, which are among the earliest uses of lime plaster technology (Goren and Goldberg, 1991; Goren *et al.*, 1993).

Central America. Although shell middens in mounds are common, massive earthen constructions are less widespread and tend to date to the last two millennia or so. One of the earliest excavations in America was conducted by Thomas Jefferson on a small (ca 3.5 m high) mound in Virginia (Sharer and Ashmore, 2003). More typical and striking examples of earthen mounds, however, can be found in the Mississippi Valley (e.g. Cahokia, Poverty Point, Watson Brake), Ohio Valley (Hopeton), southeast United States (Etowah, Moundville), and throughout Central and South America.

As in the Old World, these constructions are complex stratigraphically and socially, and overall, the same general principles of site formation occur. Again, the geoarchaeological task is to use our knowledge of the deposits and depositional/postdepositional processes to gain access to the *detailed* activities of the people who lived there and constructed them. In the New World, geoarchaeological work on mounds – with the exception of a few shell mounds and middens (Stein, 1992), and ceremonial mounds (Van Nest *et al.*, 2001) – is immature and tends to concentrate on landscape scales, such as land use, soil fertility, etc.; smaller scale examination of individual deposits, as emphasized below is not common. When one considers the size of these mounds, their abundance, it is mystifying that so little effort has gone into the detailed study of their stratigraphic makeup.

Different geoarchaeological techniques have been applied to the study of mounds, ranging from larger scale aspects and site survey, down to smaller scale studies of mound formation. The former types of investigations have employed geophysical methods (Herz and Garrison, 1998), while in the latter case, physical and chemical characterization (grain size, soil chemistry, and micromorphology) have been used (Cremeens *et al.*, 1997).

Geophysical techniques have been applied to several areas of the New World (see also Chapter 15). In eastern North America, for example, Dalan and Bevan (2002) raised the issue of how one distinguishes a natural from human-made accumulations without having to probe the sediments directly by coring. They were interested in the "lumpiness" of the geophysical data resulting from anthropogenic mixing of materials; it differs from most geophysical investigations, which concentrate on site survey and the isolation of culturally induced anomalies. They studied directionality data using seismic and magnetic susceptibility techniques at Cahokia Mounds State Historic Site, Illinois; at the Hopeton Earthwork in southern Ohio, magnetic susceptibility and seismic studies were also carried out.

While results at Cahokia were not diagnostic, those at the Hopeton Earthwork were more useful. They concluded that different earthworks have variable geophysical signatures, and that different methods and techniques are successful at different sites. Valuable recognition of anthropogenic signatures and the appropriate method to recognize them are dependent on the type of earthwork, the types of construction materials and processes employed, and on postdepositional processes. In fact, a combination of different techniques is needed to obtain significant results. It is also quite clear that the results of detailed studies of earth mounds and experimental versions in Europe, have still much to offer to New World investigations (Bell *et al.*, 1996; Gebhardt and Langohr, 1999; Langohr, 1991; Macphail *et al.*, 1987).

11.3.1 Case studies

Moundville. Some initial and very promising work was recently conducted at Moundville, a late Mississippian site in central Alabama. The site rests upon an alluvial terrace composed of sands, silts, clays, and gravels, and which is ca 45 m above the Black Warrior River. The geoarchaeological investigations aimed to explore aspects of the site structure, construction, and site formation employing relatively low impact and invasive techniques (Gage, 1999; Gage and Jones, 1999). As ground-penetrating radar (GPR) can penetrate to several

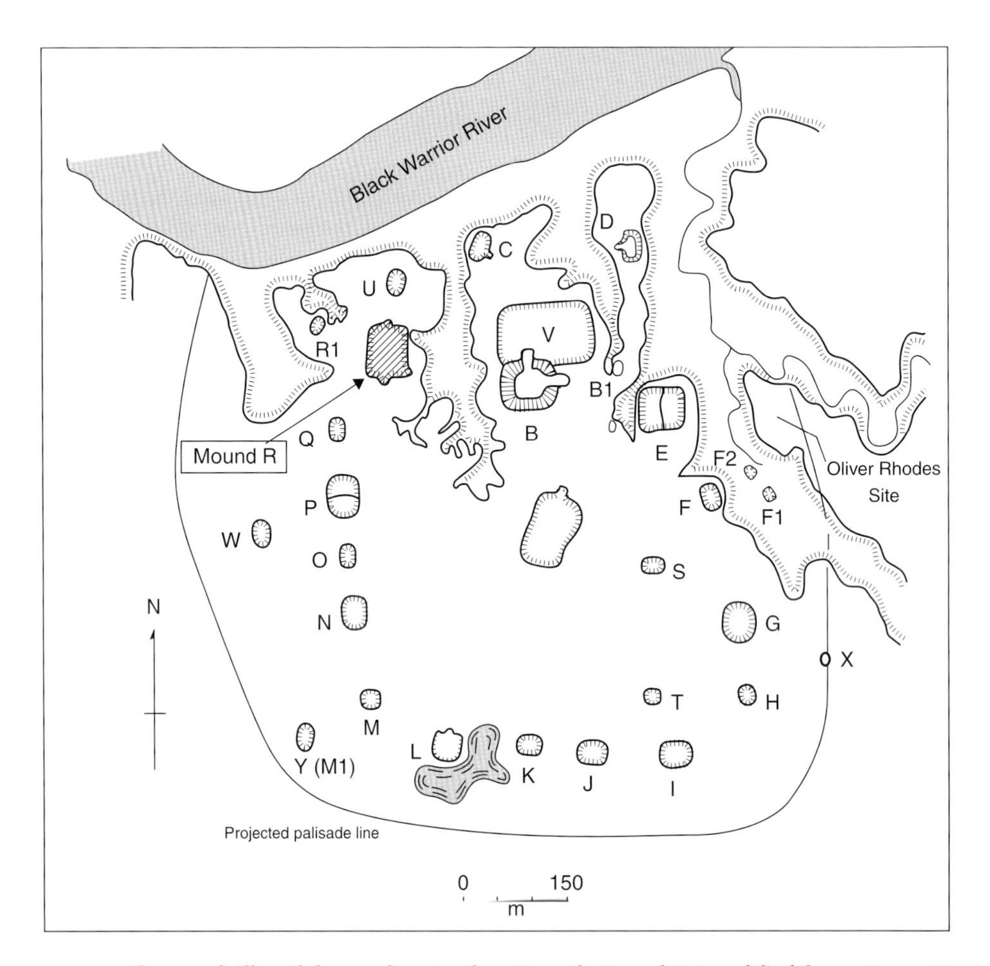

FIGURE 11.6 Map of Moundville, Alabama showing location of Mound R (modified from Gage, 2000).

meters below the surface, it was preferred to shallower techniques such as resistivity, gradiometry, and magnetometry (see Chapter 15). Furthermore, a program of coring was undertaken in order to corroborate the geophysical results (Gage, 2000). The 4 inch diameter cores were split and logged.

One of several mounds chosen was, Mound R (AD 1350 and 1450), which is the third largest structure at the Moundville complex (Fig. 11.6). It sits about 6 m above the surrounding flat area, and covers an area of ca 6375 m^2. "Mound R is not considered to be a burial mound, but rather a domiciliary platform which speculatively would have supported various elite residences, of wattle and daub construction" (Gage, 1999: 307). The geogenic sediments constituting the matrix of the mound are derived from both the alluvium as well as red, grey, and purple clays of the locally outcropping Coker and Gordo Formations. These materials are thought to have been derived from local quarries termed "borrow areas," transported with baskets, and dumped on the mound.

Both the GPR survey and coring revealed a sequence of stratified deposits that point to discrete phases of construction (Gage, 2000). Calibration of the results was achieved by comparisons with surface deposits and with previously excavated areas. It was concluded that burned organic horizons formed marker horizons that represent the end of an occupation phase (Fig. 11.7). Where no organic material was present, it appeared that the

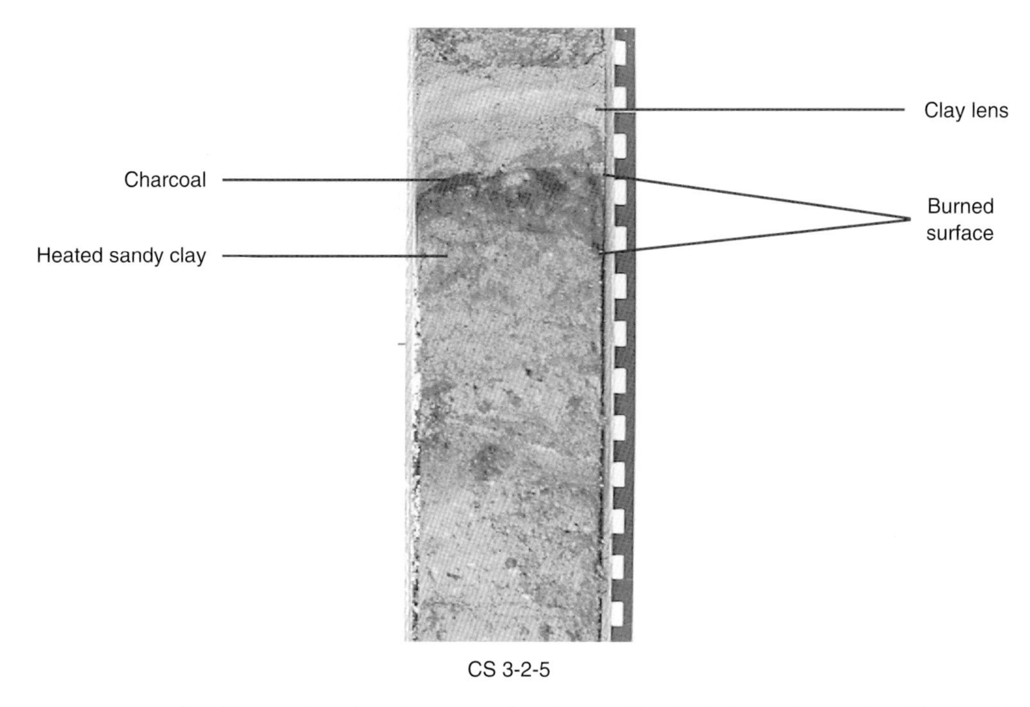

FIGURE 11.7 Example of burned surface from core 3, Moundville. Scale bar is in cm (modified, with permission from Gage, 2000).

mound summit at the time was leveled in order to produce a stable, flat surface. This leveling was accomplished by dumping and spreading basket loads followed by compaction by the site's builders and occupiers (Gage, 2000).

Interestingly, Gage related these types of data to dynamics of the population at the site (Gage, personal communication):

> *Mound R's patterns of occupation mirror the changes in social organization proposed for the site. With the changing population, successive episodes of decreased volumes of construction become apparent until probably the end of Moundville III or early Moundville IV when Mound R is abandoned.*

For whatever reason, successive episodes of construction over years, decades, or centuries was undertaken by groups of laborers in an effort to modify the existing communal landscape. The commonality of building techniques and methods leads to a similarity, notable in almost all such earthen structures. Basketload upon basketload was excavated from secondary locations, transported to a common locale, and deposited to form mounded earthen structures. As a result, profiles of these structures reveal individual piles of sediment as well as summit surfaces buried by subsequent additions of mound fill.

Cotiga mound. This is a 2100-year-old Woodland burial mound built on a Holocene alluvial terrace in West Virginia, United States that was studied by Cremeens (1995, 2005) and Cremeens *et al.* (1997) to elucidate the setting of the site in the landscape, and the mound's stratigraphy, as well as the site formation aspects of the mound. Field reconnaissance, coring and backhoe trenching, and seismic fan shooting and ground-penetrating radar (GPR) were used to provide a cross-sectional view of the mound. Laboratory analysis consisted of grain size analysis. It is one of the few studies of this kind that focuses on the mound deposits themselves (Bettis, 1988; Parsons *et al.*, 1962). Overall, the mound stratigraphy as observed in the field, consists of mostly homogeneous silt loam capped by a soil profile in the upper part, basket-loaded material, and several archaeological features (Cremeens *et al.*, 1997: 467), showing the presence of tip-lines. A very recent micromorphological study of the deposits (Cremeens, 2005) showed rapid accumulation of basket fills with no pedogenesis during construction. In addition, clay translocation, and iron/manganese formation and the development of a complex network of voids, characterize the principal types of postdepositional processses.

They found that the site was built on one of the few stable, flat pieces of land in the area, which was also free from flooding. The silty material that constitutes the bulk of the mound was not taken from the immediate proximity of the mound but from alluvium not far away; the occupants appear to have avoided sandier, deeper alluvium that would have been available in nearby exposed banks. Nine depositional lithological layers were revealed that represented distinct construction episodes. Moreover, the lack of pedogenesis between the layers indicated relatively continuous construction and occupation. Finally, variations observed between individual basket loads appear to result from variations in pedogenetic horizons at the locations from which the material was taken.

Hopeton. A detailed analysis of the stratigraphy, lithology, and pedology at the Hopeton Earthworks (ca 200 BC to 600 AD; Price and Feinman, 2005) in south central Ohio was carried out by Mandel *et al.* (2003) with the aid of exposures along a 48 m trench that cut across the South Wall of the Great Square. The principle aims were to describe the lithology and stratigraphy of the wall fill; determine whether there were temporal gaps during wall construction; and evaluate postdepositional alteration of the wall fill. The methodology consisted of detailed examination of the three profiles exposed in the trench, accompanied by laboratory analysis of bulk samples for grain size analysis and micromorphological study of undisturbed blocks.

Field observation revealed the presence of five major stratigraphic units as distinguished on the basis of color, grain size, etc. Moreover, analysis revealed that the fills consist of two different types of silty material that can be related to local soils as preliminarily mapped at the site: one comes from the local Ockley soil series that occurs across the eastern two-thirds of the site, while the second type of fill resembles the Mentor soil series that occurs over the western third of the site. The specific source locations for these materials has yet to be identified.

Both of these materials represent "loaded fills" in which individually masses of sediment or soil used in the construction of the earthwork are discernible (Van Nest *et al.*, 2001). Furthermore, the lack of soil development within any of the wall fill units shows that, as at Cotiga, wall construction seems to have been relatively rapid. Finally, only a limited degree of pedogenesis seems to have affected the earthworks since the South Wall was completed, partially mitigated by the fact that some of the South Wall was previously buried 3–4 m below the present day surface. Nevertheless, micromorphological evidence reveals the development of weak A-Bw and A-Bw&Bt horizons, including the formation of soil lamellae.

The latter, in particular soil lamellae, can form relatively rapidly, in <150 years (Thoms, 2000), but any kind of textural pedofeature in a mound formed of dumped soil must be regarded with suspicion because dumped materials can produce "instantaneous" textural pedofeatures when dumped under wet conditions. Such features have nothing to do with pedogenic argillic Bt horizon formation, as observed, for example by Romans and Robertson (Macphail *et al.*, 1987, Fig. 11.13; Romans and Robertson, 1983). At the Whitesburg Bridge Site, a Late Archaic site on the Tennessee River near Huntsville Alabama, dumped soils appeared red because wet soils collapsed on dumping and the voids became immediately infilled with ferruginous clay (Sherwood, 2005). In addition, broken or disrupted clay coatings become distributed throughout the deposit.

Meso- and South American mounds. In Central and South America, geoarchaeological investigations associated with mounds and occupational deposits have tended to focus on the regional scale. In Belize for example, much of the work has focused on issues of soil, landscape development, and settlement change (Dunning *et al.*, 1998; Pope and Dahlin, 1989), although many studies link soil and sediment analysis to architecture and site function (Smyth *et al.*, 1995). Remote sensing, satellite imagery, and geophysics are the common tools.

In Brazil, Bevan and Roosevelt (2003) used geophysical techniques and excavation to elucidate patterning of sediments and features, and to understand how they reflect on overall site history of the organization of its occupants. They employed several types of geophysical techniques, but the magnetic data were instrumental in revealing a number of cooking hearths and burial urns, and enabling the estimation of "the number of cooking hearths within the mound without being able to isolate or separately detect any one of them." (Bevan and Roosevelt, 2003: 328).

Furthermore, the geophysical analysis was accompanied by excavations, which demonstrated that baked clay hearths and their associated burned floors are the cause of the magnetic and conductivity anomalies. Before these excavations were carried out, the geophysical survey suggested that there could be as many as 8,000 hearths in the mound, if these hearths were the only magnetic objects in the mound. A reevaluation of the magnetic survey, based on the findings of the excavations, now suggests that there may be "only" about 2,200 hearths within the mound.

11.4 Urban archaeology of Western Europe

11.4.1 Introduction

Few occupation deposits were studied using geoarchaeological techniques 30 years ago, and in Europe most deposits forming urban archaeology – including dark earth (see below) – were simply machined away although its importance had been noted as early as the beginning of the twentieth century (Biddle *et al.*, 1973; Grimes, 1968; Macphail *et al.*, 2003; Norman and Reader, 1912). This section deals mainly with stratigraphy and deposits of the Roman and post-Roman early medieval world. Only occasional reference is made to upstanding structures, because they are often rare and mainly the realm of the specialists in church, historical, standing structure conservation and vernacular archaeology. Standing structures are the specialist area of building analysis where constructional material (e.g. stone) conservation and standing structure survey are key components; and essential background to any geoarchaeological investigations of the same site (Clark, 2000; Fulford and Wallace-Hadrill, 1995–6; Keevill, 2004). In Europe, the Classical World, which in this chapter is represented by the Roman Empire, produced a number of materials that were specific to this civilization. In addition to the stone-founded buildings of both Roman urban areas and villas, and their use of various types of lime-based mortar and plaster (*sensu lato*) (see Chapter 13), occupation could produce significant amounts of roadside deposits, midden accumulations and debris from domestic, industrial, and stock management practices. Numerous Roman structures, as well those from the early Medieval period, were also constructed of timber, and earth-based materials, such as clay, daub, and cob (or clunch), and these produced easily weathered deposits. As such, they became a major constituent of deposits termed dark earth (see Chapter 13).

During the Roman period there was systematic use of different constructional materials, such as lime-based mortar and the use of water proof hydraulic mortar (puzzolanic cements mixed with volcanic inclusions), rather than the use of unfired mud brick or adobe, as found in the Near East through China to the New World (see Chapters 10 and 13). Some Roman activities were carried out for the first time on

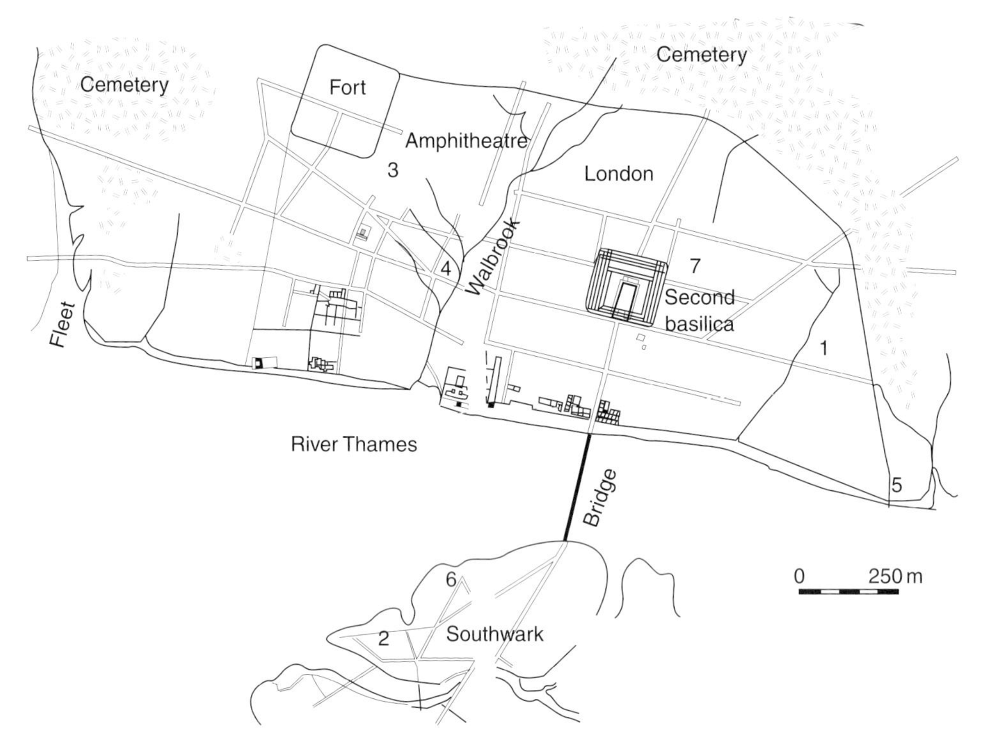

FIGURE 11.8 Roman (ca AD 50–180) London and Southwark showing locations of the extra mural cemeteries, the fort, amphitheatre and second basilica, and sites mentioned in the book: 1. Colchester House; 2. Courages Brewery; 3. London Guildhall (Amphitheatre); 4. No 1 Poultry; 5. Tower of London; 6. Winchester Palace; 7. Whittington Ave. (from Watson, 1998).

an industrial scale. Roman use of space could also be highly organized (e.g. *insula*, space organized within the street plan that are both visible in Pompeii and Roman Vienne or London). A settlement could comprise roads/tracks, structures/houses and land used for dumping/ cemeteries, the last usually on the *contemporary* town edge. It should also be noted here that Roman towns were not defined from the outset by walls. For example, towns like London and Canterbury, United Kingdom, were established in the first century AD, but were not enclosed within walls until the third century AD (Salway, 1993) (Fig. 11.8). Moreover, land use during the history of a settlement could also radically change during the lifetime of an occupation. In Southwark, London and Deansway, Worcester, for example, late Roman (fourth century AD) cemeteries were found in waste areas that were once a residential part of the first – second century town (Cowan, 2003; Dalwood and Edwards, 2004).

11.4.2 Some issues in settlement and urban archaeology

Many of the research objectives set out for the study of historic settlements in Europe can be applied worldwide. For example, there is always a need to ascertain a settlement plan in order to understand a settlement's morphology and to understand how this may have changed and developed through time. Some areas may have always been open ground where stock were kept, middens accumulated or where markets were held, whereas other parts may have had religious, commercial (warehouses, shops, and food preparation), industrial, and a (high and low status) housing land use. Space could also be divided up within structures

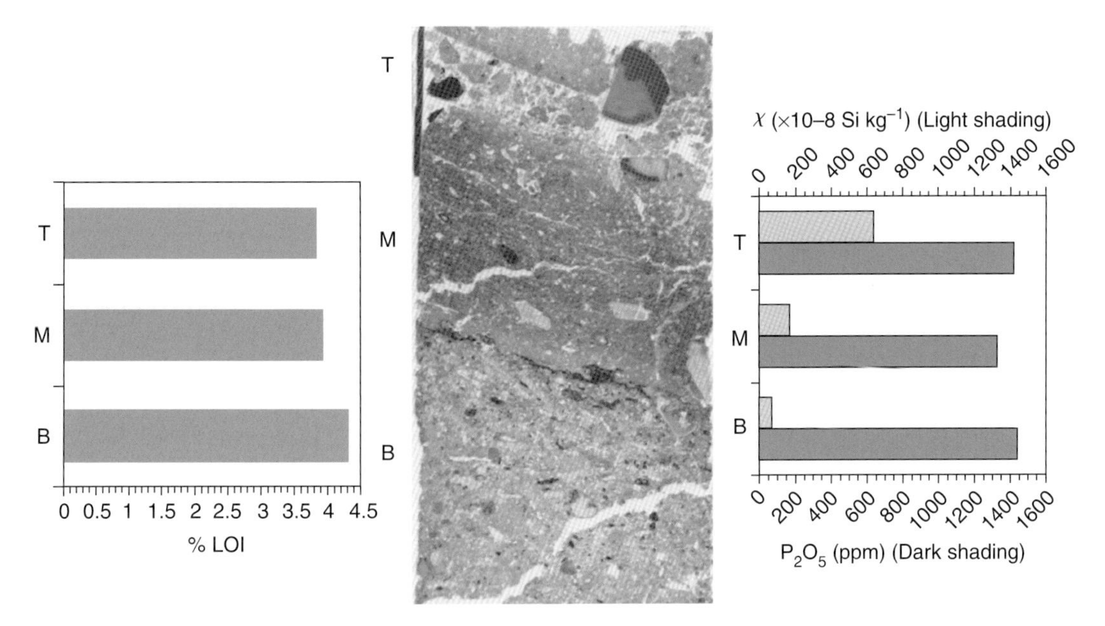

FIGURE 11.9 A scanned image of 13 cm long thin section (M968/7 No 1) from Poultry, London, United Kingdom; loss-on-ignition (LOI), magnetic susceptibility (x) and phosphate (P_2O_5); deposits at a razed (burned down) Roman domestic building forming three layers:

Top layer (T)-compacted aggregates of burned dark humified organic matter (dung?) tempered brickearth soil and few charcoal inclusions (2.3–3.8% LOI; 185–625 × 10^{-8} SI kg^{-1}, 1080–1410 ppm P_2O_5); a deposit of burned daub wall debris following the burning down of an earlier building, now represented by a:

Middle layer (M) – massive, compact, laminated, dung-(?) and straw- (longitudinal and cross-sections) tempered brickearth floor (3.9% LOI) that has undergone some burning (156 × 10^{-8} SI kg^{-1}), as well as having been influenced by secondary phosphate (vivianite)(1320 ppm P_2O_5) during its use/disuse; a carefully prepared brickearth floor that included tempering with dung and straw, and which was weakly affected by the Boudiccan fire (AD 60), and constructed over a:

Bottom layer (B) – mineralogenic (4.3% LOI) layer of building debris – some burned (56 × 10^{-8} Si kg^{-1}) – and gravel, that shows inwash of silty clay and secondary phosphate (vivianite; 1430 ppm P_2O_5), and an uppermost layer of layered charcoal; a coarse foundation dump affected by local anthropogenic wash.

themselves as noted in Chapters 10 and 12 (Cammas *et al.*, 1996; Courty *et al.*, 1994; Cruise and Macphail, 2000; Macphail *et al.*, 2004; Matthews, 1995; Matthews *et al.*, 1996). Another important aspect of settlement archaeology is *continuity* (Leone, 1998). There are important examples of brief changes to "urban" life at both Roman Colchester and London, after the towns were razed during the Boudiccan revolt (AD 59/60) (Fig. 11.9). It was at this time that the short-lived period of horticultural land use can be demonstrated before typical Roman urban construction/life ensued, which at Whittington Ave, London was the building of the basilica (Macphail, 1994).

One of the chief ways of investigating continuity of urban occupation has been through the study of road and trackway alignments, and their relationship to a settlement's contemporary land use. At Elms Farm, Essex, there appears to have been continuity between the town's Late Iron Age origins and the Roman settlement (Atkinson and Preston, 1998). At Canterbury, in contrast, metalled lanes of the seventh century Saxon town show no aligned relationship to Roman roads, even though only 200–300 years separates them (e.g. Whitefriars, Alison Hicks, and Mark Houlistan, personal communication, 2003). The Saxon roads themselves occur over post-Roman dark earth, and

Early-Middle Saxon *grubenhäuser* or sunken feature buildings of broadly Germanic origin, are present. In London the Roman amphitheatre went out of use during the fourth century, and by AD 1050 this major landscape feature had become essentially infilled and leveled. It then became the location of a farmstead on the periphery of Late Saxon London, a site that later became the medieval London Guildhall (Bateman, 1997, 2000).

11.4.3 Sediment formation

Occupation Deposits in general formed in the ways outlined in Chapter 10, but some materials are particularly characteristic of the Roman world. For example, in the amphitheatres studied from Auraurica Augusta (Basel Switzerland) and London, the floor deposits are >1 m thick and show multiple phases of repair and construction. Some layers are simply sands that were probably formed by raking between events (*arena*: sand in Latin), whereas others are rich in ashes, charcoal, and mortar that may testify to constructional history and alterations during the life of the building (Philippe Rentzel, Basel, personal communication) (see example of the Ottonian Church, Magdeburg, Chapter 13). The London amphitheatre had a final, well-constructed mortar floor. It is interesting that whereas turf was commonly used along parts of Hadrian's Wall in northern Britain, local brickearth was employed to construct the late third century ramparts at Canterbury. Similarly, at the site that later became Pevensey Castle, Sussex, the construction of the late third century rampart of the coastal fort *Anderida* led to dumps of subsoils, possibly as foundations associated with wood piling were dug. Ensuing layers of ash, charcoal, and food waste-rich material and middening, may have resulted from occupation by construction workers and the garrison (Macphail, 2002). Lastly, the experimental reconstruction of a Roman villa at Butser Ancient Farm (Chapter 12) produced large piles of constructional debris, such as, spare and rejected lime mortar and pink colored brick, dust-tempered *opus signinum*, gravel and sand, and *tesserae* (rectangular small – ca 1–2 cm ceramic "bricks" or colored rock fragments).

One of the characteristic deposits associated with Roman towns are the fills of roadside gullies and the dumps made on open ground next to roads, which also contributed to ground raising. Some examples can be given from first Century AD London (Cowan, 2003; Rowsome, 2000). Roadside gully fills are dominated by sand-dominated sediments, washed from the road surface, but there is a ubiquitous minor content of small bone fragments and human coprolitic material, as well as secondary phosphate deposition. The last occurs as crystalline vivianite ($Fe_3(PO_4)_2 \cdot 8H_2O$) or as amorphous iron phosphate (see also Box 9.1 and Simpson *et al.*, 2005). Late Roman (fourth century) Deansway, Worcester possesses exactly the same kind of roadside deposit (Dalwood and Edwards, 2004).

One enigmatic type of dumped sediment that accumulated in waste ground in Roman towns, is typically organic. Where well preserved, as a result of a high water-table as at No 1 Poultry, such deposits are characterized by a high LOI (25.9% loss-on-ignition; Figure 11.10) (see Chapters 12 and 16) (Macphail *et al.*, 2004). They are also rich in phosphate (e.g. ~4080 ppm P_2O_5). The layered nature of the organic remains and the highly humified character of some of them imply that these deposits are formed from dumped dung-rich stabling waste (see Chapter 12, Table 12.3). This geoarchaeological interpretation consistent with the results of the macrobotanical study. These deposits seem to owe their origin to the presence of many horses (evidence: faunal studies) and other stock in early first century London. At that time, the developing urban area of London acted as a frontier town on the edge of an expanding Roman Empire. Rowsome (2000) alludes to the American "western" towns of Wichita and Dodge City of the 1870s as suitable analogs. Deposits that were once dung rich, and which represent stock concentrations were also found in waste ground areas of Deansway, a suburb of Roman Worcester (Dalwood and

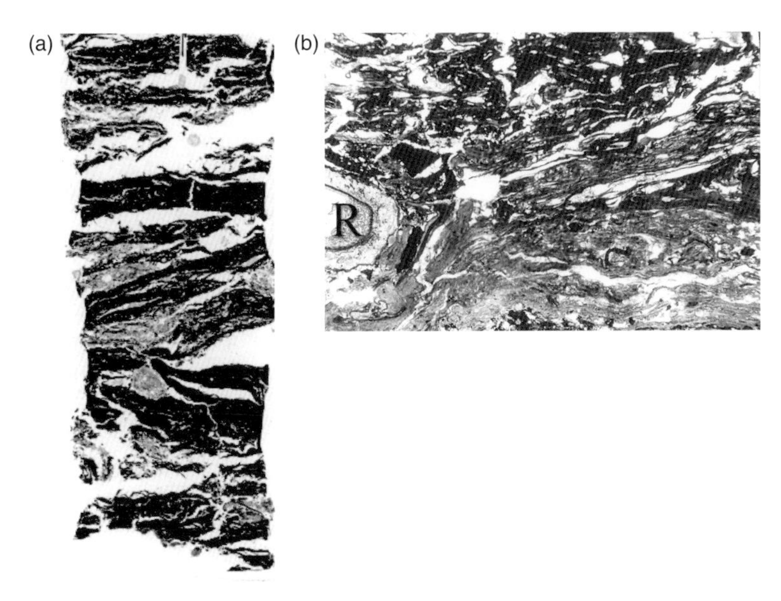

FIGURE 11.10 Dumped stabling waste at Roman No 1, Poultry, London, United Kingdom (Macphail and Linderholm, in press; Macphail *et al.*, 2004), to in which plant macrofossils point to open area dumping of waste from post-Boudiccan animal husbandry (Hill and Rowsome, in press; Rowsome, 2000) (a) Laminated organic tissue fragments, amorphous organic matter, and intercalated silts (neutral pH 6.6; high organic (content 20% LOI), with high amounts of phosphate (3950 ppm P_2O_5), and very low MS (20×10^{-8} SI kg^{-1}). Scanned image of 13 cm long thin section M891. (b) Photomicrograph of thin section M422 showing detail of well preserved stabling waste deposits; a finely bedded organic deposit of plant tissues, amorphous organic matter, commonly intercalated with silt, which is extremely humic (34.6% LOI), neutral (pH 6.6), with very high amounts of phosphate (4170 ppm P_2O_5), and very low MS (14×10^{-8} SI kg^{-1}). Note fleshy rooting (R) of this wet deposit. Plane polarized light, frame width is ~5.5 mm. (ms = magnetic susceptibility).

Edwards, 2004; Macphail, 1994). Similar open-air deposits were found at the small Late Iron Age Roman town located near the Essex coast at Elms Farm, Heybridge. Here they are associated, with animal management relating to the role of this settlement as a market center for the surrounding area (Atkinson and Preston, 1998).

The use of space and changing use of space in Roman structures has been under studied, because much more emphasis has been given to architectural remains. Increasingly, however, a less simplistic approach has been undertaken to scrutinize a variety of lifestyles. At the House of *Amarantus*, Insula 9, Pompeii, not all rooms had properly constructed floors. Room 3 in House 12 had a beaten floor, while Room 4 was used to stable a mule that succumbed to the effects of the eruption of Vesuvius in AD 79. Its skeleton rested upon compacted stabling floor deposits in which P. Wiltshire found pollen of cereals and grass of assumed to originate from fodder, hay, bedding, and dung (Fulford and Wallace-Hadrill, 1995–1996). Similarly, at 7–11, Bishopsgate, London, a room with a domestic brickearth floor was reused as a stable, the charred remains of which seem to date to the late Hadrianic fires (Macphail and Cruise, 1997; Macphail *et al.*, 2004; Sankey and McKenzie, 1998).

11.5 Early medieval settlement

The period of the classical Roman Empire was followed by the Migration Period – sometimes poetically termed the *Dark Ages* – when people moved from the east into other areas into

Scandinavia in the North, and through France, Spain, Italy, and North Africa in the south. Although the centuries following the Roman Empire were marked by the development of dark earth in urban areas, the impact of these migratory peoples was also recorded in the occupation sediments in urban areas.

Compared to the "Romans" and medieval cultures, however, their settlement features were far less substantial and have commonly been "lost." Nevertheless, there is evidence of "missing" strata within the dark earth (see Box 13.2), as shown, for example, by post third century to pre-eleventh century occupation in Tours, France (Galinié, personal communication), and from Early/Middle Saxon occupation of London (Grimes, 1968). In the latter case, however, the only "obvious" structural elements of this are often *grubenhäuser* (see below). In fact, it has often been suggested that the Saxons preferred the "countryside." In the United Kingdom, for example, many Saxon settlements can be found on river terrace locations in Bedfordshire (see below) and along the Middle Thames (Foreman *et al.*, 2002); it is quite common for Saxons to "re-use" prehistoric barrow grounds for their own, for example, West Heslerton, North Yorkshire (Haughton and Powlesland, 1999). At West Heslerton, the whole of the associated settlement was excavated, excavation showed for the first time how extensive, complex, and organized a Middle Saxon settlement could be (Tipper, forthcoming). This site included an area of rectangular house structures with assumed suspended floors, and a large number of grubenhäuser or sunken featured buildings. Two other famous grubenhäuser sites, are the royal stronghold of Tilleda (Sachsen-Anhalt, Germany), which has a very long and unequivocal "weaving shed" grubenhaus, and West Stow (Suffolk, United Kingdom), both having a number of modern reconstructed grubenhäuser (Grimm, 1968; West, 1985).

The presence of Middle Saxon populations in London's former urban areas, as characterized by Roman dark earth, may be misleading when Middle Saxon activity was focused to the west of London, along the Strand (*Lundenwic*). One recurring question is why, if "Roman" London was abandoned to become a "wildwood" as happened during the first two or three decades in parts of post world war II Berlin (Sukopp *et al.*, 1979), then where are all the tree root hollows in the dark earth? (J. Rackam, personal communication). Only a few coarse root traces have been observed. London, it seems could not have ever been totally abandoned. There are therefore, a continuing number of geoarchaeological challenges for the study of this period.

11.5.1 Grubenhäuser

Grubenhäuser are common archaeological features in early medieval settlements, and are often the only structural feature remains of occupations that date from ca AD 450 to 800 and later, in western Europe (Guélat and Federici-Schenardi, 1999; Gustavs, 1998; Tipper, 2001; West, 1985) (Fig. 11.11). Other names include pit houses, SFB's (sunken feature buildings or structures), *cabane en fosse, profonde cabane, grophus*. As they originate from the Anglo-Saxon tradition, however, the generic term *grubenhäuser* will be used here for convenience. (They should not be confused with sunken buildings with real cellars, sunken buildings with the occupation floor at the bottom of the cut feature, nor with Roman "podstal" of Belgium, which are byres in long houses; there are also real sunken floored pit houses of native American origin and dating to the Jomon culture of Japan.) Grubenhäuser are generally reckoned to have had a suspended plank floor over the cavity, and ovens and hearths found in them are often either late or secondary features of collapsed/dumped material (Tipper, 2001). It is this cavity, or what remains of it, that is most commonly found. Infrequently, however, the remains and traces of wooden structural elements have been preserved, for example near Berlin, Germany (Gustavs, 1998).

In addition to the architectural information that they may retain, grubenhäuser can be

FIGURE 11.11 (a) A field photo of Middle Saxon, grubenhaus (field photo); West Heslerton grubenhaus 12AC09507, North Yorkshire, United Kingdom.
(b) West Heslerton grubenhaus 12AC09507; bulk data: lateral control sample WH124: 6.8% LOI; χ=858 × 10^8 SI kg^{-1}; 4410 ppm P.
Bulk data of vertical sequence:
WH125: 6.2% LOI; χ = 895 × 10^{-8}SI kg^{-1}; 3230 ppm P
WH123: 6.4% LOI; χ = 816 × 10^{-8} SI kg^{-1}; 3160 ppm P
WH122: 6.7% LOI; χ = 863 × 10^{-8} SI kg^{-1}; 3870 ppm P
WH121: 7.0% LOI; χ = 619 × 10^{-8} SI kg^{-1}; 5040 ppm P.

(LOI = loss on ignition; χ= magnetic susceptibility P = phosphorus)

important repositories of occupation sediments on sites where floor deposits of rectangular houses and other structures have been lost, or where microstratigraphic information is not easy to recognize, as in the dark earth (Macphail and Linderholm, in press). They have been investigated from southern Sweden, England, France, the Low Countries, Germany, Poland, and Switzerland, and have provided important information on settlement morphology and history. Some notable examples are Tilleda (Sachsen-Anhalt, Germany), West Stow Suffolk, United Kingdom, and West Heslerton (North Yorkshire) (Grimm, 1968; Tipper, Forthcoming; West, 1985). On the other hand, the variation in preservation of grubenhäuser, the variety of their size and shape, and just as importantly, the way they have been investigated, has produced some simplistic interpretations (Tipper, 2001), for example, "all are weaving sheds." This has also led to the view that proper excavation and analysis is no longer necessary, with some being machine-excavated, and their fills completely unstudied. It has also produced the idea that all cut features resembling grubenhäuser, are in fact grubenhäuser, and that the sediments making up a site can be ignored because grubenhäuser are the only "features" making up early medieval sites. In some cases, one "official" view is that "nothing can be done with these fills." This attitude is diametrically opposite to the experience of those people who have studied them in detail.

11.5.2 Microstratigraphic analysis of grubenhäuser

A large number of grubenhäuser have now been studied through microstratigraphic techniques that combine soil micromorphology, chemistry (microchemistry, bulk analysis of loss-on-ignition and phosphate) and magnetic susceptibility (Fig. 11.11; Box 11.1). Nine have been studied in this way from West Heslerton, North Yorkshire, ten from Bedfordshire (4 sites) and four from Svågertop, Malmö, Sweden. This approach has been combined with finds recovery, plotting, and coordinated environmental techniques, such as macrofossil analysis (Macphail and Cruise, 1996; Macphail and Linderholm, in press; Tesch, 1992; Tipper, 2001, forthcoming). On rare occasions pollen has also been recovered from specific materials in grubenhäuser, such as coprolitic remains (Macphail *et al.*, forthcoming).

Box 11.2 Grubenhäuser

Grubenhäuser are sunken featured buildings of mainly Germanic origin (see main text) and record settlement during fifth–seventh century AD in northwest Europe. Their exact use is still under debate but they are important settlement features in landscapes where remains of domestic rectangular houses have often been totally lost. Too often the fills of grubenhäuser have been ignored in geoarchaeological studies but detailed analyses from England and Sweden, for example, have shown what unique information can be extracted to help identify habitation and settlement patterns during the so-called Dark Ages. The primary findings from this approach are that grubenhäuser fills can grouped into two types (Macphail and Linderholm, in press):

1) *moderately homogeneous fills* that could relate more dominantly to *in situ* formation, construction/use and disuse history.
2) *heterogeneous fills* that probably relate to secondary use and dumping, and which provide evidence of site activities.

Homogeneous fills. In England, at West Heslerton, Yorkshire, Stratton, Bedfordshire and Svågertop, Malmö (Sweden), homogeneous fills gave evidence of possible infilling by mainly mineralogenic construction material used in the construction of the grubenhaus. This was either composed of (a) local turf (for low turf walls and roofing?) from natural soils that were little affected by settlement occupation, or (b) from turf (and daub) formed during the lifetime of an occupation. In the first case (a), this gave information on the natural environment, and when grubenhäuser fills are of this kind, it could imply that either these structures were constructed early in the life of the occupation or perhaps were constructed on the expanding margin of a settlement. In the second case (b), "turf" contains large amounts of small anthropogenic inclusions, bone, charcoal, human and animal coprolites, burned material, and secondary phosphate features. All of these components indicate that "soil" has formed in an occupation-"contaminated" substrate. This finding permits the inference that this structure is unlikely to have been constructed and used early in the history of the occupation. Where dating material is so sparse, as in this specific early medieval period, such interpretations based upon fills, could clearly aid phasing.

A 3D plotting of finds of this last type (b) at West Heslerton for example, produced the archaeological interpretation of a grubenhaus with a tertiary fill (Tipper, 2001). It essentially amounts to a different way of explaining the same phenomenon. This anthropogenic material-rich grubenhaus fill also records values of organic matter (6.2– 7.0% LOI; $n = 6$), P (3160–5040 ppm P) and magnetic susceptibility (χ 619–895 $\times 10^{-8}$ SI kg^{-1}) that are all high; in the case of P, it is 3–5 times higher than pre-Saxon levels. Similarly, χ is 6 times higher. The bulk analyses totally support the interpretation of the fill using soil micromorphology.

At Svågertop Figs 11.12 and 11.13, a natural soil infilled grubenhaus also contained little evidence of human activity (Grubenhaus 433: χ mean 38.6 $\times 10^{-8}$ SI kg^{-1}; P_2O_5 mean 170 ppm; $n = 9$), compared to three others (e.g. Grubenhaus 1954: χ 81.8 $\times 10^{-8}$ SI kg^{-1}; P_2O_5 mean 2200 ppm; $n=20$). The presence of turf in Grubenhaus 433, was recognized on the basis of humic soil being present that is characterised by organic-rich excrements of soil mesofauna in a biological microfabric, as found in experimental turf roof samples in Sweden, for example (Cruise and Macphail, 2000; Macphail, 2003b).

(cont.)

Box 11.2 *(cont.)*

Heterogeneous fills. These can provide specific information on events, activities, settlement morphology, discard practices, lifestyle, cultural, and domestic economy. In addition, they can clarify the origins and character of fragile objects, such as unfired loom weights formed out of "soil". For example, at Svågertop the burned dung microfacies type is composed of a phytolith-rich matrix that includes articulated phytoliths of monocotyledonous plants, layered microfabrics (pseudomorphs of cattle dung?), charred amorphous (humified) organic matter (dung?), silt, diatoms, burned soil and rooted and iron-depleted sediment.

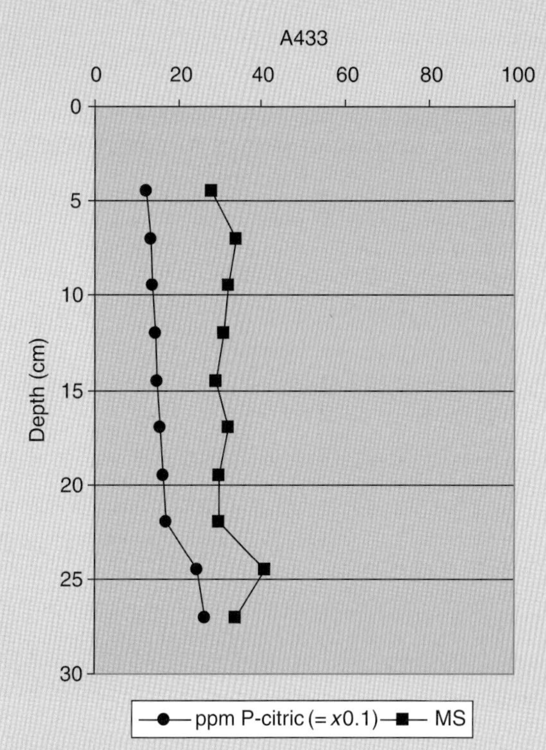

FIGURE 11.12 Grubenhaus "homogenous" fill of local soil/turf: Svågertop, grophus 433; vertical sampling column through the grubenhaus fill and underlying soil (below 23 cm), showing measurements of x (MS, $\times 10^{-8}$ SI kg^{-1}) and P_2O_5 (ppm P citric, $\times 0.1$). Note how little difference there is between the soil and the fill, supporting the soil micromorphological evidence of a natural soil-infilled grubenhaus. Such a finding implies the likelihood of little anthropogenic impact on this part of the site – possibly indicating that it is located in an area peripheral to a settlement and/or implying that it could be an *early* feature. It also may suggest that turf was a major component in its construction and that this has fallen in.

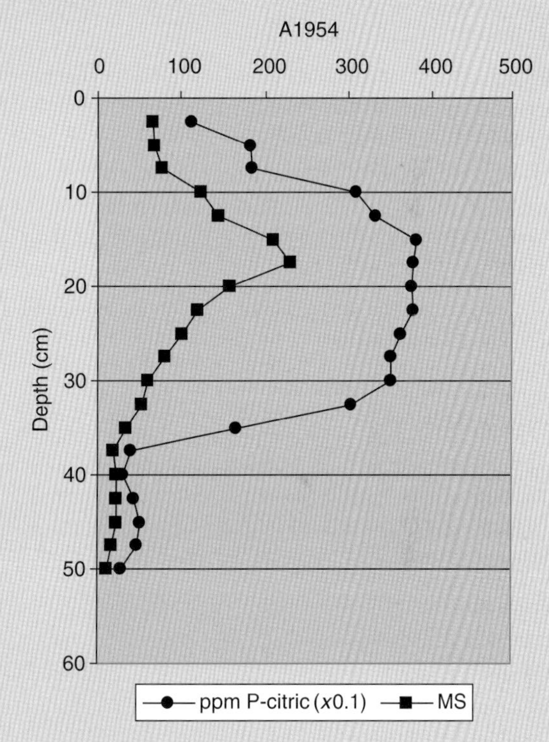

FIGURE 11.13 Grubenhaus "heterogeneous" fill of dominantly burned (cattle?) dung: Svågertop Grophus 1954: vertical sampling column through the grubenhaus fill and underlying soil (below 39 cm), showing measurements of x (Ms, $\times 10^{-8}$ SI Kg^{-1}) and P_2O_5 (ppm P citric, $\times 0.1$). Here there is a clear contrast between the high values recorded in the anthropogenic fill, which results from dumping of occupation waste, and the natural low levels in the soil.

(cont.)

Box 11.2 *(cont.)*

The effects of burning on magnetic susceptibility (χ) and phosphate concentrations induced by the burning of dung have been noted above (Figs 11.12 and 11.13). The presence of silt is consistent with the ingestion of mineral matter from grazing and drinking, diatoms and gleyed sediment fragments, being indicative of fodder/grazing originating from wetland areas. These data contribute to the analysis of the integrated economy, which included the necessity of over-wintering cattle and the use of their manure for barley production (Engelmark, 1992; Engelmark and Linderholm, 1996; Viklund, 1998).

Importantly this approach has not been applied to grubenhäuser in isolation, but has been applied on the same sites to other early medieval deposits and features, including colluvium, ditches, pits, middens, trackway deposits, waterholes, and wells. Often preceding and later archaeology are also included in the study.

The chief findings from this research are that grubenhäuser fills can grouped into two types (Macphail and Linderholm, in press) – see Box 11.2 for details:

1 heterogeneous fills that probably relate to secondary use and dumping, and which provide evidence of site activities, and
2 moderately homogeneous fills that could relate more dominantly to in situ formation, construction/use and disuse history.

Using this model it has been possible to suggest that grubenhäuser studies:

- are capable of providing unique information concerning site settlement and history, site morphology/settlement organization, contemporary soils and construction techniques.
- can provide data on activities and land use, through the analysis of their fills by artifact and ecofact recovery, the last including microstratigraphic techniques (see below).
- when combined with the analysis of pit and well fills, ditch and trackway sediments, and midden deposits, they can contribute to a holistic dataset that can be utilized when developing consensus interpretations of early medieval sites.

11.6 Medieval floors of northwest Europe

Floors and their deposits have been studied across Europe. For brevity only two main types will be described, namely:

- floor deposits formed in stables and byres
- beaten (trampled) floors present in "low" status domestic space/structures.

Prepared floors, such as brickearth domestic floors and lime-based mortar floors of "high" status space/structures, are dealt with in Chapter 13.

One of the chief sites referred to in this part of the book is the Medieval site that developed into the London Guildhall (see Figs 11.8 and 11.14). The Medieval deposits, which are studied over an approximately 10-year period, had formed over and around the Roman arena that went out of use at the end of the fourth century AD (Bateman, 2000; Porter, 1997). Moist preservation conditions permitted wood, and both micro- and macrofossils to be studied, with plant macrofossils and pollen contributing to the better understanding of the early Medieval (e.g. AD 1060–1140) settlement stratigraphy. Two major structure types were identified, namely post and hazel wattle buildings, and essentially with plank walled buildings, some of the latter with grooved posts, down which horizontal planks were slid (Damian Goodburn, personal communication). The geoarchaeology was investigated through a microstratigraphical (microfacies) approach, using soil micromorphology, chemistry, and pollen (see Chapters 16 and 17).

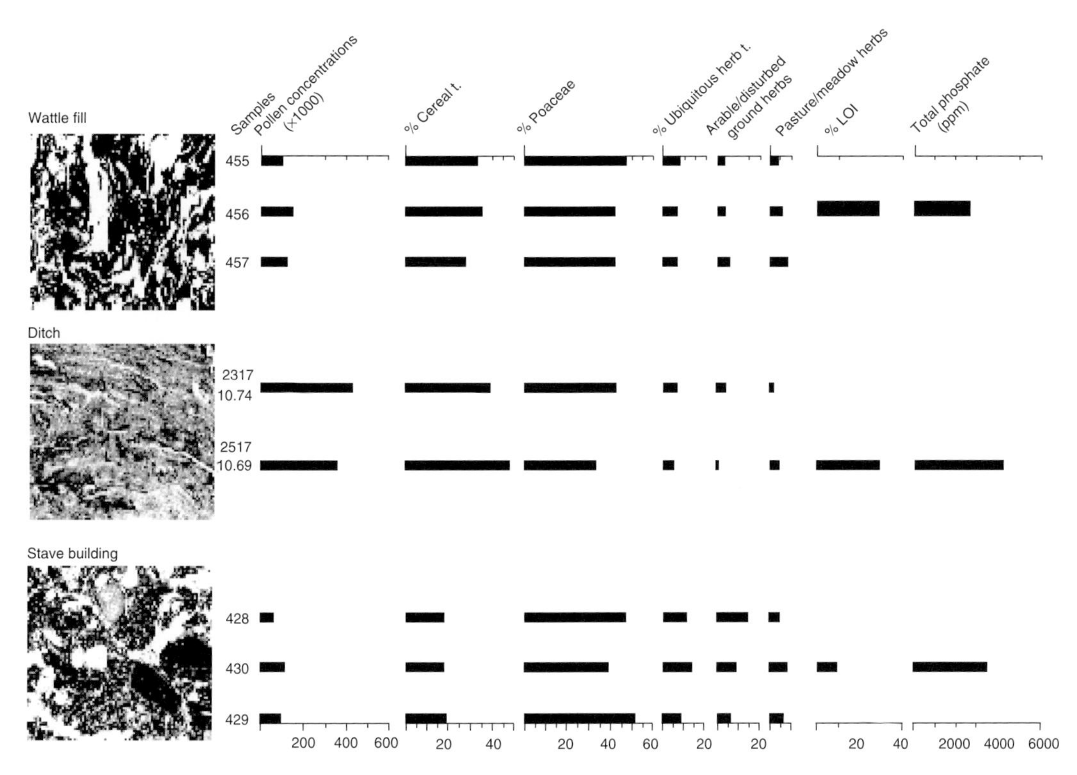

FIGURE 11.14 Microstratigraphical signatures at Early Medieval London Guildhall, United Kingdom (AD 1060–1140): wattle wall fill (highly humified probable cattle dung); ditch fill (finely laminated organic remains – cereal-rich stabling refuse); and stave building (heterogeneous mineralogenic beaten floor deposits) Selected chemistry and pollen data; all photomicrographs plane polarized light, frame width is ~5.5 mm (from Cruise and Macphail, 2000; Macphail *et al.*, 2004) (t.= type, as in, for example, Cereal type pollen).

11.6.1 Floor deposits formed in stables and byres

Although stable floor deposits have been reported from Mediterranean caves (Boschian and Montagnari-Kokelji, 2000; Wattez *et al.*, 1990), dry southern Europe (Cammas, 1994), and arid Middle Eastern (Matthews, 1995; Matthews *et al.*, 1997) open-air sites, much of this work is still fundamentally based upon the results from experiments at Butser Ancient Farm (Chapter 12).

Floor deposits and surface accumulations within and without post and wattle buildings were found to be highly organic and phosphate rich, with low magnetic susceptibility values. Typically, the low amounts of mineral material are often in the form of silt, which probably reflects ingestion of silt during drinking and eating by stock (although animals can trample-in soil clasts). Moreover, fragments of dung, insect-worked dung, and fragments of layered phosphate-embedded layered monocotyledonous (grass and cereal stems) material were identified (Fig. 11.14). Thus both the chemistry and micromorphology is consistent with the character of experimental stabling deposits found at Butser (Macphail *et al.*, 2004; see Fig. 11.12; Table 14.1).

Animal byres have also been investigated in the context of manuring generally, for example, and models have been made concerning how dung (and other domestic refuse) reaches the fields (Bakels, 1988; Carter and Davidson, 1998; Macphail, 1998; Mücher *et al.*, 1990). Patterns of phosphate distribution have also been utilized to infer the location of animal stalls for all periods (Conway, 1983; Crowther, 1996).

11.6.2 Beaten (trampled) floor deposits

Several geochemical approaches to the study of floors have been carried out (Entwhistle *et al.*, 1998; Middleton and Douglas-Price, 1996), but the best and most easily interpretable results have come from soil micromorphology. Although the last have been combined with bulk analyses, the interpretation of bulk data has to be very closely related to the thin section studies. This is because floor layers can vary dramatically over the millimetric scale. In fact, beaten (trampled) floors are characterized by trampled-in layers that vary in thickness from 200 μm to 1 mm, with each layer potentially representing trampling-in from a different environment. This is what gives beaten floor deposits (e.g. at the Stave House at the London Guildhall) their essential heterogeneity, and often mineralogenic character compared to stabling deposits (Figs 11.14 and 11.15; See Table 12.4). For example, burned soil, ashes, and charred food waste can be derived from a local hearth, while soil, dung, human coprolites, and earthworm granules can be trafficked in from outside the structure. Equally, iron slag, leather, and bark can be worked-in from the neighborhood's industrial and craft activities. These examples, taken from early Medieval London (No. 1 Poultry, the London Guildhall, Spitalfields) show that simple bulk analyses from a "floor" may not only *average* many microlayers of different use but also reflect trampling from different areas within a structure and outside that structure. This point has already been made by previous authors (Courty *et al.*, 1989; Matthews *et al.*, 1996).

FIGURE 11.15 Medieval Spitalfields, London, United Kingdom; a series of compact mineral-rich beaten (domestic) floor deposits, with coarse and fine black charcoal, fine bone, shell, ash, burned eggshell and earthworm granules; note postdepositional burrow (top right). Findings are consistent with experimental Pimperne house data from Butser Ancient Farm (Macphail *et al.*, 2004). Thin section scan, width is ~5.5 mm.

11.7 Conclusions

In this chapter, we have attempted to provide some taste of the different kinds of occupation deposits that can be encountered around the world, and the way different issues in archaeology have been tackled by a variety of approaches. It is quite clear that tells and some European Roman and Medieval stratigraphy have been investigated in great detail compared to many of the mound sites of North America. The sophistication of the developing interpretations of cultures in the Near East and Europe, based upon detailed studies is evidence of what is achievable. The utilization of ethnoarchaeological and experimental investigations as ways to improve our understanding of occupation deposits, as suggested in Chapter 10, has been demonstrated in this chapter. Some examples of experiments are given in ensuing Chapter 12.

12

Experimental geoarchaeology

12.1 Introduction

A major issue in geoarchaeology is how well is the past interpreted. It is often asked, how can morphological and micromorphological features and numerical data be understood accurately? In the sciences, empirical data from field and laboratory observations and experiments are commonly employed to help make interpretations, often based on comparisons with present day environments. For example, numerous modern sedimentary environments have been studied, and these can be utilized to elucidate fossil deposits (Reineck and Singh, 1986) (see Chapter 1). Similarly, the very many soil types that are present on the earth have been well documented (Chapter 3). Both ethnoarchaeology (Brochier *et al.*, 1992; Gifford, 1978; Goren, 1999; Gron, 1989; Shahack-Gross *et al.*, 2003) and experimental archaeology (Breuning-Madsen *et al.*, 2001b; Evans and Limbrey, 1974; Macphail *et al.*, 2003; Newcomer and Sieveking, 1980; Nielsen, 1991) have been employed to "calibrate" archaeological findings. Experiments in ancient technology can also be relevant, to iron working (Tylecote, 1979, 1986). It may not be immediately obvious that such industrial activities not only produce geoarchaeological traces (see Chapter 15), but use enormous amounts of woodland resources, and this can impact the landscape if managed woodland was not used (Cowgill, 2003).

The problem most frequently encountered in geoarchaeology, which like geology and archaeology is an historical science, is that there are no modern analogues for many past situations. For example, where does one find "virgin" soils on which to carry out palaeoagricultural experiments (Macphail *et al.*, 1990)? Many situations have not yet been fully studied or investigated at all. In addition, the effect of burial, all kinds of "aging," and other postdepositional processes alter or delete the original components, thus frustrating attempts to accurately understand the past.

There are some people of course, who would give up at this stage. We know, however, that geologists have successfully understood past environments where soils and soft sediments have been completely transformed into rock (Retallack, 2001; Wright, 1986). There is no reason, therefore, why geoarchaeologists should not attempt to elucidate ancient landscapes and human activity from archaeological soils and deposits. On the hand, the challenge for geoarchaeologists could be seen as being more complex than that for geologists who deal solely with natural processes. The human activities that the geoarchaeologist attempts to recognize can be extremely diverse and are sometimes unknown to the modern mind or experience. Human actions also commonly interact with geological/pedological processes, making the situation even more complex.

Experimental archaeology is a thriving field, and numerous experiments have been devised over the years to understand a variety of problems crucial to archaeology. Outside the realm of lithic technology where replication experiments abound (Amick and Mauldin, 1989; Stafford, 1977), issues mostly center around the integrity of archaeological assemblages and taphonomy. They include trampling studies as a means to interpret artifact distributions (Gifford-Gonzalez *et al.*, 1985; Nielsen, 1991; Villa, 1982; Villa and Courtin, 1983), or evaluations of artifact movements in natural settings (Rick, 1976; Shackley, 1978). Most recently, researchers have increased their efforts to understand Palaeolithic pyrotechnology turning to experiments in order to discern processes associated with burning of wood and bone (Stiner *et al.*, 1995; Théry-Parisot, 2002).

From the perspective of geoarchaeology, two main areas of experimentation, will be examined. Broadly, they touch on (1) the effects of burial, and (2) the processes involved in anthropogenic deposit and feature formation, and how these effects can be identified in the archaeological record. The premises underpinning these experiments are firmly based upon archaeological analogues and/or reference analogue or ethno-archaeological observations, which are our other chief sources of information for developing and checking interpretations (Macphail and Cruise, 2001). Investigations into the effects of burial on archaeological materials and old land surfaces were initiated during the early 1960s employing experimental earthworks (see Section 12.2).

Equally innovative, and in many ways more challenging as it requires a full time commitment, is the construction of some "ancient farms." Although industrial activities, such as charcoal making, construction material manufacture, and metal production and working were attempted at some locations, we focus here on reconstructions of settlement structures and palaeoagriculture (cultivation and stock management). The following locales are best known to the authors, although numerous other experimental sites occur (e.g. Leyre, Denmark; Melrand, Morbihan, France): the ancient farms at Butser, United Kingdom (Iron Age and Romano-British) and Umeå, Sweden (Iron Age). Some results from the Experimental Earthworks Project and related research are dealt with first.

12.2 Effects of burial and aging

It is absolutely crucial that any interpretation of buried soils and structures is based upon exact knowledge of preservation conditions once they are buried or abandoned. Major transformations affecting ancient soils and sediments are chemical and biochemical ones, such as oxidation of the organic content, and physical changes, for example, compaction and bioturbation. In order to replicate these processes, experimentation was deemed essential, when the Experimental Earthworks Project was set up after the Charles Darwin centenary meeting of the British Association for the Advancement of Science in 1958 (Bell *et al.*, 1996). Two British experimental earthworks were constructed in 1960 (Overton Down, Wiltshire) and 1963 (Wareham, Dorset), and these have been monitored at intervals of 1, 2, 4, 8, 16, and 32–33 years ever since, with projected final excavations in 2024–26 (Bell *et al.*, 1996) (Fig. 12.1). Both experiments were located in areas of ancient barrows (earth mounds) studying base-rich typic rendolls (rendzina) at Overton Down (Chalk), and strongly contrasting acid orthic spodosols (humo-ferric and ferric podzols) at Wareham (Tertiary sand). Comparison with modern control soils was undertaken at both sites.

Overton Down. At Overton Down (excavated 1992) the thickness of the Ah horizon had decreased from ca 180 to 90–100 mm; this took

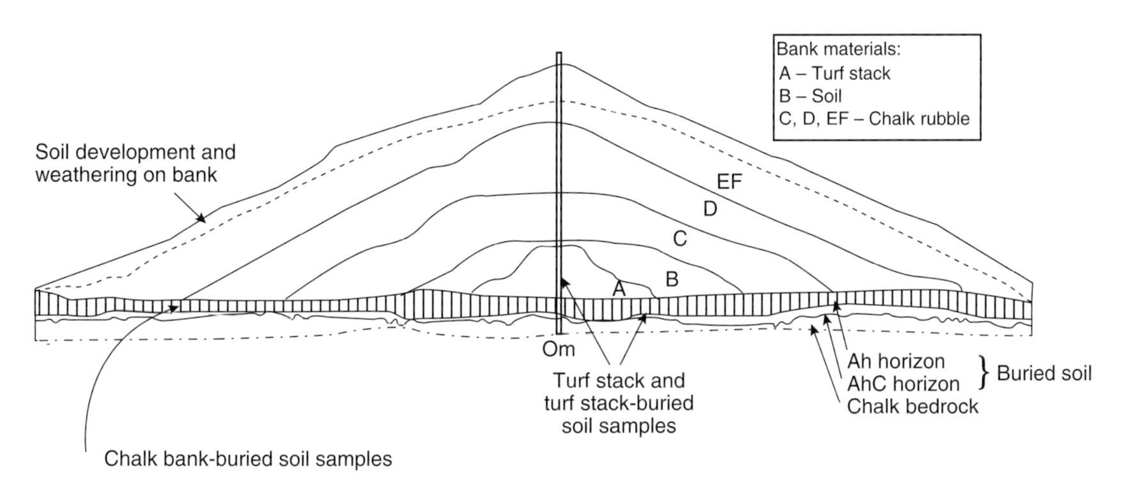

FIGURE 12.1 The Experimental Earthwork at Overton Down, Wiltshire, United Kingdom, 1992 (Bell *et al.*, 1996); section drawing and soil sampling points – the chalk bank-buried soil (2 thin sections and ~10 bulk samples from 2 profiles) and the turf stack-buried soil (9 thin sections and ~18 bulk samples from 2 profiles) (from Crowther *et al.*, 1996).

FIGURE 12.2 The humic rendzina control profile (Soil Pit 1) within the enclosure protecting the Earthwork from grazing animals. This situation has resulted in the formation of a 2 cm thick grass mat and incipient surface gleying; terrestrial diatoms are also present in this moist surface grass mat; the control profiles have an Ah horizon that varies in thickness from 15 to 22 cm.

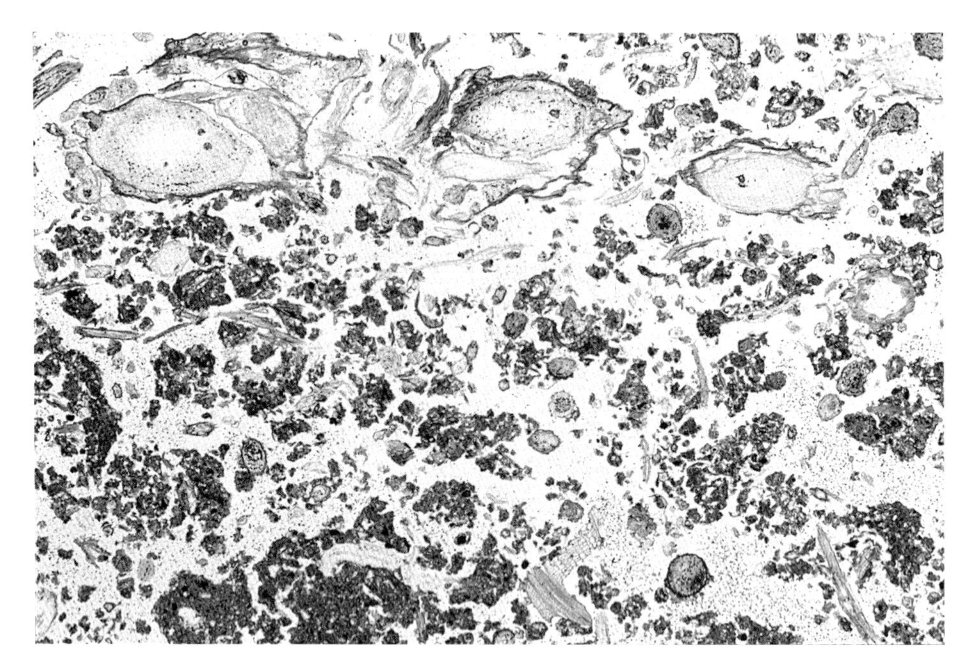

FIGURE 12.3 Photomicrograph of the Ah horizon of control profile (Soil Pit 2) outside the enclosure at 1.5 cm depth, showing a heavily rooted topsoil (note cross-sections through living grass roots), a very open mainly compound packing porosity (mean 55% voids), and very abundant humic and organo-mineral excrements. Plane polarized light (PPL), frame width is ~9 mm.

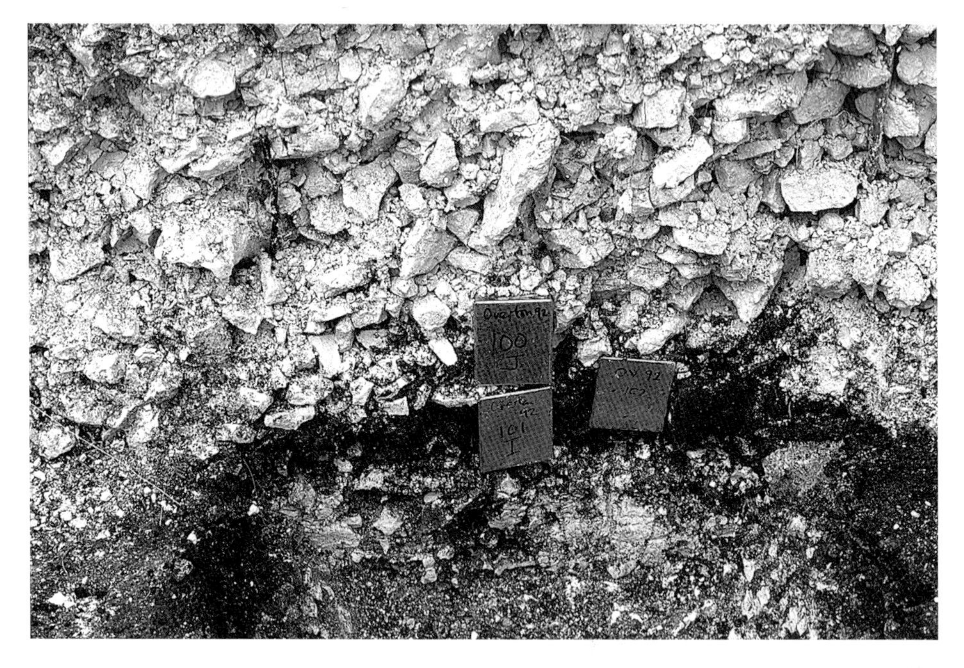

FIGURE 12.4 As in Figure 12.1 – chalk bank-buried soil and Kubiena box samples; the buried soil had become compressed (~11 cm thick), although overall the chalk bank-buried soil ranges in thickness from 5 to 18 cm.

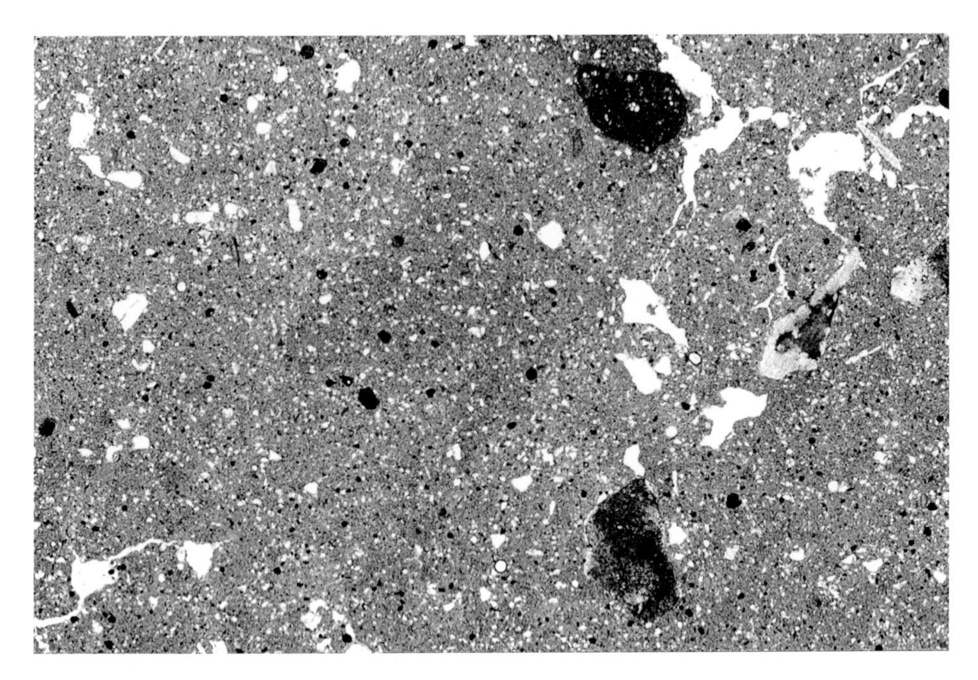

FIGURE 12.5 As in Figure 12.1, a photomicrograph of the chalk bank-buried rendzina showing changes to soil structure (now becoming massive structured through the aggregation of earthworm excrements) with concomitant loss of porosity (22.73 to 48.42% less void space compared to the control profile); organic carbon has been reduced from 11.0% to 7.88% (mean values); earthworms have also mixed chalky soil from the bank with the original decalcified soil, raising the pH (from 6.9 to 7.9). PPL, frame width is ~9 mm.

place in part through organic matter loss (ca 11.0% to 7.59–7.88% organic C) (Crowther *et al.*, 1996). An open microfabric characterized by 55% void space (Figs 12.2–12.3), that shows signs of earthworm-working and grass rooting had become compacted (minimum 14% void space (Figs 12.2–12.3) at 1 cm depth). Moderately broad (>500 μm) mammilated earthworm excrements were replaced by aggregated excrements and a spongy microfabric composed of thin (50–500 μm) cylindrical organo-mineral excrements (Figs 12.4–12.5). In addition, soil acidity increased under the turf stack (from pH 6.9 to 5.6). Where the soil was buried by chalk rubble the soil became more alkaline (pH 7.9) because here, earthworms mixed the buried soil with the overlying chalk. *Lycopodium* spores (which had been laid on the old ground surface at the time of monument construction) were mixed 90 mm upward into the overlying chalk bank (Crabtree, 1996). (The stratigraphy of land snail populations can also be affected by burial; Carter, 1990.) Micro to meso-sized nodules of Fe and Mn had also formed after burial at Overton Down.

The formation of iron pans and pseudomorphic iron-replacement of plant material are both manifestations typical of buried soils that result from localized gleying; as also noted at the 500,000 year old Boxgrove, United Kingdom (Crowther *et al.*, 1996; Limbrey, 1975; Macphail, 1999). Interestingly, the buried soil at the nearby Neolithic (e.g. ca 3400–3600 cal BC) long barrow at Easton Down (8.5 km distant) exhibited very similar soil microfabric types to those at Overton Down (Whittle *et al.*, 1993). This showed how quickly long lasting changes occurred in rendolls once they were buried.

TABLE 12.1 Wareham Experimental Earthwork – changes to the buried soil (Macphail *et al.*, 2003)

Measurement	Turf-buried	Sand-buried
Field		
LFH (thickness)	– – –	– – –
Ah (thickness)	0	0
Chemistry		
Organic C and N	– ?	– ?
C/N	– –	– –
pH	0	+ +
Alkali soluble humus	+ +	+ +
Pyrophosphate ext. C	+ +	+ +
Pyrophosphate ext. Fe and Al	0	0
χ (Magnetic susceptibility)	–	–
Phosphate-P and Available P	0	0
Available K	– – –	– – –
Soil micromorphology image analysis		
Void space	–	– – –
Organic fragments	– – –	– –
Dark organic matter	– –	– – –
Mineral	+	+ + +
Large ~2 mm size voids	–	– – –
Medium ~1–2 mm size voids	+ +	+ + (Layer C – 726)
		– – (Layer D – 781)
Small ~<1 mm size voids	+ +	+ + +
Void shape factor (irregular/elongate)	–	– – –
Soil micromorphology		
Optical counts		
Plant residues and passage features	– –	– – –
Organic excrements (total and variety)	– –	– –
Organo-mineral excrements (total)	+ +	+ +
Organo-mineral excrements (aggregated)	+ ?	+ ?
Soil micromorphology description		
LFH (laminae thickness)	– – –	– – –
LFH (Iron concentration)(Microprobe)	+ +	+ + +
Ah (microstructure and related distribution)	Small change	Change
Ah (amorphous organic matter)	0	0

Key: + + + marked increase, + + increase, + slight increase, 0 unchanged; – slight decrease, – – decrease, – – – marked decrease.

Changes that began over the first 32 years after burial are clearly recognized in soils about 5,000 years old. The ditch sediments have also been studied through both granulometry and chemistry (Bell *et al.*, 1996; 1990–1995) and in thin section, the latter showed that chalk and soil banding currently visible in the ditch fills is probably an ephemeral feature when compared to archaeological ditch fills on similar soils (Macphail and Cruise, 1996: 106).

Wareham. The buried Spodosol at Wareham, changed dramatically with a reduction in thickness of the LF(H) horizon from ca 70 to 1.6–3.6 mm between 1962 and 1980. However, by 1996 it was found that a (amorphous: H_r) humus form was still developing from an (identifiable excrement and plant fragment-rich: F_m) humus form. This occurred, despite continued "ferruginization" resulting from both gleying, induced by the poorly draining nature of basal turf stack, and continued podzolization (Babel, 1975; Macphail *et al.*, 2003) (Table 12.1). For example, there were increases in the sand-buried soil's pH, alkali soluble humus and pyrophosphate extractable C, but marked decrease in available K and change to P content. Image analysis and soil micromorphology showed that while there was an increase in the proportion of organo-mineral excrements, plant residues and organic excrements decreased. Microprobe analysis also revealed that the buried soil had been enriched in iron in some places, confirming the tenant of continued podzolization as indicated by the bulk organic chemistry.

Such monitoring of buried soils has been crucial to the understanding of processes not only affecting monument-buried soils, but also those which result from the weathering of fills in pits, ditches, treethrow holes and pit-houses, for instance (Macphail and Goldberg, 1990; Macphail and Linderholm, in press). Equally, in alluvial soil sequences (cumulic soils), the diffusion of ephemeral land surface topsoils into soils forming in the overlying alluvium – after alluviation – follows the same model of organic matter aging and bio-mixing (Catt, 1986, 1990) as identified at Overton Down. One major finding at Wareham was the concentration of charcoal along the "old land surface" produced by periodic heath fires and now found within the 70 mm thick LFH (Figs 12.6 and 12.7). Such a concentration results from the fact that charcoal is more stable compared to the other forms of organic matter. A concentration of charcoal along an old ground surface, when found under an archaeological monument, could easily be interpreted as indicating pre-construction clearance. The results from the experiment at Wareham now point to the possibility that a "taphonomic" concentration of charcoal has to be considered at some sites.

The preservation and movement of buried artifacts, including organic remains, were monitored in the Experimental Earthworks Project. Bone, for instance, was preserved at base-rich Overton Down, whereas most had been lost at acidic Wareham (Bell *et al.*, 1996). At the Historical-Archaeological Experimental Centre

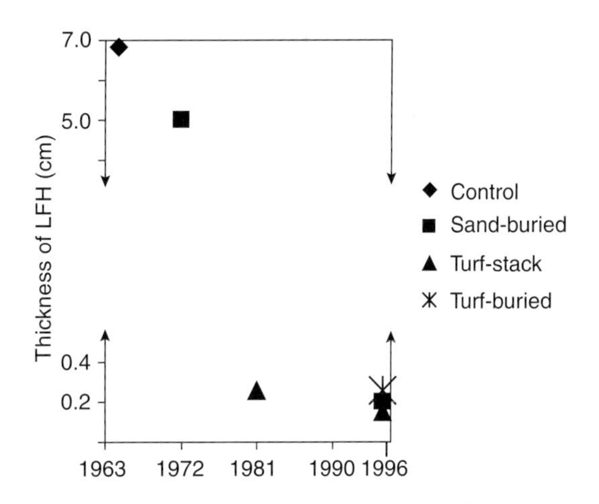

FIGURE 12.6 Wareham Experimental Earthwork 1963–1996; from the 1963 (assumed thickness from 1996 control profile), 1972 (after 9 years burial), 1980 (17 years burial), and 1996 (33 years burial) (Bell *et al.*, 1996; Evans and Limbrey, 1974; Macphail *et al.*, 2003).

FIGURE 12.7 Experimental Earthwork at Wareham, Dorset, thin section 781 sand bank-buried soil: photomicrograph showing junction of buried LFH and sands of bank, with plant remains (PR) and an artificial concentration of charcoal (C) that has been resistant to oxidation; organic matter from the bLFH has been worked by mesofauna into the overlying sand layer D. PPL, frame width is ~2.6 mm.

in Leyre, Denmark, the sealing of buried pigs in wooden coffins, either through the use of grass turfs (sods) or by compacting wet soil, led to the rapid establishment of anaerobic conditions and the cessation of meat decay (Breuning-Madsen *et al.*, 2001, 2003). Although, these experiments were of only short term (ca 3 years), the results demonstrate that the authors were able to replicate the likely conditions of burial of some well-preserved Bronze Age burials.

12.3 Experimental "Ancient Farms" at Butser and Umeå

The experimental farms at Butser (1975 onward) and Umeå (ca 1986–1998) were both initiated to examine many aspects of the Iron Age Period as a whole. At Butser, rural settlement and agricultural activities dating to the Late Iron Age to Romano-British periods (spanning the end of the first millennium BC and beginning of the first millennium AD) were reconstructed whereas selected Swedish Iron Age (ca BC 500–1000 AD) farming methods were investigated at Umeå.

12.3.1 Butser Ancient Farm

Founded in 1975, Butser Ancient Farm, near Petersfield, Hampshire, United Kingdom was founded by Peter Reynolds to investigate some archaeological questions on farming and rural settlement during the Iron Age and Romano-British Periods (Reynolds, 1979, 1981, 1987). Iron Age lifestyles, for example, often have had to be reconstructed on the basis of truncated, dry land sites, where postholes and pits are the only field features. Most importantly for geoarchaeologists, issues concerning roundhouse construction/use and palaeoagriculture, were tackled. It is beyond the scope of this book to discuss the yields from ancient crop production, or how grain was successfully stored in pits. Yet other feature-fills,

FIGURE 12.8 Butser Ancient Farm, Hampshire, United Kingdom, 1990: the 1977–1990 razed Moel-y-gar stabling roundhouse, just before sampling of the stable floor, and the burned (rubefied) and scorched daub wall material. The daub wall was pushed over and the site levelled (a frequent occurrence in Roman London, for example, after the AD 59–60 Boudiccan revolt razed the town; see Fig. 11.9) and the buried stable floor was re-sampled in 1995.

such as postholes (Engelmark, 1985), have supplied additional data used for settlement reconstruction that is crucial to geoarchaeological studies. Two important experiments concerned the study of different floor deposits: (1) the stabling of animals in a roundhouse, which was subsequently burned down as part of the experiment when the site had to be relocated (Figs 12.8 and 12.9) and (2) the development of "beaten" floors in domestic round houses (Fig. 12.9).

The now abandoned (1990) Old Demonstration Area at Butser Ancient Farm was located on a lower downland (short calcareous grassland) slope and dry valley site on Chalk, some 5 km southwest of Petersfield, Hampshire, United Kingdom. The site had been under grassland for some 100 years prior to the Demonstration Area being established. The site is broadly mapped by the Soil Survey of England and Wales as having a cover of brown rendzinas – rendolls (Andover 2 soil association; Jarvis *et al.*, 1983), although the valley bottom, where the houses and arable fields were located, has a typical (colluvial) brown calcareous earth soil cover (Coombe 1 soil association/Millington soil series; Gebhardt, 1990; Jarvis *et al.*, 1983, 1984). The large domestic (Longbridge Deveril) round house was reconstructed in the New Demonstration Area, 3 km to the south on noncalcareous and siltier (loessic) drift soils (Fig. 12.9) (palaeoargillic brown earths, Carstens soil association; Jarvis *et al.*, 1983, 1984: 116).

Posthole fills. Although feature-fills are known to include charred grain (Engelmark, 1985) and animal bones as well as artifacts, they also contain recoverable geoarchaeological information. When the Pimperne roundhouse – a large domestic structure with a diameter of 12.8 m – was dismantled in 1990 after a 15 year life, Reynolds (1994) found that most of

FIGURE 12.9 Butser Ancient Farm, Hampshire, United Kingdom, 1994: the New Demonstration area, with the newly reconstructed Moel-y-gar roundhouse (foreground) and the new large domestic Longbridge Deveril Cowdown round house (background) from which newly forming beaten floor deposits were sampled. Note small door to Moel-y-gar roundhouse, which suits the short stature of Dexter cattle used for ploughing at Butser.

the posts had rotted in the ground, sometimes leaving a post-pipe lined with bark. The postholes had also begun to silt up and already contained artifacts from the use of the roundhouse by visitors. Such a finding is crucial to archaeologists and environmentalists wishing to determine use of space from posthole fills. In fact, the tripartite division of Swedish Iron Age longhouses is based upon the recovery of macrobotanical fossils from postholes. In general these houses are thought to have been divided up into areas for: (1) fodder storage and stabling, (2) cooking and (3) grain storage (Engelmark and Viklund, 1986).

Reynolds (1995) demonstrated that posthole fills actually recorded the use of the site during its lifetime (or first 10–15 years) and thus any sediment within these postholes was a record of the structure's use, and not a post-occupation fill. On this basis, 21 posthole fills and "wall gully fills" from a rare example of a British Early Neolithic longhouse at White Horse Stone, Kent, United Kingdom were analyzed for loss on ignition (LOI – organic matter estimates at 375°C), phosphorous (P), and magnetic susceptibility (χ); these bulk analyses were complemented by three thin sections (Macphail and Crowther, 2004). This investigation was carried out in order to reveal distribution patterns of these parameters, as post-Neolithic erosion had removed all "floor" soils and other surface occupation deposits.

The results suggested that both enhanced levels of magnetic susceptibility and phosphate related to a predominant use of the long house for domestic purposes throughout its inhabitation. In addition, generally enhanced magnetic susceptibility levels, and only moderate phosphate enrichment (compared to a stable-see above), which could be explained by the presence of fine burned soil (and fine charcoal) and bone, consistent with the inclusion of trace amounts of pot and flint flakes found in thin sections. This domestic use model was consistent with the recovered finds; pottery and flints. It can also be noted that the bulk and soil micromorphological character of the fills was a close match to experimental domestic floors formed at Butser, which are soil dominated (i.e. minerogenic) with inclusions of fine charcoal and fine burned soil; these had no resemblance to stabling deposits (see below).

Reynolds (1995) also found that the ground surface between posts of the wattle wall of the Pimperne House was burrowed by rodents (rats, mice, and voles) forming a gully. This feature could easily be misidentified in the archaeological record, as a gully, produced by dripping water or wall constructional feature. We can note therefore, that at White Horse Stone (see above), the longhouse "drip gully" had soil micromorphological characteristics consistent with both burrowing (biologically worked natural soil) and infilling with soil from the beaten earth floor (fine charcoal-rich soil).

Experimental results from occupation surfaces at Butser came from the large domestic Pimperne House (1975) and the smaller (7.6 m diameter) Moel-y-Gar roundhouse (1979) when the former was dismantled and the latter burned down in 1990 (Fig. 12.8). The fire that burned down the thatched roofed Moel-y-Gar house lasted only around 20 minutes. Major scorching and rubefication of the wattle and daub walls (and magnetic susceptibility enhancement) took place above the wooden door lintel (Reynolds, personal communication, 1990). This produced an order of magnitude higher χ value (190×10^{-8} SI kg^{-1}) compared to floor deposits and soils outside ($12\text{–}17 \times 10^{-8}$ SI kg^{-1}). The daub walls were then knocked over and the debris used to bury the stable floor, which was excavated five years later in 1995 (Heathcote, 2002). It should be noted that it is not at all unusual to find a stratigraphic sequence composed of a scorched surface, overlain by a jumble of burned and rubefied daub wall material. This stratigraphy occurs after a timber and daub wall construction has been burned and the site either collapsed or was levelled on purpose. Furthermore, this has been recognized at three sites in Roman London, including structures seemingly razed during the Boudiccan revolt of AD 59–60 at No 1 Poultry (Fig. 11.5) and at Whittington Ave (Brown, 1988; Macphail, 1994; Macphail and Linderholm, in press; Rowsome, 2000).

Additional data from Butser were published in 2004, when the first results from the floor of the domestic Longbridge Deverel Cowdown roundhouse, reconstructed in 1994, also became available (Macphail *et al.*, 2004). The distribution pattern of magnetic susceptibility was also mapped across the Pimperne house floor by Peter Reynolds and Mike Allen (unpublished; Fig. 12.10). Much information has come from *failures* in experimental reconstruction, so that the number of realistic models concerning the fabrication of Iron Age structures can be reduced, this is a great achievement (Reynolds, 1979). One example of particular interest to geoarchaeologists, is how *not* to construct a turf roofed house (see Chapter 13 and Umeå, below).

Research at Butser Ancient Farm also provided a wealth of reference material, including dung from Dexter cattle (traditional draft animals used on the farm for ploughing), goats and ancient mouflon sheep; pasture soils and plough soils with varying manuring regimes provided key background material for thin section and bulk analysis studies (Gebhardt, 1990, 1992; Heathcote, 2002; Lewis, 1998; Macphail *et al.*, 2004; Reynolds, 1979, 1981, 1987). Chemical traces or "biomarkers" (5ß-stigmastanol and related 5ß-stanols) of inputs from animal dung in the manured fields were also investigated (Evershed *et al.*, 1997). Lastly, the site now contains a reconstructed a small Roman villa with hypocaust, as based upon the Sparsholt villa, Winchester, Hampshire (Reynolds and Shaw, 2000).

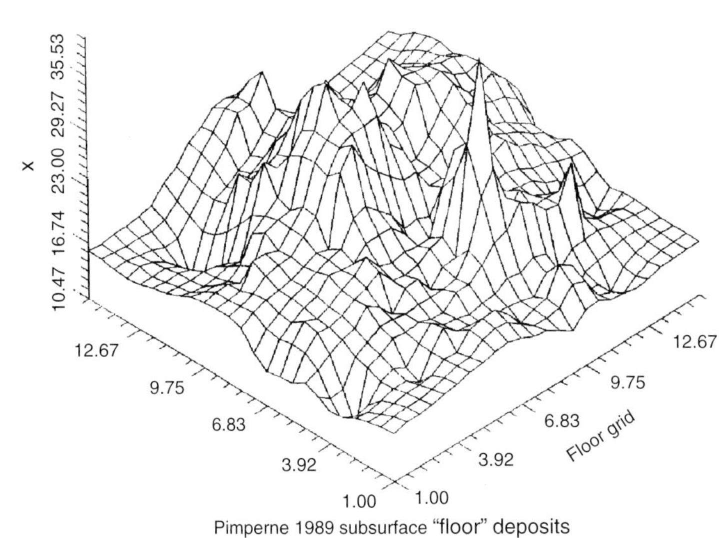

FIGURE 12.10 Butser Ancient Farm, Hampshire, United Kingdom, 1990: magnetic susceptibility distribution map of the domestic Pimperne roundhouse floor, showing the "high point" of the hearth area (M. Allen and P. Reynolds, unpublished data).

12.3.1.1 Experimental floors at Butser

Whereas for the Palaeolithic context, the notion of "floors" and "occupation surfaces" is a difficult matter to recognize and interpret, it is more clear-cut when dealing with architecture and complex societies of later time periods. Experiments have been useful in interpreting occupation surface (e.g. floor) deposits (see below), for example, in order to differentiate domestic "beaten" floors from animal stables (see Table 12.3). Ethnoarchaeological research is, published elsewhere (Boivin, 1999; Goldberg and Whitbread, 1993; Matthews *et al.*, 2000).

The chief findings from the Butser (1990, 1994, 1995) studies, which provide much of the basis for modelling occupation surface (floor and floor deposit) formation, can be summarized as follows (Macphail and Cruise, 2001; Macphail and Goldberg, 1995; Macphail *et al.*, 2004):

Stable/byre floor (Plate 12.1; Figs 12.11–12.13; Table 12.2). Although the most detailed results are derived from the Moel-y-gar stable at Butser, in all, stables holding cattle, sheep and goats, and horses have also been studied (Heathcote, 2002; Macphail *et al.*, 2004). The chief characteristics found at the Moel-y-gar have been found clearly replicated in a series of Roman and early Medieval sites in London (see the London Guildhall below), but site conditions (desiccating, waterlogging, charred/ashed, mineralizing) and cultural practices have also to be considered when studying archaeological sites. For example, only very rare phytoliths, but many woody remains are present if leaf hay was employed as fodder (Akeret and Rentzel, 2001;Macphail *et al.*, 1997; Rasmussen, 1993; Robinson and Rasmussen, 1989).

Phosphate derived from wastes reacted with the underlying buried chalky rendol (rendzina) (Plate 12.1) substrate (Macphail and Goldberg, 1995). To examine this more closely the stable floor was sampled from three locations, and three different strata types were found – (1) a stabling (cattle?) crust, (2) layered sheep/goat dung, and (3) a highly turbated and mixed layer of fragmented dung, crust and daub – essentially from the same "period" and "context" of the experiment (Figs 10.3, 12.11–12.13).

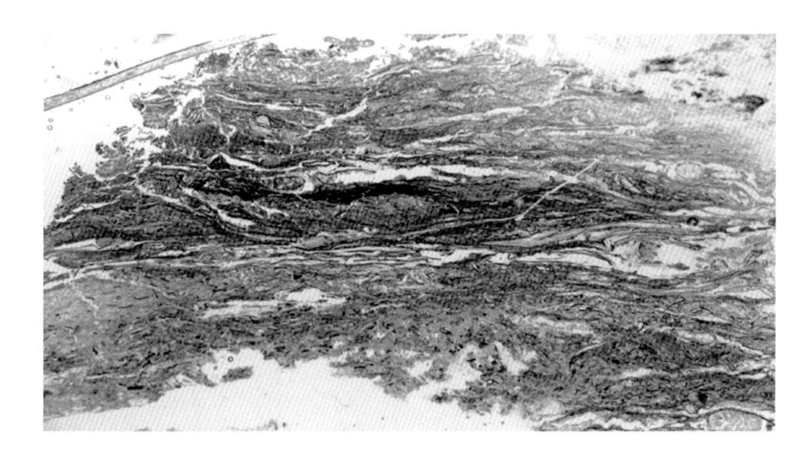

FIGURE 12.11 Moel-y-Gar center: detail of compact layered crust of dung, and fodder and bedding remains. (XRD and microprobe shows cementation by hydroxyapatite with up to 1.1–1.3% P; mean 0.5% P and 2.9% Ca. Thin section scanned image 6.5 cm wide.

FIGURE 12.12 Moel-y-Gar 0.6 m from wall; layers of compact crust and sheep/goat excrements. Thin section scanned image 6.5 cm wide.

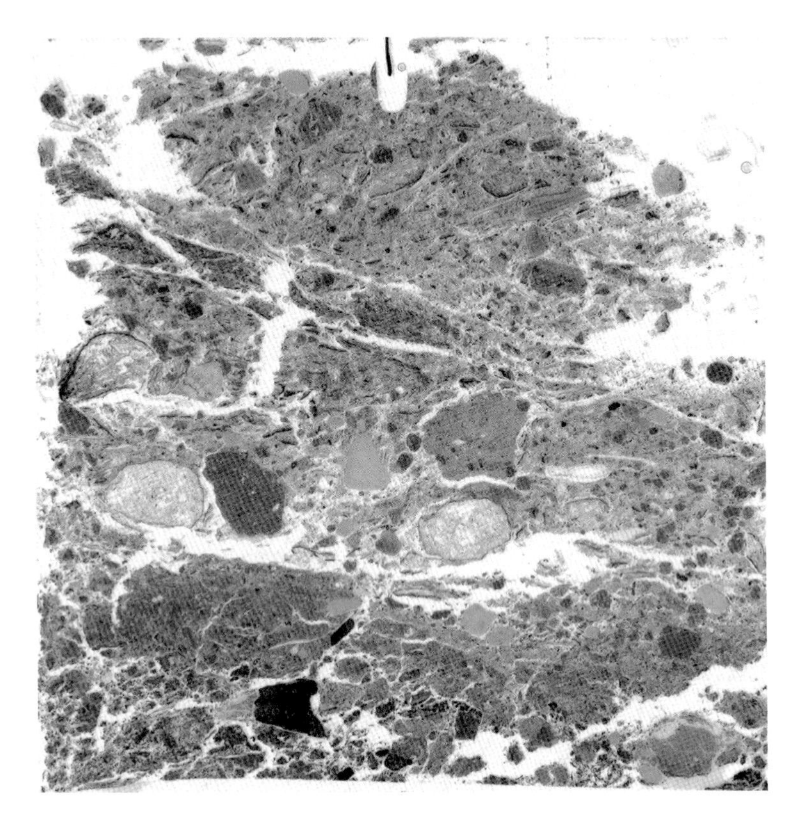

FIGURE 12.13 Moel-y-gar next to wall; fragmented and mixed dung, bedding/folder, crust, and daub (from wall). Thin scanned image 6.5 cm wide.

Butser Ancient Farm, Hampshire, United Kingdom, 1990; illustrating lateral variation across Moel-y-gar stabling floor that dates between 1977 and 1990 (see Fig. 10.3). Dung from this stable was used to manure experimental arable fields (Fig. 12.17).

Domestic "beaten" floor. As beaten floors were studied from both the Pimperne round house dismantled in 1990 (Fig. 12.14), and the newly built Longbridge Deveril roundhouse of 1994, it was possible to see how rapidly such a beaten floor formed. In fact, its massive surface structure started to develop almost as soon as the roof was finished, the grass cover died, and trampling had begun to affect the exposed surface (Reynolds, personal communication, 1994). The crumb structures were not being renewed by biological activity, nor were the blocky structures being reformed by wetting and drying.

Sampling of the 1990 Pimperne house also examined a straw matted area (Fig. 12.15), and there are many examples of archaeological floors being covered by some kind of plant "matting" which subsequently have given rise to compact floor deposits sometimes associated with planar voids relict of monocotyledonous plant "mats," long articulated phytoliths and iron staining, etc. (Cammas, 1994; Macphail *et al.*, 1997; Matthews *et al.*, 1997) (Table 12.3).

In archaeological sites beaten floors can be even more heterogeneous and contain far more bone and coprolitic waste than is allowed at sites such as Butser! Hence archaeological beaten floors may contain higher amounts of

Table 12.2 Bulk analysis of organic matter (LOI), P (2N nitric acid), phosphate (P_2O_5; 2% citric acid), and magnetic susceptibility (χ) on Butser experimental and London archaeological stabling and beaten floors. (Measured on <500 μm soil fraction; * measured on <2 mm soil fraction)

Site and context	Depth (mm) or sample no.	% LOI	P (ppm)	P_2O_5 (ppm)	χ ($\times 10^{-8}$ SI kg^{-1})
Butser 1990					
Demonstration area grassland	0–80	16.6	1770		
Animal pasture	0–80	32.6	1600	16*	
Arable field (manured)	0–80	18.1	2820	17*	
Arable field (manured)	90–170	17.5	2220	26*	
Arable field (non-manured)	0–80	19.0	2010	23*	
Arable field (non-manured)	0–80	15.8	1990		
Stabling					
Moel-y-gar 1990					
Stable crust	0–60	40.9	5960	4730	27
Stable floor	60–100	32.3	2840		18
Buried soil	100–150	22.8	1460		16
7–11, Bishopsgate (charred Roman stable)	Sample 3	9.7	9350	4080	128
	Sample 5	4.3	9220	4160	79
No 1 Poultry					
(Roman stabling)	432	28.3		4300	14
Stabling	422	34.6		4170	14
Stabling	437	36.5		4170	16
Stabling	468	16.2		3920	27
Stabling	891b	20.9		3960	25
Stabling	891a	18.9		3950	16
Stabling	Mean ($n = 6$)	*25.9*		*4080*	*19*
Stabling	Std. Dev.	*8.51*		*155*	*5.78*
Below stabling	968–1a	4.8		2450	86
Saxon stabling pit-fills	385–8	27.4		4330	75
	385–9	13.1		4140	241
Beaten Floors					
Pimperne House 1990					
Compact surface	0–30	20.2	2430		47
Buried soil	30–80	19.9	2310		28
No 1 Poultry					
(Roman beaten surfaces)	887a	4.3		1750	91
Beaten surfaces	968–3	3.6		1300	59
Beaten surfaces	968–4a	7.0		660	114
Beaten surfaces	968–4b	7.4		1050	173
Beaten surfaces	968–7a	3.9		1320	156
Beaten surfaces	973–2	13.9		3270	227
Beaten surfaces	Mean ($n = 6$)	*6.7*		*1560*	*137*
Beaten surfaces	Std. Dev.	*3.89*		*911*	*60.8*

(Cont.)

TABLE 12.2 (*Cont.*)

Site and context	Depth (mm) or sample no.	% LOI	P (ppm)	P_2O_5 (ppm)	χ ($\times 10^{-8}$ SI kg^{-1})
No 1 Poultry (Saxon beaten surfaces)	234–4a	8.1		3340	785
Beaten surfaces	234–5b	9.6		3860	779
Beaten surfaces	234–6a	17.8		3020	231
Beaten surfaces	234–6b	15.6		3110	153
Beaten surfaces	234–7a	11.5		3070	3278
Beaten surfaces	234–7b	20.7		2470	2894
Beaten surfaces	385a	5.9		2680	174
Beaten surfaces	Mean (n = 7)	*12.7*		*3080*	*1185*
Beaten surfaces	Std. Dev.	*5.43*		*449*	*1330*
Case Study					
London Guildhall					
(Early medieval)					
Wattle wall fill of dung	456	28.9	6800	2740	54
Drain fill (Stabling refuse)	251	28.6		4250	61
Stave building beaten floor surface	429	9.0		2190	84

phosphate than their experimental counter parts (Table 12.2). Also magnetic susceptibilities can be very high if buildings have been subsequently burned (see above) or when industrial activities such as metalworking are recorded in the trampled deposits.

12.3.2 Umeå Ancient Farm

The Umeå Ancient Farm, managed by Roger Engelmark and Karin Viklund of Umeå University, north Sweden, is not very far south of the Arctic Circle. It was located on acid soils formed in outwash sands, in general, originally Spodosols (iron podzols). Although primarily initiated to aid the interpretation of weed and crop remains at Swedish Iron Age sites, a number of outcomes have proved useful to archaeological soil studies. For example, it attempted to grow crops and to compare yields from "slash and burned" coniferous woodland and manured soils (Viklund, 1998). In addition, it led to the successful reconstruction of a turf-roofed building (see Fig. 13.4). It is quite clear that in a living mull horizon the roots bind the soil (see Table 3.2) and must be used; moder or mor humus types, or peat, will merely decay (Engelmark, personal communication).

As background to the cultivation experiments the different field types were analyzed for organic and phosphate content. The phosphate was also fractionated into organic and inorganic portions (see Chapter 16.4), yielding a "P ratio" of inorganic P to organic P (Table 12.4; Fig. 12.18). Fields manured with dung, which is characterized by organic P, display high P ratios. This finding was utilized to compare Iron Age field soils (with high P ratios) to long house occupation soils where P ratios are relatively low (Engelmark and Linderholm, 1996) (see Fig. 16.3). The high proportion of inorganic phosphate in occupation areas (compared to manured soils) is due to the weathered and mineralized presence of such materials as

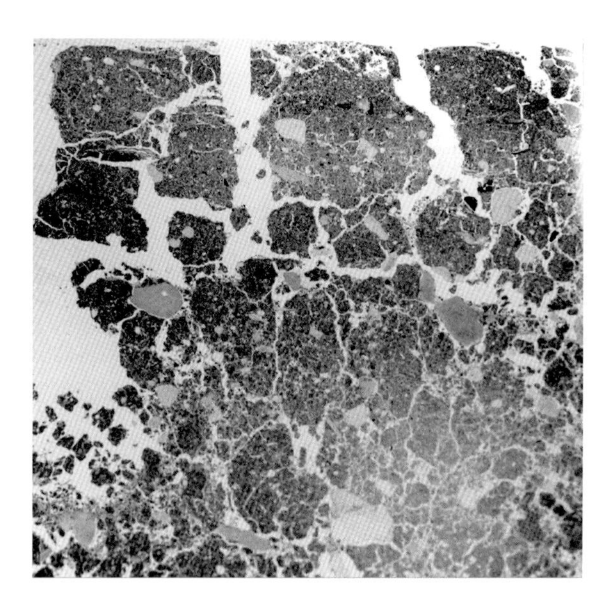

FIGURE 12.14 Butser Ancient Farm, Hampshire, United Kingdom, 1990; Domestic Pimperne Round house floor near hearth 1975–90 (width of thin section scan is 6.5 cm) 0–30 mm: Beaten floor; 2430 ppm P, 20.2% LOI, 47×10^{-8} SI kg^{-1}(note massive structure and typical vertical fissures of dried-out sample). 30^{+}mm; Buried soil; 2310 ppm P, 19.9% LOI, 28×10^{-8} SI kg^{-1} (note remains of blocky structures of original Mollisol's Ah horizon).

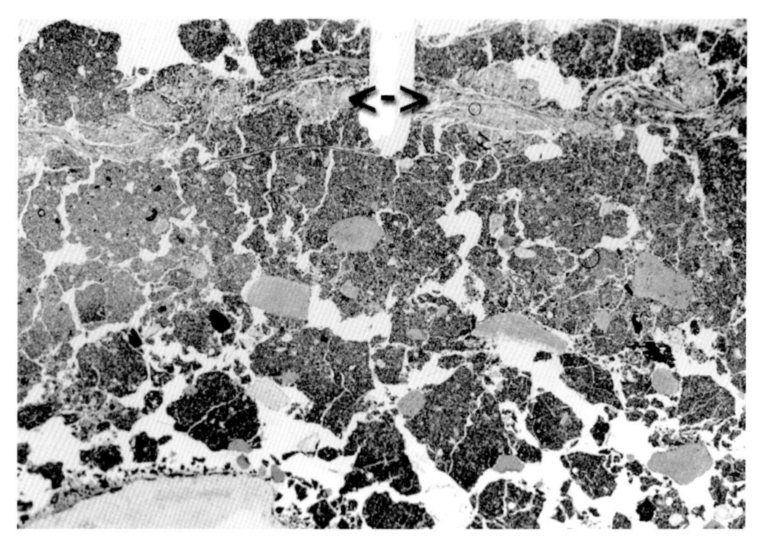

FIGURE 12.15 Pimperne house floor. The matted beaten floor area (1975–90) adjacent to the door (width of thin section scan is 6.5 cm). Detail of mat and massive structured beaten floor. Fragmented remains of pre-matted soil floor occur below (marked with <—>).

bone, ash residues, and coprolitic material, and associated secondary phosphate minerals; the last normally takes the form of iron phosphate in northern Sweden (Linderholm, personal comunication) (Fig. 12.8). The above approach was successfully applied to occupation and manured field sites on noncalcareous soils in the United Kingdom (see Chapter 16) (Macphail *et al.*, 2000).

At Umeå, the local soils – podzols, a slash-and-burn soil and a manured soil – were also examined using qualitative and quantitative soil micromorphology techniques. This analysis was carried out within a broader context of the experiments in agriculture and phosphate analysis, in order to ascertain the key micromorphological differences among the different land use types. At both the field and micro-

Table 12.3 Experimental floors, stables, and domestic space: a summary of characteristics (see also Chapter 11, Figs 11.14 and 11.15)

Floor type	Characteristics
Stable floors	Typically homogeneous where high concentrations of organic matter, phosphate, and pollen grains may be preserved. Organic matter occurs in the form of layered plant fragments which, depending on pH, are either cemented (e.g. hydroxyapatite) or stained by phosphate. If preservation conditions permit the survival of pollen grains, they are likely to be abundant and possibly highly anomalous with respect to the surrounding area
Domestic floors	Typically heterogeneous; floor deposits are comparatively mineralogenic, massive structured, and contain abundant anthropogenic and allocthonous inclusions such as burned soil, charcoal, and ash. Plant fragments are less common and may occur in single layers of organic mat remains. The palynology of domestic floors is an under-investigated subject, but available data suggest the likelihood of far more diverse weed pollen assemblages (reflective of settlement flora), lower concentrations and poorer preservation than in stabling deposits. Magnetic susceptibility is likely to be enhanced and P and LOI will reflect in situ activities and trampled-in materials

Table 12.4 Swedish experimental soil data (Macphail *et al.*, 2000); mean values of % LOI (550°C); 2% citric acid soluble P (P_{inorg}); 2% citric acid soluble P after ignition at (550°C) that converts organic P to inorganic P. (Unpublished data provided by Roger Engelmark and Johan Linderholm, Environmental Archaeology Centre, Department of Archaeology and Sami Studies, Umeå University, Sweden)

Context	%LOI	P_{inorg} ppm	P_{tot} ppm	P ratio (P_{tot}/P_{inorg})
Slash and Burn	5.3	94	423	4.4
Modern field	6.4	105	436	4.2
Permanent field (manured)	10.5	139	684	4.9
Two-field system	5.3	161	493	3.1
Three-field system	6.4	153	541	3.5

scopic level the natural noncultivated shallow podzols showed the expected horizonation (e.g. LFH/Ah/A2/Bs; see Chapter 3, Table 3.2). On the other hand different results were found in the cultivated soils as outlined below (Macphail, 1998).

Slash-and-burn. The topsoil humus and mineral horizons were mixed and characterized by coarse fragments which included wood charcoal (Table 12.4).

Manured soil. The horizons of the podzol had been completely homogenized into a new humic microfabric (Ap horizon) that extended into the subsoil Bs horizon; a fragment of cattle dung (from manuring) was found within this Ap (Fig. 12.17; Table 12.4), helping to account for the high P ratio; a large anomalous limestone clast was also present and testified to manuring and the introduction of allocthonous ("foreign") material.

Such findings are consistent with the presence of microscopic dung fragments in the manured experimental fields at Butser (Fig. 12.17), which have enhanced phosphate levels; manured archaeological soils also contain burned flint and daub, shell, grindstone, and charred coarse plant fragments (Carter and Davidson, 1998; Macphail *et al.*, 2000). Such activities have produced the thick cultisols – plaggen soils of the Netherlands (Chapters 3) – and the spreads of artifacts around Middle Eastern settlements (Wilkinson, 1990, 1997). Some of the results and applications of other experimental archaeology in palaeoagriculture (e.g. Anne Gebhardt, Helen Lewis) are discussed in Chapter 9.

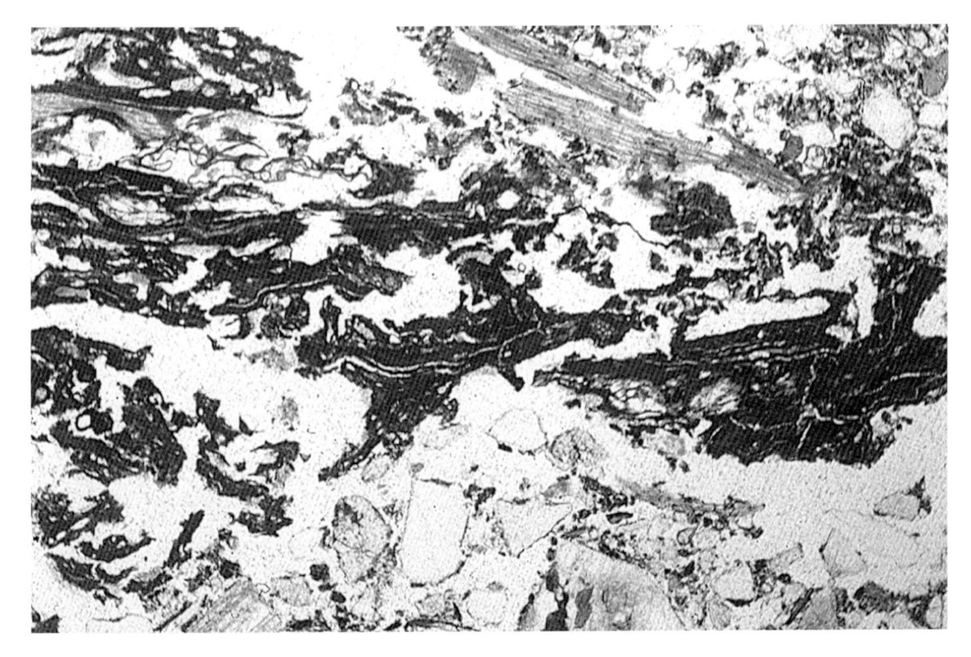

FIGURE 12.16 Umeå Ancient Farm, north Sweden: photomicrograph of experimental manured Ap horizon (formed in a shallow iron Spodosol/podzol) showing inclusion of relict dung fragments in acidic sands. PPL, frame width is ~3.35 mm.

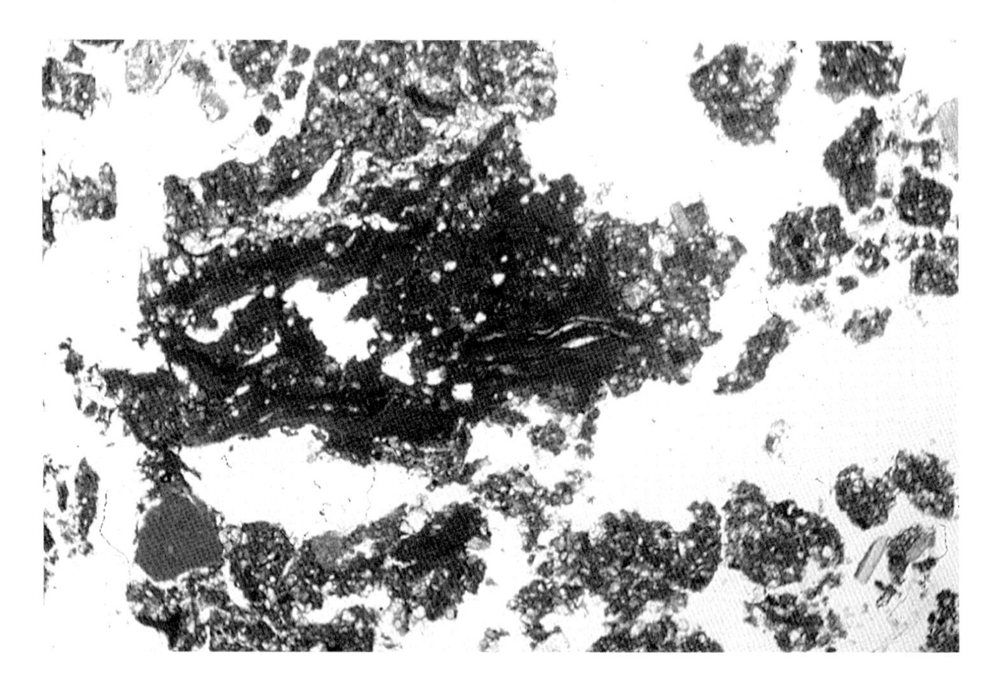

FIGURE 12.17 Butser Ancient Farm, Hampshire, United Kingdom, 1990: photomicrograph of experimental manured Ap horizon (formed in a colluvial rendoll/rendzina) showing inclusion of relict dung fragments in chalky soil. PPL, frame width is ~3.35 mm.

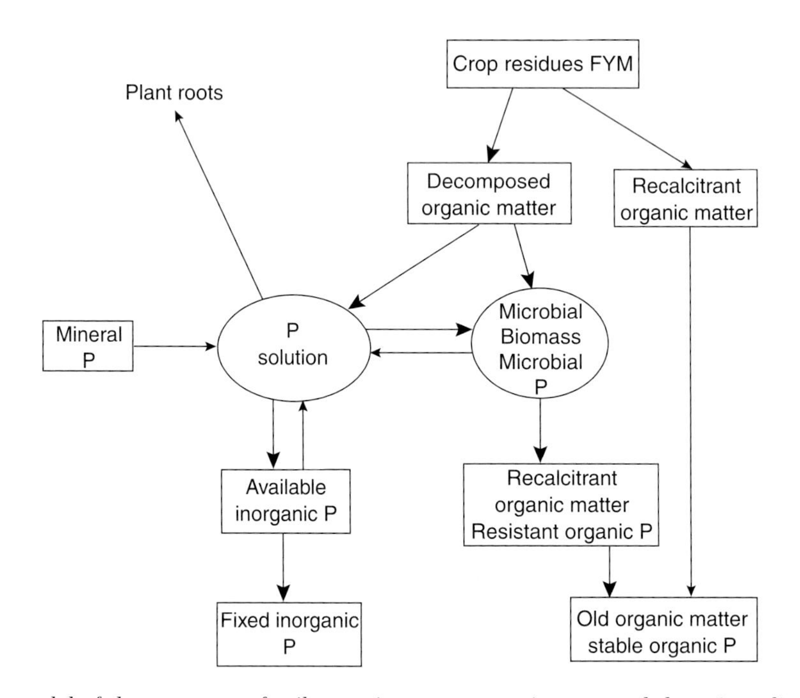

FIGURE 12.18 A model of the turnover of soil organic matter constituents and the mineralization–immobilization of phosphate in Swedish cultivated soils (Umeå). The main flow is indicated by dense arrows (redrawn from Engelmark and Linderholm, 1996) (nb: FYM = farmyard manure; recalcitrant organic matter = very long lasting organic matter).

12.4 Conclusions

The above description and discussion of experiments in geoarchaeology, and most importantly their application to real sites, has had to be brief and eclectic. Nevertheless, it has shown that if geoarchaeologists want to interrogate archaeological deposits with a realistic aim to reconstruct past human activities, one of the chief starting points is through intelligent and focused experimentation. Data on what can happen to soils and deposits when they are buried under some different regimes has proved crucial. Other experiments in industry, pyrotechnology, palaeoagriculture, and development of occupation surface deposits have all led to improved and more accurate interpretations. It is always important, however, for the limitations of experimental results to be appreciated. The example given above concerns the lack of "virgin" soils for cultivation studies. Similarly, one should be aware that the repulsive conditions of some ancient occupation deposits cannot be replicated by *modern* "ancient farms" because of health and safety reasons.

13

Human materials

13.1 Introduction

The archaeological record mainly consists of artifacts, features, and ecofacts. Among artifacts, we typically think of lithics (e.g. basalt and other volcanic rocks, limestone, and especially flint and chert) as being the principal components for prehistoric periods. These rocks are later supplanted by materials generally made with more sophisticated technologies. These include metals (copper, bronze, iron), ceramics, and construction materials, such as plasters and earth-based adobe and daub (Figs 11.2 and 11.3,). The analytical study of many of these materials is usually subsumed under "archaeometry." However, since this book is not an archaeometry text, we have decided to limit our discussion to materials that humans actually make, particularly those that are employed in construction materials. The discussion is aimed at showing that their study from the geoarchaeological point of view can furnish insights into not only how they were made but also what they repesent about site activities and formation processes. The recycling of these materials in archaeological sites as fills, dumps, for example, is considered in Chapter 10. An additional aspect of human materials and occupation deposits is what happens to them when a site is abandoned – how do they weather and become transformed by soil processes (Fig. 11.4). Examples of this from Europe (dark earth) and South America (*terra preta*) are given in Box 13.2.

13.2 Constructional materials

This section examines lime- (and gypsum-) based (plaster, mortar) and earth-based (adobe, daub, turf) building materials (Box 13.1). It can be noted that plasters, mortars, and adobe/daub commonly have both a *matrix* (fine binder) and a *temper* (coarse component). The latter can include both plant (straw, reeds) or geological/anthropogenic (rock, ceramics, and/or brick fragments) material. The terms, matrix and temper, are used henceforward.

13.2.1 Lime- (and gypsum-) based building materials

There are numerous types of lime- (gypsum-) based constructional materials, some of which have been in use for around 9,000 years (Goldberg and Arpin, 2003; Rollefson, 1990; Goren and Goldberg, 1991). The first extensive use of lime-plaster dates to the Neolithic (Kingery *et al.*, 1988) (Fig. 11.5). Moreover, although some consider that the use of gypsum as a building material infers Portland Cement or post-Medieval activity, it was in fact in use as early as 5,000 years ago by the Egyptians. The Siloam Tunnel, Jerusalem, for instance was lined with plaster, which has now been dated to around BC 700 (Amos *et al.*, 2003).

For convenience and as a very broad generalization, coarse-tempered materials are termed "mortar," such as that used to cement building

Box 13.1 Brickearth walls

Brickearth walls: The study of first century Roman contexts at the Courage Brewery site, London (see Fig 11.3 and Table 13.2), include vertical and horizontal thin sections, and a bulk sample of walls made of brickearth (Macphail, 2003) (Figs 13.1 and 13.2). Both undisturbed brickearth wall and wall material that underwent burrowing by earthworms, were examined. The brickearth building material is a coarse silt and very fine sand-dominated fine sandy silt loam (Tables 13.1 and 16.2). Its grain size and micromorphological character is typical of *in situ* brickearth argillic brown earth soils present in Roman London, as found for example, in the Leadenhall Street area of the City (see Fig. 13.3b). The Eb upper subsoil and upper Bt horizons tend to be fine sandy silt loams,

FIGURE 13.1 Courage Brewery site, Southwark, field image of upstanding remains of painted plaster coated brickearth "clay" wall.

FIGURE 13.2 Photomicrograph of typical compact fine sandy silt loam brickearth (Br) wall and humic "dark earth" burrow (Bu), Plane Polarised light, frame width is ca. 5.5 mm.

(cont.)

Box 13.1 *(cont.)*

whereas the more clay enriched lower subsoil Bt horizons are clay loams (Macphail, 1980; Macphail and Cruise, 2000). The brickearth is also typical of brickearth used elsewhere in London for constructional purposes, for example, at Whittington Avenue (Brown, 1988; Macphail, 1994). The part of the natural soil profile typically used for brickearth walls is either decalcified Bt horizon subsoil material or the more deeply quarried and still calcareous little-weathered B/C horizon material. It is probable that brickearth was imported from the nearby city of London because brickearth is not present in Southwark; south of the Thames the nearest brickearth apparently occurs some 3–4 km to the southeast (Armitage *et al.*, 1987). As noted already, the digging of Roman "silt pits" can also be cited from first/second-century Paris, France (Ciezar *et al.*, 1994).

stones (Rentzell, 1998) or as Roman *arriccio* (a coarse layer applied directly to the face of a wall), and generally finer-tempered coatings are called "plaster" (e.g. Roman *intonoco*: fine surface layer applied on top of *arriccio*; Mora *et al.*, 1984: 10; Sue Wright and MoLAS staff, personal communication). A painted, fine "plaster" over a coarse-tempered ("mortar") wall covering is well illustrated in Pye (2000/2001, figure 1) (Fig. 13.4). Plaster essentially differs from mortar by containing a temper dominated by sand rather than gravel-size material (Courty *et al.*, 1989: 121).

In order to produce a lime-based mortar or plaster, limestone (including chalk) would be burned so that $CaCO_3$ was converted into quick lime (CaO) (Table 13.1); slaked lime is formed from this through the addition of water ($Ca(OH)_2$). The hydrated lime then reacts with carbon dioxide (CO_2) in the atmosphere to produce $CaCO_3 + H_2O$ (Blake, 1947; Stoops, 1984). Essentially therefore, the composition of a mortar or plaster is the same as the original material. It occurs either as fine limey material, but more commonly, with added coarse material, which serves as an aggregate (temper) (Fig. 11.5). The formation of Quicklime and "Plaster of Paris," is as follows:

Limestone $CaCO_3 + \text{heat} \Rightarrow CaO$

Gypsum $CaSO_4 \cdot 2H_2O + \text{heat} \Rightarrow CaSO_4{\cdot}0{,}5H_2O + 1.5H_2O$

Gypsum is a ubiquitous mineral found in both marine and lake deposits. It also constitutes prominant geological strata, as for example, the gypsum mines in Germany and around Paris. "Plaster of Paris" is manufactured by heating gypsum which rehydrates in the presence of water. As a cautionary tale, the presence of gypsum crystals in some Early Medieval London occupation deposits confounded workers in conservation, but its occurrence was easily explained as the result of the importation of estuarine Thames plants and associated sediment on to the site (Kooistra, 1978).

13.2.2 Mortar

The term mortar refers to a mixture of quicklime with sand and water. It is used as a bedding and adhesive between adjacent pieces of stone, brick, or other material in masonry construction. It commonly consists of one volume of well-slaked lime to three or four volumes of sand, thoroughly mixed with sufficient water to make a uniform paste that is easily handled on a trowel (Columbia Encyclopedia, 2004). There are variations, however, in this composition of coarse/fine ratio (sand temper : fine matrix) that are delineated in thin sections of the hardened material (Macphail, 2003b; Stoops, 1984). As noted above, lime mortar hardens by absorption of carbon dioxide from

Box 13.2 *Terra Preta* and European dark earth

Amazonian dark earths. Anthrosols of Amazonia termed *terra preta* (dark earths) that are associated with precontact settlement and occupation deposits have been intriguing archaeologists and soil scientists for more than 40 years (Graham, 1998; Holliday, 2004: 320–323; Smith, 1980; Sombroek, 1966). "Indian black earth," or *terra preta do Índio,* vary in color from a dark earth that is rich in cultural artifacts (mainly ceramics with some lithic material) to a brown soil, the latter termed *terra mulata* (Sombroek, 1966); "occurrences" varying in area from 0.5 ha to >120 ha, and up to 2 m thick (McCann *et al.*, 2001; Woods and McCann, 1999). The present-day importance of *terra preta* is that they are much more fertile than the surrounding Oxisols, and are much more capable of sustainable agriculture; they are even "mined" by local people to improve land for horticulture (Bill Woods, personal communication).

Investigations around Santarém, Pará, Brazil, for example, have found that there are two types of Amazonian dark earths. *Terra preta* contains much more Ca and P compared to both *terra mulata* and the background sands and clays, although these "brown" dark earths are also equally more rich in organic carbon than background levels (McCann *et al.*, 2001; Woods and McCann, 1999). This important research also shows that sites are not the simple result of universal middening. Rather, the smaller areas of *terra preta* reflect Amerindian "habitation," whereas the more widespread *terra mulata* have resulted from "agriculture" – soil amelioration having been induced by slash and burn, mulching, and composting (Woods and McCann, 1999, figure 3). This useful soil finding can probably be best evaluated by soil micromorphology, as in the case of European Roman-Early Medieval dark earth (see below). The reader is referred to Glaser and Woods (2004) and Lehmann *et al.* (2004) for the latest research into Amazonian dark earths.

European dark earth. As urbanism declined in what had been the Roman or Classical world, a number of processes and events were recorded in deposits called dark earth. The interpretation of these dark earth deposits has not been straightforward, however (Cammas *et al.*, 1996a; Galinié, 2000; Macphail, 1981; Macphail and Courty, 1985; Sidell, 2000) (Fig. 13.3). It has also been argued that dark earth did not develop because of the total abandonment of towns and cities, but rather it resulted from a change in the use of urban space (Macphail *et al.*, 2003), and the once-stratified Roman archaeology became obscured or totally lost (Yule, 1990).

Abandoned buildings constructed of brickearth floors and plaster-coated brickearth walls (see Box 13.1), began to collapse as biological agencies destroyed supporting wood beams, or structures were robbed. In fact, a wide variety of mechanisms have been identified that could lead to dark earth forming out of constructional materials and occupation deposits (Tables 13.1 and 13.2). Some of these processes have been modeled from World War II blitzed areas of London and Berlin (Macphail, 1994, in press; Sukopp *et al.*, 1979; Wrighton, 1995).

Dark earth developed as a soil, and can reflect the original constructional materials of an abandoned building, forming in the first decades a very thin humic calcareous soil or pararendzina (Entisol) (Chapter 3). The breakdown of earth-based materials such as daub and brickearth yielded mainly clay and silt (see Table 16.2). The weathering (decarbonation) of lime-based mortar and plasters is exactly the same as soil formation

(cont.)

Box 13.2 *(cont.)*

on limestone, where pararendzinas form. This occurs through attack by humic acid from plants, and under moist conditions when calcium bicarbonate ($Ca(HCO_3)_2$) is removed in solution by rain water containing more or less dissolved CO_2 – carbonic acid (HCO_3) (Duchaufour, 1982: 74). Weathering in the form of decarbonation occurs and dark earth is thus generally strongly decarbonated. As evidence of this, dark earth that has taken 400–600 years to form often contains biogenic calcite (e.g. earthworm granules) from biological activity, but many of these granules show incipient decalcification.

The reworked remains of insubstantial buildings are believed to be one of the mechanisms contributing to dark earth and dark earth-like deposits. At Colchester House, London, there is dark earth dating to two different periods (Macphail, in press; Sankey, 1998; Sidell, 2000). There is an upper, typical dark earth above a mortar floor and the ruins of a substantial fourth-century building, which merges upward into sixteenth-century deposits. The fourth-century mortar floor itself, however, rests on an earlier truncated dark earth deposit. This lower dark earth apparently formed in first–second century deposits that date to the time when London acted as an entreport, as attested to by the very high numbers of artifacts and coins present. The chief component of this lower dark earth is weathered brickearth building debris presumably from the original structures here – a situation studied in detail at Courages Brewery, Southwark (See Box 13.1) (Cowan, 2003).

Although there are clear instances of post-Roman–Early medieval periods of major land use changes that led to dark earth formation, for example, post-fourth-century grazing at Deansway, and the formation of a pasture soil ahead of the construction of the late ninth, early tenth-century Saxon burh at Worcester (Dalwood and Edwards, 2004; Fig. 17.2), dark earth may also record continuity of occupation. For example, at Pevensey Castle (*Anderida*), East Sussex, United Kingdom, dark earth represents continuing post-Roman British, Saxon, and Norman

FIGURE 13.3 The London Guildhall site, showing the junction of the arena surface of the amphitheatre abandonment ca AD 250 or 364, and the overlying dark earth formed between the 3rd–4th century and 11th century.

(cont.)

Box 13.2 *(cont.)*

occupation (Macphail, in press; Savage, 1995). Equally, what appears to be "continuous" Late Roman and Early Medieval occupation is recorded in the French dark earth sites of St Julien, Tours and the Collège de France, Paris (Galinié, in press; Guyard, 2003). On the other hand, a rural dark earth formed at the Büraburg, Nordhessen, Germany, after the walls of the "oppidum" were raised and cultivation activity dominated the landscape (see Table 9.1b) (Henning and Macphail, 2004). Thus, dark earth sites need to be studied on a case-by-case basis. There is not one single model that explains them all, and their study is a mine of information for early medievalists.

the air (Blake, 1947). Various forms of mortar have been recorded from as early as the pre-Pottery Neolithic in the Near East (Garfinkel, 1987; Goren and Goldberg, 1991; Rollefson, 1990).

The term mortar is used here in a generic sense, for a coarse tempered "cement" that can be applied to walls, and solid floors. At Roman sites the latter is exemplified by *opus signinum* that often would have had a fine plastered surface, or be tessellated (with "coarse" ca 2 cm size ceramic tessera bricks), or covered with a mosaic (using "fine" ca 1 cm size stones, etc.). *Opus signinum* was the Roman equivalent of modern concrete. Mortar was also used to cement stones in stone buildings, although in some areas of England the stones are in fact large flints, as at Fishbourne Roman Palace, West Sussex.

At the Courage Brewery site, Southwark, London (Cowan, 2003; Dillon *et al.*, 1991), a reference fragment of first century AD *opus signinum* was found to be composed of pure microcrystalline (micritic) calcite cement with dominant gravel-size (>2 mm) clasts (temper) that include flint, burned brickearth, limestone, chalk, and pot (Courty *et al.*, 1989). It also has a fine and medium well-rounded sand-size temper likely derived from Thames alluvium ("river sand"; Blake, 1947).

Chalk fragments and other small limestone pieces may be relict of incompletely burned lime. On the other hand, coarse limestone and pounded pottery were typical tempers added for strength, and in the case of the latter produced a strong water-resistant "hydraulic" mortar (Blake, 1947: 322–323; Stoops, 1984). Here SiO_2 has been released and strengthens the cementation process. Hydraulic mortar is especially important for water basins and aqueducts (e.g. Roman Basel, Switzerland (Rentzel, 1998)). The manufacturing process for the famous pozzolanic cements used volcanic rocks because again the silicate minerals reacted with the lime to produce a hard material (Ca-Al-silicates) that even harden under water (Brown, 1990; Lechtman and Hobbs, 1983; Malinowski, 1979). Although low amounts of organic matter present in the cement may originate from accidental contamination of material from the mixing trough, a small addition (10%) of impurities was said to add to the mortar's hardness (Blake, 1947).

A typical ratio of *coarse* (gravel and sand) to *fine* (cement) material at 60 : 40 found in Roman London sites, is comparable to the soil micromorphological findings of Stoops (1984) from 15 examples of mortar from Roman Pessinus, Turkey. It also appears similar to wall cement from Pompeii, mortar in dark earth at Bath, mortar cementing a flint wall at Fishbourne Roman Palace, West Sussex (Figs 13.5 and 13.6), and in mortar floors at Colchester House, London (Macphail and Cruise, 1997; Sankey, 1998). Whereas at the Courage Brewery site burned brickearth is included as a coarse temper, oolitic limestone and chalk are the major coarse components at

TABLE 13.1 Components of urban stratigraphy and their potential for weathering (revised from Macphail, 1994)

Urban stratigraphy (and land use)	Materials and features	Components
a) Clay and timber buildings and structures	Brickearth, manufactured daub	Clay and silt
	Lime-based mortar	Calcite cement and gravel
	Lime-based "plaster"	Calcite cement, silt, and sand
	Timbers	Wood (charcoal if burned)
	Roofs	Tile, thatch, or turf (the last containing organic matter and phytoliths, and a focus of biologicalactivity)
b) Backyard middens and manure heaps	Middens	Human and dog coprolites, bone, ash (calcium carbonate), organic matter, phytoliths, oyster shell, wood, and Gramineae charcoal (focus of biological activity)
	Manure	Herbivore/omnivore (e.g. pig) coprolites, organic matter, phytoliths, diatoms, Gramineae charcoal (focus of biological activity)
c) Cultivated plots	Cultivated soil	Finely fragmented "urban stratigraphy" of building debris, midden, and manure material, plant organic matter, some times mixed with natural soils (focus of biological activity)
	Gardens	As above, and including ground-raising activity and soil importation
d) Pit areas and dumps	Pit fills, dumps, and cess pits	Mixed natural soils and underlying geology with all kinds of urban materials of mineral and organic origin, including industrial waste, cess, and nightsoil (sometimes the focus of biological activity)
e) Destruction leveling and fires	Dumps, collapses, and razing	Mixed (sometimes burned) building materials, often with charcoal and other urban materials (see above)
	Robber trenches	Infills over wall foundations, and so on
f) Abandonment and fires	Unoccupied buildings and building shells	Accumulated collapsed and weathered building debris, organic waste from visiting humans, animals, and nesting birds (focus of biological activity)
g) Rural	Grassland, shrubs, and woodland	Originally heterogeneous components (as above) becoming increasingly homogenized into dark earth

Bath and at Fishbourne, respectively. Similarly, local building materials are found in mortars used at Roman Augst, Switzerland (Rentzel, 1998). It is significant that these lime-based building materials are characterized in terms of their soil micromorphology, because deposits such as "dark earth," limestone building stone, chalk, and other natural geological materials need to be differentiated from manufactured materials – and this is not always immediately obvious, even in thin section.

TABLE 13.2 Phases of dark earth formation elucidated from an analogue (postwar Berlin) and three selected Roman/Early Medieval sites (Macphail *et al.*, 2003)

Site	Dark Earth phase	Interpretation
Berlin, Germany Monitored pedogenesis on postwar (1945–1978) Berlin's wastelands	Pararendzina formed an poorly weathered concrete and brick-rich building debris	*Total site abandonment* Very shallow (100 mm) base-rich soil formation after 20–30 years (closed woodland canopy formed after 28 years)
Courage Brewery, Southwark, London, United Kingdom 1st – 2nd century Late 5th century Roman London; dark earth below post-Medieval dumps	Dark earth Phase 3: calcareous brown earth development	3 – *Essentially abandoned waste ground ca* AD *400–1050* (Mature 300+ mm thick decarbonated but still base-rich soil formation)
	Dark earth Phase 2: calcareous brown earth developing in accreting ash-rich midden deposits	2 – *low intensity urban activity – middening/ shrub and ruderal vegetation in abandoned house plots during 200–300 years, developing into wasteland (late Roman cemetery)*
	Dark earth Phase 1: pararendzina formed in brickearth and lime-based mortar/plaster debris over (sometimes partially robbed out) brickearth floors	1 – *Extremely short-lived (1–5 years?) total site abandonment* Extremely thin (10–50 mm) base-rich soil formation with likely moss/lichens/ fungi and ruderal vegetation cover
London Guildhall, United Kingdom – Roman Arena Mid-third century–Late 5th Roman/mid 11th century Saxon and Early Medieval; dark earth sealed by Early Medieval occupation deposits and floors	Dark earth Phase 4: immature brown calcareous earth formation	4 – *High intensity middening (Increasing Late Saxon urban use of adjacent land)* (150 mm poorly weathered midden dumps)
	Dark earth Phase 3: Mature pararendzina development	3 – *Essentially abandoned waste ground* ca *AD 400–1050* (40 mm mainly decarbonated base-rich soil)
	Dark earth Phase 2: Pararendzina forming in ash-rich midden/butchery/ nightsoil dumps	2 – *Low intensity urban activity (middening) (50–100 years?); ca.* AD *364–400* (50 mm A/C soil profile)
	Dark earth Phase 1: pararendzina formed in and over weathering Arena mortar floor	1 – *Short-lived (10–100 years?) total site abandonment* ca AD *250 or 364*
Deansway, Worcester, United Kingdom–Late Roman to Saxon (fourth–ninth century); dark earth sealed by Saxon deposits and rampart of burh (ca AD 890)	Dark earth Phase 2: mature brown earth development	2 – *Very low intensity occupation* (grazing?) *during ca 300–500 years (eighth century Saxon hearth)*
	Dark earth Phase 1: brown earth forming in accreting dung-rich midden deposits (stock area?)	1 – *Short-lived (100 years?) nontypical urban landuse, developing into wasteland (late Roman cemetery)* (250–300 mm Ah horizon)

See Table 3.2, for soil types (Bowsher *et al.*, in press; Cowan, 2003; Dalwood and Edwards, 2004; Sukopp *et al.*, 1979).

As described in Box 13.2, weathering can break down the lime-based cements, hence the need for "water-proof" hydraulic mortar. Cements that were used for building foundations standing below water level had to be particularly resistant to water. One way to offset this problem was achieved during the mid-thirteenth-century building of the (short-lived) barbican of Henry III at the Tower of London (Keevill, 2004). Here, the stones were "cemented" with molten lead, hence the anomalously high levels of Pb found during heavy metal analysis of the juxtaposed and newly formed moat sediments (Macphail and Crowther, 2004).

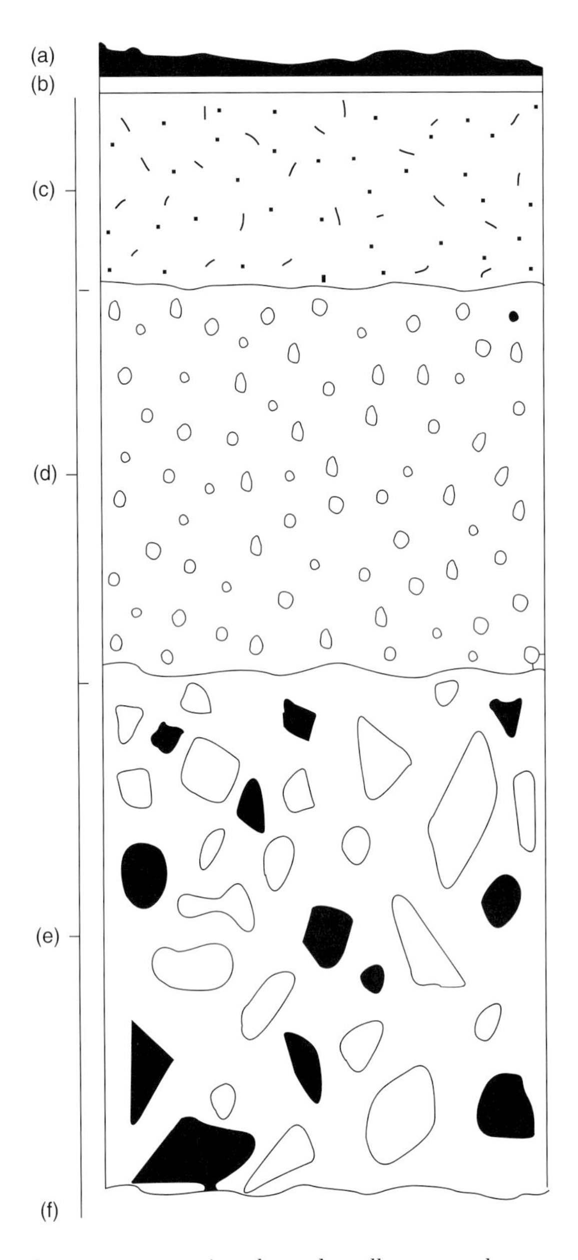

FIGURE 13.4 Section through wall mortar, plaster, and painted surface of the Roman Empire (from Pye, 2000/2001,) showing layered structure of painted wall plaster; (a) the painted surface; (b) layer of pure lime usually 1–2 mm thick, but which is not always present; (c) fine plaster, often<10 mm thick and sometimes mixed with ground marble or other white filler (temper); (d) coarse plaster, often>10 mm thick; (e) preparation layer of plaster (or mortar) of varying thickness, mixed with coarse filler (temper) and applied direct to wall; (f) wall below the plaster layers.

13.2.2.1 Mortar floors

At the Norman site of Dragons Hall, Norwich, a well-made mortar floor composed of flint gravel temper and a calcium carbonate cement matrix was investigated (Shelley, 1998; Macphail, 2003b). Here, the mortar floor had a fine plaster surface (see below). Microprobe mapping picked out the siliceous (Si) nature of the flint gravel and quartz sands present in the mortar and plaster screed surface, and the dominance of calcium (Ca) in the fine matrix of micritic calcium carbonate. More siliceous sands are present in the floor and postabandonment deposits, which are more phosphate-rich than the floor makeup. At the Ottonian (tenth century) Church site, Magdeburg, Germany several "mortar floors" were separated by beaten floor deposits. On closer inspection (see Fig. 10.4c,d), however, the mortar – composed of gravel-size limestone and a finely crushed limestone-rich cement – was shown to have micritic textural infillings that indicated that these were layers of mortar "settling" *in situ*; they appear to be relict(s) of building work rather than in-place mortar floors (Macphail and Crowther, in press). At the Roman theatre at Basel, some mortar-rich strata were likewise identified as relict constructional debris rather than proper floors (P. Rentzell, University of Basel, personal communication).

13.2.3 Plaster

Plaster here refers to wall and floor plasters, made of either lime or gypsum (see Section 13.2 above). Plasters have been observed from sites as early as the PPN (Pre-Pottery Neolithic) of the Near East, and include painted plasters (as reviewed in Arpin, 2004; Matthews, 1995; Matthews *et al.*, 1996; Spensely, 2004). In the Classical World, however, where frescoes are a major concern for conservators, plasters are classed as a fine lime-based coating (Roman *intonoco*) applied to prepared surfaces (Fig. 13.4), such as "combed" rough mortar (Roman *arriccio*) covered walls (as replicated at the reconstructed villa at Butser Ancient Farm; Reynolds and Shaw,

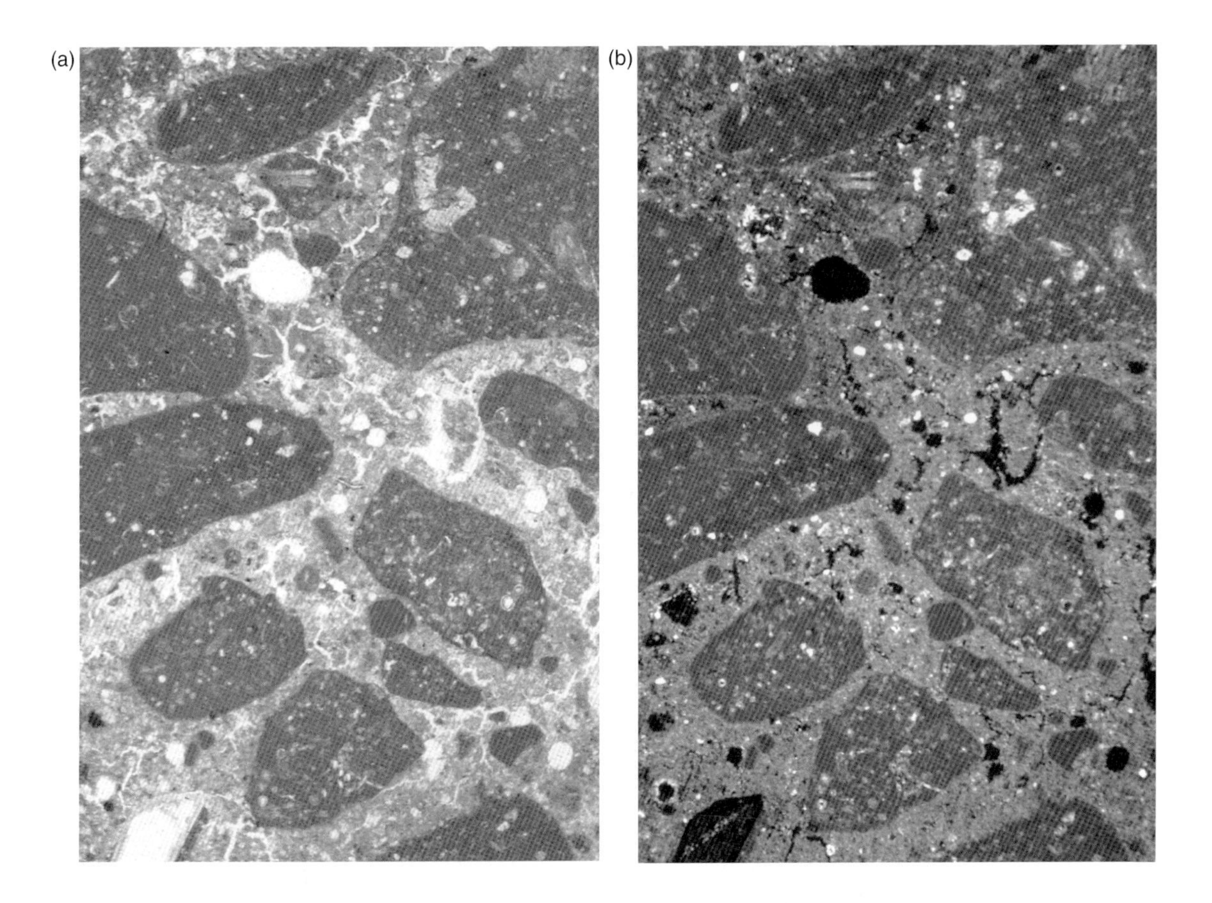

FIGURE 13.5 (a) Fishbourne Roman Palace, West Sussex: photomicrograph of Roman mortar (cementing flint wall) composed of a dark grayish fine micritic lime mainly tempered with coarse sand-size to gravel-size subrounded chalk clasts. Plane polarized light, frame width is 5.5 mm, (b) As Figure 3.5, but under crossed polarized light; lime mortar cement has midorder interference colors, chalk with rather low interference colors, and a scatter of quartz silt.

2000). Plaster was also directly applied to "clay" walls, or to Roman "concrete" floor surfaces, such as *opus signinum*, and to the walls and surfaces of many complex societies both in the old and new world. Some Roman examples are given first.

Plaster was examined from an *opus signinum* floor sample and from an *in situ* standing brick-earth clay wall at the Roman Courage Brewery site (Cowan, 2003; Dillon *et al.*, 1991) (Fig. 13.1). Both plaster samples differ from mortar by containing a temper dominated by sand rather than gravel-size material (Macphail, 2003c). In the case of the *opus signinum* sample the sand-size material is poorly sorted and the cement is very similar to the cement of the attached mortar. On the other hand, the plaster coating the brickearth clay wall is finely (3–5 mm) layered (Fig. 13.6), reflecting several applications of plaster to the wall. It has a well-sorted fine and medium sand-size temper set in a weakly organic micritic cement (producing a coarse : fine ratio of 60 : 40, cf. Stoops, 1984). Some of the sand-size material could be from "clean" local alluvial sand (Table 13.2), whereas other sandy material has clay coatings associated with it. The last suggests the use of local argillic brown sandy subsoil B(t) horizon material (a pedologically clay-enriched "forest" soil horizon; Avery, 1990; Duchaufour,

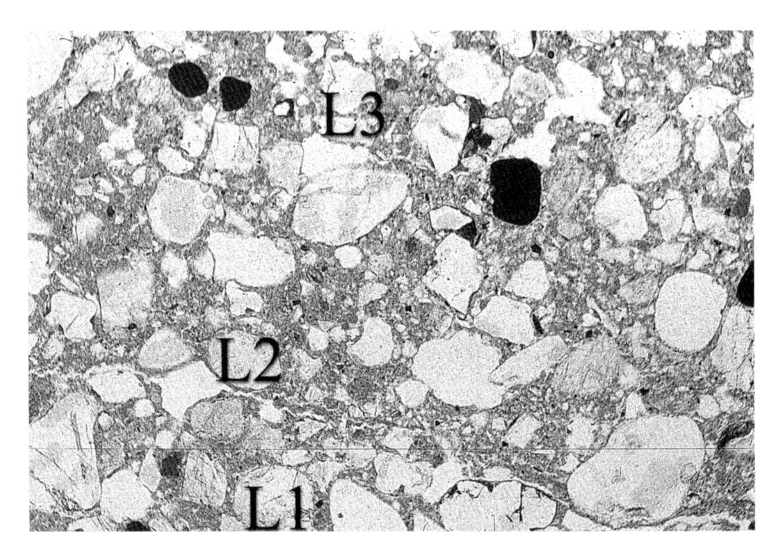

FIGURE 13.6 Courage Brewery site, Southwark, photomicrograph of first–second century Romano-British plaster layers (on brickearth clay wall); three distinct layers (L1, L2, and L3) can be distinguished reflecting how the plaster was applied. Frame length is ~5.5 mm, plane polarized light.

1982) (see Chapter 3 and below). The higher amount of organic matter in comparison with the mortar cement, could again relate to contamination from the mixing trough, the addition of weakly humic soil, and/or the possible addition of oil lees (residues from olive oil pressings) or equivalent material (Blake, 1947: 318). Stoops (1984) suggests the use of *sieved* "river sand" for plaster, although at Southwark the local sandy subsoil is already sufficiently well sorted to produce the kind of plaster temper studied and would not have needed to be sieved/prepared. Although a coarse : fine ratio of 60 : 40 seems common during antiquity, M. Madella (University of Cambridge, personal communication) points out that in the Greco-Roman catacombs some 30 m beneath Naples, Italy, the plaster lining the reservoirs and aqueducts is much richer in fine matrix material than normal, in order to aid its water-resistant qualities.

Plasters are particularly common in sites with complex societies, such as Maya mounds (Spensely, 2004) and Middle Eastern tells. Micromorphological analysis of stratigraphic sequences within well-dated buildings is enabling the resolution of much smaller timescales than that achieved by routine examination of building levels and phases routinely identified during excavation and analysis of bulk wet-sieving and flotation samples. At Çatal Höyük and other Near Eastern sites, a range of small-scale, seasonal, annual, and life timecycles of activities have been identified (see Table 10.1) (Matthews, 1995; Matthews *et al.*, 1996, 2000). Some surfaces, such as walls in the main rooms, were replastered in the order of ca. monthly–seasonally, as well as at annual intervals, whereas others such as the adjacent storeroom floors and walls, were only plastered every ca 20–40 years. Ovens were rebuilt in the order of every ca 10–25 years. These differences in the duration of surfaces have major implications for interpretation and dating of the residues and artifacts found in different rooms and areas with in a site. At other tell sites, floors and surfaces may be renewed less frequently. The maximum number yet recorded is at Tell Brak, which is 25. The application to floors and walls of nonlime based "plasters" composed of red soil, at times of birth, death, and marriage, was observed in modern India by Boivin (1999), who drew inferences concerning the replastering of surfaces at Çatal Höyük.

Such geoarchaeologically based findings have strong ramifications in interpreting the role of buildings, and the social status of its users and owners at the site. Clearly a site that is replastered monthly/yearly reflects a markedly different status from one replastered much less frequently. Equally, when examining plastered surfaces and the occupation deposits on them, the question asked is, how can these be interpreted in terms of cultural activity?

13.2.4 Earth-based constructional materials

The amount of earth-based structures in archaeology is often grossly underestimated apart from the most obvious upstanding ramparts, earth mounds (Old World barrows and *tumuli*, New World mounds; Van Nest *et al.*, 2001), and tells, and the like (see Chapter 10). Across the globe, adobe, mud brick, rammed earth, daub, clay, turf, and local variants, have been used for construction. Even famous stone-built linear features such as Hadrian's Wall, England and the Great Wall of China, have lengths once composed of earth-based ramparts. Furthermore, although the remains of Roman masonry structures are admired today, Roman period populations across Europe employed local, imported, and processed soil and geological materials for constructional purposes (Blake, 1947). The remains of these structures are often much more difficult to study and reconstruct, because of various human and natural reworking processes (Box 13.2) (Macphail, 1994). It is therefore important to understand the types of earth-based building materials that have been used, and are currently in use in the world.

13.2.4.1 Turf

Turf used for construction is generally mineralogenic topsoil. It is the living grass-covered and grass-rooted mull humus ("prairie") Ah soil horizon (see Chapter 3). Peat (Histosols), and acid moder and mor humus horizons can be used, for example, in ramparts and walls, but their high organic content can be highly susceptible to oxidation, leading to shrinkage; they are totally unsuitable for turf roofs where they are exposed to oxidation (see below). The Experimental Earthwork Project examined the postdepositional transformations affecting turf mound constructions on both base-rich (rendols) and acid (Spodosols) soils; findings are described in Chapter 12.

Turf was widely employed as a building material, and until the twentieth century was used regularly in Scotland and Ireland (sod-house), and by early settlers of the prairie regions of the "West" in northwest America – "soddies" (Evans, 1957; Fenton, 1968). It is still used in Scandinavia and Iceland, and experimental studies in its use have been carried out (see below).

Characteristics of turf. Experiments, information on natural soils, and archaeological studies allow the identification of the past use of turf. Turf can be recognized from its relative high organic content (e.g. LOI; and as seen in thin section), its topsoil biological microfabric – roots, root channels, excrements of mesofauna (Babel, 1975; Bal, 1982), although these can be altered through burial and ageing processes (see Chapter 12). Supporting data can also come from palynology, magnetic susceptibility, phosphate, and microchemical analyses (see below). It must be remembered, however, that turf, like any other earth-based building material, is not always a pristine natural material, but may be a mull Ah horizon formed in soils containing many anthropogenic components, and this occurs where settlement occupations are long lasting. The use of turf from both "natural" soils and "occupation" soils has been recognized in grubenhäuser, for example (Macphail and Linderholm, in press) (Box 12.2). Turf used to construct thick walls at Viking long houses in Iceland can also be distinctly humic with relict evidence of high biological activity in the topsoils of the local andosols formed on volcanic deposits.

Some archaeological examples. A large number of turf barrows and associated environmental studies have been reviewed for England (Macphail, 1987). A more recent study, the Romano-British kingly burial at Folly Lane, St Albans, concerns the burial chamber sealed by a putative collapsed turf tumulus (Niblett, 1999). Turf used for construction was

FIGURE 13.7 Umeå University ancient farm, north Sweden, 1994; turf roofed wooden house constructed by Roger Engelmark and Karin Viklund in birch and pine woodland. Note grass covered "living" turf (Mull humus) roof. Turves were cut from a local grassland that had been ameliorated during the period 1850–1950; boreal iron Spodosols are the natural soils (Cruise and Macphail, 2000; Engelmark, personal communication).

recognized on the basis of its soil micromorphology and microchemical (microprobe) characteristics. These data suggest the use of topsoils (Ah horizon) from natural "woodland soils" (acidic argillic brown earths – Alfisols) and dung-enriched topsoils that had been likely associated with animal management (Macphail *et al.*, 1998). This finding was consistent with pollen analysis of the same samples and macrofossil remains at the site (Murphy and Fryer, 1999; Wiltshire, 1999).

Experimental turf roof. Pollen analysis and soil micromorphology were also combined in order to characterize the experimental turf roof used on a wooden structure built by Roger Engelmark on the experimental "ancient farm" at Umeå (Bagböle) in northern Sweden (University of Umeå) (Chapter 12). Analyses of two roof turves showed that the characteristics of the turf used were still recognizable, although somewhat transformed by its use. It can be noted that the 140 mm thick turf roof comprised two turves over a birch bark roof liner, with the bottom turf being face down and top turf facing upward and displaying a living grass sward (Figs 13.7 and 13.8) (Cruise and Macphail, 2000; Figs 22.1 and 22.2). It can also be noted that the successful turf roof at Umeå has only a very gentle pitch; the turf roof constructed on a steeply pitched round house roof at Butser failed (Reynolds, 1979).

Ancient turf has been identified through pollen and macrofossil analysis, but even when strongly transformed by the effects of burial, the biological traits of the turf employed should still be recognizable through soil micromorphology (e.g. turf mounds, turf roofs, turf ramparts; Babel, 1975; Bal, 1982; Crowther *et al.*, 1996; Dickson, 1999; Dimbleby, 1985; Macphail *et al.*, 2003; Scaife and Macphail, 1983; Wiltshire, 1997). Lastly, mineralogenic peat and the humic turf of Spodosols (podzols) have been, and still are, used for fuel. In the archaeological record, "peat ash" accumulations have been found at settlements, and as residues of manuring in soils (Carter, 1998a,b; Carter and Davidson, 1998; Simpson, 1997).

13.2.4.2 Building clay

Building clay (e.g. brickearth) is a local or imported soil (usually subsoil) or sediment that is employed for construction, either in its natural unaltered form, or mixed with a plant temper (see adobe, daub, and mud brick, below). It can be used for ground raising/leveling, to make clay floors, and wooden framed and supported walls, often as cut slabs. Some

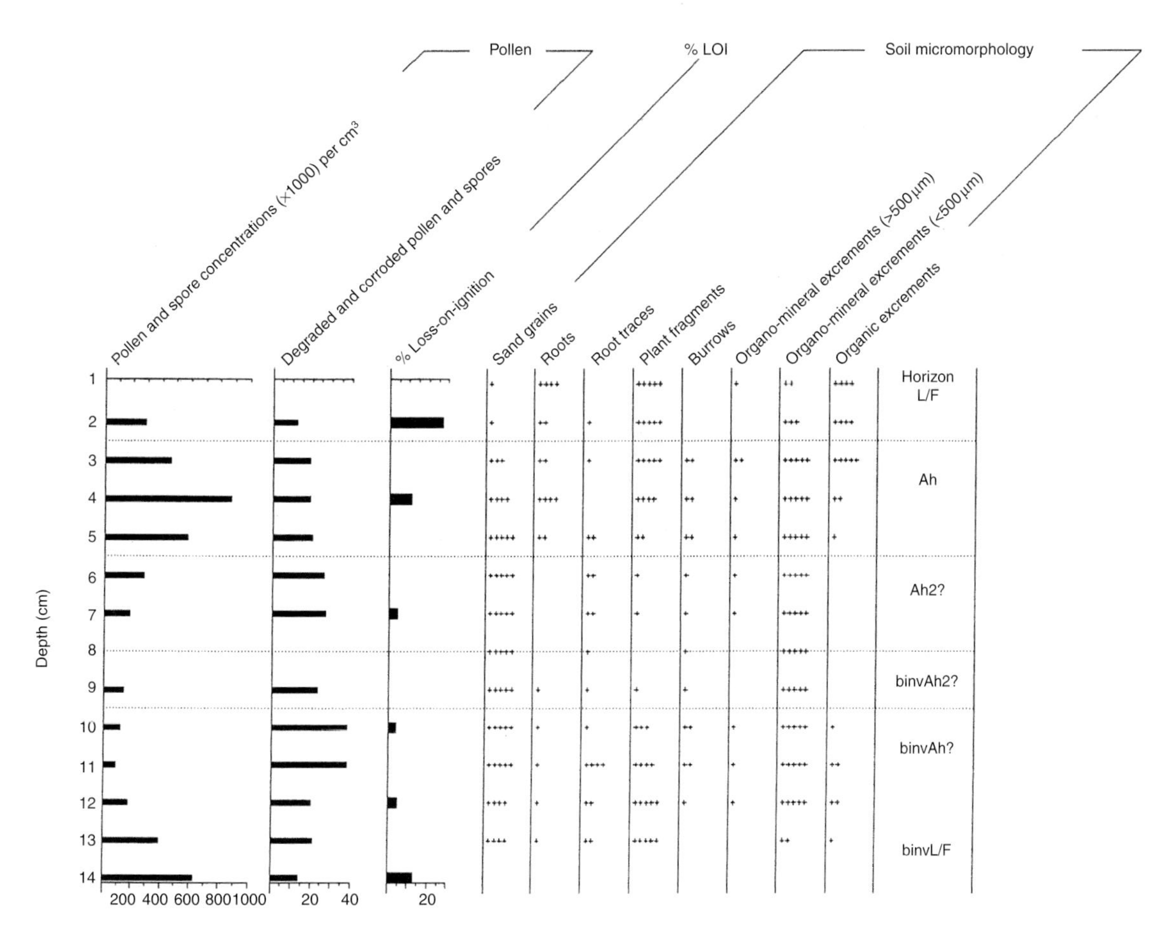

FIGURE 13.8 Diagrammatic section through the Umeå experimental turf roof (courtesy of Roger Engelmark and Johan Linderholm), showing an inverted (face down) turf at the base (binvAh – buried inverted Ah horizon) and "right-way-up" upper turf (e.g. Ah); note pollen concentrations are highest in the binvL/F (Litter and Fermentation layer) and the Ah horizons, with LOI being highest in the "modern" L/F; although organic excrements are concentrated in the "modern" L/F, much of the micromorphology of the upper turf is mirrored in the buried inverted turf (from Cruise and Macphail, 2000). The way-up of turfs in earth mounds is also often deduced from the pollen (Dimbleby, 1962; Wiltshire, 1999), although these data can be combined with measurements of organic matter and micromorphology (Scaife and Macphail, 1983; Macphail *et al.*, 2003).

examples from northern Europe include loess (a silt loam), and quarrying produced "silt pits" at the Roman and Medieval Louvre site, Paris (Ciezar *et al.*, 1994). In southern England, a fine sandy silt loam called brickearth, was employed for building (see below). These are building clay examples from 'complex societies', but the Early Neolithic (Körös culture) site of Ecsegfalva (County Békés, Hungary) is characterized by much burned daub, selectively made from very fine and fine river worked loess, the exact connotations of which are yet to be elucidated (Whittle, 2000).

The example of brickearth. This is a naturally occurring superficial deposit that commonly dates to the Devensian (Pleistocene), and is formed from loess and reworked loess, with fluvial activity often adding a fine and medium sand content to the silt-dominated loess (Avery, 1990: 13; see below). It can be noted that during the Roman and Saxon periods, people used local silt-dominated loessic brickearth at Canterbury although in London the locally employed brickearth has a marked sand-size component (see Box 13.1). Findings from London so far suggest that during the Roman period, the more clay-rich subsoils and natural (Bt and C[t]) horizons of Alfisols were utilized. During later periods (Saxon and Medieval) however, the compact and clay depleted upper subsoil material (Eb or A2) horizon upper subsoil seems to have been employed with equal frequency (Table 13.2). The obvious difference between brickearth building clay and turf, is that turf is humic and biologically worked, whereas brickearth building clay is mineralogenic and may have textural/argillic pedofeatures. Nevertheless, as these are imported soil, materials pedofeatures such as clay coatings may not be oriented to the present way-up.

13.2.4.3 Worked and transformed earth-based building materials (adobe, daub, and mud brick)

Adobe, daub, and mud brick are soil-sediment based materials, which have been mixed with organic matter, often in the form of a temper (e.g. straw and reeds) or other organic binding agents, such as dung (Figs 11.2 and 11.3). The soil/sediment and organic matter are mixed with water, then puddled (e.g. by trampling in a puddling pit), a process where by the mixture loses all structure and most of the air is removed. On drying, the mixture is massive (i.e. without structure), with no gas voids. Mud bricks and adobe may be shaped in a wooden "former," and when the former is removed, they are left to dry and harden in the sun. Daub is more commonly applied to, and built up on, wattle walls and trellis-like frameworks, which themselves are supported by more substantial wooden structures. Different forms of rammed earth (e.g. pisé) are built up within wooden frames, sometimes with wooden "reinforcing" supports and sometimes with an oblique horizontal boundary to the next horizontal layer, in order to impart structural strength (Pierre Poupet, CNRS, personal communication). One of the early geoarchaeological investigations into the study of tells and mounds, including pisé structures is that of Bullard (1970) at Tell Gezer, Israel.

Other variants of daub constructions include local English names such as cob, where the daub contains a high amount of chalky material, such as from East Anglian boulder clay, as seen at the reconstructed Anglo-Saxon Village at West Stow, Suffolk, England (West, 1985), Medieval Stanstead (Suffolk), and Cressing Temple (Essex; Macphail, 1995). In comparison, clunch is a hand-puddled chalky soil that is employed to build up a thick (20–30 cm) freestanding wall (e.g. Butser Ancient Farm, England). In the latter case, the idea is again to remove all the air. This process may be surprising to those who have examined ancient adobe, daub, and mud brick, because these materials from archaeological sites appear to contain many voids. These voids, however, are in fact the spaces left by the oxidized plant temper; long articulated phytoliths are sometimes the only remains visible (Courty *et al.*, 1989) (Plate 13.1). Plant material can also sometimes be identified from molds of these voids, or plant impressions. It can be noted that when small amounts of pounded chalk or other limestone are added, when the mixture dries, a small amount of secondary micrite forms and this helps the "cementation" process.

Adobe is the common term applied to earth-based building materials in Africa, the Far East, and the New World. In the United States it is defined as an unburnt brick dried in the sun, the term being adopted during expeditions to ("Spanish") Mexico during the early nineteenth century, although the Spanish word itself has origins from the Arab world (Oxford English Dictionary, second edition, online http://dictionary.oed.com/). One example of such is noted in a nineteenth-century report of the houses in Costa Rica being built of adobes or undried bricks two feet long and one broad, made of clay mixed with straw to give adhesion. The term can also be employed to describe a "clay" used as a "cement." Both adobe and rammed earth (e.g. as used in China, Jing *et al.*, 1995) preserve well in dry climates, but may collapse during a particularly wet season in some countries. Adobe/rammed earth walls therefore may have a thatch or tile top (for rain protection), and/or be built upon a masonry foundation (protection from ground water) to offset slumping (Pierre Poupet, CNRS, personal communication).

Morris (1944) conducted one of the earliest studies of adobe buildings in pre-Conquest North America, and compared mud brick building techniques of the New World with those of 3500 BC Sumeria. He also made one of the first identifications of an adobe mud brick kiva wall at a Pueblo site near Aztec, New Mexico. Here the site, dated to AD 1090–1110, featured hand-formed bricks around 5–12 inches long, 3–5 inches wide, and 2–3 inches thick, but without a temper. It was estimated that the construction of ca 28 m high and 342 × 159 m size Huaca del Sol (Peru) required more than 143 million adobes (mud bricks), with the Luna Platforms requiring more than 50 million (Hastings and Moseley, 1975). The adobe bricks were manufactured variously from local water lain silt, coarse desert soils, and organic "sumps" within the valley. They were formed in molds and varied in size according to stage of construction, changing from 8–11 cm to 12–18 cm thick; a study of the size range, mold markings, and maker's marks also revealed chronological implications.

In the Near East, geoarchaeological studies of adobe constructions from tells revealed information on the source of mud bricks, as well as decay processes acting at the sites, such as Tell Lachish (Goldberg, 1979; Rosen, 1986) (see Box 11.1).

Daub. As will be discussed below, earth-based materials are subject to rapid breakdown and weathering, even from rain and rain-splash; hence the need for protection in the form of eaves or an overhanging thatched roof. It is therefore obvious that very little daub survives in the soil, unless it has been burned, which essentially fossilizes it; burned and rubefied daub has been studied from Butser. Occasional fragments of burned daub are ubiquitous to most sites, and are to be expected at known settlement sites. As noted above, reed-tempered burned daub was found in unexpectedly high amounts at Early Neolithic Ecsegfalva (Plate 13.1). Plant-tempered burned daub formed the base of a hearth within Middle Neolithic stratigraphy at Arene Candide, but only rare fragments of burned daub were found of a relict structure within the buried soil at Neolithic Hazleton, Gloucestershire, United Kingdom (Macphail, 1990).

Relict daub is found more commonly in settlements. Several types of daub, including a chalky cob-like material, was recovered from the Middle Saxon village at West Heslerton, North Yorkshire, United Kingdom (Macphail *et al.*, forthcoming). It can also be noted that in Roman London, brickearth was mixed with a straw-like temper and used to make both floors and walls (see Fig. 11.5), and daub has been identified at Mayan sites (Wauchope, 1938). On both continents, the reworking of earth-based building material produces totally transformed deposits (Box 13.2).

13.3 Metal working

Archaeo-metallurgy is a specialist field and has to be treated by laypeople, such as geoarchaeologists, with caution. Iron processing sites of hearths and kilns, however, can often be indicated both by magnetic survey (Chapter 15) and analysis of magnetic susceptibility (Section 16.5). In addition, the geoarchaeologist should be aware of how to recognize metal working materials in their samples. Just a few examples of studies are given below.

Iron working and slags. In the first instance, iron-working debris can be found in loose bulk samples by the simple use of a hand magnet (J. Cowgill, personal communication). In thin section, hammerscale, the material that is produced when the hot iron is hit with a hammer, is opaque, but under oblique incident light can show multiple fine layering, picked out by metallic and reddish brown colors according to preservation. Iron slag is often globular with a vesicular porosity, and is a mixture of both newly formed silicate minerals (olivine group) and iron that shows a dendritic pattern (Macphail, 2003b). Further work may be carried out using a metallurgical incident microscope, and SEM, EDAX, and microprobe, although this is a specialist field (e.g. Kresten and Hjärthner-Holdar, 2001; Tylecote, 1986).

The presence of iron slag, which is highly magnetic, will raise the magnetic susceptibility of bulk sample measurements (see Fig. 16.4). It is therefore crucial that their presence is identified in thin section where it may indicate local iron working or the trampling-in of iron working debris (Dalwood and Edwards, 2004; Macphail and Cruise, 2000). In fact, the inclusion of iron or other materials within micromorphological samples offers the twin benefits: to examine the metallographic properties of the metal and to determine how it was made, as well as detailed information about its microstratigraphic context. If iron slag undergoes hydromorphic weathering ("rusting") at wet sites, the iron loses much of its magnetic susceptibility (J. Cowgill, Environmental Archaeology Consultancy, United Kingdom, personal communication; cf. Crowther, 2003). Reference thin sections of archaeological iron slag and iron slag that had undergone hydromorphic weathering (both slags supplied by J. Cowgill), were passed through a magnetic susceptibility coil, and a marked qualitative difference in magnetic susceptibility was measured (Macphail, 2003b).

Other evidence of high temperatures/burning can be seen in mineralogenic soil where the quartz has lost its birefringence, after losing its crystal structure and at higher temperatures, even showing signs of melting (Courty *et al.*, 1989). The opal silica of phytoliths may however, melt at the lower temperatures of domestic fires. Clinker, the porous remains of burned coal, may also be a residue of domestic fires, and in London, is indicative of late and post-Medieval deposits.

An example of bronze working. An important question facing the team investigating the twelfth century levels at Magdeburg, Germany, in an area that soon became the New Market in front of the Gothic Cathedral was, is there evidence of bronze bell manufacture? The deposits were studied using both thin sections and bulk samples (Macphail and Crowther, in press). The bulk sample testified to marked burning/heating (magnetic susceptibility: 31.0% χ_{conv}), and to heavy metal enrichment, especially by copper (156 $\mu g\ g^{-1}$ Pb, 40.8 $\mu g\ g^{-1}$ Zn, and 441 $\mu g\ g^{-1}$ Cu). In thin section, two 3.0–3.5 mm size metal fragments were noted, which were opaque, but with greenish staining of the surrounding ashy, calcitic sediment. In oblique incident light the fragments were a patchy metallic gray and metallic red. Using the microprobe, a "large-scale" map of the whole thin section showed the metallic fragments to include copper, and copper was also present in smaller concentrations around these fragments, indicating the dispersal of corrosion products (copper salts) into the sediment (Plate 13.2). Detailed microprobe mapping and analysis of one metal fragment found the gray metallic areas to be bronze (copper and tin), the red areas to be copper corrosion products, one of the bronzes having an average chemistry of 29.27% CuO, 17.77% SnO_2, 0.97% PbO ($n = 4$) (Plate 13.3). Specialist metallurgical incident microscope observations by Thilo Rehren at the Institute of Archaeology, University College London, identified various features (e.g. dendritic metallic pattern) that showed that these were bronze casting droplets, and that the low amount of lead (Pb) present, was indicative of bronze bell manufacture; the amount of lead in the bronze affects the sound of the bell (T. Rehren, personal communication; Tylecote, 1986).

13.4 Conclusions

In this chapter we have tried to describe some human materials of geoarchaeological interest. These are also important components of Occupation Deposits (Chapters 10 and 11). They include constructional materials that are lime-based (plaster and mortar) and earth- and clay-based (adobe, mud brick, and turf). When postdepositional processes affect constructions and occupation deposits, these can be transformed into "soils." Examples of this phenomenon are given in Box 13.2 (*Terra preta* and European dark earth). Lastly, as an example of industrial activity, some indicators of metal working were detailed.

14

Applications of geoarchaeology to forensic science

14.1 Introduction

At one time forensic science teams commonly consisted of a pathologist, a toxicologist, and a finger print expert. However, as crime scenes, such as clandestine graves and weapons caches, have been investigated more thoroughly, the techniques of archaeological excavation have been applied more frequently (Hunter *et al.*, 1997). More recently, the pioneering use of palynology at crime sites by Patricia Wiltshire has led her to develop a whole new applied discipline, that of environmental profiling (Wiltshire, 2002, in press). This approach entails the application of plant and animal ecology, geology, and pedology to produce circumstantial evidence to help support the case, either for the prosecution or defense.

There is a very real and important role for the forensic geoarchaeologist to play in modern forensic science (Pye and Croft, 2004; Petraco and Kubic, 2004). This participation may involve simply suggesting to search teams where a clandestine grave may *not* be located, for example, because the soil is too shallow (e.g. rendzinas on chalk). Or it might help link a suspect to a crime site, when the mud on the suspect's shoe matches that found at the grave site (Pirrie *et al.*, 2004). Environmental profiling is, however, a holistic approach, where both seeds (and other macro plant remains) and palynomorphs (pollen, plant spores, and other microscopic entities) from specific ecologies could only come from a specific environment with particular soil types and geology (Horrocks *et al.*, 1999). Equally, if certain plant remains have been found on a suspect's clothing and/or vehicle, they could link a suspect to a specific environment, which could then be searched for a clandestine grave.

14.2 Soils and clandestine graves

In any given landscape, vegetation and soil patterns reflect geology and geomorphology (e.g. the catena; Fig. 3.2), which when understood, can provide useful guides to where a clandestine grave may be located. An essential tool is the soil map (see Chapter 15). The legend that goes with the map will also yield important information. For instance, some soil types that are, or have been affected by freezing and thawing (e.g. through periglacial activity) are described as having "patterned ground." The presence of stone stripes, soil stripes, ice wedges, and the like, needs to be known before ground is probed for graves, because on some shallow soils, these features provide deep soils that could be mistaken for clandestine graves. Ancient treethrow subsoil features and solution hollows in limestone are other natural features that need to be differentiated from *bona fide* archaeological features or clandestine graves.

In addition to soil maps a good idea about landscape morphology can be gained by studying stereo pairs of air photographs, which provides a three-dimensional view (see Chapter 15). Although these also give information on the vegetation cover to be expected, unfortunately most aerial photographs are "out of date" in terms of the vegetation, which could have changed dramatically since the photos were taken, but again this may be important if "cold" cases are reviewed.

Soil maps and guides provide information on soil types, soil depths, and soil horizons that can be expected in any one location or region. Areas with shallow soils can therefore be eliminated from any searches. On the other hand, areas covered by colluvium can be targeted because of greater soil depth. In addition, disturbed ground can be identified by the presence on the surface of anomalous subsurface and subsoil material. For example, in areas of Spodosols (podzols) where there are very strong color contrasts between dark humic topsoils (Ah horizons), gray leached A2 horizons, and ochreous illuvial Bs horizons, the presence of bright reddish and yellow Bs material on the surface is anomalous (see Chapter 3). Equally, the identification of mottled reddish brown and gray gleyed subsoils (Bg horizons) and stony and rocky parent material should be an obvious clue to soil disturbance if these are found on the surface.

A correct understanding of pedogenesis and soil horizons at a crime scene is therefore extremely useful, being basic information for the forensic scientist and their "scene of crime officer" (United Kingdom – SOCO; United States – CSI) colleagues. If a spade belonging to the suspect exhibits traces of soil, and a link to the crime scene grave is sought, it is critical that more than one soil sample is collected. Someone untrained in soil science may not realize that soils are commonly made up of different soil horizons, and that sometimes these have strongly contrasting chemical and grain size characteristics. A suite of soil analyses can be expensive, but if only a topsoil sample is tested against soil on the spade that originates from digging into subsoil, it is money wasted.

Again, the local soil guide will provide information on the types of soil horizon to be found, and the basic chemistry and grain size of these. For example, the Chiltern Hills, Hertfordshire, United Kingdom are formed on Cretaceous Chalk. Their soil cover has been mapped as palaeoargillic brown earths and these are bi-sequal in character, that is, they have formed out of two parent materials – silty loess over clayey drift (Avery, 1964; Jarvis *et al.*, 1983, 1984). The deep subsoil (Ctg horizon) parent material that rests on the chalk, sometimes in solution features, is called Clay-with-Flints, a stony clay partly derived from long-weathered Tertiary Reading Beds. Topsoils and the upper subsoils, however, are formed in a fine loamy loess (aeolian coarse silt and very fine sand – ca 40–100 μm in size). Thus the Ah, A2 (Eb), and Bt horizons have strongly contrasting grain size and chemistry compared to that of the Ctg lower subsoil horizon. It can be seen then that sampling of all these horizons is needed (both individual samples and a bulk sample of all horizons), if the soil on the spade is to be properly compared to that of the crime site (Fig. 14.1). The same principle applies to studies of valuable artifacts (treasure), which are from unknown provenances. Any soil attached to them has to be compared to possible clandestine excavations just as thoughtfully.

Soil type and its relationship to drainage and groundwater levels, for example as modeled in a catena, can also give some insights into likely body preservation. Base-rich, well aerated soils generally encourage rapid decomposition of a corpse through the activity of earthworms, snails, slugs, and other invertebrates such as beetles and flies (e.g. coffin flies) (Hopkins *et al.*,

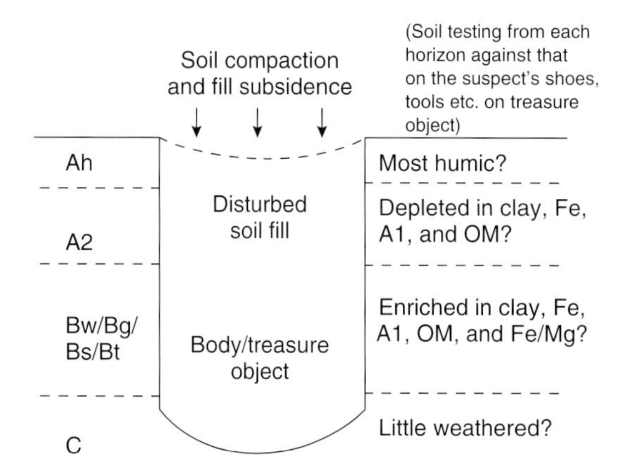

FIGURE 14.1 Variables affecting testing of clandestine burial/excavation sites. Different soil horizons have different characteristics such as chemistry, clay mineralogy, and microfossil remains (e.g. pollen) – a simple "soil sample" may not be specific enough to link soil on the suspect's shoes/tools or soil on a treasure object, to the clandestine grave or treasure site.

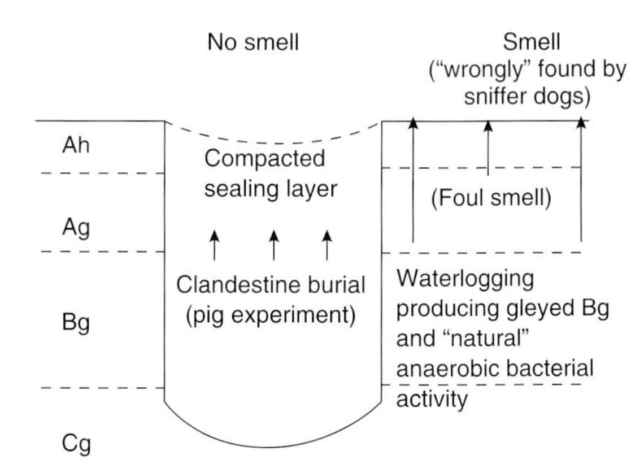

FIGURE 14.2 Soils, burials, and canine sensing. Natural waterlogging and gleying can produce foul smells (sulfides and alcohols) that may indicate to a dog that a rotting corpse is present. On the other hand, a real clandestine burial may yield no identifiable smell either to dogs or foxes if a sealing soil layer is present – buried pig experiments (Wiltshire, UK forensic palynologist, personal communication; Wiltshire and Turner, 1999; cf Breuning-Madsen *et al.*, 2001).

2000; Turner, 1987). Poorly drained, cool, clay-rich soils, in contrast, may permit human remains to undergo less rapid decomposition, especially if well sealed. As such they are less liable to be dug up by scavenging animals such as foxes (Turner and Wiltshire, 1999). Dogs are commonly employed by search teams seeking clandestine graves. Likely locations are probed, using a narrow gauge auger, for example, to produce a small opening in the soil. These are left open so that a dog can locate a grave through the smell of decomposition that is generated by a corpse (Fig. 14.2).

The presence of poorly drained soils can have a detrimental influence on the effectiveness of dogs searching for clandestine graves (Wiltshire, UK forensic palynologist, personal communication 2002). Heavy textured (fine silty and clayey) soils with poor draining characteristics can be seasonally waterlogged. Under such conditions, animals and plants can die and then may well undergo anoxic decomposition by obligate anaerobic bacteria, and these give rise to a suite of chemicals (e.g. sulfides and alcohols) that are both foul-smelling and indicative to canines that rotting corpses are present (Turner and Wiltshire, 1999). The study of a soil map will readily help identify locations where seasonal waterlogging may occur, as for example, with the presence of stagnogleys (United Kingdom) or pseudogley/gleysols (FAO/ Europe)(see Chapter 3). In the American soil classification system, waterlogging is suggested by the formative element "Aqu." Aquepts, for example, can include periodically wet brown soils (Soil Survey Staff, 1999). It is worth remembering that these soils have horizons with the subscript "g" (Ag, Bg, Cg), which indicates that soils are mottled due to the effect of "hydromorphisim" (waterlogging), and ochreous, gray to blue-green colors may be encountered.

It also has to be remembered that the excavation of a clandestine grave has to be carried out meticulously, as in archaeology. Both samples and finds should be even more carefully recorded, as they become potential "exhibits" in any ensuing trial. The proper documentation of bulk and monolith samples is therefore absolutely essential, and should follow the protocols of the police force involved.

14.3 Provenancing and obtaining geoarchaeological information from crime scenes

The analysis of soil materials has been used as circumstantial evidence in the investigation of crimes. It is difficult to link a suspect absolutely to a crime scene through soil analysis, and more commonly soil data is employed to demonstrate that a suspect is lying. For example, the suspect may claim that the mud on a shoe or in the wheel arch of the getaway car came from a certain location, but soil analysis may show that this is unlikely to be true. There are numerous geoarchaeological methods utilized in this kind of work. One common major constraint is the small quantity of sample that may be available to work on. In the case of palynology, this is not a problem because palynomorphs may be present in thousands of grains per gram of soil. On the other hand, a muddy stain on the trouser knee of the suspect does not yield 40 g of soil for standard grain size analysis or 10 g for bulk chemistry (see Chapter 16). Techniques that require very small amounts of sample (e.g. 0.5 g or less) that permit essential replicate analyses to be carried out, are therefore much more commonly employed (Wilson, 2004). Replicates are required to offset both natural heterogeneity and allow statistical testing, as well as ultimately providing persuasive and defendable evidence in the witness box (Jarvis *et al.*, 2004).

Methods include many that examine nonlabile/stabile characteristics in soils. Calcium carbonate content ($CaCO_3$), for example, is liable to change because of weathering, but a total chemical analysis of elements (including calcium) may well yield reproducible results. Jarvis *et al.* (2004) recommend the combined use of two different types of ICP analysis (see Chapter 16), Inductively Coupled Plasma-Atomic Emission Spectrometry (ICP-AES), and Inductively Coupled Plasma-Mass Emission Spectrometry (ICP-MS). ICP-AES is efficient in measuring oxides (SiO_2, Al_2O_3, MgO, CaO, etc.) whereas ICP-MS is able to measure trace elements at extremely low concentrations, that is, in parts per million (ppm) rather than as percents. Individual mineral grains can be identified according to their mineralogy, as done traditionally using the petrological microscope, or just as often, now employing X-ray techniques (EDAX, microprobe, *QemSCAN*; Pirrie *et al.*, 2004)(Chapter 16). Grains can thus be provenanced in the same way as pollen suites.

Grain size analysis on very small samples is carried out using Coulter counting or laser diffraction methods that now can provide amounts of clay-(<2 μm) to sand-(<2000 μm) size material. Equally, the study of surface texture of quartz grains under the SEM has been successful, because both very small quantities of sample are needed, and because assemblages of surface texture types can be unique to different geological and soil environments (Bull and Goldberg, 1985)(Chapter 2). *QemSCAN* has the advantage of both identifying the mineralogical composition of grains (and soil particles) and recording their shape and size – all potentially specific to a geological area (Pirrie *et al.*, 2004).

14.4 Other potential methods

As yet, soil micromorphology, the study of thin sections of undisturbed soil (see Chapter 16), has been underused in forensic geoarchaeology. This probably stems both from the conservative nature of police and their forensic advisers (who have to justify funding of such work), and the common paucity of sufficient amounts of soil material to be analyzed, especially undisturbed samples. Some interpretations of thin sections can also be equivocal and perhaps not stand up in a court of law, but the clear identification of micro-clues and/or sequencing of grave-fills would probably be sustainable as circumstantial evidence (see below). A gravesite can be investigated in the same way as an archaeological feature-fill is studied. Anomalous soil heterogeneity and poorly sorted textural pedofeatures (e.g. silt and dusty clay void infills) are indicative of disturbed ground, as investigated from treethrow holes (Macphail and Goldberg, 1990). Equally, the effects of burial on soils has been investigated experimentally on both base-rich and acid soil types (rendzinas and podzols) from southern England, and it is now known how long it takes for both meso- and micro- soil fauna to begin to transform their buried environment at these specific locations and soil types (Bell *et al.*, 1996; Crowther *et al.*, 1996; Macphail *et al.*, 2003). Unfortunately, no experimental soil micromorphological results are as yet available from the burial of bodies or similar organic remains, although soils from archaeo-logical cremations and cemeteries have been investigated (Fig. 14.3).

Lastly, soil micromorphology in archaeology has been successful in identifying trampling of microscopic materials ("micro-clues") into

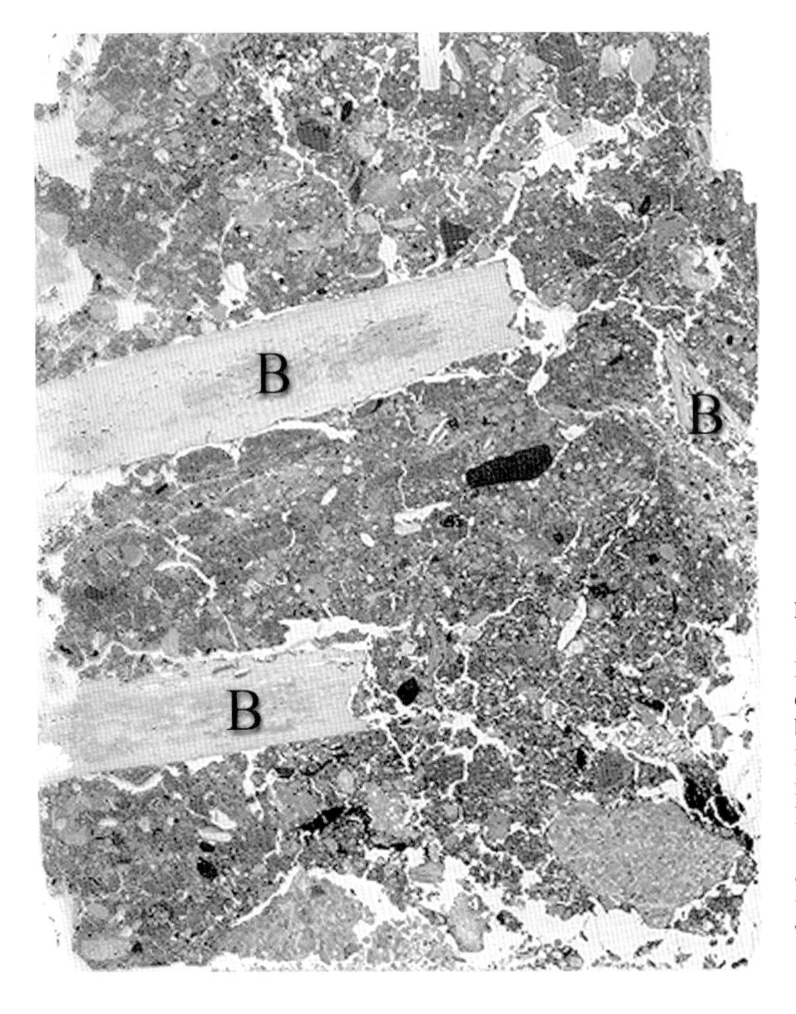

FIGURE 14.3 Buried bones (B) – Dark Age nunnery soil, Bath, United Kingdom. Dark earth soil formed in post Roman building debris (Jurassic Oolitic Limestone building stone fragments, mortar tempered with limestone clasts, and charcoal). "Ancient" bones in this cemetery are set in mature earthworm worked and homogenized soil. Thin section scan; height is ~80 mm.

"beaten floors," from a wide variety of anthropogenic and natural sources, and this could be used as a form of provenancing. Examples from Butser Ancient Farm, Hampshire and late Saxon and Norman London sites, show how fragments of industrial (e.g. iron slag, burned soil), domestic (e.g. burned bone, cereal processing and toilet waste), and animal management activities (e.g. dung and stabling crust) can all be recorded (Macphail and Cruise, 2001). Many crimes occur in environments strongly influenced by human activity, and just as in archaeological soil micromorphological investigations, there is the potential for these environments to be reconstructed. This is especially the case when carried out in parallel with other techniques employed in environmental profiling such as palynology.

14.5 Practical approaches to forensic soil sampling and potential for soil micromorphology

If for example, a grave site was being investigated, there are a number of locations and features that could be usefully investigated through soil micromorphology and other techniques (bulk chemistry, mineralogy, grain size, and palynology)(Jarvis *et al.*, 2004; Wiltshire, 2002). Firstly, potential pathways to the crime scene site could be sampled. Here, only the surface 1 mm or so could be of interest, if the ground was hard. A sample of loose soil crumbs would be collected from each sampling location, to complement soil for pollen, chemistry, and macrofossils. This loose soil would then be dried and impregnated and then made into a thin section, so that intact soil could be examined at the microscale, and used to try and match intact soil material gathered from the suspects clothing, shoes, car, and tools. If the crime scene is muddy, however, it may be worthwhile taking an undisturbed sample through the uppermost, homogenized trampled layer, including shoe print impressions. This would be the same as taking a Kubiena box through a tire track where material from the uppermost 10–30 mm track deposit could conceivably have been collected in the suspect's car – on the tires, in the wheel arches, and/or even trampled by the suspect into the car's foot wells (Figs 14.4a,b). Obviously, these samples would also need to be compared to samples from the suspect's normal home surroundings and areas of movement.

At the gravesite itself, loose soil samples should be taken from both the grave as well as the surrounding margins of the grave cut, probably the most trampled part of the crime scene (see Bodziac, 2000). During the excavation of the grave, the most important locations to sample, are (given in Fig. 14.5):

1 The complete soil profile including the topsoil (A horizon), subsoil (B horizon), and parent material (C horizon). This can be accomplished by taking an undisturbed soil monolith, to provide a control profile of the undisturbed soil at the site (Goldberg and Macphail, 2003).
2 The grave fill, from one or more locations.
3 At the junction between the fill and the undisturbed subsoil/parent material. This could permit tool marked-soil to be compared to material on tools used to excavate the grave. It would also allow estimations of weather conditions at the time of grave cutting (e.g. soil wash would be greater if dug when raining). Finally such sampling would help establish the condition of the corpse when buried and possible age of the burial, because body fluids would have likely produced soil staining (e.g. phosphate, iron, and manganese features) (Breuning-Madsen *et al.*, 2003); and
4 At the top of the grave fill/new soil surface, where trampling is likely to have occurred.

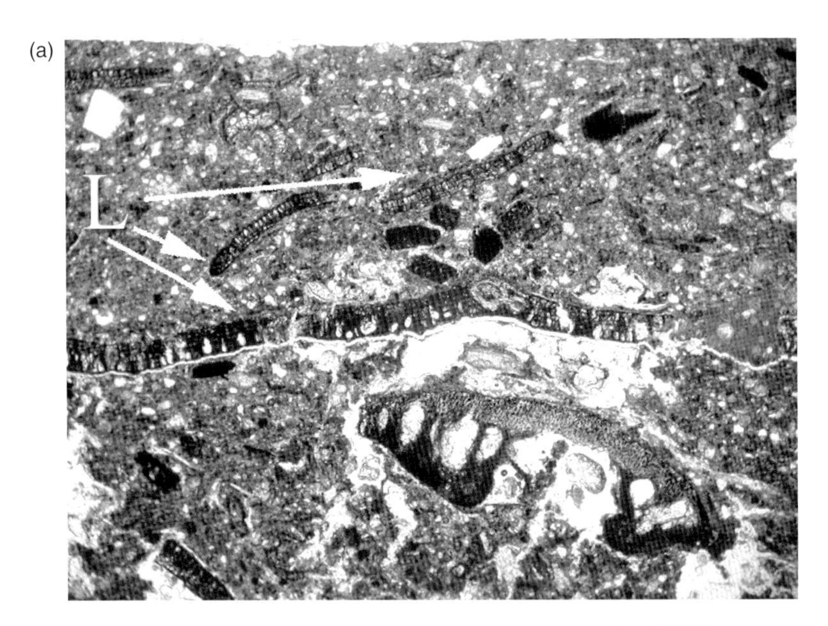

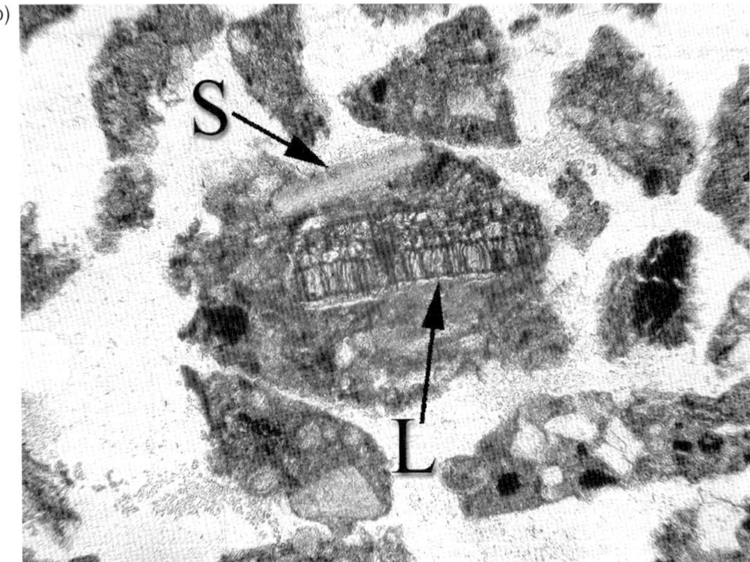

FIGURE 14.4 Experiment in forensic geoscience; comparison of soil on a muddy track that had been driven over by a *suspect's* car (Fig. 14.4a) and soil particles recovered from the *suspect's* car tyres and footwells. (a) Soil from uppermost muddy track deposits that featured tires tread marks; a compact calcareous chalky soil featuring embedded grass and tree leaves (L). Plane polarized light (PPL), frame width is ~3.2 mm.(b) detail of chalky soil fragments collected from the *suspect's* car, with inclusions of landsnail shell (S; ubiquitous in muddy track deposits) leaf fragment (L). PPL, frame width is ~1.6 mm. Both humic soil and plant material are potential sources of pollen.

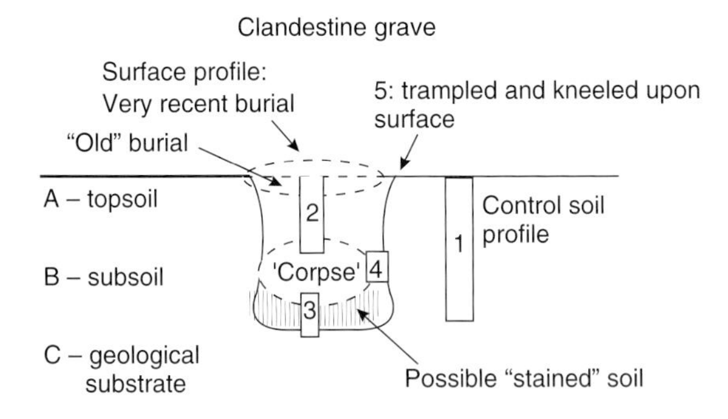

FIGURE 14.5 How to sample a clandestine grave for geoarchaeological evidence. Five potential sampling locations identified; 1 – control profile; 2–4 – soil micromorphology sample locations to help reconstruct "time" and "climatic" conditions of burial, evidence of trampling, postburial weathering, and to differentiate "exotic" materials brought in by "suspect"; 5 – surface soil materials picked up on footwear or on trousers (kneeled-upon surface). All samples to be tested (palynology, macrofossils, mineralogy, and chemistry) against tools, clothing, and the like, in possession of suspect (Wilson, 2004; Wiltshire, personal communication).

14.6 Conclusions

Geosciences are increasingly being employed in forensic investigations of crime sites. If bulk and soil micromorphological analyses are to be useful in this context, however, there are two essential caveats to be remembered. First, it is important to have a grounding in the types of natural geological, pedological, and anthropogenic materials of the crime site and its environs, and secondly, because findings have to stand up in court – hence the term forensic – sufficient replicates and comparisons need to be made.

Acknowledgments

The authors thank P.E.J. Wiltshire, E. Wilson, and Hertfordshire Constabulary for their comments on an earlier version of this chapter.

III

Field and laboratory methods, data, and reporting

Introduction to Part III

In the previous sections (Parts I and II) different geoarchaeological contexts were explored and discussed in light of their field and laboratory data. We also drew upon unpublished reports, as well as articles in journals and books. In this section, we present field and laboratory methodologies employed widely in geoarchaeological studies. Those chosen are meant to provide guidelines so that techniques can be employed appropriately. Techniques vary from aerial photography and ground-truthing by coring, to instrumental analysis of individual samples by electron microprobe. We have also suggested ways by which fieldwork and laboratory studies can be reported, from a field evaluation through to a published article. Lastly, we discuss some of the directions of development that we anticipate for geoarchaeological research in different parts of the world.

15

Field-based methods

15.1 Introduction

Geoarchaeology begins in the field, for it is there where observations are made and initial data are generated. Missed opportunities in making sound field observations are very difficult to make up for, and since the archaeological record is extracted from the "geological" context, it is important that the geoarchaeologist sets out in the field with appropriate strategies for the task at hand. Such strategies may entail simple regional mapping of sites or features, that is, be confined to the landscape scale of operations, or they may focus on documenting the microstratigraphy at a site and then seeing how that relates to the geology of the immediate surroundings of the site location (e.g. fluvial terrace). Obviously, the field protocols will be somewhat different for each one of these endeavors. Below we present some of the most common types of tools and strategies that are needed to carry out proper geoarchaeological fieldwork. These approaches range from regional-scale methods (e.g. aerial photography) down to those used at the microstratigraphic scale. These methods are not meant to constitute a cookbook or comprise a series of rigid protocols that should be followed religiously, because every site delivers a different challenge. As always, common sense should be used.

Nevertheless, as we point out in Chapter 17 on report writing, geoarchaeologists should be involved in all phases of the research project: from initial planning, through survey and excavation phases, and ultimately to the write-up of the results. This continued presence operates at all levels, from understanding the site in its landscape setting, down to the understanding of individual stratigraphic units, along with their contents (e.g. artifacts and construction materials).

15.2 Regional-scale methods

Several methods provide a geoarchaeological record for landscape scale investigations. So, for example, in attempting to locate sources of raw materials for building, or for pottery temper, it would be useful to consult a geological map of the region. Similarly, to locate present and past sources of water, large scale aerial photographs, or satellite images should be examined in conjunction with geological and topographic maps. Below, we summarize some of the most well-known regional-scale methods currently employed in geoarchaeology.

15.2.1 Satellite imagery and aerial photos

Satellite images from space and aerial photographs provide a direct visualization of landscapes, vegetation, and resources, and represent the most regional scale of remotely sensed data. Images can span a huge variety of scales, from that of the planet, down to an aerial photograph of a site and its immediate surroundings at 1:1,000, for example.

15.2.2 Satellite images

Satellite images are acquired from spacecraft at elevations greater than 60,000 feet (US Geological Survey, USGS: http://erg.usgs.gov/isb/pubs/booklets/aerial/aerial.html), and as such, they provide information on a regional scale, and generally in less detail than that acquired from an aerial photograph taken at a lower altitude. They employ diverse types of sensors that are responsive to different parts of the electromagnetic spectrum. The electromagnetic spectrum represents types of energy that is characterized by differences in wavelength (in meters) and frequency (waves per second) (Fig. 15.1). It spans the longer wavelengths of radio waves (e.g. 10^3 m), down to very short wavelengths of "hard X-rays" (10^{-10} m) used in X-ray diffraction techniques, for example; visible light, infrared (IR), and ultraviolet light (UV) are somewhat in the middle of the spectrum with wavelengths centering between 10^{-5} to 10^{-7} m.

Remote sensing from satellites, makes use of the fact that energy derived from the sun travels through the atmosphere and strikes the earth's surface or target (Fig. 15.2: A, B and C) (Canada, 2004). Much of this energy is reemitted to the atmosphere as different wavelengths (and frequencies) where it is collected by different sensors within a satellite (or by different satellites) that detect the different types of electromagnetic radiation (Fig. 15.2: D). Energy collected by the sensor is transmitted electronically to a facility (Fig. 15.2: E) that processes the data into an image that can be obtained as digital data or a hardcopy printout (Fig. 15.2: F). These images are then used as is, or enhanced in order to make observations and interpretations about the objects being recorded (Fig. 15.2: G). Certain factors influence this path of energy transmittal and detection, including, for example, the energy source, scattering, and absorption in the atmosphere, energy/matter interactions at the earth's surface (e.g. scattering and reflection), and the type of sensor (e.g. whether passive or active) (Canada, 2004; Lillesand and Kiefer, 1994).

Data in digital images are represented as small equisized areas (picture elements or *pixels*) in which the relative brightness values of each pixel are represented by a number (Canada, 2004). Thus, the smaller the pixel size, the greater is the resolution and our ability to distinguish individual objects on the ground surface. It is essentially the size of the smallest object detectable by the sensor.

In the United States, satellite images have been taken from both manned space flights and from satellites. The early Landsat satellites contained a multispectral scanner (MSS) that measured four different wavelengths, with a resolution of ca 80 m (http://erg.usgs.gov/isb/pubs/booklets/aerial/aerial.html#Satellite). The Thematic Mapper, aboard later satellites, measured a greater number of spectral bands and has a greater resolution of ca 30 m. Table 15.1 shows the spectral bands of the Thematic Mapper and their principal applications.

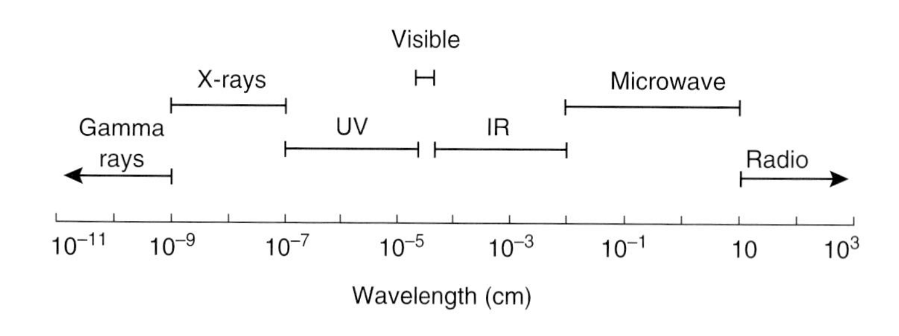

FIGURE 15.1 The electromagnetic spectrum.

In Europe, satellite SPOT Images (*Système Probatoire d'Observation de la Terre*) from France, for example, can record panchromatic images (60 × 60 km with spatial resolution of 10 m) or multispectral ones. In the latter case, images are recorded in green, red, and reflected IR, with cells of 20 × 20 m. Color composite images can be made by mixing two spectra (Sabins, 1997). Stereopairs can also be produced for three-dimensional visualization (see aerial photos below).

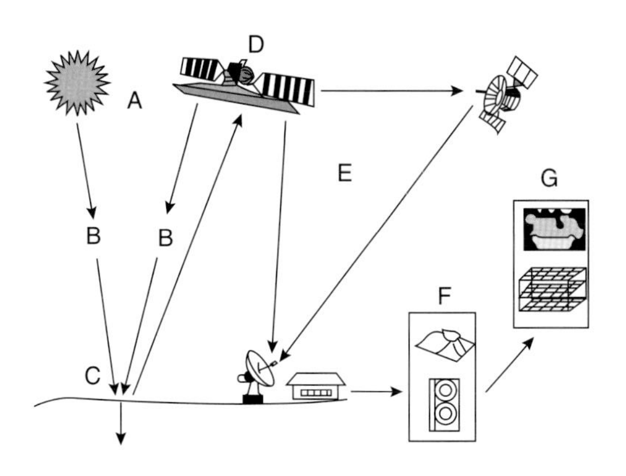

FIGURE 15.2 Generalized scheme of remote sensing detection using a satellite (from 2004. National Resources Canada: *Fundamentals of Remote Sensing*. http://www.ccrs.nrcan.gc.ca/ccrs/learn/tutorials/tutorials_e.html).

Examples of the use of satellite imagery in geoarchaeology are numerous. These include archaeological site prospection in England in which satellite data provided complementary information to those found in aerial photos (Fowler, 2002) and detection and surveying of former patterns of land use in Spain (Montufo, 1997). In Yemen, Harrower *et al.* (2002) used LANDSAT satellite imagery along with GPS data to document the geomorphological, palaeoecological, and archaeological evidence of mid-Holocene environments, as well as changes in land use. Archaeological features like mounds and trackways in Norway have been successfully idendifeid from multispectral images; identifications were later tested by soil chemistry (Grøn *et al.*, 2005).

TABLE 15.1 Spectral bands used in the Thematic Mapper of Landsat satellites (modified from Lillesand and Kiefer, 1994, table 6.4)

Band	Wavelength (μm)	Nominal spectral location	Principal applications
1	0.45–0.52	Blue	Water body penetration and coastal water mapping. Discrimination of soils and vegetation; forest-type mapping; identification of cultural features
2	0.52–0.60	Green	Green reflectance peak of vegetation. Assessment of vegetation and vigor of plants; also useful for identification of cultural features
3	0.63–0.69	Red	Chlorophyll absorption region. Differentiation of plant species and identification of cultural features
4	0.76–0.90	Near IR	Determination of vegetation types, vigor, and biomass content; delineation of water bodies; soil moisture identification
5	1.55–1.75	Mid-IR	Indicative of soil mixture content and soil moisture; differentiation of snow from clouds
6	10.4–12.5	Thermal IR	Vegetation stress analysis; soil moisture discrimination; thermal mapping
7	2.08–2.35	Mid-IR	Discrimination of types of minerals and rocks; sensitive to vegetation moisture content

15.2.3 Aerial photographs

In contrast to satellite images, aerial photographs are taken at lower altitudes, either from airplanes or from small balloons or photo towers constructed at the site (Whittlesey *et al.*, 1977). Aerial photographs have been used for over a century to depict landscapes, and these early photographs have been valuable in documenting recent landscape modifications and their rates of change, even if they are over relatively short time spans.

Aerial photographs can be classed as either oblique or vertical (Figs 15.3–15.5). In *oblique* photographs, the axis of the camera is tilted from the vertical, generally 20° or more (Lattman and Ray, 1965). This provides somewhat an artistic feeling for the landscape and relief (the difference between high and low points on a map) and landscape features – both cultural and natural – are enhanced; the downside is that features in the foreground appear larger than those in the background, with concomitant differences in scale. This type of photograph is common in the United Kindom to show features (e.g. ditches, walls, and barrows) associated with Bronze Age, Iron Age, Roman, and Medieval sites (Riley, 1987).

Vertical aerial photographs are generally more useful in archaeology and particularly in disciplines, such as geology (Hamblin, 1996; Lattman and Ray, 1965; Marcolongo and Mantovani, 1997; Mekel, 1978) and botany, where field mapping is done. This is because the stereographic properties of overlapping vertical photographs can be used to view the ground surface in three dimensions. An airplane, flying at a specific altitude will take a series of photographs so that there is about 60% overlap from one photograph to the next, and about 25% sidelap of photographs from successive adjacent flight paths (Fig. 15.4). When two of the overlapping pairs are viewed with a stereoscope, a three-dimensional image

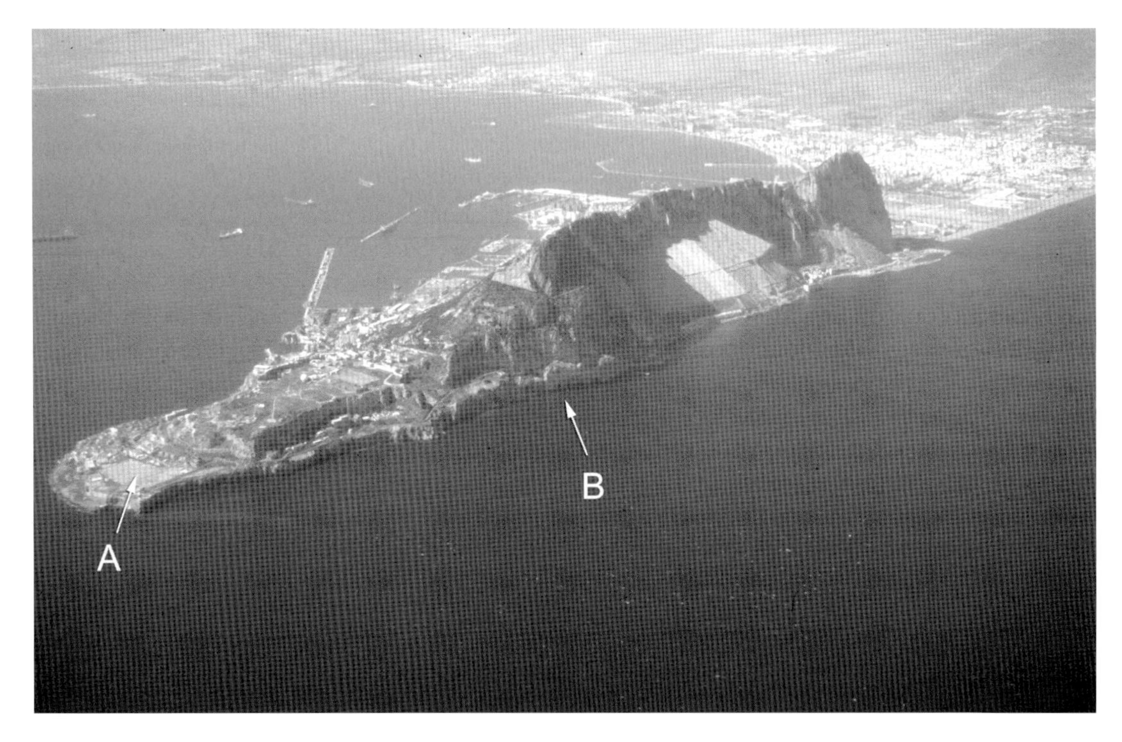

FIGURE 15.3 An oblique aerial photograph of Gibraltar showing a marine Pleistocene abrasion platform (A) and Gorham's and Vanguard Caves (B) along the eastern side of the peninsula (See Chapter 7).

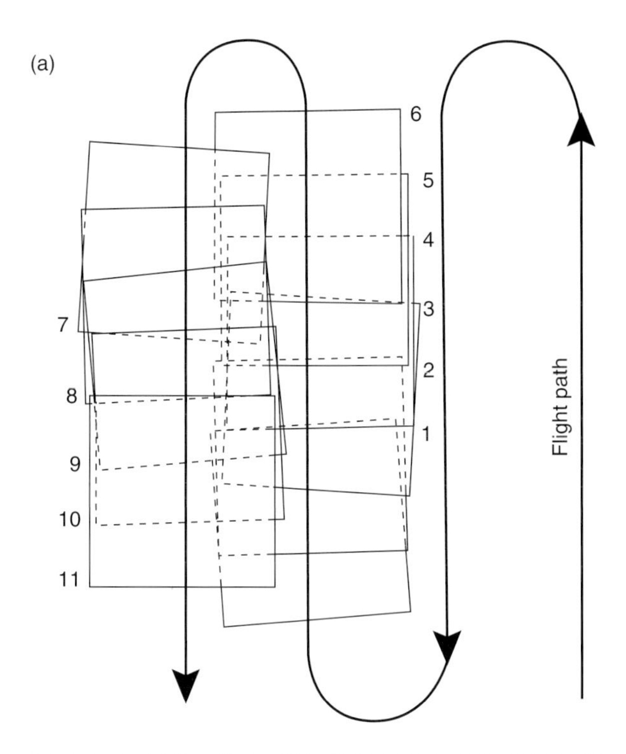

Figure 15.4 (a) Stylized view of flight path of airplane for vertical aerial photos that can be used for stereoscopic viewing in three dimensions. Note the overlap along the flight path and the sidelap between flight lines (from Lattman and Ray, 1965; figure 1-8). (b) One of the authors viewing a set of aerial photographs with a stereoscope.

of the landscape can be seen (Lattman and Ray, 1965).

These stereo images provide a much clearer view of the landscape, including the vertical and spatial distribution of rocks and strata, vegetation, springs, and other important features. However, images in simple stereoscopic viewing are vertically exaggerated by an amount that is a function of the degree of overlapping of the photo pairs. The overall scale of the photo depends on height of the aircraft above the ground, as well as the focal length of the

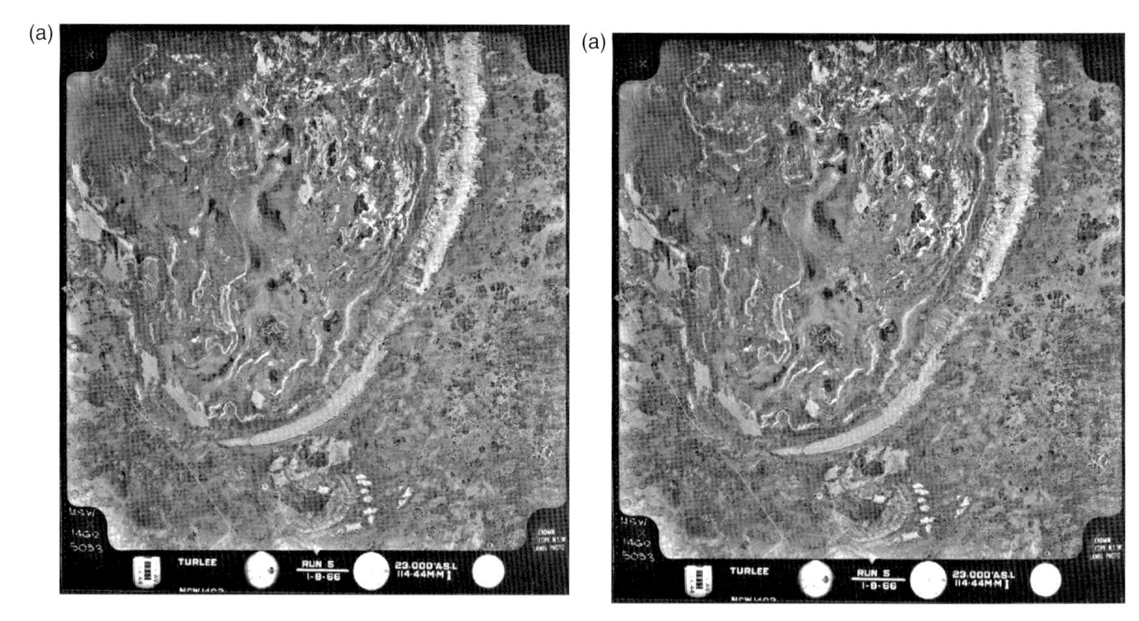

FIGURE 15.5 Stereopair of aerial photographs. These are normally viewed with a stereoscope but it is possible to train oneself to view the photographs stereoscopically without the device. (a) Shown here is a portion of the eastern side of Lake Mungo in southeast Australia. The banana-shaped ridge on the right is called a lunette and is composed of a sequence of aeolian sands (both quartz grains and sand-sized pellets of clay blown up from the lake bottom) and paleosols that formed during the late Pleistocene. In the right-hand side of photo (b) weakly developed, vegetated linear dunes can be seen downwind of the lunettes; left of the lunette, the floor present-day lake surface is hummocky. Within the lunette deposits are prehistoric sites, including the earliest use of ritual ochre in a burial (Mungo III), and a cremation (Mungo I); dating is controversial, but recent estimates place both burials at about 40,000 years ago (Bowler *et al.*, 2003). Aerial photographs used with permission, © Department of Lands, Panorama Avenue, Bathurst 2795, NSW, Australia, www.lands.nsw.gov.au

camera. In addition, within each photo the scale can be somewhat variable, since objects at higher elevations will occur at a different scale than those at lower elevations. This variability is of only slight consequence in areas of low relief but can be significant in highly mountainous terrains with great relief. Thus, because of the lateral displacement of objects in an aerial photo, depictions of objects differ from those of maps, which are orthographic projections. Consequently, linear distances on aerial photos are not equal as would be the case for maps (Lillesand and Kiefer, 1994).

Both black and white and color films, either panchromatic or infrared-sensitive, are used in aerial photography, from which contact sheets are made (usually 9 × 9 in. or ca 23 × 23 cm). Panchromatic film is most common and is used in constructing topographic maps, geological surveys, and in agriculture (Sabins, 1997). Black and white IR film is useful for penetrating through haze and exploiting the fact that the maximum reflection by vegetation takes place in the region of photographic IR, and the fact that IR is totally absorbed by water (Sabins, 1997). Similarly color IR film is useful in viewing different terrain signatures, such as vegetation (healthy and stressed), clear versus silty water, damp ground, shadows, water penetration, and contacts between land and water (Lillesand and Kiefer, 1994, table 2.3). Consequently, they are used in a variety of applications in basic and applied research that include geological and soil mapping, agriculture and forestry (types of crops and diseases), water pollution and wetland management, and urban development.

Interpretation of photographs translates the visual patterns on the photograph into information that becomes useable to the observer. The typical characteristics that are observed include the following (Lillesand and Kiefer, 1994):

- Shape – morphology of objects, some are characteristic, some not.
- Size, both absolute and relative, particularly with regard to the scale of the photograph.
- Pattern – the spatial arrangement of individual entities, including repetition, layout, and ordering.
- Tone and hue – relative brightness or color of objects or materials; different types of soils will have different tones.
- Texture – frequency of change of tones and their arrangement on the photograph, typically by objects too small to be seen individually, as for example, a smooth lawn versus a coarse forested canopy.
- Shadows – provide outline and relief, particularly if the photograph is taken at a low angle to the sun.
- Site – topographic or geographic location, reflecting, for example, peaty soils in lowland locations *versus.* coniferous vegetation on well drained upland acidic sands.

Awareness, knowledge, and experience with the above can be applied to geoarchaeological situations, with particular emphasis on geology and soils. In geological mapping, for example (see below), landforms (e.g. alluvial fans), lithology (e.g. basalt versus rhyolite), and structure (folds and faults) can be identified, as well as mineral resources (Rowan and Mars, 2003). Similarly, drainage networks (both on the surface and partly buried as relicts) and topography can be clearly seen, the latter often a reflection of structure and lithology. A horizontally bedded limestone in an aerial photograph may have the characteristics as presented in Table 15.2, (Lillesand and Kiefer, 1994).

Similarly, deposits associated with glacial, aeolian, lacustrine, etc. landforms can be identified by associations of characteristics. Those for loess deposits, for example, are shown in Table 15.3.

Soil mapping (see below) can be made more efficient and can cover larger areas with aerial photos. Most soil survey maps (see Section 15.2.6 below) in the United Stated are presented on an aerial photo base. In addition, the use of photos provide insights into moisture characteristics of the soil. With the use of stereo pairs (see above) topographic boundaries, such as breaks in slope, can then be drawn directly onto the photograph, and plateaus, slopes, and valleys demarcated. The vegetation cover also seems to "stand up," and be identifiable. Such a desktop reconnaissance can be a vital step before working "on the ground," whether it be for archaeological survey or for a search for a clandestine burial (Chapter 14).

15.2.3.1 Air photos in archaeology

Aerial photos are often useful for seeing things that are simply invisible on the ground. Aerial

Table 15.2 Characteristics of horizontally bedded limestone in aerial photographs (modified from Lillesand and Kiefer, 1994: 251)

Topography	Gently rolling with sinkholes varying from 3 to 15 m deep by 5 to 50 m wide
Drainage	Centripetal drainage into sinkholes; few surface streams; sinking streams
Erosion	Gullies with gently rounded cross-sections are developed into clayey residual soils
Photo tone	Mottled due to numerous sinkholes
Vegetation and land use	Typically farmed, except for sink holes, which are usually inundated with standing water

Table 15.3 Characteristics of loess in aerial photographs (modified from Lillesand and Kiefer, 1994: 261)

Topography	Undulating mantle over bedrock or unconsolidated materials; tends to fill in preexisting relief
Drainage and erosion	Gullies with dendritic pattern and broad, flat bottoms and steep sides with piping
Photo tone	Light, resulting from good internal drainage; tonal contrast demarcated by gullies and associated vegetation
Vegetation and land use	Farming

photography in archaeology has a long history, going back to photographs of the well-known site of Stonehenge in England, which was photographed from a military balloon in 1906 (Bradford and González, 1960). This tradition continued into the 1920s, and up to the present, with photographs of Bronze Age and Roman sites (Bradford and González, 1960). A classical example of early work in archaeology is that of J. Bradford, who mapped over 2,000 Etruscan graves at the site of Tarquinia in Central Italy; most of these tombs were not visible at the surface, although some had been previously discovered centuries before (Bradford and González, 1960). Photos can be used not only to discover sites – both on land and submerged – but also used for making measurements, constructing maps, and three-dimensional renditions of the site and surroundings.

For sites with architectural features, a number of these show up on photos that aid in their recognition. *Shadow marks*, for example, are produced by differences in surface relief that are augmented by oblique illumination of the sun's rays and are particularly enhanced with low sun angle; they result in shadows being cast on the surface. Such marks can be produced by ditches or embankments, or the construction of monuments, such as the Poverty Point site in Louisiana or mound sites in the central and southeast United States.

Crop marks and *soil marks* (Evans and Jones, 1977; Jones and Evans, 1975; Riley, 1987) pick out crop responses to soil variations relating to archaeological features. They are best developed:

- in arable areas where there can be a marked summer water deficit (e.g. in the United Kingdom, eastern and southern England and eastern Scotland);
- where bare soil-marks relate to the occurrence of light colored subsoils (e.g. on chalk); or
- where shallow loamy soils have rooting depths between 30 and 60 cm and crop marks are caused by soil moisture deficit acting with nutrient supply and soil depth.

Generally, crop marks are better recorded in cereals than grasses, appearing first in shallow soils when the potential moisture deficit, which occurs when water transpired by a crop exceeds precipitation, is greater than the amount of water in the soil available to the plant. More faint crop marks also occur in wet years, probably because of waterlogging, which may lead to nitrogen shortage, or archaeological features may cut through drainage impedance levels, also producing a crop mark (Bellhouse, 1982). Soil conditions, crop marks, and potential soil moisture deficit have been linked to better plant growth over a ditch, because deeper soil allowed plants to reach greater available water (Evans and Jones, 1977; Jones and Evans, 1975). Bellhouse (personal communication) also noted that crop marks could also occur through the poaching of grass, a phenomenon caused by stock trampling topsoil, which inhibits the growth of grass.

Crop marks are also produced by differences in vegetation related to soil moisture conditions and controlled by subsurface features. Ditches filled with organic-rich material, for example, will tend to favor vegetation growth, resulting in slightly more rigorous vegetation growth than outside the ditch area, which may be rocky and with less available moisture.

Similarly, above a buried wall, soil cover may be thinner and consequently vegetation will be slightly stunted in comparison to adjoining areas, where soil cover is thicker. Vegetation may be grass, crops, weeds, or woods. Crop marks are abundant in the United Kingdom where many sites have been mapped (Riley, 1987).

Soil marks show up on photographs as differences in tone, texture, and color, and moisture of the soils resulting from human features such as ditches and heaping of soil. A ditch, for example, may promote the accumulation of darker humus within the depression, serving as a contrast for the surrounding, lighter, less humus-rich soils (Bradford and González, 1960).

Finally, other types of constructional features resulting in changes of the landscape or surface coatings on rocks can be visible. Among the most visible and striking are the famous Nazca lines in Peru, whose origins are still unknown.

15.2.4 Topographic maps

Topographic maps are an invaluable resource both in and out of the field, as they place an area, sites, and their surroundings in a well-defined spatial context. In turn, this contextual information can be used to understand site distribution, proximity to certain landscape features, such as plateaus, sheltered areas (e.g. canyons), water sources, types of vegetation, and mineral resources (information available from USGS: http://erg.usgs.gov/isb/pubs/booklets/topo/ topo .html; http://erg.usgs.gov/isb/pubs/booklets/usgsmaps/usgsmaps.html#Topographic%20Maps).

Generally, the most useful aspect of topographic maps is the elevational data. This is expressed as individual spot elevations (bench marks), but usually with contour lines of equal elevation. Their spacing reflects the type of terrain, whether steep, with closely spaced lines, or more gently sloping, with large spaces between contours. Latitude and longitude, as well as UTM (Universal Transverse Mercator) coordinates can also be read off the map directly, or interpolated for other areas. Depending on scale of the map (see below) it can also portray a number of useful features that could be of geoarchaeological interest. These, for example, include location of mines, buildings, and roads, as well as important surface features (e.g. levees, dunes, moraines, channels, beaches, reefs, and ship wrecks). Swamps and types of rivers are also indicated, whether they are perennial or intermittent. The location of springs in arid environments can be particularly valuable for locating potential sites.

Maps are available at a variety of scales. In the United States, they can be found at scales of 1 : 24,000 (7.5-minute maps; 1 in. = 2,000 feet; occasionally at 1 : 25,000 in metric units), the 15-minute maps (1 in. = 63,360; 1 in. = 1 mile), 1 : 100,000, and 1 : 250,000. The larger the scale, the less detailed information is portrayed, but the tradeoff is in viewing a larger area at one time. Also available in the United States are other types of maps, such as Orthophotomaps, which are multicolored, distortion-free, photographic image maps. They are produced in standard 7.5-minute quadrangle format from aerial photographs, and they show a limited number of the names, symbols, and patterns found on 7.5-minute topographic quad-rangle maps. Most of these maps are at a scale of 1 : 24,000, but some are at 1 : 25,000 (http://topomaps.usgs.gov/). They are particularly useful in showing areas of low relief.

15.2.5 Geological maps

The use of geological maps is an essential part of geoarchaeological investigation, and as such should be consulted before, during, and after fieldwork has been carried out. A remarkable amount of potential valuable information is available on geological maps, which are useful to both archaeologists and geologists.

First and foremost, a geological map displays the types of rocks and sediments exposed at the surface. Such materials include both lithified (consolidated) materials, such as bedrock, or

FIGURE 15.6 Maps showing the topography around the Palaeolithic sites of Pech de l'Azé I, II, and IV, in the Dordogne region of SW France. (a) topographic map produced by the mapping of over 2000 data points shot in with a Total Station by H. Dibble and S. McPherron (www.oldstoneage.com). Contour lines are 1 m and the grid markings are 10 m. The sites of Pech de l'Azé I, II, and Pech de l'Azé IV are shown on the map. The saddle like object just above 940 grid line is an abandoned railway trestle shown in (c). (b) a computer generated surface made with these same data, showing somewhat more realistically the morphology of the landscape, particularly the cliff just behind both sites that leads up to a plateau surface behind them and shown by closely packed contours in (a). (c) Photograph of Pech de l'Azé II from just above the trestle depicted in (a). Pech de l'Azé IV is just below and to the right of the photo frame (maps in a and b, Courtesy of S. McPherron).

looser – likely younger – materials such as alluvium, colluvium, or aeolian deposits. Information about existing rock and sediment types can be useful in locating possible sources of raw materials for lithic (e.g. chert, volcanic rocks) or ceramic production (e.g. clays for plaster and coarser materials for temper). In turn, knowledge of their distribution can be important in determining distances and energies of expense of procuring such resources, and be instrumental in guiding the collection of samples of materials for instrumental analysis (e.g. X-ray fluorescence – XRF; see Chapter 16), as used in obsidian (a black volcanic glass) sourcing, for example. In the case of the very large collection of obsidian tools at Arene Candide, Liguria, Italy, neutron activation analysis was carried out and it showed that during the important Mediterranean "colonization" period of the Early Neolithic, people relied on obsidian from Sardinia and Palmarola, whereas by the Late Neolithic there was shift to Lipari (Ammerman and Polgase, 1997). Equally, a study of the local geology at Arene Candide permitted the identification of sources of ochre (Ferraris, 1997).

Outcrop patterns of different types of bedrock and sediments in an area also provide information about structures, such as folds and faults, reflecting the tectonic history of an area (Fig. 15.7). The presence of faults or lithological discontinuities (breaks), coupled with the lithological distributions, can indicate the location of water sources (springs) or zones of mineralization, which could be used, for example, to determine provenance of ores used in smelting. Furthermore, orientation of strata and structures – indicated with strike and dip symbols – can be used to project the occurrence of strata with depth. Similarly, patterns of more recent Quaternary sediments, when used in conjunction with soil maps (see below) can reflect relative ages of the deposits and the landscapes associated with them. Most geological maps provide structural cross-sections that portray the earth's fabric beneath the surface.

Included in many maps is auxiliary information such as a booklet that provides a variety of geological details. These items include, a geological overview of the area, complete descriptions of the rocks and deposits, earthquake history, tectonic structure and history, regional raw materials, hydrology, and soils.

Geological maps are obtainable from national and local geological surveys at a variety of scales that typically range from 1 : 25,000 to 1 : 1,000,000 (Fig. 15.7). Some are available on the web, as for example a UGSG Open-File report from Connecticut, United States (http://www2.nature.nps.gov/geology/usgsnps/gmap/gmap1.html). In many countries, however, most maps are available only for purchase, if they can be found at all.

15.2.6 Soil survey maps and soil mapping

As described in Chapter 3, different types of soils develop in response to the five soil-forming factors. Although soil maps can be drawn from remotely sensed data and previously studied areas, in reality they have to be checked and refined in the field, usually by a governmental agency. These can be form national, regional, or agricultural agencies (e.g. British Forestry Commission, US Department of Agriculture). Countries with their own soil classifications have been listed (Avery, 1990; Bridges, 1990). Other countries have used guidelines from the FAO (Food and Agriculture Organization) or *Soil Taxonomy* (United States; http://soils.usda.gov/technical/manual/; http://soils.usda.gov/technical/fieldbook/) (FAO-UNESCO, 1988; Soil-Survey-Staff, 1999); also some countries have had soil maps drawn up by external commercial companies (King *et al.*, 1992). Finally, private, large-scale farming operations may also have small-scale soil surveys done over their property (e.g. kibbutzim in Israel).

The chief role of a soil survey is to draw up soil maps using the techniques of soil profile (pedon) description, sampling, and laboratory analysis; to produce a soil classification; and then map soils through an auger survey (see Section 15.4). A pedon is the smallest identifiable three-dimensional soil unit and a polypedon is composed of one or more contiguous pedons, that is, a three-dimensional soil body that makes up a specific "soil series" (Buol *et al.*, 1973). In both the United Kingdom and the United States, soil series are named after the locality where it was first studied or where it is most extensively present (Avery, 1990); they are defined according to "intrinsic properties of practical significance" (Jarvis *et al.*, 1984: 54). Apart from being defined according to soil type (Table 3.2) they can also be defined in relation to the substrate type (geology, including organic soils), texture (clay, silt, or sand, or sand over silt, etc.) and mineralogy and condition, (e.g.) "ferruginous" or "saline," presented weakly or strongly swelling clay, and if stony, composition of the stones, as such criteria used for the present soil map(s) of England and Wales (Clayden and Hollis, 1984).

Once a soil series has been identified and classified it is then surveyed and mapped by surveying with auger or trenching (Section 14.4). Because of financial and time constraints, a soil type may have only been mapped by two auger

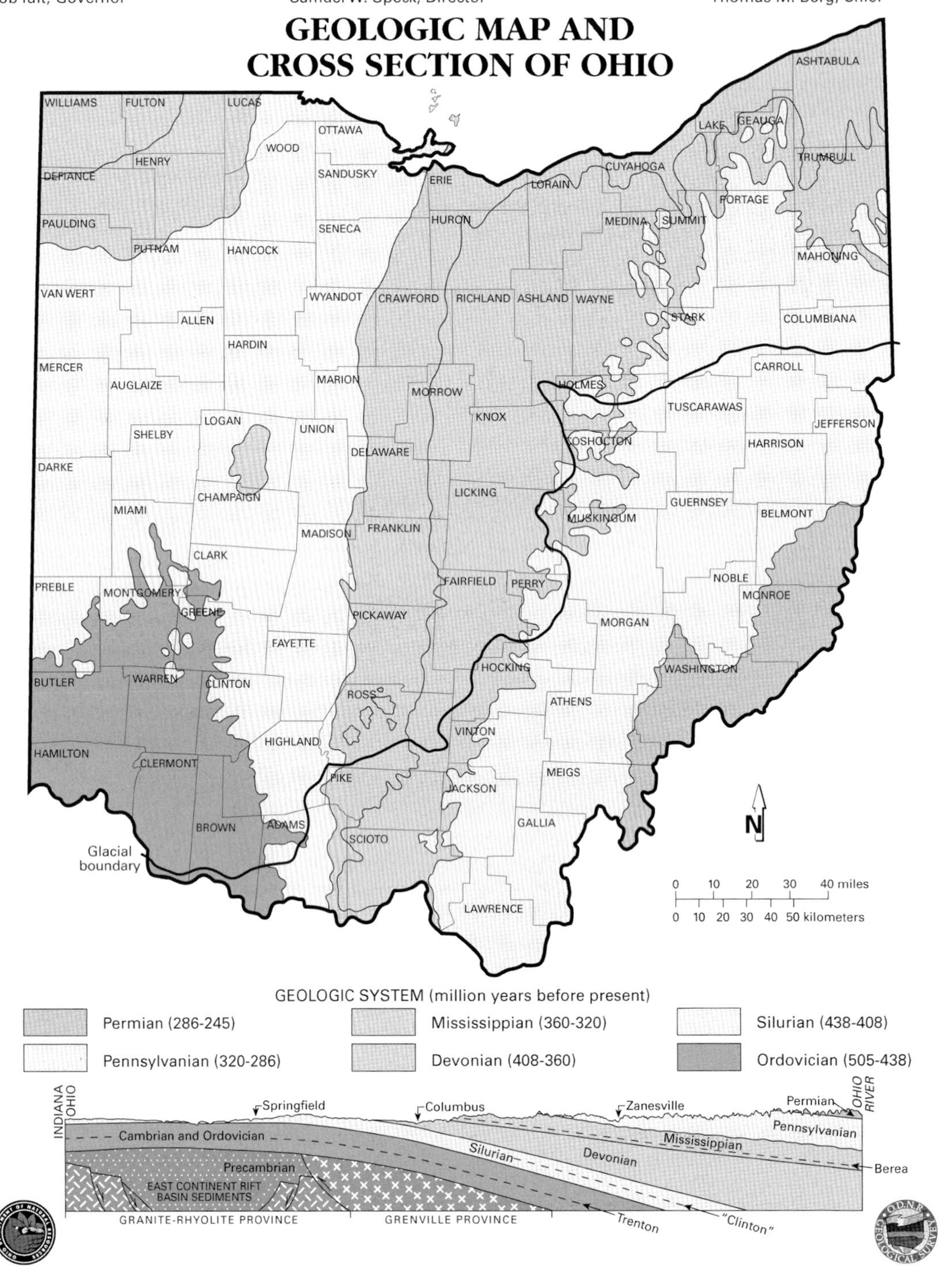

FIGURE 15.7 Geological map of the state of Ohio, USA. This simplified map of the state shows a series of broad bands of limestone and clastic sediments that dip toward the east, as shown in the cross-section below. The black squiggly line represents the limit of glaciation during the Quaternary resulting in a more exposed and eroded terrain in the SE quarter of the state. (Used with permission from the State of Ohio, Division of Geological Survey).

borings per square kilometer. As a result, many soil maps are composed of a great deal of interpolation. For example, the 1983 (and still current) 1 : 250,000 soil map of England and Wales (e.g. Jarvis *et al.*, 1983) was compiled as a combination of an earlier detailed map (1:63,360 – 1 in. map), which gave a 20% coverage of these two regions, and new information based upon small pit and auger borings at an average frequency of 250 per 100 km^2 (Avery, 1990: 45).

According to the details required (e.g. local versus regional study), different scale soil maps can be used. In the first instance, the soil surveyor would probably use a 1 : 10,000 base map. The detail from this would be transcribed (and simplified) onto a lower scale map, for example 1 : 50,000 for a "county" soil map. In North America, each county has its own soil survey at different scales, and many can be downloaded as pdf files, or queried online (e.g. http://soils .usda. gov/survey/online_surveys/ alabama/russell/ al_russell.pdf; see also http://soils.usda.gov/). Procedures for how these maps were made are discussed within each county report. Information from these individual reports has been used to create global soil maps at a scale of 1 : 5,000,000, and only some 18 sheets are needed to cover most of the world (FAO-UNESCO, 1988).

The amount of detail and accuracy required to produce a soil map for a geoarchaeological survey is very much greater of course, and is more likely to be 10–20 borings per km^2. Two types of survey can be carried out, namely *free survey* and *grid survey* (Avery, 1990: 37). The first is employed by experienced soil scientists who use variations in topography and vegetation from aerial photography (see Section 15.2.3) to infer the location of different soil types that can then be later checked by augering. Soil map boundaries can be indicated in along a break of slope, for example, and then inspected by augering. Within this kind of survey, it may be important to know the variation in depth of the soil profile from plateau top to valley bottom, that along catena profile (see Chapter 3), and an auger transect can then be carried out.

The second type of survey, the grid survey, is very useful when recording a single parameter, such as soil depth. This can be as detailed as required, from 1 auger per 10 m, 50 m, or 100 m, etc. Depth to gleyed (mottled or leached soil) calcium carbonate-enriched horizon (K horizons), can be readily measured or to a buried landscape. In the same way it is possible to map an extensive archaeological layer, such as the several hectare size Late Bronze Age/Early Iron Age "deposit" at Potterne (Lawson, 2000).

Such soil information along with that from a geophysical survey can then be incorporated into an overall GIS database. Two words of caution need to be mentioned. First, the use of soil survey maps must be done with the caveats: although this is a good start for a desktop study, the scale is probably not detailed enough to answer critical geoarchaeological questions and the soils will need to be checked during trenching or excavation. Also, the way surveys were made and the information that they contain may not always be suitable for archaeological purposes. These aspects are dealt with in detail in Holliday (2004, chapter 4). Second, the depth, chemistry, and physical properties of modern soils may have very little in common with those of the ancient soil cover, since many of the soil forming factors may have been different in the past. This is shown by a podzol cover that dates to the Bronze Age or Iron Age: eairlier soils had the attributes of bioactive brown soils (see Chapter 3, Soils). Also, if the modern topsoil appears to be highly fertile and produces good crops of strawberries, such information is only extrapolated to past landscapes and cultures at great peril to the worker's reputation.

In any case, soil survey information is very useful in geoarchaeological contexts and should be consulted at the beginning of most geoarchaeological projects. One of the major uses is its predictive value in evaluating site distributions and site survey. For example, we used county soil maps to help guide the survey of a small valley/river system in northern Alabama, United States. The occurrence of Inceptisols (soils with scant profile differentiation) on the flood plain of tributaries to the Tennessee River indicated that older sites (e.g. Archaic or Paleoindian, i.e. >2 ka) would not be present on the soil surface, but could possibly be found at a depth, buried by the young alluvium on which this soil developed. Similarly, remnants of more developed fluvial soils in the valley (mapped as Ultisols) indicate older soils and underlying deposits, suggesting that more ancient sites might be found within these fluvial deposits; relatively young sites could naturally occur on the surfaces of these older soils.

In addition, state and county soil surveys [in cooperation with the US Natural Resources Conservation Service (NRCS, as part of the US Dept of Agriculture)] provide a wealth of potentially useful information, including temperature and precipitation data; lengths of growing season; wildlife habitats; and chemical, physical, and engineering properties of the soils. Thus, the occurrence of different types of soils can point to certain subsurface conditions, which in the past might have influenced, where former settlements might have occurred. For example, the presence of aquic soils on a map indicates waterlogging at least for part of the year, and thus might reasonably be an unlikely locus for past occupation. On the other hand, soil units with good drainage would be much more favorable locales for both past and present settlement. A relationship between soil survey map units and Neolithic settlements was such an instance noted in Belgium (Ampe and Langohr, 1996). Nevertheless, one must realize that in this latter case, present-day drainage characteristics might not necessarily represent those in the past: modifications can come about by past climate changes, or tree clearance, which have the effect of reducing evapotranspiration, and thus raising groundwater levels.

15.3 Shallow geophysical methods (resistivity, palaeomagnetism, seismology, ground penetrating radar)

This section describes a number of methods that employ geophysical techniques to examine and reveal subsurface archaeological sites and features, while placing them in a regional landscape. Geophysical techniques have been applied since the middle of the twentieth century, but their use – particularly in England – has become more routine since the 1980s (Gaffney and Gater, 2003). Modern instruments permit large areas (several hectares) to be investigated at a time, thus revealing extensive and detailed information on site structure well before excavation is carried out. With the sophistication and ease of use of the equipment ever increasing, and costs per area surveyed decreasing, geophysical approaches are very efficient means to reveal material buried within the upper few meters of the surface. Details on these techniques can be found in several books and web sites [(Clark, 1996; Conyers and Goodman, 1997; Gaffney and Gater, 2003; Garrison, 2003; Herz and Garrison, 1998; Kvamme, 2001); the journal, *Archaeological Prospection*, http://www.wileyeurope.com/ WileyCDA/WileyTitle/productCd-ARP.html; http://www.cast.uark.edu/~kkvamme/projects.htm].

As discussed by Kvamme (2001), however, not every technique is appropriate for all situations, and commonly more than one is used at a given site, as each may be more effective in detecting different details. Several factors can influence the effectiveness (e.g. speed, volume, and usefulness of data) of a geophysical survey, including vegetation, bedrock, soil type, and local conditions. Vegetation, for example, cannot only directly impede survey with the presence of trees and tall grasses, but it can also indirectly influence soil moisture, which in turn can affect electrical methods (see below and Table 15.4).

TABLE 15.4 Some common geophysical techniques used in archaeology and associated attributes (adapted from Kvamme, 2001; Gaffney and Gater, 2003)

Attribute	Magnetometry	Electrical resistance	Ground penetrating radar (GPR)
Type	Passive	Active	Active
Common depth	<1.5 m	0.25–2 m	0.5–9 m, depending on antenna and soil properties
Survey time (for 20 m grid of 20 lines)	20 to 30 minutes	45 minutes	60 minutes
Situations to avoid	Metallic debris, igneous areas	Surface very dry; saturated earth; shallow bedrock	Highly conductive clays, salts, rocky glacial deposits (e.g. moraines)
Tree and vegetation effects	Trees and high grass impede survey, invisible in data	Trees and high grass impede survey, positive anomaly	Trees and high grass impede survey, roots produce anomalies
Advantages	Speed, hearths, burned areas detectable	Good feature definition, specific depth settings	Vertical profiles, stratigraphy, results in real time
Disadvantages	Restricted depth, need open parkland for speed, iron clutter detrimental, constant pace of movement, high cost, must process data for results	Probe contact slow, must deal with cables, must process data for results	Equipment bulky, difficult data processing, interpretations difficult, constant speed of movement, high cost
Daily data volume	High	Low	High
Data processing complexity	Moderate	Low	High

Geophysical techniques are usually classified as either "passive" or "active" (Table 15.4). In the former case, an ambient property (e.g. magnetic field) is measured (eg. magnetometry), whereas the latter measures a response to energy put into the ground (e.g. ground penetrating radar – GPR).

15.3.1 Magnetometry

Magnetometry techniques measure the magnetic field (across the ground surface), part of which is contributed to by buried archaeological remains and features (e.g. kilns, pits, and ditches). Thus, traversing a surface with a magnetometer can pick up irregularities in the earth's magnetic field (anomalies) produced by these objects. This is particularly so for those anomalies produced by heating, which enhances the *magnetic susceptibility* (a measure of the magnetism of an object when it is placed in a magnetic field) of the objects; magnetometers are not effective in detecting features such as walls, unless they have been fired (Gaffney and Gater, 2003). Because heating increases the magnetic susceptibility of an object or sediments, ditches filled with anthropogenic material (e.g. ashes, wastes from domestic and industrial use) can be detected.

Although several types of magnetometers exist, the most commonly employed is the fluxgate magnetometer (Gaffney and Gater, 2003). With this handheld instrument (Kvamme, 2001), two sensors are mounted vertically in the same instrument, separated by ~50 cm. The upper sensor measures the earth's magnetic field while the lower one does the same, but

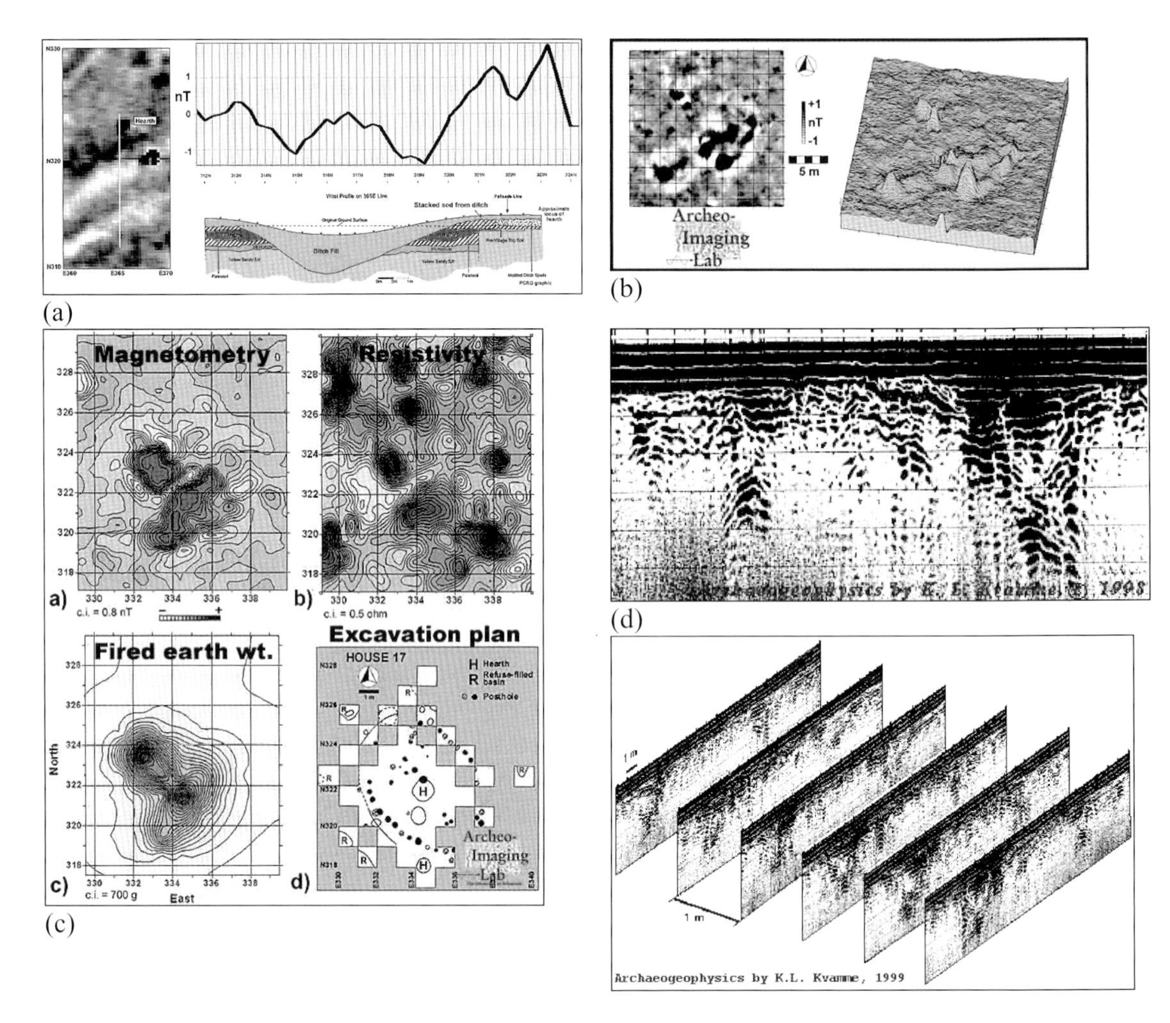

FIGURE 15.8 Different graphical techniques to display geophysical data. These illustrations are from research carried out by Dr. Ken Kvamme, University of Arkansas. Parts a–c are from the Menokan Village site, North Dakota, a late Woodland fortified earth lodge village; d is from the Revolutionary War site of Breed's Hill (Bunker Hill), Boston, USA. Shown here in (a) is a stratigraphic profile with overlying magnetometry scan across the area (values are in nanoTeslas): "an increased magnetic response is seen over the stacked sod, while a magnetic void is indicated in the area of the ditch where topsoil was removed." (b) Grayscale and wireframe presentation of magnetic data showing: "An oval shaped burned house, may reveal an interior entryway ramp and a central hearth." (c) Various types of contoured data (magnetometry, resistivity, weight of fired earth) and excavation plan, are shown from a nearby house showing a clear correspondence between the various types of data. (From Kvamme: http://www.cast.uark.edu/%7Ekkvamme/geop/ menoken.htm). (d) The upper part shows a single ground penetrating radar (GPR) profile across part of the site, whereas below it are a number of sequential profiles: "The 20 meter long profiles show possible fortification ditches from the Revolutionary War battle, or subsequent modifications made by the British. Two features about a meter deep are indicated that continue in adjacent transects over a span of at least 26 m. The interior of the fortification would have been to the right in the first figure, which illustrates 6 of the transects" (with permission from Kvamme: http://www.cast.uark.edu/%7Ekkvamme/geop/bunker_ h.htm).

being closer to the ground surface, and potentially buried objects will receive a slightly higher and measurable signal. Large areas (commonly over 100 m) on a side can be readily surveyed by pacing over a gridded area (usually 1 × 1 m intervals), and by using modern instruments that record thousands of points at a time before they must be downloaded. Once recorded, the data points can be displayed and processed (if need be) in a number of ways. Displays include XY plots of values (Fig. 15.8a), wireframe traces along an axis (XY), grayscale renderings (Fig. 15.8b), and contoured data plots (Fig. 15.8c), dot density plots. Interpretation of magnetic and other geophysical data depends on experience, as well as a knowledge of theory, a facet of investigation that is true for most disciplines (Gaffney and Gater, 2003).

Magnetometry surveys are numerous and widely published. For example, a geophysical survey carried out *after* topsoil stripping at the early medieval (Anglian) settlement at West Heslerton, North Yorkshire, showed a much higher resolution "map" of features compared to the prestripping survey (Tipper, forthcoming).

Electrical Resistance is one of several active geophysical techniques in which an electrical current is placed in the ground and the resistance is measured. Differences in resistance above or below the background are called anomalies, which can be produced by objects or features that provide greater or lesser resistance to flow of the current. High and low resistance anomalies can be produced by various objects and features (Table 15.5; modified from Gaffney and Gater, 2003):

Table 15.5 High and low electrical resistance anomalies (modified from Gaffney and Gater, 2003)

High resistance anomalies	Low resistance anomalies
Stone walls	Ditches, pits, gullies
Rubble and hardcore	Drains
Constructed surfaces (e.g. plasters)	Graves
Roads	Metal pipes and installations
Stone coffins, lined cisterns	

The general approach is that soil can conduct a current, which is enhanced by moisture content and salts within it. As a consequence of the former, resistance surveys are best conducted in wetter seasons to enhance contrast among objects and features. Thus, if the soil is too dry even large stone features may not show up (Gaffney and Gater, 2003).

Resistance measurements take into account the nature of the soil/sediment and objects, as well as how much of it is present. A more realistic measure is to determine the resistivity, which takes into account just the intrinsic nature of the materials themselves. It is expressed as Ohm-meters, as opposed to resistance, which is given in Ohms (Gaffney and Gater, 2003).

Resistivity measurements are made by inserting electrodes (usually four) into the ground, through which an electric current is passed, and the drop in voltage is measured. Different spacings (arrays) of the electrodes can be used (e.g. Wenner, Schlumberger, Twin-Probe), but regardless, the wider the distance between the electrodes, the greater is the ability to probe to greater depths. A spacing of 0.5 m with a Twin-Probe array can detect features up to a depth of about 0.75 m (Gaffney and Gater, 2003).

As with magnetometry, a resistivity survey is conducted by gridding off the area, hopefully well mapped and positioned on the ground, and making systematic measurements over a given linear distance, such as 1 m or 0.5 m. Again, the area and interval are dictated by time and monetary budgets, as well as the objectives of the research. An example of a resistance survey is illustrated in Figure 15.8.

Ground Penetrating Radar (GPR) has been increasingly used over the past two decades. Essentially, radio waves are transmitted downward into the earth and these can be reflected back if they strike a material with contrasting *dielectric constant* (property of an electrical insulating material (a dielectric) equal to the ratio of the *capacitance* of a capacitor filled with the given material to the capacitance of an identical capacitors in a vacuum without the dielectric material. "dielectric constant" *Encyclopaedia Britannica,* 2004, Encyclopaedia Britannica Premium Service. September 28, 2004 <htpp://www.britannica.com/eb/article?tocId=9030383>), such as a different type of sediment, features, floors, or other prepared surfaces. The depth of the feature is measured by the time it takes for the signal to return to the receiving antenna.

Different frequencies of antennas can be used, and the higher the frequency, the smaller the instrument, and poorer the depth of penetration due to dissipation, but higher is the resolution and the ability to recognize small-scale features. Lower frequency antennas, unfortunately, tend to be quite large and difficult to move around (Gaffney and Gater, 2003). Different considerations govern which type of antenna should be chosen, including electrical and magnetic properties of the materials at the site; desired depth of penetration; size of morphology of the features; access to the site; and possible external interference from radar (Conyers, 1995; Conyers and Goodman, 1997).

Initially, a typical GPR survey consists of dragging the antenna across the gridded area, while marking the distance traveled, say every meter. More typically nowadays, especially for large antennas, they are carried along on wheeled support with distances being measured automatically. Results are recorded digitally and either plotted directly as a radargram or be treated with software to remove artifacts. Nevertheless, "It is fair to say that the interpretation of GPR data is even more dependent upon the skill and experience of the operator than other geophysical techniques" (Gaffney and Gater, 2003: 113), and there are several factors that can affect the results and their interpretations. These include moisture conditions and the lack of regularity of archaeological features, which hinder the ability to set up reference patterns (Gaffney and Gater, 2003).

GPR has been very effectively used to reveal the extent of sites buried by masses of volcanic ash, such as at Ceren, in San Salvador (Conyers, 1995; Conyers and Goodman, 1997). In Boston, GPR was employed to aid the National Park Service in reconstructing the battleground of Bunker Hill (Breed's Hill) (http://www.cast.uark.edu/%7Ekkvamme/geop/bunker_h.htm). Here the question was to attempt to define the battleground between the American and British forces. Several GPR transects across selective parts of the hill (Fig. 15.8d), a ditch, and rampart structure, which appeared to have been used as a fortification for firing down at the advancing enemy forces. Archaeological excavations indeed exposed such a structure but it was incapable of revealing who was firing upon whom.

15.4 Coring and trenching techniques

The methods described above, whether passive or active geophysical techniques, or remote sensing from the air, can only provide information about the general nature of features, objects, and deposits at or just beneath the ground surface. Such information enables us to make inferences about depth of burial of a site, its lateral extent, size, and distribution of objects, and features. They also provide some information about the stratigraphic setting and context, which can enable physical correlations to be made (see Chapter 2). For the most part, however, these techniques are somewhat limited in their ability to provide details about the deposits themselves, including texture, composition, and fabric. Many probing techniques exist that permit direct observation of the geological context of sediments and soils without or before the need to carry out full-scale excavation. These techniques include augering, coring, and trenching by hand or

(a)

(b)

FIGURE 15.9 Different types of augering and coring equipment. (a) a large, truck-mounted auger during fieldwork at Zilker Park, Austin, Texas; the auger has a diameter of about 1 m. Material is scraped off the blade, described, sampled, and processed by sieving for artifacts or ecofacts; (b) a post-hole digger is illustrated in the foreground. This is useful in collecting samples relatively near the surface (~upper 25–30 cm; (c) a hand bucket soil auger used for examining surface material or for starting deeper boreholes; (d) a split-spoon core is laid open showing sandy material at the top overlying a darker,

Figure 15.9 *(Cont.)*
organic-rich buried horizon associated with the archaeological material dating to the late prehistoric period; (e) a mechanical corer (GeoProbe®) taking a sample next to the Tennessee River in Knoxville (photo courtesy of S.C. Sherwood).

machine. All these approaches can be used to determine vertical and lateral extent of archaeological and natural deposits and their total thickness. Such data can be integrated on a regional level to make palaeogeographic reconstructions of landscapes, such as positions of former floodplains or lake shores (see Stein, 1986, for details and examples).

Augering techniques involve inserting a screw-like or gouge-like auger into the ground and bringing material to the surface (Rowell, 1994). The material is loose and accumulates in the auger head, so original stratigraphic features such as bedding are not preserved; in the case of a screw auger, soil and sediment adhere to the blades of the auger. Parameters such as color, composition, and texture can be recorded in rough stratigraphic order (see Section 15.5) by removing the soil retained on the auger blade (Fig. 15.9a). Each soil sample can either be laid on the ground, in the order it was removed from the ground, or put into a labeled sample bag. Moreover, since it is possible to measure the depth that the auger has reached by inserting a ruler to the bottom of the auger hole, one can reasonably accurately estimate the depth beneath the surface of a given material. It is evident that these loose materials, however, are unsuitable for thin sectioning techniques that employ intact blocks (see below).

Several types of augers are commonly used. The most common, and cheapest, is a

(e)

FIGURE 15.9 *(Cont.)*

posthole digger or hand auger – usually a bucket auger (or open 'Jarrett'/posthole auger) (Fig. 15.9b,c), – which can be purchased as part of a soil sampling kit. In this case, the auger head is attached to a vertical pipe-like stem with horizontal handle. The vertical part is commonly composed of threaded steel pipes that permit extending the augering to depths of several meters. Hand augers are usually ~35–100 mm (1.5 to 4 in.) in diameter. (A video of the procedure can be found at http://www.ams-samplers.com/main.shtm?PageName=Regular_Auger_ 5-8in_NC_ Connection3.shtm.) Screw-type augers are handy, but can be difficult to remove from the ground (Stein, 1986), and it is important to remember that in heavy clay soils it is preferable to try to take a smaller – yet retrievable – sample than is done in more loose sandy soils, where larger samples are more easily taken (Appendix 15A.1).

Gouge-type augers can be as narrow as 10 mm – for soil work – or 50 mm for deeper coring. The benefit of the gouge auger is that it is inserted into the ground, and after being rotated will remove an intact profile, if the ground is not too loose, wet, or stony. This profile can then be recorded (photographed or drawn), and the material can be usually safely used to subsample for bulk chemical analysis. This is a common method used when carrying out a phosphate survey, for example, especially when the subsoil horizon is the focus of archaeological soil phosphate (Linderholm, 2003). On the other hand, because there is a risk of contamination, gouge augering is very seldom used in palynological studies (see below).

A more robust type of augering involves a larger mechanically driven corkscrew-type blade (up to 900 mm/36 in.) mounted on the back of a truck (Fig. 15.9a). With this technique much larger volumes of material can be brought up to the surface, and greater depths can be achieved. Furthermore, material can be sieved for artifacts and ecofacts, permitting the documentation of potential sites at given depths. This larger auger is more costly to operate, and is usually rented by the day. Again, its use is a balance between budget, as well as surface area and depth to be evaluated.

Coring is a more prevalent technique for obtaining intact subsurface information (Stein, 1986; Schuldenrein, 1991), and like augering, hand and machine coring devices are common. Either technique involves pushing a hollow pipe beneath the surface to a measured depth, and then removing the corer and extracting the material from the pipe. Unlike augering, the stratigraphic integrity of the deposits is conserved, although sediment compression and slumping are common problems. Similarly, coring can be ineffective or be stopped when stiff clays and gravelly deposits are encountered.

Usually, the pipe is lined with a plastic tube enabling it to be capped at both ends and stored for later analysis (e.g. granulometry, mineralogy, pollen, diatoms, and soil micromorphology). Depending on the type of device, core lengths can be up to 1 m; the longer the core, however, the more difficult it is to bring it up from the subsurface. To achieve greater depths, a weighted slide hammer is used to pound the core into the ground.

When sampling, peat bogs piston corers or a "Russian" corer can be used. The latter has a half-circle section and a thin blade to hold it in position in soft sediment such as peat. The meter or half-meter long closed chamber is inserted, then the shaft rotated 180°. This ingenious tool then opens and cuts a half section out of the soft sediment at the same time as the cutting edge recloses the chamber. Hence when the corer is withdrawn there is minimal contamination of the sample from nearer surface deposits. This procedure is very helpful when striving to avoid contaminating pieces of peat that may contain thousands to millions of pollen grains. The half section of retrieved soft sediment core is then commonly slid into a labeled length of semi-circular drainpipe. The soft sediments of lakes are sometimes very sloppy and difficult to core. In the boreal region, these are best cored when frozen in winter (Roger Engelmark, Pers. comm.).

Several other types of corers are available, depending on type of sediment/soil, intended depth of sampling, and budget. A hand corer is usually 25–50 mm (1–2 in.) in diameter, and the coring head can fit on to the same pipe stems used in hand augering. The coring head can be of several types depending on type of sediment. A common type is the split-core sampler, in which the pipe of the coring head is split in half and held together by a ring at the tip of the sampler. After removal from the ground, the ring is unscrewed from the core barrel and the plastic liner can be lifted out (Fig. 15.9d). The Livingston piston corer is used extensively in coring aquatic environments and peat for the recovery of pollen.

More ambitious – and expensive – coring strategies involve mechanical coring rather than manual coring. The smallest is probably a one/two person held machine driven by a two-stroke motor that "vibrates/rams" the corer into the ground. As with augering, the coring device can also be part of a self-propelled vehicle (GeoProbe®(http://geoprobe.com/products/machines/machinedream.htm), or mounted on the back of a dedicated truck or trailer (Giddings corer) or larger drilling rig (Fig. 15.9e). In these cases, hydraulic pressure is used to either push or drill the core into the ground, reaching depths of tens of meters, and thus alleviating the difficulty of removing manual cores from several meters' depth. The obligatory use of plastic linings in some corers is somewhat limiting: it requires buying many tubes, and transporting them in and out of the field. Mechanical coring enables a large area or transect to be sampled during the course of a day. In addition, it may justify the expense of renting a corer, which as of 2004 can amount to thousands of dollars per day.

The first step after coring is to "log" the cores, that is, identify the different strata, measure their thickness and describe them (see Section 15.5). Cores that are between 50 and 100 mm wide are also suitable for selective or continuous sampling for thin sections and soil/sediment micromorphology. A few extra logging techniques can be carried out after cleaning, but before the core is subsampled:

- prepare a photographic record of the core. This can be done rapidly and cheaply with digital camera;
- run the core through a magnetic susceptibility meter coil, recording χ (Chapter 16) at regular intervals or at the location of visible stratigraphic units;
- employ X-radiography to record details of the fine-scale stratigraphy resulting from deposits rich in "opaques," such as alluvium (Barham, 1995).

Measuring of χ and making X-radiographs can only be carried out on cores within plastic containers. Sometimes metal monoliths are used to collect samples, but these can have plastic inserts that are then employed for these techniques (see Section 15.6 on Sampling).

15.4.2 Trenching

Trenching follows along the lines of augering and coring in revealing the detailed aspects of subsurface soils and sediments. Trenching can be part of test excavations before full excavations are started, or part of an exploratory program to evaluate site limits, the extent of certain types of deposits for palaeogeographic reconstructions, or simple reconnaissance work.

The simplest trenching involves digging simple 0.50 × 0.50 × 0.50 m test pits. The material that is removed can be systematically sieved for recovered artifacts, and the exposed profiles can be studied, although their small size might limit the degree of detailed observations. Larger trenches of different sizes can also be undertaken, such as the 1 × 1 m "telephone booth" or 2.0 0 × 0.50 m "slit trench." For safety concerns, these types of trenches can only reach depths of about 1.5 m without needing shoring-up. Such small trench excavations are sometimes termed *sondages*.

In mitigation archaeology (CRM in the United States), where time is critical, mechanical trenching is the method of choice, and many trenches can be dug in a day or two, thus justifying the costs in renting the heavy machinery. A backhoe is usually employed for most trenching, although bulldozers and grade-alls are also used. In the United States, often the machine can be borrowed from the agency that is contracting the work, such as a State Department of Transportation, which has access to this equipment. In any case, mechanical trenches can range in length from a few to tens of meters in length, and can reach depths of up to 3–4 m. However, with greater depths, the trench must be stepped out for safety reasons to avoid collapse, thus adding expense and time to the trenching effort. In restricted areas, such as on urban sites or in sensitive areas, where wide excavations are not permitted, such as the famous hillfort of Maiden Castle, Dorset, United Kingdom, deep trenches are stabilized by steel shuttering. At the 7 m deep Early Iron Age ditch at Maiden Castle, the ditch fill deposits could be accessed only through gaps between the shuttering. During site evaluations, perhaps along a proposed road corridor, archaeology can be evaluated through trial trenching, with the protocol being 2–5% trenching of the landscape producing a 2 m wide trench, which is then hand cleaned. Site investigation by trenching and excavation of these trenches, is carried out in a more intensive way. For example, sampling of the landscape using 30 m long, 2 m wide trenches at 30 m horizontal intervals provides an approximate 8% coverage (e.g. A41 corridor between the M25 and Tring, Hertfordshire, England; McDonald, 1995).

Although it is the most aggressive technique and leads to the greatest disturbance of the landscape and site, mechanical trenching provides clear and relatively large windows into the subsurface. Thus, it differs from coring, which is spatially punctual and can provide microfacies correlations over large areas that can be difficult, if not incorrect. With trenching, it is possible to observe lateral and vertical changes of deposits, soils, and archaeological features directly and place them in their proper stratigraphic context, as is done in site excavation. It is perhaps the best way to obtain detailed stratigraphic information from several places over large areas.

15.5 Describing sections: soils and sediments in the field

The next issue is to record and describe the stratigraphic information on soils and sediments or features that were revealed by augering, coring, and trenching, as well as from sections exposed by full-scale excavations. A large degree of variability exists in the way that stratigraphic information is recorded (Courty *et al.*, 1989), and this arises from fundamental

FIGURE 15.10 A backhoe digging a trench at the western end of the mound of Titris Höyük, Turkey. Excavation director Dr. Guillermo Algaze on the left oversees the operation that attempted to establish the stratigraphic relationship of the anthropogenic mound fill on the left with the flat alluvial floodplain sediments at the right and background. Part of the mound is overlapped by the alluvium.

differences in ways that profiles are viewed. Pedologists, for example, may view a section in terms of its pedological horizons, and their pedogenic nature. Geologists, on the other hand, are trained to observe lithological differences recording lithostratigraphic units (e.g. color, texture, composition, and sedimentary structures – see Chapters 1–3). Emphasis is subliminally placed by geologists on depositional criteria, although postdepositional effects (e.g. mottling, rooting, and diagenesis/pedogenesis) are included. They tend to be less concerned with soil genesis, although many of the same descriptive criteria are employed. Geologists may also tend to look at lateral facies or microfacies, which are critical in understanding site histories (Courty, 2001). Obviously, proper observation and description should include all aspects of the soil/sediment. In any case, when dealing with non-open-air sites (e.g. caves, buildings, features) a pedostratigraphic approach may be less practical and a simple lithostratigraphic description may be more appropriate and pragmatic.

Soil scientists, on the other hand, tend to view profiles in soil stratigraphic units that center around soil horizons, whether they are actually forming or are relict features that have been buried, but still preserved. They see the section in terms of smaller scale pedological processes, which for example, strongly affect "surfaces"/topsoils by subaerial weathering or subsoils by down-profile translocation of particles and solutes, or by groundwater movement (see Chapter 3).

Plate 15.1 shows a profile from the southern end of a profile at the Hopeton Earthworks site, Ohio (see Chapter 10; Occupation Deposits). This is a Mississippian period site (ca AD 200) in which soil material was mounded above the floodplain. The figure shows the soil stratigraphic descriptions done by Rolfe Mandel, Kansas Geological Survey (Mandel *et al.*, 2003). In this strategy, which is representative of soil stratigraphers, the stratigraphic breakdown, shown in the photo and accompanying table, tends to focus on soil horizons. Note that in the

Table 15.6 Soil stratigraphic description by Rolfe Mandel of profile from Hopeton Earthworks site, Ohio, USA. The profile illustrated in Plate 15.1

Depth (cm)	Stratigraphic unit	Soil horizon	Description
0–10	Wall fill B2	Ap	Light yellowish brown (10YR 6/4) silt loam, yellowish brown to dark yellowish brown (10YR 5/4 to 4/4) moist; weak medium and fine platy structure parting to moderate fine and medium granular; soft, very friable; many fine and very fine roots; abrupt wavy boundary
10–30	Wall fill B1	Ep	Very pale brown (10YR 7/4) to light yellowish brown (10YR 6/4) silt loam, yellowish brown (10YR 5/4) moist; weak medium platy structure parting to weak fine and very fine subangular blocky; soft, very friable; common siliceous granules; common siliceous pebbles 30–40 mm in diameter; common fine and very fine roots; common fine, medium, and coarse pores; abrupt wavy boundary
30–55	Wall fill A2	2Ab	Brown to dark yellowish brown (10YR 4/3 to 4/4) silt loam, dark brown to dark yellowish brown (10YR 3/3 to 3/4) moist; weak fine subangular blocky parting to weak fine and very fine granular; slightly hard, friable; many fine, medium, and coarse pores; common worm casts and open worm burrows; common siliceous granules; common siliceous pebbles 30–40 mm in diameter; common fine and very fine and few medium roots; common fine, medium, and coarse pores; gradual smooth boundary
55–83	Wall fill A2	2Bw1b	Brown (7.5YR 5/3) loam, brown (7.5YR 4/3) moist; weak fine subangular blocky structure; slightly hard, friable; common pale brown to very pale brown (10YR 6/3 to 7/3) silt coats on ped faces and in pores; common worm casts and open worm burrows; common siliceous granules; common siliceous pebbles 30–40 mm in diameter; few fine and very fine roots; common fine, medium, and coarse pores; clear smooth boundary
83–103	Wall fill A1	2Bw2b	Brown to dark yellowish brown (10YR 4/3 to 4/4) loam, dark yellowish brown (10YR 3/3 to 3/4) moist; weak fine subangular blocky structure; slightly hard, friable; common pale brown to very pale brown (10YR 6/3 to 7/3) silt coats on ped faces and in pores; common worm casts and open worm burrows; many siliceous granules; many siliceous pebbles 2–70 mm in diameter; few fine and very fine roots; common fine, medium, and coarse pores; abrupt wavy boundary
103–120	Subwall soil	3Btb	Yellowish brown (10YR 5/4) silty clay, dark yellowish brown (10YR 4/4) moist; few fine prominent red (2.5YR 4/6) mottles; weak medium and fine prismatic structure parting to moderate fine and very fine subangular blocky; very hard, extremely firm; common distinct brown (10YR 5/3) clay films on ped faces; few round pebbles 30–50 mm in diameter in lower 5 cm of horizon

table, the following criteria are systematically entered:

- color, with Munsell designation (the standard is to record the moist color, but dry colors may also be recorded if useful contrasts are noted);
- texture, using soil nomenclatural terms (a geologist might use "clayey silt" for "silt loam" or similar term where appropriate);
- moisture content (e.g. moist);
- structure (e.g. platy; subangular blocky);
- cohesion, compactness, firmness (e.g. soft, friable);

TABLE 15.7 Sediment and profile description of profile 2 from Hohle Fels Cave (Schelklingen, Southern Germany) (cf. Fig. 15.11). Note that the "Geological Units" are described separately and independently from "Archaeological Layers" (modified from Goldberg *et al.*, 2003)

Geological units	Munsell color	Field observations	Archaeological layers	Archaeological features and comments	Uncalibrated ^{14}C age (AMS in italics)
Modern stratigraphic complex					
a	2.5Y 7/4	Cave pavement (Höhlenfestschichten) consisting of alternating banded layers up to 2 cm in diameter; coarse yellow, rectangular gravel, and blackish fine sediment-bearing pebbles	0	—	Post-1958
b	5Y 2.5/1	Slag-bearing layer containing pebble chips and few 3–4 cm size pebbles mixed with some finer clayey sediment	0	—	1944
C	7.5YR 3/4	Thin gravel "leveling layer" with rounded pebbles up to 2 cm across	0	—	1944
D1	2.5Y 5/6	Sandy deposit containing limestone gravel with boulders up to 45 cm in diameter. Gravel displays preferred subhorizontal orientation. Yellow-ochre	– –	— —	
Stratigraphic A-complex (Holocene; below Unit 1 k, pure and typical Magdalenian)					
0c	5Y 2.5/1	Slightly rounded to well rounded 2–5 cm size limestone fragments lacking preferred orientation and occurring with in a dark gray/black argillaceous site with fine-gained calcareous sand	0/1	Mixed horizon: containing metal age and Neolithic ceramics and Magdalenian artifacts; disturbed Holocene surface caused by clearing in 1944	
Magdalenian					
1k	2.5Y 2/0	Clayey silt with 1–2 mm size, irregularly to slightly rounded limestone gravel, slightly elongated and without recognizable orientation	I	Magdalenian lithic and 0rganic artifacts	*13,240±110* 13,085±95 12,770±220
Pit filling of Qu 76/77/86/87					
1gb	10YR 5/2	Medium gray, very dense clay, with homogeneous distribution of rather big-charcoal fragments. Inclusions of medium sized, angular to slightly rounded limestone	0	Contains black fired ceramics	

Stratigraphic B-complex (moonmilk-sediments and limestone gravel)					
1s	10YR 8/2	White, fine calcitic sand with slight amount of interstitial clayely silt; subangular limestone inclusions, 2–5 cm, mostly without orientation		Essentially sterile	
3as	2.5Y 5/6	Very heterogeneous layer. In the northern part, a marked unconformity separates a matrix-rich area from a matrix-poor one Matrix-rich area: Matrix is pale yellow/brown, pale gray/brown, partly darker brown silt with little clay and fine sand. High abundance of 1–8 cm, angular limestone gravel inclusions displaying irregular orientation, occurring along with very well rounded, 1 cm diameter clasts Matrix-poor area: Matrix is virtually lacking, and clasts consist mostly of 2–8 cm, and some 10–13 cm of limestone fragments/gravel; these are angular to rounded, and irregular to slightly elongated shape, without orientation			
Stratigraphic C-complex (mainly Gravettian)					
3ad	10YR3/2–4/4	Heterogeneous layer with pockets of variable brightness (generated by washing out of finer components) and containing different types of sediments; the darker sediment possibly exhibit an anthropogenic component. The lighter matrix consists of dark gray-brown silt with small amounts of clay and calcareous sand. The dark matrix is blackish silt with small amounts of calcareous sand and a somewhat higher clay component compared to areas of the bright matrix. Inclusions (2–7 cm) are relatively abundant, mostly angular, but some partly rounded; they dip northward	I/IIb	Mixed Magdalenian/Gravettian-horizon (reworked and displaced Gravettian)	
3b-complex	7.5YR4/6	Reddish silt with little clay and fine calcareous sand. Inclusions of angular to slightly rounded limestone gravel occur in variable amounts without orientation	II b	Gravettian lithic and organic artifacts	*27,150±600*
3bt	2.5Y2/3	Dark gray, partly black silt, with an abundance of sand-sized burnt bone (*Knochenkohle*), and a small amount of clay. Inclusions consist of some stones, bones, and bone ash (>1 cm)	II b	Gravettian lithic and organic artifacts	

Table 15.7

Geological units	Munsell color	Field observations	Archaeological layers	Archaeological features and Comments	Uncalibrated ^{14}C age (AMS in italics)
3c	10YR5/6	Very moist clayey silt with many limestone clasts of variable size. They range from angular to rounded, and dip toward the northeast	II c	Gravettian lithic and organic artifacts	*29,550±650*
3cf	5Y 2.5/1	Black clayey silt with a high proportion of bone ash that underlies a limestone-rich gravel layer	II c	Gravettian lithic and organic artifacts	*28,920±400*
Stratigraphic D-complex (mainly Aurignacian)					
3d	10YR 4/4	Red-brown argillaceous silt with ca 60% of 4–8 cm, slightly rounded limestone gravel inclusions. Matrix only slightly compacted, contains abundant small pores and cavities	II d	Aurignacian?	29,560+240/—230 *30,010±220*
5	7.5YR 4/4	Generally, loose mixture of angular limestone gravel and moist clayey silt with a minor component of coarse calcareous sand	II e	Aurignacian lithic and organic artifacts	
6	7.5YR 5/6	Rounded limestone gravel with little interstitial matrix. Mostly coarse-grained calcareous sand, downward-facing edges, or fracture planes of limestone fragments frequently bear gray clay crusts	III	Aurignacian lithic and organic artifacts	*30,550±550* *31,100±600*

Notes on descriptions:

1. These are provisional descriptions based on unpublished work by J. Hahn (12.09.1994); Waiblinger (1997); N.J. Conard and T. Prindiville (02.09.1997); and P. Russell and N.J. Conard (09.98). The geological units (GH in original publication) and archaeological layers (AH in original publication) serve as preliminary field descriptions and are subject to modification during future excavations.
2. Colors were determined using the Munsell Soil Color Chart on moist samples, both in the cave under artificial light conditions and outside the cave in natural light.
3. Differentiation of individual sediment units on the basis of their calcium carbonate contents was not possible due to the large number of limestone clasts.
4. ^{14}C ages are uncalibrated and compiled from Conard and Bolus (2003); Conard and Floss (2000); Hofreiter *et al.* (2002); and Housley *et al.* (1997).

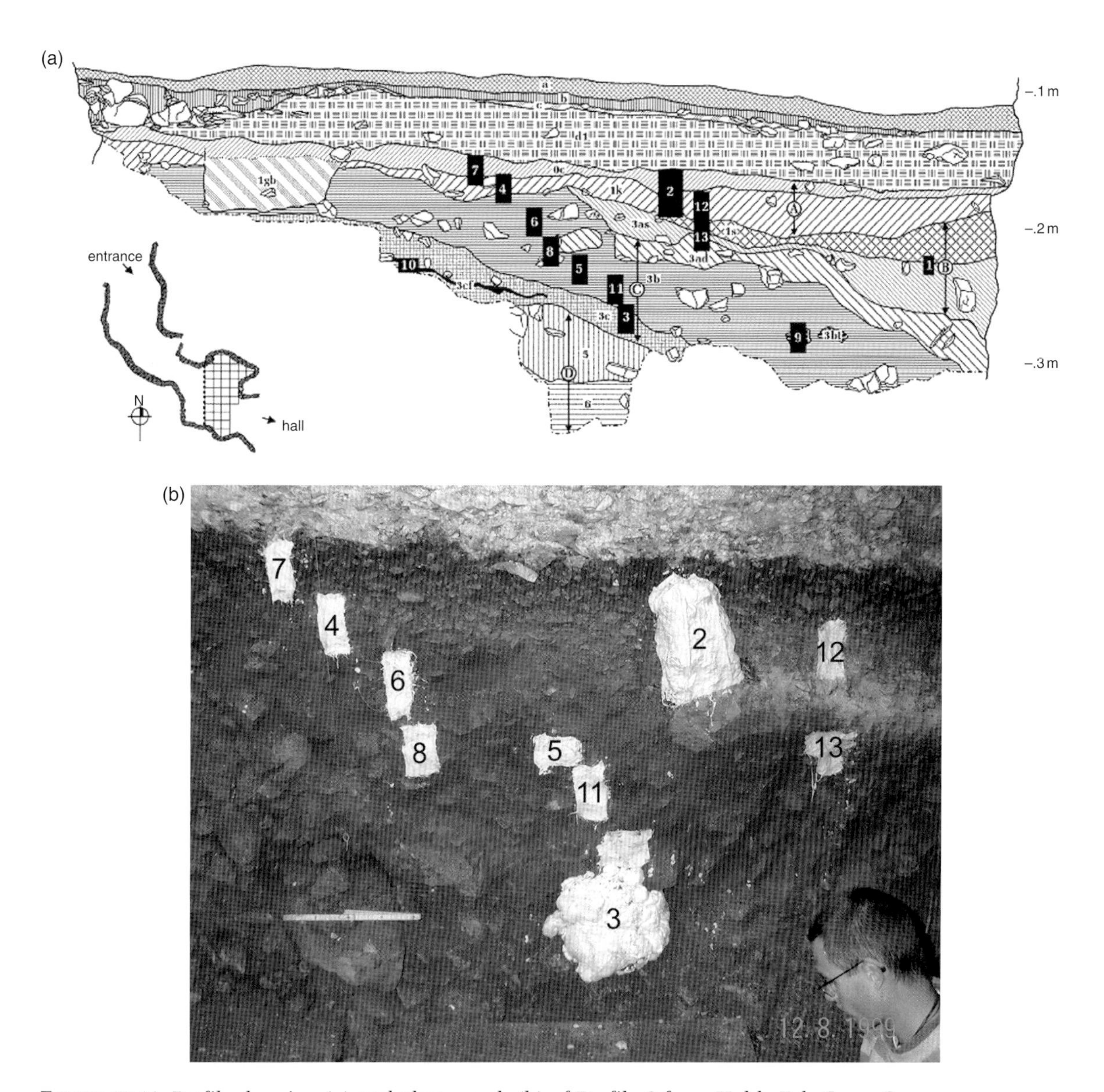

FIGURE 15.11 Profile drawing (a) and photograph (b) of Profile 2 from Hohle Fels Cave, Germany, showing deposits and locations of micromorphological samples. Due to the stony and crumbly nature of the deposits, it was necessary to jacket them in plaster in order to remove intact samples for micromorphological sampling (see Table 15.6 for systematic descriptions of deposit). Shown are Gravettian and Magdalenian silts and clays rich in éboulis (the lighter layer at the top dates from the 1940s). The white squares and blocks are pieces of gauze soaked in Plaster of Paris. In samples 4, 6, and 8 for example, a swath of gauze is placed on the section to stabilize it. After it dries, the sediment is partly excavated around it, leaving a boss of sediment which is then covered with more gauze as shown in samples 2 and 3. This results in an area of sediment that is stabilized on the front and sides. It is then removed from the back and the exposed part is sealed with a final layer of plaster/gauze; for other sampling methods see Goldberg and Macphail (2003).

- composition with sizes and abundance (e.g. "common siliceous pebbles");
- roots;
- pores and porosity;
- lower boundary, with degree of transition and morphology (e.g. "abrupt wavy").

It should be pointed out that with this type of descriptive approach, there is a distinct genetic flavor, such the use of "3Btb" which indicates that this is a "buried illuvial horizon."

At the same time that the deposits are described, key profiles are photographed and drawn (Hester *et al.*, 1997; Hodgson, 1997). Common in CRM/survey work is to employ data recording sheets, which permit rapid sketches of key strata and features (Fig. 15.12). Sampling locations for pollen and sediments can be clearly shown as well. In either case, other than what to describe and register, how one does it is often a matter of personal taste, or a function of the CRM firm that employs the person doing the description.

15.6 Collecting samples

After exposures have been described, it is possible to proceed with sampling. Many different strategies for collecting samples exist. These depend on the overall goals of the project, the specific questions being asked, and on the budget(s).

Two types of strategies are generally used. The most typical strategy involves the collection of bulk samples for analysis by a variety of physical, chemical, and even biological techniques. Many of these samples are for classical pedo-geological analyses, including granulometry (grain size analysis), pH, carbonate content, heavy minerals, various forms of iron and phosphates (see Chapter 16; Courty *et al.*, 1989). Random and systematic sampling techniques can be employed. For example, it is very common to sample systematically an exposed section from bottom to top, collecting representative samples from the principal stratigraphic units. Samples are taken from the bottom upward, because all material disturbed by the sampling process, and a source for sample contamination, affects only the previously sampled section below. Sampling can also be done where deposits or features are scattered throughout the site area, as for instance taking spot samples from the major units. Common sense should be employed at all times. In any case, with these procedures, contacts between stratigraphic units are to be avoided to preclude mixing of deposits of different lithologies and depositional events. This is particularly important when collecting samples of microfossils, heavy minerals, or for chemical elemental analysis, where just a trace of "soil contamination" may produce misleading results. Thus the sampling trowel or other tool must be carefully cleaned for each sample. In some case, it is recommended that plastic trowel be used for sampling for magnetic properties.

The other sampling strategy involves collecting intact, undisturbed samples for micromorphological or microprobe analysis (for details see Courty *et al.*, 1989; Goldberg and Macphail, 2003). Such samples can be problematic to collect, as commonly archaeological sediments are loose, poorly sorted mixtures of coarse and fine material. Imagine trying to sample a shell midden in one piece.

Fortunately, years of experience – one of the benefits of getting older – have led to several successful approaches in the removal of undisturbed samples. Strategies of sampling vary mostly according to type of sediment, but considerations of leaving unsightly and unstable large holes in witness sections must also be made. The easiest sampling involves compact sediments or soils, such as loess. In this case, blocks of sediment can be cut out of the profile using a knife and trowel, and then wrapped with toilet paper/paper toweling and tape, labeling the sample number and way-up with an indelible marker. Similarly, metal or plastic

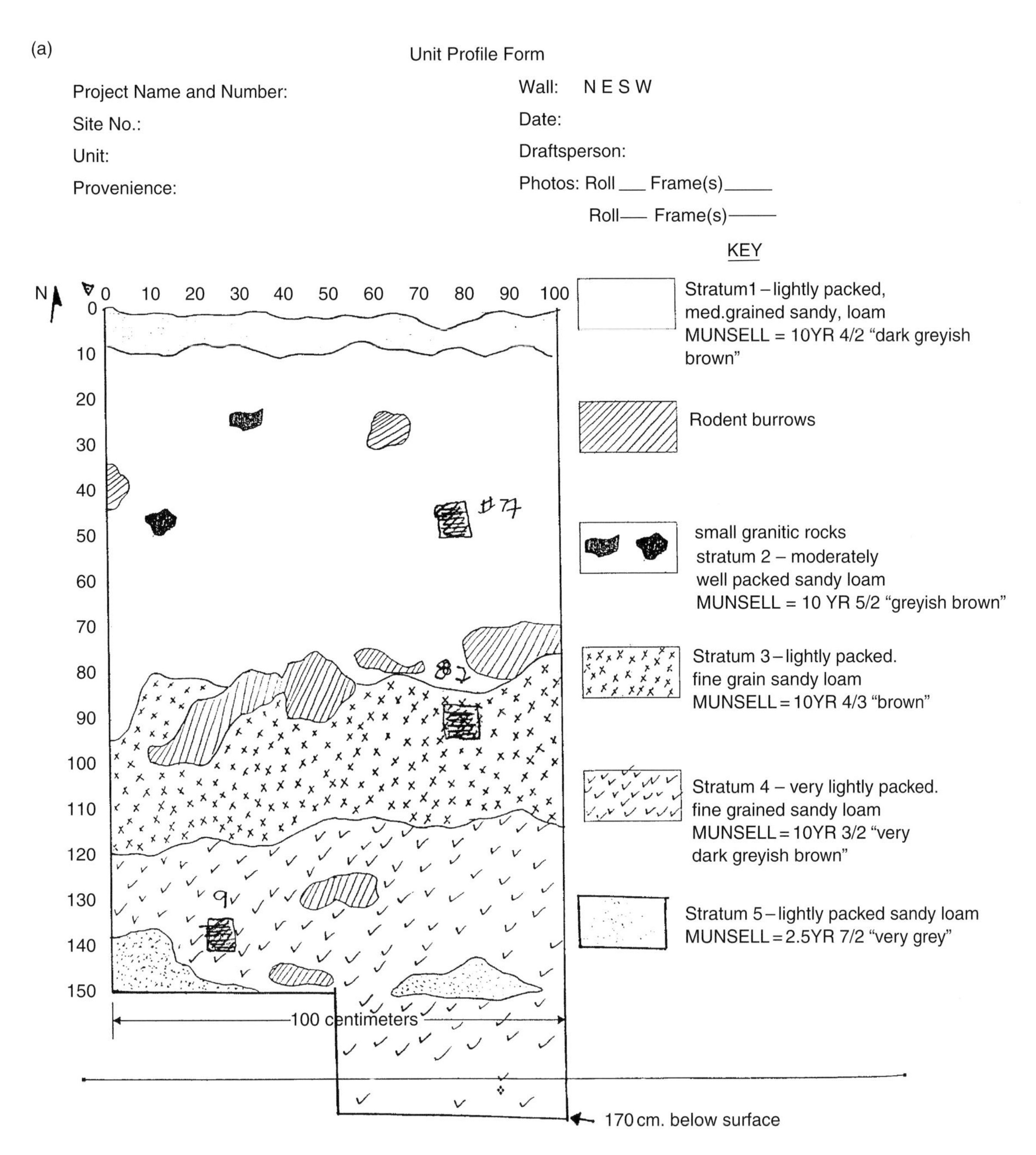

FIGURE 15.12 Two examples of field recording sheets. (a) Form is used by field staff in CRM work. Shown is a sketch of the stratigraphy with legend that describes the major stratigraphic units; (b) a field recording sheet provided by geoarchaeologist Charles Frederick, Texas. Although, overall, similar information is recorded, the geoarchaeological and stratigraphic detail are greater than that in (a), and more suitably match the type of information that interests geoarchaeologists.

(b)

page __ of __

Field Exposure Description Form

Profile Designation ______________________ Project: ____________
Latitude: ________ Longitude: ________ UTM zone ____ E ________ N ________ Datum ______
Topo Quad: ________ Landform: ________ Geomorphic Surface: ________ Elevation: ________
Described by: ________ Date: ______ Geologic Unit: ________ Slope: ________
Remarks: ____

Zone: ____ Depth: ______ Texture: ____________ Consistence: ________
Lower Boundary: ________ Structure: ____________ Reaction: ________
$CaCO_3$ Morphology: ________ Pores: ____________ Roots: ________
Coats: ____________ Color moist: ____________ dry: ________
Coarse Fragments(%): ______ Comments: ____________
Horizon: ________

Zone: ____ Depth: ______ Texture: ____________ Consistence: ________
Lower Boundary: ________ Structure: ____________ Reaction: ________
$CaCO_3$ Morphology: ________ Pores: ____________ Roots: ________
Coats: ____________ Color moist: ____________ dry: ________
Coarse Fragments(%): ______ Comments: ____________
Horizon: ________

Zone: ____ Depth: ______ Texture: ____________ Consistence: ________
Lower Boundary: ________ Structure: ____________ Reaction: ________
$CaCO_3$ Morphology: ________ Pores: ____________ Roots: ________
Coats: ____________ Color moist: ____________ dry: ________
Coarse Fragments(%): ______ Comments: ____________
Horizon: ________

Zone: ____ Depth: ______ Texture: ____________ Consistence: ________
Lower Boundary: ________ Structure: ____________ Reaction: ________
$CaCO_3$ Morphology: ________ Pores: ____________ Roots: ________
Coats: ____________ Color moist: ____________ dry: ________
Coarse Fragments(%): ______ Comments: ____________
Horizon: ________

Zone: ____ Depth: ______ Texture: ____________ Consistence: ________
Lower Boundary: ________ Structure: ____________ Reaction: ________
$CaCO_3$ Morphology: ________ Pores: ____________ Roots: ________
Coats: ____________ Color moist: ____________ dry: ________
Coarse Fragments(%): ______ Comments: ____________
Horizon: ________

Zone: ____ Depth: ______ Texture: ____________ Consistence: ________
Lower Boundary: ________ Structure: ____________ Reaction: ________
$CaCO_3$ Morphology: ________ Pores: ____________ Roots: ________
Coats: ____________ Color moist: ____________ dry: ________
Coarse Fragments(%): ______ Comments: ____________
Horizon: ________

Zone: ____ Depth: ______ Texture: ____________ Consistence: ________
Lower Boundary: ________ Structure: ____________ Reaction: ________
$CaCO_3$ Morphology: ________ Pores: ____________ Roots: ________
Coats: ____________ Color moist: ____________ dry: ________
Coarse Fragments(%): ______ Comments: ____________
Horizon: ________

Zone: ____ Depth: ______ Texture: ____________ Consistence: ________
Lower Boundary: ________ Structure: ____________ Reaction: ________
$CaCO_3$ Morphology: ________ Pores: ____________ Roots: ________
Coats: ____________ Color moist: ____________ dry: ________
Coarse Fragments(%): ______ Comments: ____________
Horizon: ________

Zone: ____ Depth: ______ Texture: ____________ Consistence: ________
Lower Boundary: ________ Structure: ____________ Reaction: ________
$CaCO_3$ Morphology: ________ Pores: ____________ Roots: ________
Coats: ____________ Color moist: ____________ dry: ________
Coarse Fragments(%): ______ Comments: ____________
Horizon: ________

FIGURE 15.12 (*Cont.*)

boxes (so-called Kubiena tins) can be inserted into the profile, removed, and covered to ensure that they remain intact. It may look neat to have a carefully lidded box, but it may be better just to have the lid on the outside face (where it can be photographed *in situ*), and plenty of soil/sediment on the inside face after digging out, as this can then be subsampled later for bulk samples, exactly relating to the box monolith thin section. The Kubiena-box sample, with one lid on, and protruding soil on the "back" can easily be wrapped with plastic film. Sometimes, an unexpected large stone or artifact may also protrude out of the "back" of the sample, and any attempt to remove this will destroy the sample. Again such an unwieldy sample can be wrapped, and then supported by additional wrapping in a couple of plastic bags. Obviously, such samples have to be transported and stored carefully. If fine-grained cohesive sediments are being sampled, such as alluvium, several overlapping samples can be taken.

More difficult to sample are loose or stony materials, such as sands, éboulis-rich cave sediments, and midden material (Fig. 15.13). For finer grained, nonstony materials, Kubiena tins can be readily used, as well as plastic drain spouting/down-pipe (Fig. 15.13). The latter was effective in sampling sandy deposits from Gibraltar and South African caves (Goldberg, 2000; Goldberg and Macphail, 2003; Macphail and Goldberg, 2000). One side of drain spouting can be removed to extract monolith, which can be as long as the height of the profile and the length of the drain spouting (usually sold in 3 m lengths); however, the greater the length, the greater the operational difficulty in removing them. If the monoliths are then going to be used for X-radiography, however, metal down spouting should be avoided. Alternatively, the square down spouting can be introduced vertically into the sediment (similar to coring), which can then be removed, sealed, and then ready for impregnation back at the laboratory. Again, one should remember to photograph/record the location of the core(s), marking it with sample number and way up.

The most demanding sediments to sample are coarse, poorly sorted, and loose deposits. Here, the best strategy is to encase the sample in a plaster jacket (Goldberg and Macphail, 2003) (Fig. 15.13). Cheese cloth or burlap (jute) is infused with plaster of Paris and applied to the exposed surface to stabilize it; a more expensive but more convenient alternative is to use pre-plastered gauze that is used to set broken bones (available at most pharmacies). Once the bandage is hardened, deposits can be excavated from around the sides, leaving a raised bulge above the profile surface, which can also be covered with plaster bandage. The sample, now surrounded on its three sides, is detached from the profile, and the remaining surface is wrapped, and when dry, can be transported.

Finally, some issues that have arisen during our geoarchaeological experience over the past decades should be considered.

1 Apart from of what technique is actually used, the *exact* context of the sample must be noted: its position in the field should be indicated on a photograph, as well as on a section drawing. If total station recording of artifacts and features is being done at the site, the ID number of the sample (or equivalent unique identifier) should be noted by the sampler and the archaeologist. For micromorphological samples it is useful to mark the position of stratigraphic boundaries or other features on the surface of the sampled block/Kubiena box/down spouting. Very commonly, samples are given a site code but when the specialist has a cooler filled with samples from many sites over the years awaiting the go-ahead for analysis, such site codes become meaningless. It is therefore useful to at least append a recognizable site name to samples. Similarly, during sampling, good communication must exist with the project director of the site in order to ensure that samples will address archaeological issues. Such a strategy also

FIGURE 15.13 Strategies for collecting micromorphological samples. (a) Consolidated or firm samples can be simply cut out of an exposed profile and wrapped with paper and packaging tape. A column of stratified burned material (bones, charcoal, ashes) from the Palaeolithic site of Pech de l'Azé IV is being wrapped up by the American author with paper towel and tape. (b) Loose samples such as these sandy deposits from Gorham's Cave can be sampled with 7 mm square drain pipe, cut along one side and inserted into the profile, similar to that of a soil monolith.

ensures that sampling for radiocarbon dating, snails, pollen, seeds, etc., can be taken in the same place as micromorphological or bulk samples.

2 When visiting a site, the geoarchaeologist should go on the cautious, even if somewhat pessimistic assumption, that this will be his/her last visit to the site, whether the site visit lasts 1 day or 2 weeks. Thus, every effort should be made to collect samples from the major stratigraphic units, features, off-site soils, or any other specific deposits, with the notion that with these samples, a basic understanding of site formation history is attainable. There may not be a tomorrow or next season to collect further samples. While working in Gibraltar, we had planned to collect samples from a critical profile that spanned Middle Palaeolithic and Upper Palaeolithic boundary, but decided to put it off until the following year. That winter, the highest rain falls in a century fell and the profile had slumped during these heavy rains. It took several field seasons to get back to the original profile.

3 Samples represent an invaluable resource and archive. Depending on the project and expenses of getting to the field area, they can have in effect high price tags attached to them. A field project in China, for example, involves a costly plane ticket, as well as field per diem (subsistence costs). In addition, samples are often collected from exposures that will be excavated at a later time, and thus represent an archive of deposits that are no longer visible or available for study or observation. Thus, on the advice of a close colleague (O. Bar-Yosef), where and when possible, it is best to collect as many samples as possible, even if there is no current budget to analyze them. They can always sit on a shelf until the money or scientific questions arise, but at least they are there and should be treated as invaluable documentation of the archaeological record, no differentive from lithics or faunal remains.

Impregnated samples have a virtually indefinite shelf life. Air-dried bulk samples also possess good longevity. Undisturbed samples, however, if stored for a long time can suffer deterioration. Metal tins may rust, calcium carbonate or other salts in the sample can migrate toward the outer part of the sample; the sample may also be affected by biological activity. Probably the best storage method is in a cooler, at about 2–3°C. This is particularly true for organic deposits. Freezing imparts artifacts on soils and sediments, and is normally only used for ice cores, or for lake sediments taken during winter in boreal areas.

15.7 Sample and data correlation

It is crucial to correlate all geoarchaeological sampling with any other environmental sampling or recovery of artifacts. The exact relationships among the geoarchaeological samples and the archaeological contexts need to be recorded (Macphail and Cruise, 2001). It is useful to repeat the fact here, that if samples are too far apart they may well tell different stories even though from the *same context*. Although this may be appreciated better on occupation sites, because of the obvious heterogeneity and numerous lateral changes in microfacies (Courty, 2001), it can also be true of many "natural" deposits and soils. For example, pedogenesis is influenced by small-scale differences in drainage or vegetation cover, and by subtle, localized anthropogenic impacts, and by variations in rates and types of sedimentation.

Single samples can be split between different disciplines. Thus, a core or monolith sample can provide samples for microfossil investigation (pollen, diatoms, ostracods, mollusks), bulk analyses (physical and chemical studies), and enough remaining undisturbed material for soil micromorphology. Moreover, archaeological material and macrofossil samples should be collected in the same way from a sampled section in order to permit exact correlation at a later date. If such an approach is carried out, information from all the disciplines can be viewed together and enhance a *real* interdisciplinary investigation.

While setting up an interdisciplinary program, it may be worthwhile to undertake further analytical methods in addition to those done routinely. The palynologist, for example, may wish to know the ratios between organic and mineral deposition in a peat deposit, and so high-resolution organic matter analysis could be carried out to match the pollen sampling intervals. In turn, a palynologist may be willing to look at mineralogenic layers that are of more interest to the geoarchaeologist, rather than simply concentrating on the most organic and probably most polleniferous horizons.

In some cases, the potential of samples for an interdisciplinary study is not realized until a thin section is observed and fossils are found, overcoming to some extent the absence of a bulk sample for macro- or micro-fossil analysis, which was not collected through some oversight. In fact, charcoal, diatoms, nematode egg, and palynological data have been readily revealed from the thin sections themselves (Macphail *et al.*, 1998; Goldberg *et al.*, 1994). Although such observations of fossil remains commonly have no statistical value *sensu stricto*, they nevertheless provide extremely valuable data on the sedimentary environment of the specific fossil materials. In traditional sedimentology *all* fossil assemblages would be analyzed in terms of their depositional environment (Reineck and Singh, 1980). This happens very much less often in archaeology, where fossils are often recovered (along with artifacts) and analyzed without close regard to the geoarchaeological origin of the sediments in which they occur. Thin section analysis early on in a project can therefore sometimes guide ensuing environmental studies.

15.8 Conclusions

In this chapter we have attempted to make the reader acquainted with some of the basic methods that are needed to carry out proper geoarchaeological fieldwork, including prefield desktop studies, remote sensing, and fieldwork *sensu stricto*. Not all the techniques will be applicable or suitable in all instances, and many types exist for each one. Nevertheless, the reader should be able to follow the general strategy of how to tackle fieldwork, regardless of the specific strategies and tactics employed. This pathway starts with gathering data over a large area, with aerial photos, geological maps, satellite and geophysical techniques, down to the scrutiny of soils, sediments, and archaeological deposits through the examination of cores or other exposures – whether by excavation or backhoe. The last leads to detailed and careful description of the profile and appropriate sampling. We describe how to take intact blocks for micromorphology and appropriate bulk samples for physical or chemical laboratory analyses. (Analyses of soil micromorphological and bulk samples are described in Chapter 16.) We also emphasize how sampling has to be correlated for all the site team members, and make the point that without accurate field observations and proper contextualization of samples, profiles, and landscapes, any analytical work carried out subsequently in the laboratory may be of limited use, be totally uninterpretable, or even provide inaccurate results, regardless of the degree of laboratory precision.

16

Laboratory techniques

16.1 Introduction

Geoarchaeological research comprises three essential parts, and each is necessary in order to achieve the best results. So while early planning in the office and detailed fieldwork can often represent the major part of a geoarchaeological study, work in the laboratory is necessary in order to clarify issues and test hypotheses that have arisen. Below we provide the basics for some of the most important and common laboratory techniques that are used in geoarchaeology. We concentrate on why a technique should be employed, rather than the details of the laboratory procedures themselves. The latter can be found in Herz and Garrison (1998), Gale and Hoare (1991), Pollard and Heron (1996), Stoops (2003) Courty *et al.*, (1989) and Garrison (2003), as well as books and articles on soil science and sedimentary petrography that are cited throughout the text.

16.1.1 General overview and considerations

The decision concerning which laboratory studies to undertake is generally first decided during fieldwork (see Chapter 15). This is an important point, since it is possible that the wrong type or quantity of sample will be taken during the excavation. Also, as more data become available, for example from thin sections or other analyses, additional procedures may be deemed necessary. Most important, however, is to remember that quality is better than quantity. As noted in Chapter 15, small, well-focused samples (e.g. organic matter, phosphate and magnetic studies, ca 10–20 g), are better than large "mixed" samples, which include mixed components. Any data from the last will be far more difficult to interpret, or relate to microstratigraphic details in the thin sections. It is crucial when deciding upon what methods to employ, that the following are clearly thought through:

1 All techniques are time consuming, and many are expensive. It is therefore essential to ask: is this analysis really necessary, what questions are we trying to answer, and what will the results add to the site's interpretive dataset? Simply, are the costs of time, labor, and cash worth it?
2 In cases when a laboratory always carries out a suite of analyses, are they all appropriate to the study, and will all the data be useful?
3 When dealing with ancient soils that are either on the surface or buried, is the property that is being measured now applicable to the understanding of past conditions? This is especially important when comparing buried soils and so-called control profiles, particularly if the latter are present day soils. Similarly, does the measurement of soil acidity (pH) for a palaeosol really represent the conditions at the time of its formation and burial (see Chapter 12)?

The amount of required "accuracy" and "precision" must also be thought through. Accuracy, the closeness of the measurement to true value, is constrained by the equipment being employed. For example, if a balance measures only to three decimal places, then it may be better to report only two decimal places, by rounding up. If a mean value is calculated, the answer may be given by the calculator, to 6 decimal places, which of course is nonsense, and "inaccurate," when the balance weighs up to three decimal places. Again, rounding up to two decimal places is necessary. More importantly, "precision" refers to how reproducible are the results. If several measurements of pH are taken, the time the sample is in water can affect pH, and the pH reading can theoretically shift (although not demonstrated in practice; Gale and Hoare, 1991: 278). Pragmatic decisions need to be taken concerning how representative the sample is and how many replicate analyses/measurements are required to ensure that the results are robust. Statistical testing cannot be carried out on small sample suites.

When carrying out laboratory analyses, one should strive to produce data for each sample that is effective in answering a common problem. Thus, if we are trying to establish the presence of a buried argillic palaeosol, organic matter estimates, phosphate, and clay content might be suitable sets of analyses to choose. In any case, all material should be analyzed from a split of the homogenized sample (16A.1). Every report should say which methods were used, citing an authority and detailing any variations that were carried out. Different bulk analytical methods, for any given technique, will give different results (Holliday, 2004, appendix 3). If studies are aimed at making comparisons with work by others, or replicating investigations, exactly the same methods need to be undertaken. With regard to these points, when European dark earth (Chapter 13) was first investigated in detail, a multiple technique approach was used (11 methods, Macphail, 1981). Some methods proved more useful than others in answering the specific questions on the dark earth, and these, with a further six techniques were applied on subsequent sites (Macphail, 1994; Macphail and Courty, 1985). Techniques to study dark earth are selected on a case-by-case basis, and several more analytical methods have recently been used. The characterization of the hydroxyapatite cemented stabling crust at the Moel-y-gar stable (Butser Ancient Farm, Chapter 12) also underwent a mixed geoarchaeological approach (Canti, 1995), which involved field identification, pollen and macro-botanical studies, bulk chemistry (organic matter, magnetic susceptibility, and phosphate analyses), soil micromorphology, microprobe, and XRD (Macphail *et al.*, 2004). We advocate a flexible approach that provides both continuity with earlier studies, and also takes advantage of useful new techniques as they emerge.

16.2 Physical and chemical techniques

16.2.1 Grain size analysis

Grain size or particle size analysis is a routine study in geoarchaeology and other earth sciences. When studying sediments it is carried out in order to determine, for instance, energy of deposition and environment of deposition (see Chapter 1). The uniformity of grain size (sorting) may provide information on the type of sediment source; is a deposit totally aeolian in character or are there admixtures of fluvial material, as when differentiating between a loess and a brickearth (see Chapter 6)? Alternatively, in soil science, there may be more than one parent material, producing a soil profile formed in two sequential sediments; or clay translocation can be inferred from grain size data.

As discussed in Chapter 1, grain sizes form a size range continuum, from fine (clay, generally <2 μm) through granules (gravel and stones), coarse cobbles, and boulders. For practical purposes, however, they are broken down into four main categories. From coarsest to finest these are (see Table 1.1):

1 The coarse or stony fraction (granules/ gravel, stones/pebbles, cobbles, and boulders),
2 sand,
3 silt, and
4 clay.

Any particles <2 mm in size are classed as "fine earth," and usually estimated as clay, silt, and sand-size fractions. The word "estimation" is employed because it is difficult to measure exactly all the different size ranges. Material >2 mm is termed the coarse fraction or stones, and is often determined less frequently. If the deposit contains large amounts of material that is <63 μm in size (i.e. silt and clay) as in loams and/or organic matter, the deposit/soil needs to be pretreated (Avery and Bascomb, 1974; Carver, 1971; Soil Conservation Service, 1994). The first pretreatment removes all nonmineralogenic material (organic matter), and separates the nonsandy elements – usually by "wet sieving" (see next paragraph). It may also be desirable to eliminate the calcium carbonate, to produce carbonate-free grain size estimations (Catt, 1990).

The most basic method for estimating the sand and stone size ranges is through sieving. In pedology, each horizon is described in the field according to its "texture" (grain size) through "finger texturing," and its stone content or stoniness through visual estimation (see Chapter 15). The precise nature of this stoniness may be very important, and stones can be individually measured or sieved into 2–4 mm, 4–8 mm, 8–16 mm, 16–32 mm, etc., fractions. If sediments are totally dominated by sand, the different size ranges are analyzed by dry sieving the <2 mm fraction, normally employing a minimum of 63 μm, 125 μm, 250 μm, 500 μm, and 1,000 μm sieves (Table 16.1). In sedimentology, a settling tube is used in which sediment is introduced into a column of water and the rate of sedimentation is measured, with coarser grains settling first (Pye, 1987). This method more accurately depicts the sedimentary dynamics associated with transport and sedimentation. The equipment is not standardized, however, and sieving is a much more straightforward procedure, hence its popularity.

Since the finer silt and clay sizes are too small to measure directly by sieving, their size and abundance must be measured indirectly. There are many ways to estimate clay and silt fractions that first include wet sieving (to exclude the sand component), the use of a sedimentation column, and the "high tech" methods of using a laser counter or sedigraph. Loams (mixtures of sand silt and clay, Fig. 1.1), silts and clay materials are pretreated to remove organic matter and to disaggregate the soil and disperse clay particles. (The importance of "clasts" of mixed grain sized material is noted below.)

The standard procedures outlined here, broadly follow those in Avery and Bascomb (1974). Organic matter is removed from the sample using hydrogen peroxide (H_2O_2) mixed with deionized water. (This loss of material needs to be deducted from the original soil total weight, and should be consistent with the measured organic matter.) In addition, carbonate can be removed using HCl. The procedure then continues with the now mineralogenic part of the sample that should be disaggregated using a disaggregating agent such as sodium hexametaphosphate (Calgon). The sample is then mechanically stirred or shaken, and when totally disaggregated the suspended sediment can be analyzed in a sedigraph or laser counter to measure grain size, although normally only silt to very fine sand size are counted in the latter. Traditionally, however, the sample is washed through a 63 μm sieve into a 500 or 1000 ml cylinder, in order to separate silt and clay from the sand content (>63 μm). The sample then undergoes testing for silt and clay content, either by the pipette or hydrometer method. Essentially the "sediment column" is "paddled" until all the particles are evenly distributed, and the particles begin to settle out, according to

Table 16.1a Example of grain size analyses – from different sedimentary environments at Lower Palaeolithic Boxgrove, United Kingdom: % $CaCO_3$, clay, silt, and sand; grain size on a decalcified basis (from Catt, 1999) (See Fig. 16.1a)

Sample	$CaCO_3$	Clay <2 µm	Silt (Sizes in µm)					Sand (Sizes in µm)				
			2–4	4–8	8–16	16–31	31–63	63–125	125–250	250–500	500–1000	1000–2000
Unit 6	51.9	44.4	4.0	5.8	7.2	17.3	13.0	3.0	0.7	0.1	0.1	0.1
Unit 4c	1.7	40.0	5.1	6.0	9.0	12.1	22.6	5.0	0.2	0.1	0.1	0.1
Unit 4b	25.7	39.2	4.4	4.3	7.0	10.2	25.6	6.8	0.2	0.1	0.1	0.1
Unit 3	13.0	3.4	2.0	0.4	0.3	1.2	22.7	50.6	14.2	1.1	0.1	0.1

GTP 13: Unit 3 = Beach sand; Unit 4b = Estuarine silt; Unit 4c = Old land surface (ripened "soil" formed in partially decalcified uppermost Unit 4b).
GTP 10: Unit 6 = Colluvium (brickearth colluvium).

Table 16.1b Example of grain size analyses – from different soil horizons at prehistoric Raunds, and including the overlying medieval alluvium; data from Turf Mound and Barrow 5 Raunds, United Kingdom; % clay, silt, and sand (from Macphail, forthcoming)

Sample and soil horizon	Clay <2 µm	Silt (Sizes in µm)			Sand (Sizes in µm)					Texture
		2–16	16–31	31–63	63–125	125–250	250–500	500–1000	1000–2000	
22 (Ap/ alluvium)	36	12	10	11	4	10	14	2	1	Clay
24 (bA)	9	6	5	15	6	17	37	4	1	Sandy loam
26 (bB2)	12	3	2	13	9	25	32	3	1	Loamy sand
27 (bBt)	24	2	3	9	8	17	27	7	3	Silty clay loam
28 (bC)	3	5	2	10	7	20	33	10	10	Loamy sand

Note: bC = Nene river terrace sands and gravels; bBt = clay-enriched argillic horizon; bB2 = upper subsoil which also includes anomalous argillic features; bA = barrow buried topsoil which also includes anomalous argillic features; Ap = medieval ploughsoil formed in post-prehistoric fine alluvium which buries the barrow.

Stoke's law, at a given rate that is temperature dependant. Coarse silt falls quickest, while clay stays in suspension for the longest time. At specific time intervals (e.g. 30 seconds to 8 hours), therefore, sediment is drawn off from a given depth, and the amount of sediment present is dried and weighed. This permits the calculation of the different percentages of different grain sizes present. The hydrometer method allows the measurement of the density of sediment, again giving information on the nature of the sediment in suspension at different times.

In both methods, how much fine ($<$1 μm) and coarse ($<$2 μm) clay, fine silt, medium silt, and coarse silt is present can be calculated (Table 16.1; see 16A.2).

The measured amounts of the different fractions are then plotted to produce a grain size curve (Figs 16.1 and 16.2). The different grain sizes have already been presented (Table 1.1). In sedimentology, grain sizes are normally expressed on the phi (ϕ) scale (e.g. 0.25 mm= 2.00 ϕ; 0.50 mm=1.00 ϕ; 1.00 mm=0.00 ϕ; 2.00 mm=−1.00 ϕ, etc.) (see Chapter 1). A number of examples of grain size curves, and estimated amounts of the different size factions, are shown here. It is useful to compare a loess with a brickearth, the latter being less well sorted because the well sorted coarse silt and very fine sand-size aeolian content is mixed with medium size sand of fluvial origin. Both have been utilized for building clay and bricks (Chapters 10, 11, and 13).

In the Chiltern Hills, a range of chalk hills in southern England, clay-rich subsoils were covered by a more silty loessic drift (Avery, 1964, 1990). This was strongly eroded in prehistoric times after clearance, and now is mainly found in subsoil treethrow hollows and early ditchfills (Chapter 9). Across many parts of the United States and Eurasia, Alfisols (Luvisols) formed under woodland are differentiated in part from Mollisols developed under prairie vegetation, by measured increased amounts of clay in their subsoil (Bt) horizons (Table 16.1b). In coastal areas, there are enormous differences between beach sandy sediments, and silt and clay-rich estuarine deposits (Table 16.1a; see Chapter 7). Lastly, anthropogenic deposits such as dark earth are extremely poorly sorted because of dumping and the common earlier presence of buildings (Table 16.2; Fig. 16.2). It is clear that fine sandy silt loam brickearth (building "clay"), early Holocene river sand, and dumped material have contributed to the overall grain size of dark earth (Macphail, 2003c). It can also be noted that the grain size of Roman alluvium is comparable to dumps of silty Roman street sweepings. Hence when carrying out granulometry on anthropogenic deposits the resulting data may well be uninterpretable without complementary soil micromorphology.

Grain size analysis was also carried out on the Late Bronze Age/Early Iron Age site of Potterne and produced essentially bimodal distribution curves. The sandsize material originated from the local Greensand soils, but the silt size material was largely composed of ashes and phytoliths, which characterized this midden (Macphail, 2000). Hence, grain size data from occupation deposits need to be interpreted with a large measure of caution. Natural, poorly sorted deposits include till, deposited by ice.

Like any technique, grain size has its limitations and pitfalls. For example, when a sediment contains aggregates composed of sand silt and clay, disaggregation of these clasts will not reflect the depositional energies or conditions of the *clast itself*, because the clast is the grain size that was transported (Table 16.1a – Boxgrove Unit 6). Also, there may be problems when deposits are rich in mica, because lath shaped silt size mica takes longer to settle than round silt particles. Deposits rich in diatoms or phytoliths also settle at anomalous rates. Finally, the use of grain size analysis to characterize anthropogenic deposits should be strongly questioned for many of the reasons mentioned above: what is its value when studying a sediment composed of aggregates of clay, brick, ashes, bone fragments, and organic matter (see Table 16.2)? Certainly the sedimentary dynamics cannot be determined as is in the case of geogenic sediments. A relationship between grain size data and soil micromorphology also needs to be noted (16A.3).

16.2.2 Soil characterization

There is insufficient space here to itemize and explain all the many soil analyses that can be carried out and also describe all the different techniques employed. We have therefore

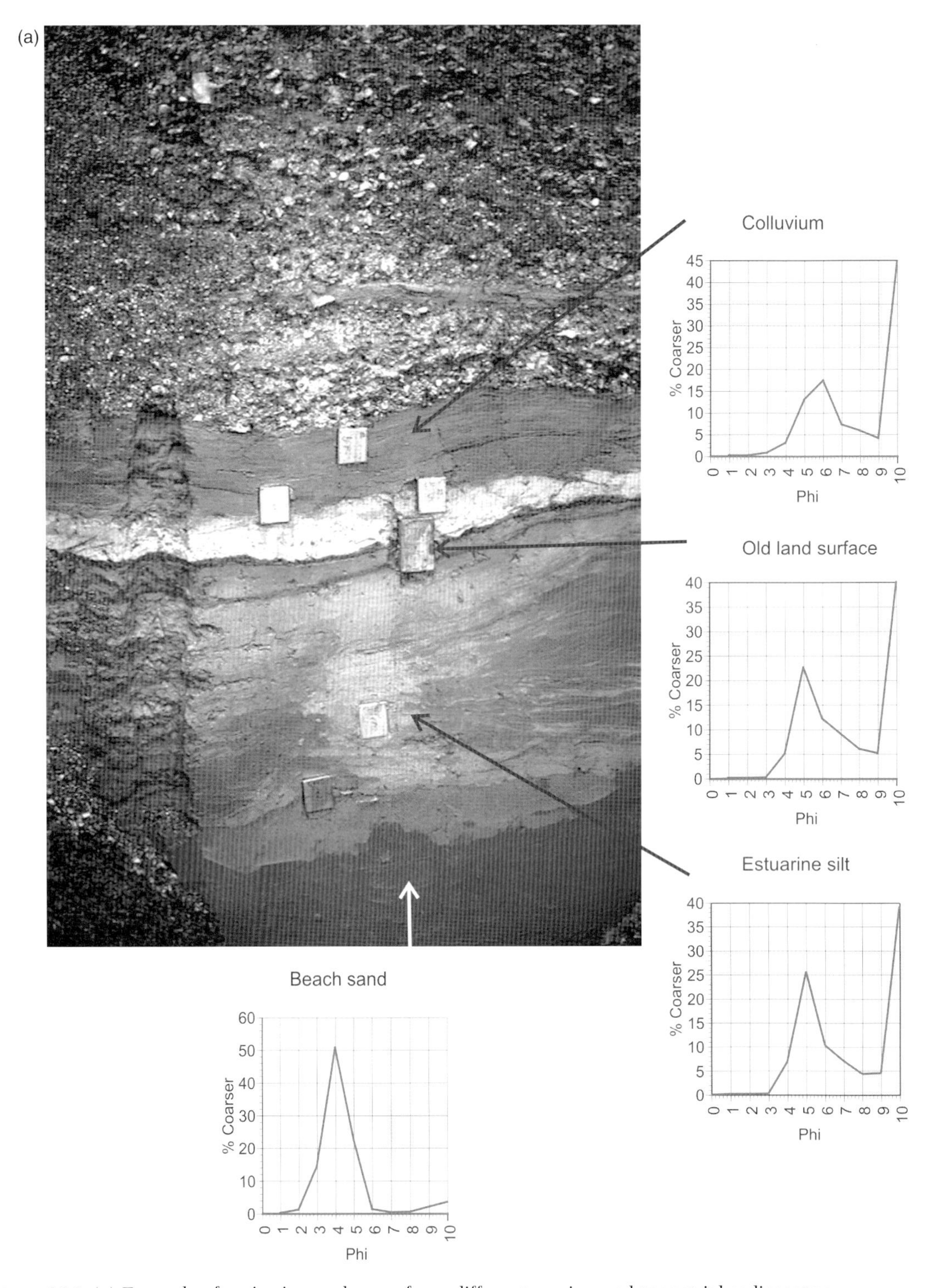

FIGURE 16.1 (a) Example of grain size analyses – from different marine and terrestrial sedimentary environments at Lower Palaeolithic Boxgrove, United Kingdom (Data from Catt, 1999); well sorted fine beach sand, less well sorted estuarine silts" and clay, the Lower Palaeothic old land surface "soil" formed through the weathering of the uppermost estaurine silts, and the least well sorted brickearth colluvium (cool climate terrestrial deposits). (b) Example of grain size analyses – from different soil horizons at prehistoric Raunds, and including the overlying medieval alluvium; data from Barrow 5 Raunds; *bC*=Nene river terrace sands

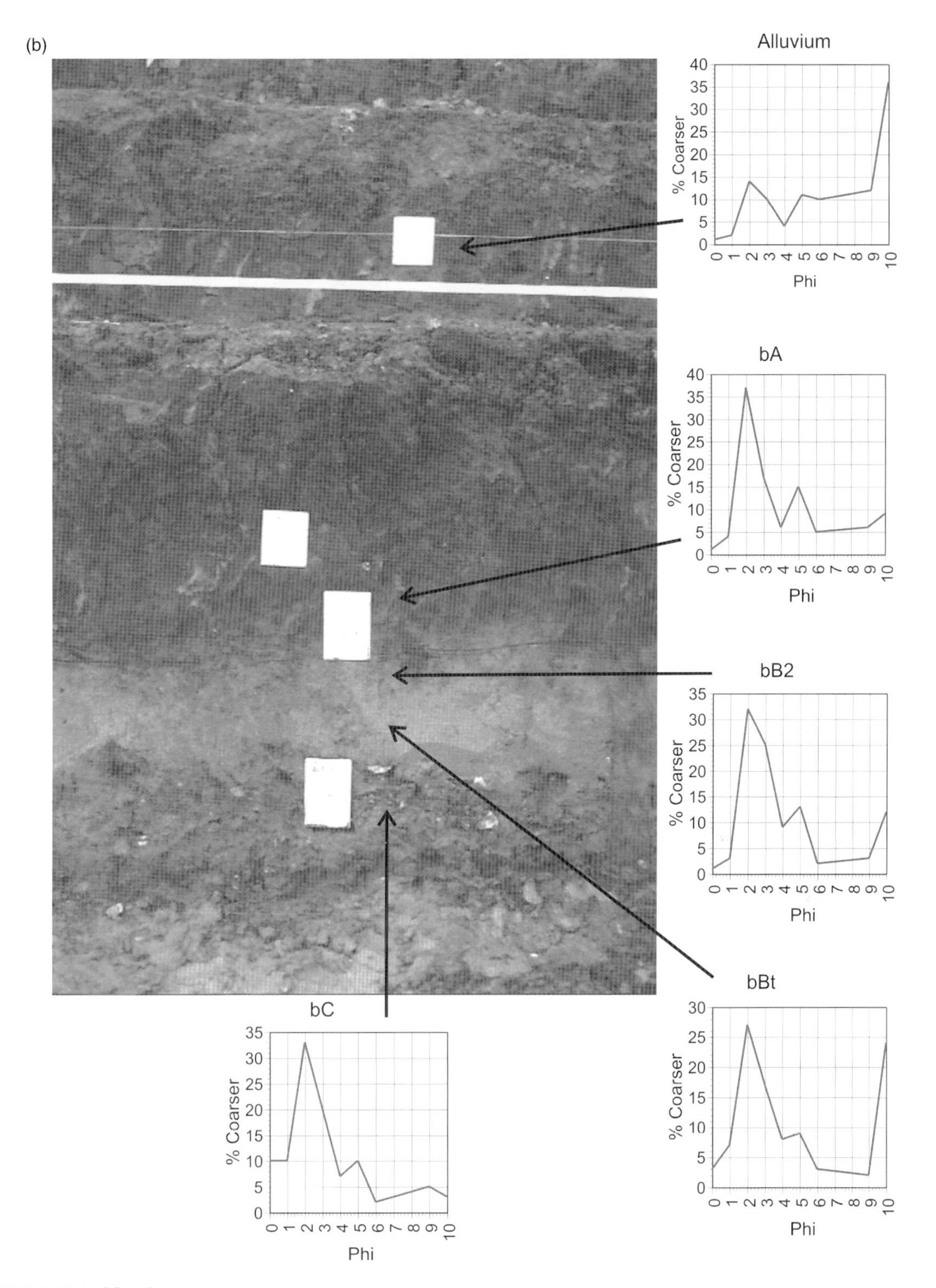

Figure 16.1 (*Cont.*)
and gravels; *bBt*=clay-enriched argillic horizon; *bB2*=upper subsoil which also includes anomalous argillic features; *bA*=barrow buried topsoil which also includes anomalous argillic features; *Ap*=medieval plough-soil formed in post-prehistoric fine clayey alluvium which buries the barrow.

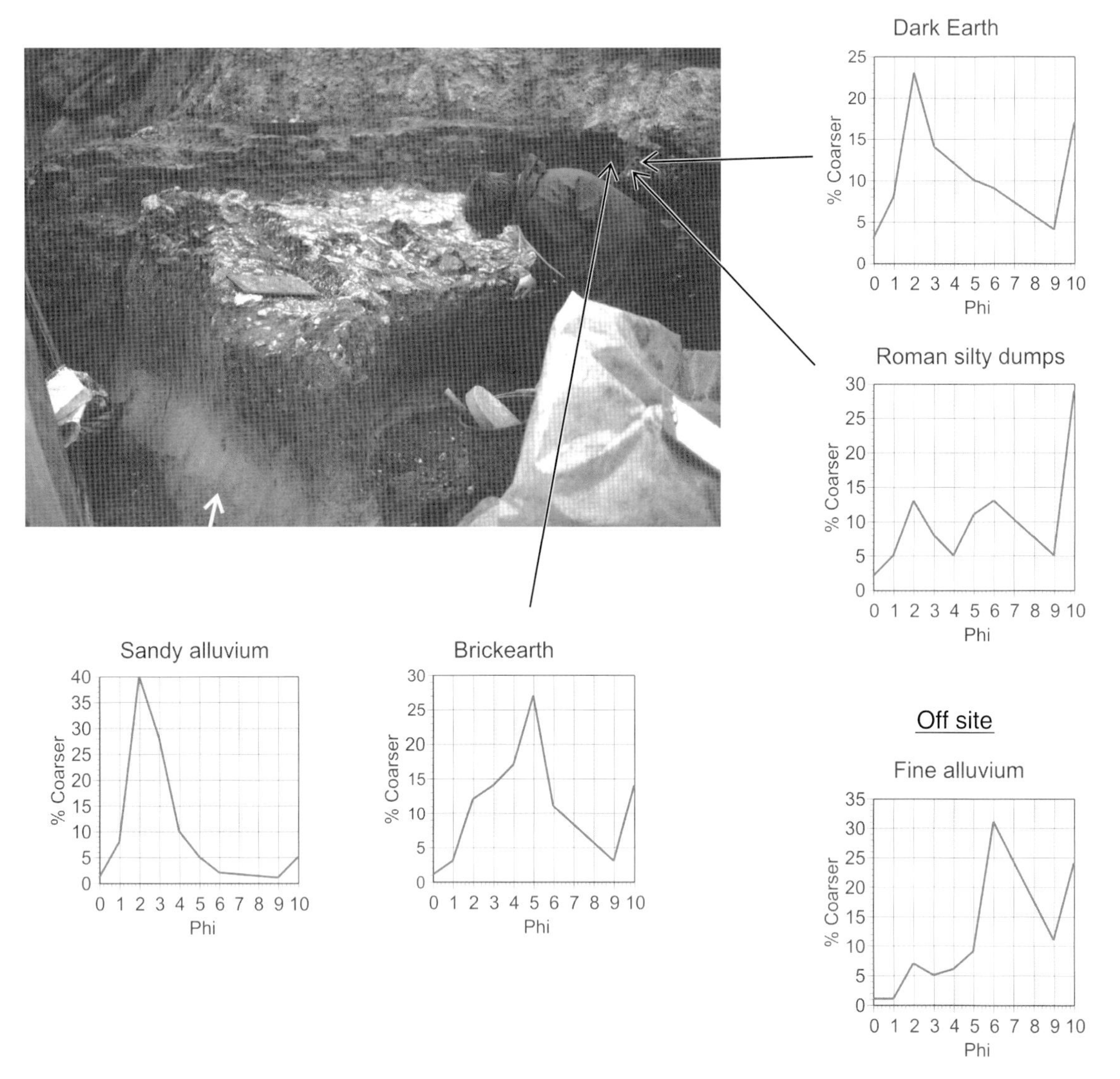

FIGURE 16.2 Example of grain size analyses – from different soils, sediments, building materials, and the dark earth at Courages Brewery, Southwark, London, United Kingdom. Dark earth ($n = 4$) includes high amounts of local sandy alluvium (used in all kinds of building activities, mortar manufacture, etc. see Chapter 13), with contributions from brickearth building "clay" and anthropogenic dumping. Such interpretations were only possible through parallel soil micromorphology analyses.

chosen to present a short guide to the most commonly employed techniques. The chief methods that are available in most laboratories, especially those in soil or earth science departments, are: measurement of water content, total organic matter (e.g. by loss on ignition, LOI), pH, organic carbon, nitrogen, carbonate, iron (Fe), and aluminum (Al). Other elements, such as heavy metals (e.g. copper [Cu], lead [Pb]; and zinc [Zn]; see Chapter 17) may be analyzed, although now with the greater availability of X-ray fluorescence and ICP, most elements can be measured easily and in bulk (see section 16.5), but again there are caveats to the indiscriminate use of these methods.

TABLE 16.2 Example of grain size analyses – from different soils, sediments, building materials, and dark earth at Courages Brewery, Southwark, London, United Kingdom: % clay, silt, and sand (from Macphail, 2003; see Fig. 16.2)

Sample	Clay <2 μm	Silt (Sizes in μm)			Sand (Sizes in μm)					Texture
		2–16	16–31	31–63	63–125	125–250	250–500	500–1000	1000–2000	
Southwark Roman										
1. Brickearth (clay wall)	14	3	11	27	17	14	12	3	<1	Fine sandy silt loam
2. Sandy Thames alluvium (preRoman subsoil)	5	1	2	5	10	28	40	8	1	Loamy sand
3. Fine Thames alluvium (Roman)	24	11	31	9	6	5	7	1	1	Clayey loam
4. Roman fine dumps ("sweepings")	29	5	13	11	5	8	13	5	2	Clayey loam
5. Dark Earth ($n = 4$)	17	4	9	10	12	14	23	8	3	Sandy loam

Note: 1. Brickearth (a loess-like fine sandy alluvium that was quarried for building clay); 2. Thames river sands (early Holocene alluvium); 3. Roman fine alluvium; 4. Roman fine dumps (contains abundant fine anthropogenic inclusions); 5. Dark earth (a mixture of brickearth (1) and sands (2) derived from constructional raw materials [e.g. sand-tempered mortar], as well as finely fragmented building debris [mortar and brick], and occupation dumps (4) that include ashes, phytoliths, fine bone, shell, and coprolites etc.)

The measurement of water content at 105°C is useful as samples are all analyzed on a water-free weight basis, and all that is needed is a simple drying oven.

Low temperature LOI (loss on ignition) is an important measure of the amount of organic matter in the soil or sediment (e.g. Ball, 1964). Its measurement can be an indicator of the intensity of human activity or the degree of oxidation associated with soil formation (see Chapter 3). LOI is carried out using an oven-dried sample in a crucible, and a furnace; a desiccator and desiccant are to make sure that samples do not take up moisture between heating and weighing (for further details on LOI and associated soil techniques see 16A.4).

Organic matter is also often determined by the analyses of C and N, and these are measured by acid digestion. This is only normally carried out in properly equipped soil laboratories, but provides very useful information on bioactivity and biodegradation, in the form of the C : N ratio. Similarly, this kind of "wet chemistry" is used to analyze different important pedological fractions of Fe and Al (Bascomb, 1968; Bruckert, 1982; Macphail *et al.*, 2003). For example, it may be important to know the proportions of "mobile" and "crystalline" forms of these components in a soil. Spodic/podzolic subsoils (Bh, Bs) have accumulated mobile Fe and Al (sesquioxides), whereas brown soils only feature crystalline forms of these (see Table 16.3). CEC – cation exchange capacity is another useful fundamental test, measuring the amounts of Ca^{++}, K^{+}, Mg^{+}, and Na^{+}, along with H^{+}. The amounts of these are related to pH and degree of leaching (see Table 3.4). In the same way, C : N results from the degree of bioactivity and bio-decay in soils, with low C : N (~5–10) being found in base rich bioactive soils (e.g. high earthworm or termite populations), 10–20 in acid or humus horizons of Spodosols.

Other analyses include measures of plant-available K and available P. These measurements are in reality mainly used to determine present-day soil fertility, and are not suitable for archaeological deposits. They have usefully been carried out, however, at Experimental Earthworks (e.g. Crowther, 1996; Macphail *et al.*, 2003) because these studies are engaged in monitoring modern or short-term soil changes after burial (Chapter 12). In addition, a common question when studying soils of historic gardens is, how fertile are the garden soils and do they require amendments to the soil when being reconstructed, as was undertaken at Hampton Court Privy Garden, London, United Kingdom (Macphail *et al.*, 1995).

When studying ancient soils, and under normal conditions, a number of characteristics are labile or easily transformed (Table 16.3). This was also demonstrated at the two Experimental Earthworks in the United Kingdom (see Chapter 12). First, pH can be affected by cations leaching out of the overburden, or through down wash of unstable fine material, or by organic matter decay. Organic matter itself is highly affected by postdepositional oxidation, and frequently cannot be compared to amounts of organic matter in present-day "control" soils. C : N ratios may also change through time, and it is worth noting that archaeological deposits with high amounts of charred organic matter, have anomalously high C : N ratios, despite being once bioactive (Courty and Fedoroff, 1982; Macphail and Courty, 1985).

16.2.3 Phosphate analysis

This technique (Arrhenius, 1934) is one of the most commonly employed in geoarchaeology, having an archaeological pedigree going back at least to the 1950s (Arrhenius, 1934, 1955; Proudfoot, 1976). In fact, Holliday (2004, Appendix 2) notes more than 50 extraction methods, more than 30 of which have been used in archaeology. From the outset it is useful to differentiate between phosphate (as organic, and inorganic iron and calcium compounds, for example), which are reported as an oxide – such as P_2O_5, and elemental phosphorus (P).

TABLE 16.3 Data on soil chemistry and magnetic properties for the L, F, and Ah horizons of the control soils and buried soils at Wareham Experimental Earthwark (Bell *et al.*, 1996; see chapter 12). Data from Macphail *et al.* (2003).

	Control soils				**Buried soils**			
					Turf buried		Sand buried	
Soil profile (pit)	1(1)	2(1)	3(2)	4(3)	5(8)	6(8)	7(4)	8(6)
Thin sections	600/601		676/677	681	879/880/882a–e		726/727	781/782
Sample series	746	747	740	759	881	883	730	783
Litter (L) layer								
Organic C (%)	31.5	40.6	42.6	40.7				
C/N ratio	29.7	47.8	54.6	31.3				
Alkali sol. humus (trans)	12.7	7.29	4.99	3.13				
pH in water (saturated)	4.2	4.2	3.8	3.9				
Fermentation (F) layer								
Organic C (%)	33.3	35.0	20.3	31.0				
C/N ratio	35.8	36.1	30.8	27.4				
Alkali sol. humus (trans)	9.79	7.70	12.8	5.33				
pH in water (saturated)	3.6	3.6	3.7	3.6				
Top 20 mm of Ah horizon[a]								
Organic C (%)	7.55	9.79	10.9	7.32	8·41	8·61	8.19	6.43
C/N ratio	30.2	29.7	29.5	24.4	21.6	22.7	22.8	20.7
Pyro. ext. C/organic C	0.128	0.126	0.207	0.142	0.293	0.237	0.282	0.244
Alkali sol. humus (trans)	8.88	10.8	9.85	8.84	6.97	6.42	6.39	6.45
pH in water (1:2.5)	3.9	3.8	3.9	3.9	3.8	4.0	4.0	4.7
Pyro. ext. Fe (%)	0.03	0.03	0.03	0.03	0.04	0.04	0.06	0.04
Total ext. Fe (%)	0.04	0.04	0.04	0.04	0.05	0.05	0.07	0.05
χ (10^{-8} SI Kg^{-1})	1.1	1.7	1.4	2.2	0.4	0.5	4.3	2.5
χ_{max} (10^{-8} SI kg^{-1})	10.0	15.3	17.6	16.1	10.7	9.0	27.0	29.4
χ_{conv} (%)	11.0	11.1	8.0	13.7	3.7	5.6	15.9	8.5
Ah horizon								
Thickness (mm)	40	70	100	60	100	110	90	70
Predicted organic C[b] (%)	nd	nd	nd	nd	8.91	9.57	8.25	6.92
Organic C (%)	5.04	7.02	8.94	6.02	6.37	6.96	6.87	5.04
C/N ratio	28.0	22.6	31.9	23.2	17.7	19.3	27.5	21.9
Pyro. ext. C/organic C	0.171	0.148	0.225	0.135	0.350	0.302	0.316	0.256
Alkali sol. humus (trans)	7.94	9.13	5.66	7.64	5.23	6.40	5.14	5.69
pH in water (1:2.5)	4.0	3.9	4.0	4.0	3.9	4.0	4.1	4.7
Pyro. ext. Fe (%)	0.02	0.02	0.03	0.02	0.04	0.03	0.03	0.03
Total ext. Fe (%)	0.03	0.03	0.04	0.03	0.04	0.04	0.04	0.04
χ (10^{-8} SI kg^{-1})	0.8	0.8	0.5	1.4	0.2	0.2	1.8	1.5
χ_{max} (10^{-8} SI kg^{-1})	9.1	11.3	8.8	17.0	8.3	7.3	14.8	16.7
χ_{conv} (%)	8.8	7.1	5.7	8.2	2.4	2.7	12.2	9.0
Phosphate-P (μg g^{-1})	94	115	83	75	70	65	86	74
Avail phosphate-P (mg g^{-1})	16.0	23.6	14.5	18.4	26.0	29.0	20.1	21.9
Avail K (mg l^{-1})	34.8	53.8	53.1	31.3	10.8	7.9	16.6	14.8

[a] Top 30 mm in case of buried soils – i.e. includes the 10 mm sample which contains the bLFH (see text).
[b] Expected concentration present in Ah horizons of this thickness, based on regression line ($r = 0.995$, $p < 0.01$) through the plot of organic C and Ah horizon depth for the control soils.
Analyses carried out at the University of Wales, Lampeter.

A well-known example of a phosphate mineral is vivianite (e.g. $Fe_3(PO_4)_2\ 8H_2O$). Both *phosphate* and *P* can be measured by wet chemistry and/or through microprobe and XRF analyses, but amounts of reported phosphate will always be *more* than reported P. The ratio between P_2O_5 and P is 2.2916, and this figure is used to calculate P from P_2O_5. Amounts of P can be reported as ppm (parts per million), $mg\ g^{-1}$, or as percent: 1,000 ppm P = 1.00 $mg\ g^{-1}$ = 0.10%; 10,000 ppm P = 10.0 $mg\ g^{-1}$ or 1.00% P. Often the amount of *phosphate* is measured chemically, but the amount of *P* is then reported after a calculation, as *phosphate-P* (Crowther, University of Wales, Lampeter, personal communication).

Phosphate analysis is important because it can give one important measure (amongst several) of the intensity of human occupation. As noted above, there are many ways of extracting phosphate, and it is up to the worker to choose the most appropriate method (Bethell and Máté, 1989; Holliday, 2004, Appendix 2). It is necessary, however, to reaffirm that in normal circumstances, spot testing and measurements of plant available phosphate are not useful, because in the former it is a nonquantitative test, and on many sites phosphate will be *present*; and in the latter the extraction method is far too weak usually to provide a useful signal. Even in impoverished Spodosol topsoils Crowther found the presence of phosphate (Ah horizons: 0.827 $mg\ g^{-1}$ P; 21.2 $mg\ l^{-1}$ available P) but with much less available phosphate compared to the Privy Garden soil at Hampton Court (plate-bande fills: 0.942 $mg\ g^{-1}$ P; 95.8 $mg\ l^{-1}$ available P) (Macphail and Crowther, 2002; Macphail *et al.*, 2003).

Two principal approaches for phosphate analyses are preferred. The first relates to available phosphate testing as is commonly used in agronomy: a sample is subjected to a mild leaching in which the more easily solubilized phosphate is removed. The second employs measurement of the total phosphate in which the entire sample is dissolved. In the latter it may be misleading to dissolve the whole sample in a very strong acid (e.g. hydrofluoric) before measuring phosphate, as then P within the crystal lattice and fossil P (e.g. apatite within rock fragments such as chalk) will be included, giving confusing and probably regionally specific data. The study of the Hampton Court soils, however, is a special case where the measurement of available phosphate was important to the reconstruction and replanting of the garden, in order to record present-day soil fertility.

Nevertheless, in geoarchaeology generally, a total phosphate method is required that extracts phosphate from (1) organic matter, including dung and immobilized recalcitrant/aged organic material, (2) bone and coprolites (inorganic "apatite"), and (3) inorganic secondary minerals, such as vivianite and amorphous Fe/Ca forms. It is immediately obvious that it is useful to be able to differentiate or fractionate organic and inorganic phosphate, which was the strategy of Eidt (1984) who attempted to infer soil use by measuring the various types of fractionated phosphate. Determination of phosphate fractions is commonly carried out by analyzing a sample twice. The first measures inorganic phosphate (*before* ignition/oxidation), while the second measures any residual inorganic phosphate (*after* ignition/oxidation) that had previously been in an organic form.

Two schools of thought can be identified. The first school uses a relatively weak acid (2% citric acid) in order to be able to *gently* differentiate inorganic from organic phosphate, a method used across Europe since the 1950s and one with a long track record for finding archaeological sites in Scandinavia (Arrhenius, 1934, 1955), and identifying anthrosols/plaggen soils (Driessen *et al.*, 2001; Pape, 1970; see Holliday, 2004, Table A2.1). It is still used successfully in Sweden (see below). It was, however, designed for acid and leached soils of this boreal region, which are carbonate free and rather naturally low in phosphate (Engelmark and Linderholm, 1996; Macphail *et al.*, 2000). It therefore has a disadvantage in that its extraction capacity is weakened if large amounts of free calcium carbonate are present, as in calcareous soils. This, however, can be offset if a few drops of HCl are added to the method, as this removes the buffering effect of the carbonate (Linderholm, University of Umeå, personal communication). Generally speaking, this method has produced good results for samples containing up to around 4500 ppm P_2O_5 (Linderholm, 2003; Macphail, 2003; Macphail *et al.*, 2004) (see Box 11.1).

The second school uses stronger acid extraction, for example, 2N nitric acid, again pretreating the sample with HCl to offset any calcium carbonate, or oxidation methods (Bethell and Máté, 1989; Dick and Tabatabai, 1977). Eidt (1984) differentiated numbers of P types in his work in South America. More recently Kerr (1995), Schuldenrein (1995), and Holliday (2004) reviewed the subject of phosphate analysis, with Kerr and Schuldenrein also carrying out their own fractionation methods on prehistoric North American sites. The presence of P at sites has also been linked to other elemental concentrations, for example in body stains (Keeley *et al.*, 1977), floors (Entwhistle *et al.*, 1998; Middleton and Douglas-Price, 1996) and campsites (Linderholm and Lundberg, 1994).

Movement of P in archaeological sites is a key issue. Considerable amounts of P can be moved down-profile through pedogenic effects, perhaps to a depth of 1 m over 1,000 years (Baker, 1976). Hence when studying the phosphate distribution in an area of boreal Spodosols in Sweden, the illuvial Bs horizon was sampled (Linderholm, 2003). On the other hand, Romans and Robertson working at Strathallan, Scotland, suggested that P could move significantly within archaeological time, that is, 10 mm per 20–40 years (Barclay, 1983). At Raunds, Northamptonshire, P that likely originated from prehistoric livestock concentrations was carried down-profile through clay translocation during the life of the sites (Macphail, 2003) (Table 16.4; see Fig. 16.5c microprobe). Movement of P in caves through time is well documented (Courty *et al.*, 1989; Jenkins, 1994; Macphail and Goldberg, 1999; Schiegl *et al.*, 1996), but short-term movement has also been reported within an experimental stable (during 15 year use) (Macphail and Goldberg, 1995) and where bone was buried at the Experimental Earthwork at Overton Down (Crowther, 1996). Certainly at the London Guildhall, the early medieval *in situ* stabling of animals markedly impacted upon the underlying dark earth and at Worcester, United Kingdom, Saxon cess pits led to the phosphate contamination of the earlier formed dark earth (Macphail, 1994; Macphail *et al.*, 2003).

The analysis of total phosphate has been applied in many ways, varying from simply measuring amounts of phosphate in an archaeological section or soil profile, or to identify different areas of phosphate distribution through transects, or to map phosphate distribution using samples taken on a grid (Crowther, 1997). Phosphate has also been commonly mapped alongside geophysical measurements, such as magnetic susceptibility (e.g. Maiden Castle; Balaam *et al.*, 1991). High phosphate levels have been found in association with anthropic agricultural soils (Barnes, 1990; Eidt, 1984; Liversage *et al.*, 1987; Pape, 1970; Sandor, 1992), cemeteries (Faul and Smith, 1980), assumed stock areas and drove-ways, structures housing animals (Crowther, 1996; Conway, 1983), camps (Balaam *et al.*, 1991; Barnes, 1990; Bell, 1990; Crowther, 1996; Schuldenrein, 1995), and occupation mounds and middens (Kerr, 1995; Lawson, 2000).

Differential distributions of mainly organic phosphate and predominantly inorganic phosphate have been commonly found in central Sweden. The values coincide with the remains of long house structures (large amounts of dominantly inorganic P) as identified archaeologically, and arable fields (small concentrations of mainly organic P) as also indicated by the presence of cereal pollen (Engelmark, 1992; Engelmark and Linderholm, 1996; Sergestrom, 1991). Years of experimental agriculture at Umeå Ancient Farm (Chapter 12) established that manure (organic P) was used to promote the growth of barley in late prehistoric (AD 0–1,000) Sweden where

Table 16.4 Example of soil data table – selected prehistoric treethrow soil, barrow-buried soils, barrow makeup, and medieval alluvial overburden (Raunds, United Kingdom; Macphail, 2003); soil samples, magnetic susceptibility (χ), and chemistry (from Macphail, forthcoming)

Sample number	Relative depth (m)	% Org C	% $CaCO_3$	pH ($1:2.5H_2O$)	pH (CaCl2)	% LOI	Mag sus. $\chi \times 10^{-8}$ $SI-kg^{-1}$	P (ppm nitric acid) *Italics= calculated*	P_2O_5 (ppm citric acid) [Inorganic P]	$P_2O_{5\text{-ignited}}$ (ppm citric acid) [organic and inorganic P]	P ratio
Turf mound (S end)											
3 (medieval Ap/alluvium,)	0.15–0.73	1.0	2.1	7.6	—	9.3	40	*1050*	170	460	2.7
4 (Ap ridge and furrow)	0.73–1.32	0.4	<0.1	7.4	—	4.0	77	*1070*	130	470	3.6
5 (mound)	1.32–1.64	0.5	0.1	7.3	—	4.2	134	*1010*	130	440	3.4
6 (bB)	1.64–1.72	0.3	—	7.6	—	3.0	81	*790*	110	330	3.0
7 (bB2)	1.72–2.04	0.1	—	7.5	—	2.1	33	*600*	90	230	2.5
8 (bBt)	2.04–2.25	0.2	—	7.5	—	3.3	48	*830*	130	350	2.7
9 (bC)	2.25+	<0.1	—	8.6	—	—	—	—	—	—	—
Long mound (E end)											
1 (mound)		1.0	—	7.4	6.9	10.2	218	1380	340	490	1.4
2 (bA&B)		0.3	—	7.2	6.8	2.3	64	780	290	330	1.1
Barrow 5											
22 (medieval Ap/alluvium)	0.20–0.52	1.2	<0.1	6.7	—	8.0	32	*1160*	130	520	3.5
23 (mound)	0.80–0.86	0.6	<0.1	6.8	—	4.6	78	*1010*	130	440	3.4
24 (bA)	0.86–	0.2	<0.1	—	—	2.0	34	*600*	90	230	2.5

	0.99										
25 (bB)	0.90–1.06	0.2	<0.1	6.8	—	2.1	30	*620*	100	240	2.4
26 (bB2)	1.06–1.25	0.1	<0.1	7.3	—	2.1	24	*640*	100	250	2.5
27 (bBt)	1.25–1.32	0.2	0.2	7.5	—	4.4	24	*900*	150	380	2.5
28 (bC)	1.32+	<0.1	9.6	8.1	—			—			
F62119 (treethrow hole 3, B140)											
41b (bBtg [main dark fill])	0.035–0.115			7.1		6.5	80	*810*	130	340	2.6
42 (bBtg [centre fill)]	0.04–0.12			7.0		5.5	14	*550*	140	260	1.9
43 (bBtg [lower main dark fill])	0.35–0.43			7.0	6.2	53	*780*	120	320		2.7
44 (bB(t)g [outside hole])	0.20–0.28			7.2	2.1	5	*460*	50	160		3.2
45				6.9	4.5	9	*660*	110	260		2.4
46				8.6	—	—					

Note: Two types of phosphate analysis were carried out – strong combined nitric and hydrochloric acid extraction and more gentle citric acid extraction; the last to separate inorganic from organic phosphate. The relationship between nitric and citric acid extraction was shown to have an acceptable 79% relationship (e.g. $R^2 = 0.79$), which allowed amounts of nitric acid extractable P to be *calculated*. Nevertheless as citric acid only extracted ca 25% as much phosphate as extracted by nitric acid, these results should be treated with caution. Statistical relationships between organic matter and phosphate are presented in Table 17.1.

impoverished boreal podzols are common; such fields develop an organic P signature (Engelmark and Linderholm, 1996). On the other hand, in settlement/long house areas, inputs of inorganic P, for example, from bone and human waste in the form of now-mineralized cess, predominated (Fig. 16.3; see Chapters 10 and 11). As should be expected, when organic matter such as dung is burned or even just charred, its organic P component is changed into an inorganic form (as in the fractionation method). At the settlement locations, hearths areas (combustion zones) are foci of inorganic P concentrations from ashes and their weathered residues. Amounts of P can also give proxy estimates of numbers of humans present for given periods under some circumstances (Linderholm, 2003).

16.2.4 Magnetic susceptibility

Together with organic matter estimations and phosphate analysis, low frequency mass-specific magnetic susceptibility (expressed as χ_{LF}) is a third key laboratory technique commonly employed to provide bulk data at archaeological sites (Crowther, 2003; Thompson and Oldfield, 1986). The magnetic susceptibility of a soil or sediment is generated by the amount of magnetic minerals that are present (Gale and Hoare, 1991: 201ff.). The first influence on magnetic susceptibility enhancement that is of interest to archaeologists, is biological activity forming the magnetic mineral maghaemite, especially in topsoils through the "fermentation processes" associated with alternating oxidation/ reduction conditions (Longworth *et al.*, 1979; Tite and Mullins, 1971). The second is burning, which also mainly affects topsoils, and this enhances χ because iron minerals become aligned (a phenomenon also used for palaeomagnetic dating of hearths). It is therefore clear that if a soil/sediment is low in iron (Fe) it is likely that it will not develop a high χ value, and Fe content is dependant on geology and soil factors (Tite, 1972). Examples of deposits and soils that are low in Fe, are (nonferruginous) peats, or an albic horizon. In studies of Roman and European dark earth sites it was found that the dark earth, which contains large amounts of burned material and especially relict Roman iron slag at Deansway,

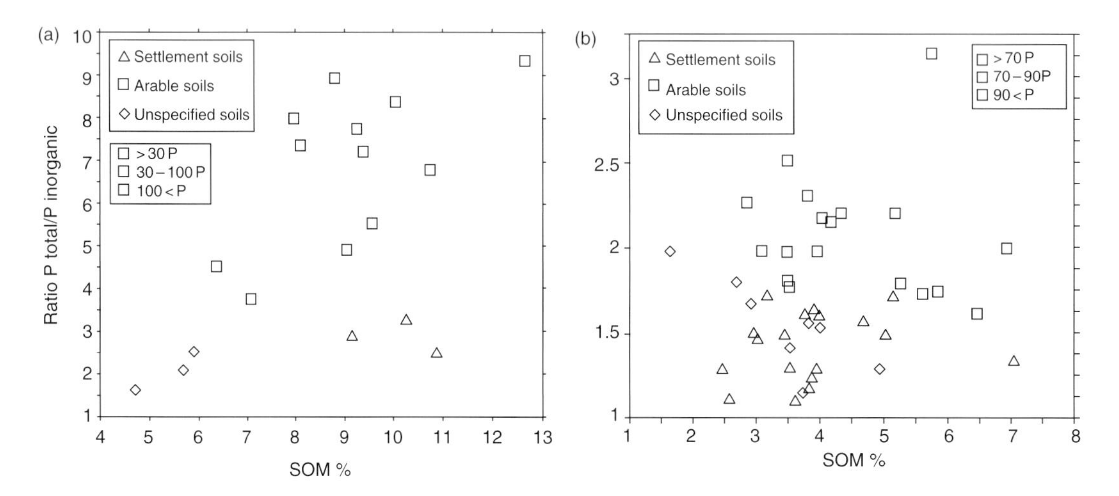

FIGURE 16.3 Scattergrams of Swedish phosphate site analyses (from Engelmark and Linderholm, 1996); (a) scatter-plot showing relationship between soil organic matter (som) and the ratio of total P/inorganic P from the site of Vässingstugan (Medieval farmstead in Småland, south Sweden); (b) scatter-plot showing relationship between soil organic matter (som) and the ratio of total P/inorganic P from the site of Gallsätter (Iron Age settlement in Ångermanland, North Sweden).

Worcester, displays much higher levels of χ compared to earlier Romano-British soils (Fig. 16.4).

Measures of χ_{LF}, are useful and are the most common measurement carried out in archaeology. It is also valuable to know what the maximum potential value for χ is in a given sample (i.e. χ_{max}) (Crowther, 2003; Crowther and Barker, 1995). A simple, qualitative measure can be gained by testing the sample a second time after it has been ignited for estimating LOI (Macphail *et al.*, 2000). Such measurements also provide proxy information on amounts of iron present. Waterlogged sediments often have a very low χ, but when ignited at 550°C can produce very high values (χ_{550}) because of the presence of bog iron (cf. Roman peat at Warren Villas, Bedfordshire, United Kingdom: $\chi = 5 \times 10^{-8}$ Si kg^{-1}; $\chi_{550} = 6098 \times 10^{-8}$ Si kg^{-1}). Equally, when deposits become waterlogged, they are affected by gleying (hydromorphic leaching) processes and the formation of mobile ferrous iron. This can lead to a decrease in the value of χ.

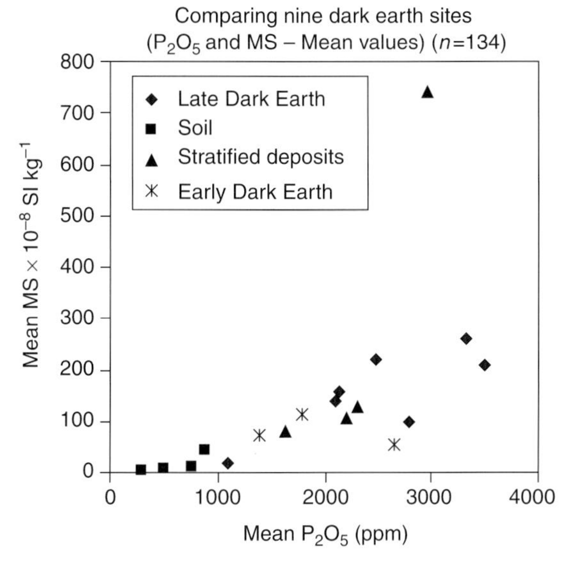

Figure 16.4 Scattergram of phosphate and magnetic susceptibility of nine English "dark earth sites" (mean values $n = 134$) comprising Roman to early medieval deposits; samples have been grouped into Soil (pre-Roman and Romano-British soils), Early Dark Earth (pre-third century Roman dark earth formed in first–second century stratigraphy), Stratified Deposits (Roman and early medieval occupation deposits, domestic and stabling floors, etc.), and Late Dark Earth (fourth to tenth century dark earth); the highest magnetic susceptibility values for late dark earth occur at Deansway, Worcester and result from relict Roman iron slag, while stratified deposits include Late Saxon floors containing slag concentrations (No. 1, Poultry, London; Macphail, in press; see Chapter 13).

Quantitative potential susceptibility (χ_{max}) can be determined by heating a sample under reducing, followed by oxidizing, conditions; by mixing samples with household flour and adding lids to create reducing conditions (Crowther, 2003; Crowther and Barker, 1995; Graham and Scollar, 1976). Crowther (2003) found that he could measure fractional conversion – χ_{conv} (χ_{LF}/χ_{max}) this way and has created a database from 30 sites. Some major findings are that magnetic susceptibility results may be problematic where the χ_{max} is naturally low (e.g. in poorly ferruginous soils) and where the present χ_{max} is no longer representative of the original circumstances in which the soil/sediment formed (e.g. through gleying). Leaching under conditions of podzolization has a similar effect (see Table 16.3). Also, he demonstrated that χ_{conv} does provide a measure of enhancement in samples that are known to have been affected by heating/burning, with values of ≥30% being recorded in certain contexts (e.g. burned daub from Early Neolithic Ecsegfalva, Hungary: $\chi_{LF} = 717 \times 10^{-8}$ m^3kg^{-1}; $\chi_{conv} = 48.1\%$) (cf Plate 13.1).

Magnetic susceptibility has also been used in larger-scale endeavors, as for example, the examination of lake cores, alluvium, and other long sequences (Needham and Macklin, 1992, e.g. figure 2.4; Oldfield *et al.*, 1985). In China, the technique has been employed extensively to recognize paleosols and associated palaeoclimates (Tang *et al.*, 2003). It appears that the amounts of magnetite and maghaemite are enhanced during times of pedogenesis. Thus when thick loess sections are studied, fluctuations in magnetic susceptibility values are clearly higher in palaeosols than the intervening values for aeolian dust (loess). Although the exact mechanisms involved in the enhancement process are not exactly known, they do clearly point to environmental changes, which can be linked with oxygen isotope stages found in deep sea sediments. Consequently, many loess sections can be correlated on the basis of their stratigraphies and magnetic susceptibility changes.

In another application of the technique, Ellwood *et al.* (1997, 2004) has measured the magnetic susceptibility of numerous cave sediments around the world, and particularly in the Mediterranean region. They maintain for example that "magnetic susceptibility (MS) data from Scladina Cave, Belgium, provide a time–depth–climate relationship that is correlated to the marine oxygen isotopic record and thus yields a high-resolution relative dating method for sediments recovered from many archaeological sites" (Elwood *et al.*, 2004: 283). At that site they suggest a date of 90,000 ± 7000 years for Neanderthal skeletal remains found at the site.

As noted above, techniques such as these, while important, should be supplemented with other analytical data and should not be taken at face value. Other measurements of χ, including χ_{550}, χ_{max}, and amounts of total iron present, are all useful additional data when employing magnetic susceptibility to reconstruct past environments (Allen and Macphail, 1987; Clark, 1990; Crowther, 2003; Gale and Hoare, 1991; Longworth *et al.*, 1979; Macphail *et al.*, 2001).

16.2.5 Sourcing (provenancing)

Sourcing raw materials of manufactured and constructional materials, and establishing the provenance of sediments and archaeological deposits, and determining the origin of a soil's parent material, is a common first step in site investigations. Methods to do this are discussed below where we deal with Instrumentation Techniques (Section 16.5), and the analysis of minerals (including "heavy minerals") through optical microscope and X-ray methods (XRD), and chemical and microchemical studies (XRF, NAA, ICP, microprobe, SEM-EDAX, "Quemscan," Mass-spectrometry). These methods are also used in the investigation of sediment diagenesis and determination of the type and maturity of soil formation.

16.3 Microscopic methods and mineralogy

16.3.1 Binocular and other macro-methods

Binocular examinations of loose samples, thin sections, and polished blocks usually range from × 4 to × 20(40). Such low magnifications provide information about the larger-scale features of the sample that might not be evident with higher magnifications alone. In the case of thin sections, useful macroscopic investigations can be carried out using binocular microscopes with both plane- and crossed polarized light, as in normal petrological microscopy (see section 16.4). Even if binocular microscopes only have an incident light source, polished blocks (i.e. impregnated blocks with a resin-coated or spray coated "cover slip"), can be studied to view macrostructure, inclusions, coarse voids, and bedding.

The coarse fraction (>2 mm) of loose samples can be analyzed for composition, texture, shape, and roundness. This technique is often used on coarse natural sediments, such as river gravels. It has also been applied to anthropogenic deposits such as dark earth and other occupation deposits, in which stratigraphic units are difficult to determine, but where features and units may be indicated by concentrations of coarse inclusions (Sherwood, 2001). This approach has been carried out to establish amounts of Roman mortar, tile and brick, stones, bone, charcoal, and food waste in the form of mussel shells in deposits from sites in London and more recently at St Julien, Tours,

France (Macphail, 1981; Galinié, University of Tours, personal and communication; Sidell, 2000). Middens have also been studied in this way, in order to identify stones, botanical (seeds, wood, and charcoal fragments), and faunal (shells, animal and fish bones) remains (Balaam *et al.*, 1987); (see Fig. 7.5). Lastly, "invisible" stratigraphy has been elucidated through the use of X-radiography, where undisturbed cores are collected in 450 mm long, 100 mm diameter plastic tubes, as standard U4/U100 samples (see Chapter 15) and then X-rayed (Barham, 1995: 164).

16.3.2 Petrography – soil micromorphology

Soil micromorphology is considered by some to be one of the new techniques in geoarchaeology. In fact, the employment of soil micromorphology in archaeology has a history almost as old as the subject itself; Kubiena first developed the subject from geological petrography in the mid-twentieth century (1930s–1940s) in Germany (Kubiena, 1938, 1953), It was almost immediately taken up by Cornwall at the Institute of Archaeology, University of London who produced reports up to the 1960s (Evans, 1971, 1972; Macphail, 1987). The Scottish soil scientists Roman and Robertson also applied this pedological technique to archaeological sites during the 1970s and early 1980s (Romans and Roberston, 1975, 1983a,b). The subject also developed internationally during this period (Courty and Fedoroff, 1982; Courty and Nornberg, 1985; Courty *et al.*, 1989; Goldberg, 1979, 1981).

Since these early efforts archaeological soil micromorphology has taken off as a mainstay of geoarchaeological investigations, especially in Europe (International Soil Micromorphology Working Group, initiated in 1990 by Macphail), with numerous articles in conference proceedings, *Geoarchaeology* and *Journal of Archaeological Science*. There has always been a strong tradition of applying soil micromorphology in the United states, especially for soil classification and even for the study of NASA's moon rock (Douglas, 1990; Soil Conservation Service, 1994). Despite the long-term major contribution made by Goldberg (initially at University of Texas, now at Boston University), this technique, however, has been completely underused by North American geoarchaeologists (Collins *et al.*, 1995; Holliday, 1992). It is only now that soil micromorphology is recognized as a geoarchaeological component in North America (Goldberg *et al.*, 2001); hence the sole micromorphy workshop held at Boston University's Sergeant Centre in the fall of 2003, as opposed to workshops held across Europe since the 1990s (e.g. chiefly at Cambridge and London in the United Kingdom, and in Belgium, France, Italy, Switzerland (Arpin *et al.*, 1998; Boschian *et al.*, 2003)(Bulletins from London, Pisa, and Ghent are on http://www.gre.ac.uk/~at05/micro/soilmain/soilpage.htm).

16.3.2.1 Protocols in soil micromorphology

A number of protocols for the scientific description of thin sections that can be applied to archaeological soils over the past several decades have been formulated. The earliest ones were introduced by Kubiena (1993) and Brewer (1964), and the most recent and most commonly used internationally, is by Bullock *et al.* (1985) with recent revisions by Stoops (2003a). Much can also be learned from FitzPatrick (1984, 1996) although much of his terminology is not in mainstream use. These descriptive systems, combined with those from sedimentary geology (e.g. Tucker, 2001) have been utilized to describe many kinds of archaeological materials and situations. For example, archaeological cave sediments, Pleistocene sediments, floor deposits, "manufactured" materials (e.g. daub and mortar), pit fills, and soils *sensu stricto* have been described (Courty *et al.*, 1989, 1994). This has permitted the accurate characterization and interpretation of archaeological deposits, but understandably as more complex and completely novel materials are investigated, the use of the petrological microscope has had to be supplemented by a whole range of complementary techniques (e.g. fluorescence microscopy, cathodoluminescence, SEM-EDAX/microprobe, Fourier Transform Infrared Spectrometry/ FTIR) (Schiegl *et al.*, 1996; Weiner *et al.*, 1995; see below).

16.3.2.2 Sample processing methods

Essentially, the undisturbed sample that was retrieved in the field (Chapter 15) needs to remain undisturbed during processing, which requires the removal of any soil water and impregnation of the sample with a resin that will eventually produce a hard block after curing (FitzPatrick, 1984; Murphy, 1986). Further details of sample processing methods for thin section manufacture, and some guidelines concerning the handling of thin sections are given in 16A.5 and 16A.6.

16.4 Thin section analysis

Essentially, petrological studies are carried out using plane polarized light (PPL), crossed polarized light (XPL), and oblique incident light (OIL). The use of these has been detailed in textbooks on petrology, and soil micromorphology (Bullock *et al.*, 1985; FitzPatrick, 1984; Stoops, 2003a). Another commonly used technique is fluorescence microscopy (van Vliet *et al.*, 1983a), which can be carried out using an attachment to the petrological microscope (see below). A multiple method approach employing the above optical methods, and microprobe is given in Figure 16.5 (see Table 16.4 and Fig. 16.1b, for complementary bulk data).

The most important aspect of the procedure is to have a continuum of observation from the lowest to the highest magnifications, that is, from ×1 (hand held thin section), through ×25, ×40, ×100, and ×200/400. Thus, there will be no observational break between what has been seen in the field and what is observed under the microscope. Macrostructure, field layers, coarse inclusions cannot be recognized at the highest magnifications. Equally, the composition of the fine fabric cannot be elucidated at low magnifications. A study protocol should therefore include the following six steps:

1 *Preliminary examination* of the thin section by eye (×1), by using a hand lens, microfiche reader, or binocular microscope (×4–×20) – the last may also provide both PPL and XPL observation (see above). If the thin section has also been scanned, the scanned image can be examined using a software package such as Adobe Photoshop®; and different levels of lighting and contrast, and color balance, can be utilized to see what may be present but not immediately obvious (Arpin *et al.*, 2002). To repeat: the chief role of this macro-study is to relate what was seen and recorded in the field (photos, images, section drawings, contexts sheets, notebook), to the laboratory study (whole contexts, layers within contexts, etc.) (Courty, 2001). At this stage, macrostructure and heterogeneity, for example, can be identified and examined more closely during stage 2.
2 *Thin section scanning* using the polarizing microscope can be carried out at ×25–×40 magnifications. Areas/features/inclusions of interest can be identified, and information on these can also be related back to the field observations. In some cases, it may be desirable to scan all the thin sections at this stage, to see if field and macro-morphological identification of similar features is correct. It is important at this early stage to switch rapidly between PPL, XPL, and OIL. For example, variations in amounts of organic matter may show up well in PPL, while calcium carbonate-rich areas (e.g. patches of ash, secondary carbonate) become strongly evident under XPL because of their high interference colors (birefringence = 0.172). Equally, OIL will make apparent iron and manganese stained features, while fragments of "black" charcoal and "red" burned (e.g. rubefied) soil become obvious. This may also be a good time to scan

using blue light (BL), so that residual roots, bone, and calcium phosphate-enriched materials (some coprolites; phosphatized soil/sediment), and other autofluorescent materials can be noted (see below). Areas and features can be denoted on the slide by drawing on the cover slip with a marking pen, or on the back of the thin section, noting a two dimensional coordinate, and/or drawing on a scanned image.

3 *Detailed observation and description* of areas/features/inclusions should be carried out carefully *without* any interpretative weighting. If interpretation is introduced at this stage it can undermine the systematic gathering of soil micromorphological data. It is recommended that any new worker or student *sensu lato* should employ Stoops (2003a) in conjunction with Courty *et al.* (1989), although Bullock *et al.* (1985) and FitzPatrick (1996) contain useful basic information on soil materials and secondary minerals. It is also suggested that as skills and knowledge develop, specialized articles can be sought (Miedema and Mermut, 1990) including those in the conference proceedings of the International Working-Meetings on Soil Micromorphology (International Society of Soil Science Commission B; e.g. Bullock and Murphy, 1983; Douglas, 1990; Fedoroff *et al.*, 1987; Stoops, 2003b). Although all new workers are strongly urged to undergo proper training in soil micromorphology and to record all their information systematically, it is important that the results are presented in such a way as to be rapidly accessible to both specialist and lay colleagues alike. Examples of description are given by Bullock *et al.* (1985: 142–145) and Stoops (2003a: 137) also includes a tabulated presentation. In addition, we present a suggested order of description and a number of worked examples from a range of site types (Appendix 16A, Tables 16A.3–16A.6). Ways to estimate approximate percentages are repeated in these books, after FitzPatrick (1984). With experience and much time spent estimating percentages, great accuracy can be achieved. One experiment showed the differences between image analysis of thirteen counted 0.5 mm wide transects as varying between 0 and 5%, but here manual counting took some 8 h per slide (Acott *et al.*, 1997; Crowther *et al.*, 1996). Individual characterization and identification of features and materials may also require instrumental (e.g. EDAX, microprobe) and other analyses.

4 *Presentation of data* needs to be carried out in ways best suited to the findings, type of site and chosen audience (see Chapter 17). Although systematic presentation is useful, it must be remembered that each site and audience varies and the presentation of data should be flexible. Very few readers are going to wade through a mass of descriptive information for its own sake, and so presentation of material has to be very carefully balanced to be both inclusive and relevant. In Appendix 16A we present a standard format (Table 16A.3), and how it can be employed to describe a freshwater sediment (Table 16A.4; see Figs 16.6e–h), a barrow buried soil (Table 16A.5; see Figs 16.6a–d), and an anthropogenic deposit (Table 16A.6). An additional approach has been to present the semiquantitative data in the form of count tables (Macphail and Cruise, 2001; Simpson, 1997; Simpson *et al.*, 2005). We believe that a primary step in the interpretational process is the characterization and identification of soil/sediment microfabric types (SMTs) and soil microfacies types (MFTs) (as Table 16A.6) (Courty, 2001; Macphail and Cruise, 2001; Macphail and Crowther, in press).

5 *Interpretations* need to be carefully explained and justified. They should first follow "identifications" of SMTs, inclusions, and (*pedo*) features (i.e. to include all sedimentological, pedological, and anthropogenic processes), to reach a first level of interpretation (e.g. homogeneous material, mixed material, illuvial or gleyed character, high or low biological activity). At this time *primary*

materials such as sediments can be differentiated from *secondary* materials such as soils. This stage is then followed by a second level of interpretation (e.g. identifying the presence of a specific type of deposit – e.g. colluvium, soil horizon, occupation layer, or cultivation soil; French, 2003: 47). This involves ascribing a MFT (microfacies type). The third level of interpretation attempts to place the slide in a broader interpretive context, such as overall soil type, mass-movement deposit, type of land use, stable, house floor. It should be recognized, however, that it may not always be possible to reach the third level of interpretation and one should avoid jumping to this level too soon during thin section analysis, as this biases the observation process. Moreover, crucial identifications of anomalies may be erroneously disregarded if descriptions are carried out with an "agenda" that weighs descriptions in favor of a predetermined interpretation.

6 *Testing of interpretations* is done by reviewing the first five stages, and comparing findings with those from independent studies in other disciplines, such as bulk chemistry, magnetic studies, and palynology. Such combined datasets can also be presented together in tables, so that the sceptical reader may be persuaded by consistent findings from different methods. As in all geoarchaeological studies interpretations can also be reviewed against field/excavation results, as well as the interpretations presented from zooarchaeology, archaeobotany, artifact recovery, and dating. In the examples presented in Appendix 1 (Tables 16A.3–16A.6), the interpretation of freshwater sediment sequence at Boxgrove was not only supported by thin section analysis of calcite root pseudomorphs, but also by bulk counts of slug plates and earthworm granules and the presence of relict saline foraminifera washed in from the eroded marine substrate (Roberts and Parfitt, 1999; Roberts *et al.*, forthcoming). At Raunds, the interpretation of soils becoming both phosphate-enriched and poached through stock concentrations was consistent with findings from macrofossils (cattle bone, grass tuber, dung beetle) and archaeological findings (Healy and Harding, forthcoming; Macphail, 2003). Cave sediment diagenesis was substantiated by detailed mineralogical analyses using FTIR (Karkanas *et al.*, 1999, 2000; Weiner *et al.*, 2002; see Chapter 8). At the London Guildhall, interpretations of soil micromorphology and complementary chemistry and pollen were based upon experiments (Macphail *et al.*, 2004), as well as the well-preserved timber building remains (Bateman, 2000). Many other examples are provided elsewhere in this book.

Finally, the first to third level interpretations of thin sections can be based upon pedofeatures (see Tables 16A.3 and 16A.4; Figs 16.5a–d). When these are described, they often show an order based upon the "law of superposition," and as such, one has to evaluate which pedofeatures came first, second, third (etc.), and last. At Raunds, a whole range of clay coating types was observed in the prehistoric soil. A yellow brown clay was always superimposed on other clay coating types and soil structures, and moreover could be traced upward into the medieval alluvium-enriched ploughsoil. This relationship therefore permitted the identification of the yellow brown clay as the *last* phase of textural pedofeatures formation (Fig. 16.5d). In the example from Boxgrove, meltwater flow and *sedimentation* produced the polyconcave and closed vughy porosity and associated textural "pedofeatures" of intercalations and void coatings, as a final phase of pedofeature formation (Fig. 16.6a; see Table 16A.5).

Furthermore it is crucial to continually bear in mind the principle of *equifinality* when interpreting soil micromorphology. So many different natural and human-induced processes can lead to the similar microfabrics and pedofeature(s). A sequence of limpid clay coatings being succeeded by dusty clay

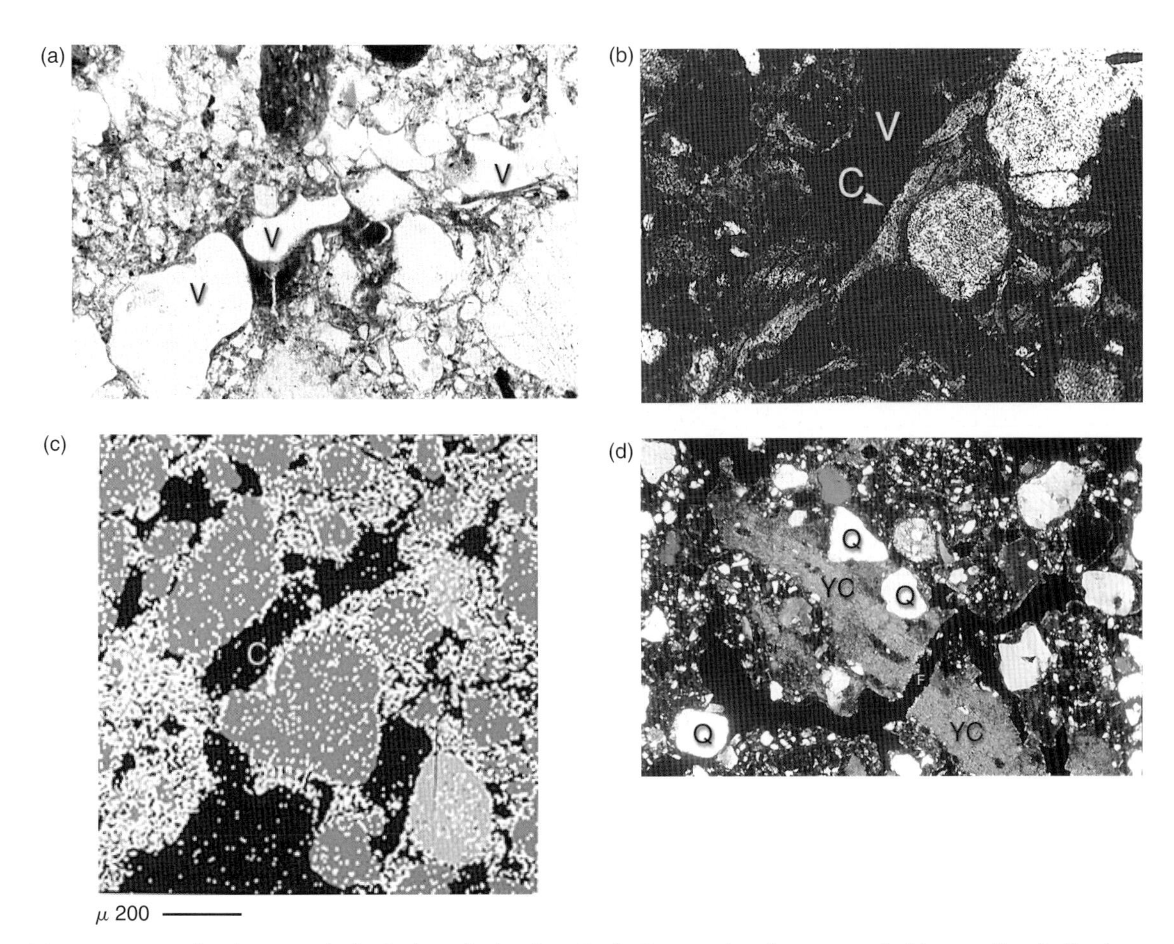

FIGURE 16.5 Soil micromorphological analysis of an Early Bronze Age barrow sealed by medieval alluvium at Raunds, Northamptonshire, United Kingdom, including complementary microprobe investigations (see also Tables 16.4 and 16A.5) (a) Bronze Age barrow buried topsoil (A horizon) with anomalous dark red (humic) void (V) clay coatings in organic phosphate-enriched soil, indicative of soil poaching and a proxy indicator of stock concentrations. PPL, frame width is 3.3 mm. (b) Bronze Age barrow buried upper subsoil (Eb or A2 horizon) with anomalous dark red (humic) void (V) and grain clay coatings (C). Uncovered thin section used for microprobe line analysis across clay coating (0.25% P; n = 9; length of transect = 114 μm). XPL, frame length is 0.33 mm. (c) As Figure 16.6b, microprobe map of Si (grey) and P (white), showing, in addition to relict phosphate-rich iron nodules, P-rich clay void/grain coating (C). (Bar equals 200 μm.) (d) Bronze Age turf mound, showing major inwash of yellow clay (YC) from postdepositional burial of the site by Saxon and medieval fine alluvium, which contrasts with the quartz (Q) sandy loam prehistoric soil. Drying of the sample leads to the formation of a sloping fissure (F) seen across the image. XPL, image width is 3.3 mm.

coatings, for example, can be formed in more than one way (Courty *et al.*, 1989):

1 Limpid clay coatings formed under undisturbed woodland, followed by clearance-induced dusty clay coatings; e.g. *woodland → woodland clearance;*
2 Limpid clay coatings formed from the dissolution of K-rich ash resulting from a campfire of burned peat, succeeded by surface freeze and thaw-induced dusty clay coatings, all in a nonwooded environment; e.g. *campfire on a prairie/tundra site.*

3 Limpid clay coatings resulting from the down-profile fine sorting of (soil surface formed) once-dusty clay caused by ploughing were followed by dusty clay that remains unsorted because erosion has produced a shallower soil profile; e.g. *plough erosion*.
4 Limpid clay coatings formed through fine alluviation (inundation of a flood plain), followed by animal trampling disturbance; e.g. *natural alluviation → grazing by stock or wild herds*.
5 Limpid clay coatings formed through groundwater movement (for instance from weathering basalt), succeeded by dusty clay coatings resulting from an overlying mud-flow; e.g. *geologically active volcano*.
6 Successive limpid clay and dusty coatings formed in pit, grave, or treethrow/clearance subsoil hollow, induced by consecutive inwash of soil according to lightness and heaviness of rain; e.g. *single disturbance event*.

Lastly, to the untrained eye, once-limpid clay coatings can take on a less limpid appearance if they suffer from partial breakdown under hydromorphic (e.g. loss of Fe) or podzolizing (acid breakdown of clay) conditions; the "outside" or latest clay laminae will obviously be affected first by such processes. Thus caution needs to be exercised during description; otherwise results may simply follow wishful thinking to attain a "finding."

We hope that neophytes do not find the above presentation too daunting, but it is exactly what the authors have to do at each site, in order to make objective observations and the best possible interpretations according to site circumstances. In fact, if objective observations are made and integrated with other geoarchaeological data (bulk and other instrumental analyses) and the site's archaeological and geological context, reasonable interpretations should be forthcoming if the six-step protocol outlined above is adhere to.

16.4.1 Fluorescence microscopy

Standard petrography is usefully combined with the use of fluorescence microscopy by adding a fluorescence attachment to the petrological microscope. Observations can be made, either using ultraviolet light (UV), blue violet light (BV), and blue light (BL). The attachment combines an exciting filter between the light source (sometimes as a form of incident light) and the thin section, and a suppression filter between the object being studied and the eyepiece (Stoops, 2003a: 24–27). A number of identifications can be made using UV, BV, and BL, although in pedology fluoresence microscopy was also commonly employed to study void space (through image analysis) made apparent by adding fluorescent dyes to the infrequating resin (Jongerius, 1983).

"Free" aluminum can produce strong autofluorescence in some soils, such as spodic horizons (Van Vliet, 1980), and may cause some burned phytoliths to be autofluorescent because they also have become coated with "free" Al (Van Vliet-Lanoë, University of Lille, personal communication). It is also likely that it is the chemical composition of amorphous organic matter in Spodosols, Andosols, and Planosols – especially aluminum complexes – that affects autofluorescence (Altemuller and Van Vliet-Lanoe, 1990; Van Vliet *et al.*, 1983). Organic material such as cellulose, roots, fungi, pollen, and spores, often become more visible using fluorescence microscopy (Courty *et al.*, 1989: 48–50, plate Id; Davidson *et al.*, 1999; Stoops, 2003a). Lastly, the presence of calcium phosphate as various types of apatite can be inferred from their autofluorescence under blue light. Guano-induced phosphatized limestone (in caves), bone, and coprolites of birds, dogs, carnivores, and humans can be identified (Courty *et al.*, 1994; Macphail and Goldberg, 1999). In areas of intense anthropogenic impact, high amounts of phosphate can be liberated from weathered ash and bone, or be excreted directly from animals. Here, occupation sediments can become saturated with calcium phosphate, but the resulting hypocoatings are difficult to see under normal lighting conditions. Although fabrics become isotropic ("black" under XPL), they are best seen using fluorescence microscopy, as well illustrated by the studies of Late Bronze Age/Early Iron Age Potterne, Wiltshire and an experimental stable at Butser Ancient Farm, Hampshire, United Kingdom (Macphail, 2000; Macphail, *et al.*, 2004; see Chapter 12). X-ray diffraction, EDAX, and microprobe studies (see below) confirmed the identification of calcium phosphate in those investigations.

16.4.2 Image analysis of thin sections

Image analysis was used extensively in soil science, especially to measure void space. A fluorescent dye in the resin made voids "visible" and were studied by a system called "Quantimet" (Jongerius, 1983; Murphy *et al.*, 1977). More recently, with the widespread use of PCs and quantification software, such measurements can be carried out quickly (McBratney *et al.*, 1992, 2000). Image analysis techniques have now been extended from the analysis of voids and the size, shape, orientation of constituents, into three-dimensional studies of soil using X-ray computed tomography (Bresson and Moran, 1998; Terrible *et al.*, 1997; Tovey and Sokolov, 1997; Francus, 2004).

Some attempts have also been made in archaeology to quantify more accurately soil micromorphology using image analysis. Image analysis data cannot be interpreted alone, however, without standard soil micromorphological descriptions and accurate identifications (Acott, University of Greenwich, 2000; Murphy, Soil Survey of England and Wales, personal communication, 1985). In the latter, manual semi-quantitative and quantitative estimates of features and inclusions that are difficult or near impossible to discriminate in practical terms by image analysis, provide a dataset on a par with those produced by image analysis.

Some examples of combined studies are the investigation of a Bedouin floor and Experimental Earthworks Project (Acott *et al.*, 1997; Goldberg and Whitbread, 1993; Macphail *et al.*, 2003). At the Experimental Earthwork at Wareham, Dorset, a number of pilot studies were carried out, and eventually nine types of mesofauna excrements were manually counted; image analysis was employed mainly to accurately measure void space, organic fragments, dark (amorphous) organic matter, and mineral grains of sampled areas at 1 mm intervals. The possible compression of the buried Ah horizon in comparison to unburied (control) soils was investigated by comparing void space. Image analysis data on amounts of "organic fragments" and "dark organic matter" was measured against bulk chemical estimations of organic matter (low temperature LOI and organic carbon). Results showed a consistent match between the two datasets (Macphail *et al.*, 2003: see Chapter 12).

Standard soil micromorphology was also combined successfully with image analysis of Boxgrove's freshwater deposits (Roberts *et al.*, 1994). Here void space and void size were analyzed employing circular polarized light (Stoops, 2003a: 21), because voids are black compared to the predominantly anisotropic calcareous silt sediment. This technique was crucial in appreciating the amount of fine void space associated with packing voids (Unit 4*) compared to fewer but much more obvious and larger void spaces (closed vughs) in overlying Unit 4d1 (Figs 16.6a–d; see Table 16A.4).

Despite the above useful combined studies and the large number of cautionary tales and caveats regularly published by specialists in image analysis (Acott *et al.*, 1997; Terrible *et al.*, 1997), there has been a very worrying trend for work and grant applications to play down the role of standard soil micromorphology in favor of techniques such as image analysis and

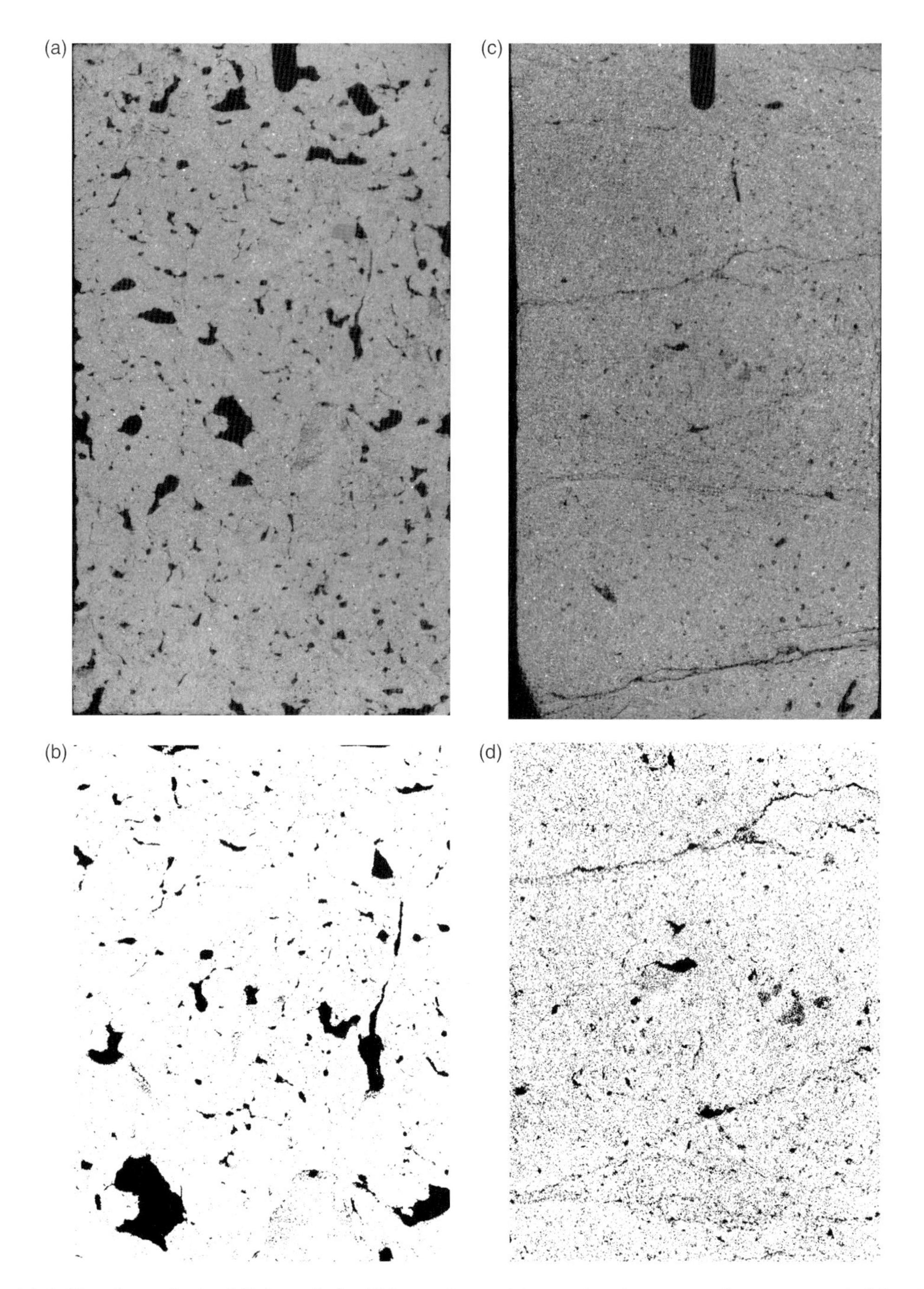

FIGURE 16.6 Circular polarized light; whole 130 mm long thin section images and areas sampled for image analysis of void space and void shape (binary images); Boxgrove thin sections 14c and 13c. (Analyses by Dr Tim Acott, Greenwich University; Roberts *et al.*, forthcoming). (a) and (b) Soil Micromorphology: massive with dominant fine to medium, coarse meso to macro-channels, showing partial collapse, with polyconcave and closed vughs, and few complex packing voids. Image Analysis: 8% voids; dominant fine and medium (61.9% – 101–200 μm^2), frequent coarse (24.1% – 201–500 μm^2) meso voids, few fine

X-ray tomography. The thinking is that soil micromorphological identifications and interpretations are too difficult and *unreliable*, especially as they are "non-quantitative." Features and materials in thin section have to be identified and clearly understood through standard soil micromorphology first, before image analysis can be used to quantify them. The suggestion that image analysis eliminates the bias of the observer and their interpretation, may sound scientific but is in fact, nonsensical. Investigators need to *know* what they are "counting." Blind use of instrumental methods are the very antithesis of holistic microfacies approaches that have contributed so much to our advances in geoarchaeology (Courty, 2001; Matthews *et al.*, 1997).

16.5 Minerals and heavy minerals

In sedimentary geology, the composition and texture of deposits can serve as an important indicator of source, transport, and depositional processes (see Chapter 1). For example, in large sedimentary basins that drain different types of bedrock, petrographic analysis of different grains can reflect the relative contributions and evolution of different source areas that furnish sediment into the basins (Pettijohn, 1975).

Similar strategies can be employed in geoarchaeology in different ways. Perhaps the foremost is the petrographic analysis of pottery in which mineralogy of pottery is examined (e.g. Rice, 1987; Whitbread, 1995; Williams and Jenkins, 1976). Information gleaned from the examination of mineralogy and texture, can be used to infer the source of ceramic temper. In the New World (southwest United States), A.O. Shepard (1985) was a pioneer in ceramic petrography. Numerous other studies in North and Central America show the value of the technique (Iceland and Goldberg, 1999; Miksa and Heidke, 2001; Shepard, 1939). Peacock (1982) made extensive analyses of Roman ceramics. Such work was expanded by Wieder and Adan-Bayewitz (2002) who were able, by employing micromorphology of the local soils in Galilee (Israel), to identify specific raw material sources for Roman pottery. Elsewhere, raw materials and trade routes have been located (Beynon *et al.*, 1986; Cornwall and Hodges, 1964; Goldberg *et al.*, 1986; Goren and Goldberg, 1993; Kapur *et al.*, 1992; Peacock, 1969).

In addition to thin sections, the mineralogy of an object or deposits can be studied by analyzing individual grains under the microscope (Carver, 1971; Tucker, 1981). Loose sand-size grains, for example, can be mounted on a glass slide and examined under the petrological microscope. Mineralogical composition can be determined optically using characteristics such as shape, cleavage, fracture, and birefringence.

Particularly useful in these types of single grain mounts are the so-called heavy minerals, which are those with specific gravities >2.9. These minerals are usually separated from the "light minerals" (e.g. quartz and feldspar) either with an electromagnetic separator or by using a heavy liquid. With regard to the latter the sediment is put into a test tube with the

FIGURE 16.6 *(Cont.)*

(6.5% – 501–1000 μm^2) macro voids and very few medium and coarse macro and mega-voids (2.5% – 1000–>2000 μm^2). (c) and (d) Soil Micromorphology: *Structure and voids*: massive, burrow, and channel microstructure; common very fine to fine (0.1–0.4 mm), frequent medium (0.8 mm), and coarse (1 mm), and few patches of very fine packing pores. Image Analysis: 15% voids; very dominant fine and medium (78.5% – 101–200 μm^2), frequent coarse (20.3% – 201–500 μm^2) meso voids, very few fine (0.9% – 501–1000 μm^2) macro voids and very few medium and coarse macro and mega-voids (0.3% – 1000–>2000 μm^2).

heavy liquid and the lighter grains float, while the heavy minerals sink to the bottom. Originally, bromoform or tetrabromethane was used as the liquid, but as these proved to be toxic, they were replaced by sodium polytungstate solution ($Na_6(H_2W_{12}O_{40}) \cdot H_2O$). Once the minerals are separated and mounted on the slide, they can be counted by scanning the slide systematically, using any of a number of point counting techniques (Tucker, 1981). The percentages of the different minerals can be tallied and then interpreted (Catt, 1986, 1990).

Normally the suite of heavy minerals can point to a sediment source, that is its provenance, or help to identify possible locations of raw material used to manufacture ceramics. In the United Kingdom, mineralogical analyses were carried out to identify loess in southern England, such as at the Neolithic site at Pegwell Bay, Kent (Weir *et al.*, 1971). At Boxgrove, mineralogy also demonstrated that there was no relationship between the marine sediments and the terrestrial "brickearth," showing that they were not derived from a local "loess" (Catt, 1999).

Alternatively, heavy minerals can be used as weathering indicators, because some minerals are more susceptible to dissolution than others, and this follows the opposite path to Bowen's reaction series (Pettijohn, 1975). Weathering indices have been used to estimate amounts of clay that can be formed in soils (Catt and Bronger, 1998; Catt and Staines, 1998; Weir *et al.*, 1971). At Tabun Cave the lower two-thirds of the sedimentary column consisted almost entirely of zircon, tourmaline, and rutile, which are very resistant minerals (Goldberg, 1973). The upper part (Layers B and C) contained a suite of minerals that matches those found in modern coastal sediments in Israel: hornblende, pyroxenes, staurolite, with relatively minor amounts of zircon, tourmaline, and rutile. The inference, corroborated by other analyses, is that these lower sediments had been subjected to intense diagenetic alteration, in which the less stable heavy minerals were removed, leaving a concentrate of the more stable ones. Traditional heavy mineral analysis through petrology is highly skilled. Specialists prefer to study grains that are not cemented in a mount, but which can be turned around to see their three-dimensional characteristics (J. Catt, ex-Rothamsted Experimental Station, personal communication). More commonly nowadays difficult grains will be placed in a SEM so that its chemical composition can be identified through EDAX (X-Ray analysis), allowing quite rapid mineralogical identifications (see QuemSCAN below).

16.6 Scanning Electron Microscope (SEM), EDAX, and microprobe

A number of additional instrumental methods are employed to analyze thin sections (Courty *et al.*, 1989: 50). Of these only Scanning Electron Methods (SEM) and the similar electron microprobe are briefly explored here (Fig. 16.5d). These techniques are capable of examining materials that are too small (<2 μm) to be resolved by light microscopy, and also provide punctuated (spot) microchemical data on uniform grains, heterogeneous rock fragments and soils, coatings, and other pedofeatures. It should not be forgotten that although the very highest magnifications can be used (e.g. up to ×60,000), it is useful to commence a SEM study at ×100 to provide a natural transition between observations with the light microscope and those with the SEM.

Different types of samples can be employed (see below). These can include an individual object, such as a piece of pottery, or several objects (e.g. mineral grains, plant or seed fragments). In soil work, it is common to examine loose aggregates of soil or sediment. In either case, the object(s) are mounted on a metal disc and coated with a nonconductive material such as gold or carbon. Alternatively, an uncovered thin section or impregnated block can be the object of study. Both these must be polished, particularly if microchemical analyses are to be performed, either with the SEM or the electron microprobe.

Scanning electron microscopy utilizes electrons to gather an image rather than light (Herz and Garrison, 1998; Pollard and Heron, 1996). Essentially, an electron source is focused on and scans across the sample by using electromagnetic field lenses. When the electrons strike the object, both electrons and X-rays are given off, and each form of energy can be detected and each provides different sorts of information.

Secondary electrons are low energy electrons produced close to the surface by inelastic collisions with the beam (Ponting, 2004). The images produced by secondary electrons are grey scale photographs with high definition and depth of field, and because the electron beam scans across the sample, it is possible to obtain three-dimensional view of the material. In this mode, it is possible to observe dissolution or precipitation of crystals, as well as their relative formation history (e.g. gypsum crystals overlying calcite). In the 1970s SEM analyses in geology were committed to studying the surface textures of quartz grains with the aim of inferring their depositional and diagenetic histories (Bull, 1981a,b; Bull *et al.*, 1987; Krinsley and Doornkamp, 1973; Smart and Tovey, 1981, 1982).

Such an approach was taken in the case of Tabun Cave, Israel, a Lower Palaeolithic site that is infilled with substantial amounts of sand. Bull and Goldberg (1985) analyzed the surface features of the quartz grains to reveal depositional and diagenetic environments of the deposits. Layers F and G (Garrod and Bate, 1937), for example, displayed primarily wind-blown characteristics but with diagenetic alterations, probably because they sit on top of bedrock where ground and sediment water would tend to be concentrated. Quartz grains from Layer E on the other hand, showed aeolian traits, but also the effects of "high energy subaqueous buffeting action" in a sublittoral and littoral environment (Bull and Goldberg, 1985: 181). The uppermost layers (D, C, and B) show mostly aeolian features, above the middle part of Layer D. These results supported the hypothesis that Layer D represented markedly wetter conditions (Goldberg, 1978). Similar types of studies can be conducted on sediments from a variety of environments (e.g. Bull *et al.*, 1987).

Backscattered electrons (BSE) are a result of elastic contact with the electron beam. With BSE, the energy emission is proportional to the atomic number of the element so that the higher the atomic number the brighter the image. The employment of BSE serves as a complement to secondary electron images for it is a relatively simple means to reveal variations of composition within a sample and thus disclose some of the very fine-scale fabric elements (Karkanas *et al.*, 2000). It is typically utilized to examine pottery, metals, soils, and sediments. The exact elemental composition of materials cannot be judged from the BSE images alone, although shapes of components coupled with micromorphological data can often point to the correct mineralogical identification in the BSE image. To observe elemental abundance secondary electrons are employed. The BSE image is often used to identify areas for microprobe analysis (Fig. 16.7).

X-rays are produced in the process of secondary electron emission when it is replaced by an electron from a different part of the atom. What is important is that the X-rays that are emitted, have energies and wavelengths that are characteristic of each element and its energy shells (Pollard and Heron, 1996; Ponting, 2004). Two techniques are used to measure the X-rays and thus infer elemental composition: energy-dispersive X-ray analysis (commonly written as EDX, EDAX, or EDAXRA) and wave-length dispersive X-ray analysis (WDX). EDAX analyses can be performed on either loose grains or aggregates, or more efficiently in conjunction with micromorphological analysis, on a polished thin section. The latter approach, allows the researcher to expand the continuum of observation from field through macro, micromorphological, and ultimately, elemental scale. An instrument that is similar to the SEM and is specifically oriented toward analysis of polished blocks or thin sections is the electron microprobe (Pollard and Heron, 1996; see below).

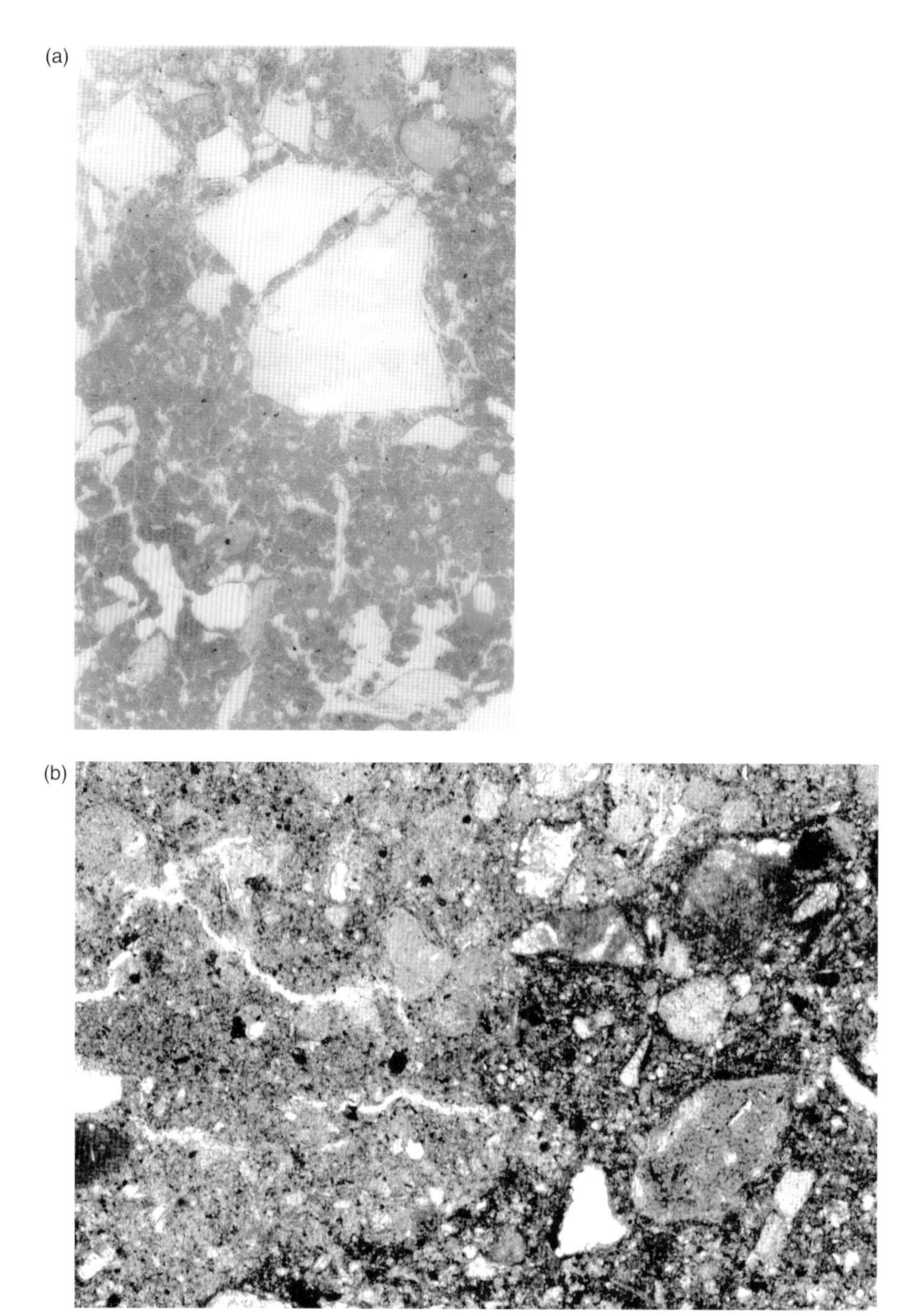

FIGURE 16.7 Hohle Fels Cave, Swabian Jura, Germany. A field photograph showing part of an excavated profile is shown in fig. 15.11b (a) A macro image of sample 13 [see Fig 15.11b for location] produced on a flat-bed scanner (Arpin *et al.*, 2002) reveals a more nuanced view of the composition and fabric of the sediment. Shown here are angular and rounded éboulis and the matrix, which exhibits considerable void space. The rounding is produced by cryoturbation under cold, glacial conditions. (b) The porosity and nature of the matrix is better shown here in this photomicrograph taken in plane-polarized light (PPL). Visible here are numerous rounded.

With EDAX it is possible to analyze the composition of specific grains, even those that are only a few micrometers in diameter because the electron beam can be focused to 1–2 μm across. This allows for the determination of different elements, which in turn is an indicator of the mineral being observed. Such analyses are presented as spectra in which energies correspond to different elements. These spectra also indicate relative abundances of the different elements, which in turn can be utilized to infer the mineralogical composition of the material or sample and ultimately determine its history (Karkanas *et al.*, 2000; Weiner *et al.*, 2002) (see Plates 3.1 and 3.2).

In addition, maps or distribution plots of selected elements, are made on specific areas of the thin section or block as shown in Figures 16.5c, 16.7c, and 16.8. In this last figure, the combination of elements, such as Ca and P, permit the identification of guano. Clasts or materials composed of calcium phosphate, at Hohle Fels Cave, are probably representative of

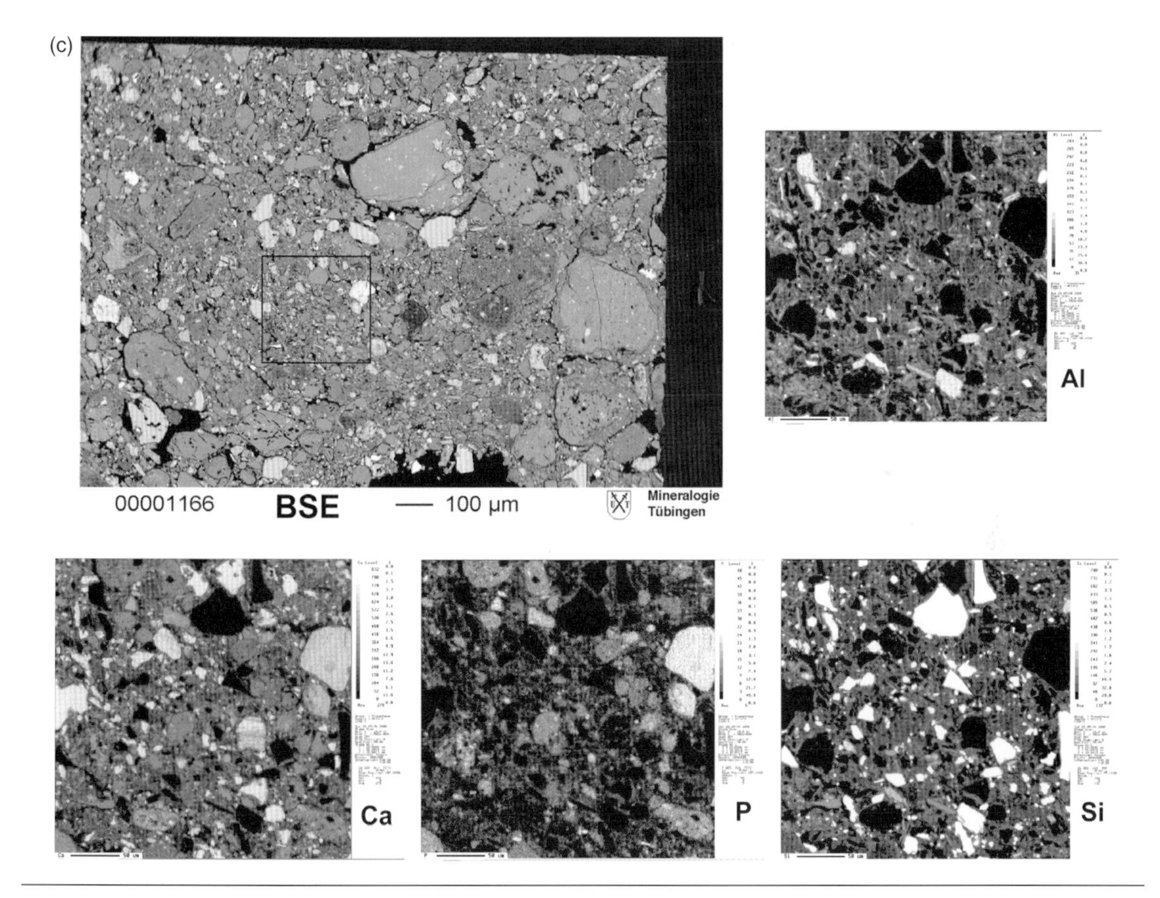

FIGURE 16.7 (*Cont.*) sand-sized grains composed of clay and phosphatized clay, quartz sand, and a large coprolite fragment in the center-right of the photograph (this cave was extensively inhabited by bears during the Palaeolithic). (c) BSE image and elemental maps of Al, Ca, P, and Si from sample 13 [not from the thin section in (b); note that these elemental images are from only a portion of the BSE image as shown by the square]. These similar components are visible here. For example the subrounded "illuminated" grain with Ca and P is piece of coprolite; the bright white areas in the Si image are either grains of quartz grains, or where the same grain is also rich in Al, they are clasts of clay. (see Goldberg *et al.*, 2003 and Schiegl *et al.*, 2003 for details).

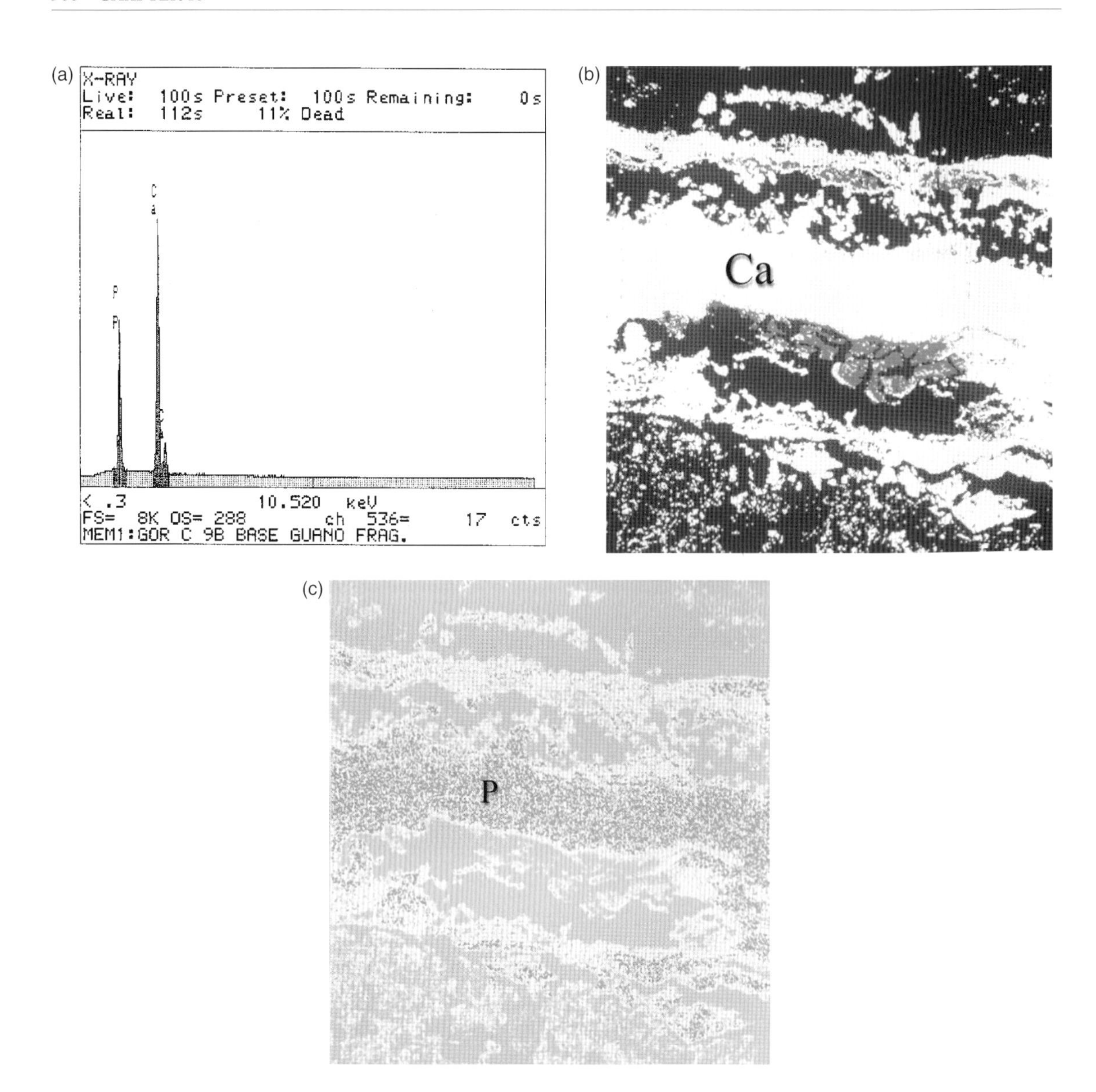

FIGURE 16.8 Examples of results from EDAX and microprobe analyses of calcium phosphate-rich deposits – from Gorham's (Middle Palaeolithic levels) and Vanguard (Middle Palaeolithic) Caves, Gibraltar (Macphail and Goldberg, 2000). These illustrate (a) EDAX analysis of suspected guano (hand collected sample 9B, Gorham's Cave; Ca and P identified:), and microprobe analyses (scale = 1 mm) of a thin section showing layered guano deposits (thin section 11, Vanguard Cave); (b) Ca (max. 36%); and (c) P (14%).

bone, bear coprolites, or grains of secondary phosphatization of the bedrock (see also Weiner *et al.*, 2002). Similarly, grains with high Si and Al contents, can be ascribed to various feldspar or clay minerals; refinement of mineral identification is accomplished by observation of grain/object shape, cleavage, texture, and overall appearance.

There are also one or two machines worldwide carrying out QEMSCAN, which is a fully automated microprobe technique. The sample is scanned by the computer automatically and *all* particles are subjected to many analyses (as many as millions overall). The software then converts the elemental data into mineralogical data. It knows how much of each element is there and it is intelligent enough to convert this in to mineral configurations. The program also provides particle size and shape of particles automatically because it has scanned the whole particle yielding the mineralogical profile, which will identify grains as small as 1 μm in size. Instrument time is very expensive (but 100,000 analyses can be made per hour; Pirrie *et al.*, 2004), and may only be used in important forensic cases (see Chapter 14).

16.7 Conclusions

This chapter presented some of the most commonly used laboratory techniques and strategies that are used in geoarchaeology. Our chief concern here has not been to describe techniques in detail – as these can be researched from papers and textbooks – but to discuss *why* certain techniques are useful or even essential to address geoarchaeological questions. There are almost an infinite number of methods that have been applied, and we have focused on the most important ones. We have outlined the methods for both bulk analyses and those employing instruments and have provided some examples of results. We have also noted some important caveats, to caution against the blind use of techniques. Thus, we have highlighted the fact that granulometry of anthropogenic deposits can produce potentially meaningless results leading to absurd interpretations, if other techniques such as soil micromorphology are not also carried out. Equally, soil micromorphological interpretations are undoubtedly enhanced if complemented by such approaches as microprobe and microfossil analyses and bulk chemistry.

17

Reporting and publishing

17.1 Introduction

A student or professional worker may be highly talented and have at their command a large personal database, but if this material is not presented in a clear and logical way, much of the work and effort could be wasted. It is also important to report information which is relevant to the project and to the questions being asked. Similarly, it is not necessary to include all data that were collected but only those that are pertinent. Reports need to be concise, and well illustrated and documented, with as little jargon as possible. The reader should be able to follow the logic of data presentation, interpretation, and ultimately *its relevance to the archaeological goals of the project.* Good reporting skills, therefore, need to be learned and practiced. Working to a timetable is also a necessary discipline; reliability is another professional requirement. Writing an article for publication or review for one's peers, requires even further effort because of space restrictions. The need to balance the requirements of an international readership and still cover a subject in a focused way, must also all be weighed. Few people can create more than one such article a year, despite pressures from the academic management to produce more, especially given that the research may have taken 3–10 years or more.

In this chapter, we make some suggestions about reporting and publishing based on our experience and those of our colleagues. We cannot, however, provide a universal guide, as different organizations, countries, and journals have different requirements according to their national laws, local authority guidelines, and the specifications of a single contract. After a few brief examples of types of site management, some ways to construct geoarchaeological reports are described including some protocols for fieldwork report and an example of a fully integrated report (Boxes 17.1 and 17.2). Also we supplied some of the ways data can be best presented, including some illustrations of the use of statistical methods, and geographical information systems (GISs) to provide. Lastly, some examples about the ways of sites can be interpreted from the gathered data in order to produce stand-alone and consensus interpretations, and the ways that these results can be published, are given.

17.2 Management of sites and reporting

Both of us have long experience of the ways to manage the geoarchaeological information of

Box 17.1 How to write a report – a suggested fieldwork report protocol

Introduction. This should contain name of the site and its exact geographical location; when was it visited; with whom was the site discussed, and who supplied information – and their affiliations (archaeological company, university); period(s) and type of site (urban, rural, tel); what is the type of exposure (open excavation, back-hoe trench, sondage, and building cellar); why was the site visited/aim of the geoarchaeological study (geoarchaeological questions); what was done (field evaluation, section description, survey, and sampling); associated work by other specialists, and overall context of the work (ancient erosion, early tillage, urban morphology, and human versus animal occupation).

Methods. This section should include a list of techniques carried out and authorities employed for field description (Munsell color chart), sampling, on-site tests (e.g. HCl for carbonate, finger texturing for grain size, coring, and survey); any specific equipment employed (different augers, geophysical equipment, types and lengths of monoliths); and follow-up laboratory techniques (e.g. pH, see Chapter 16).

Samples. This requires a tabulated *list of samples* (type of sample: undisturbed monolith, with measurements [Ordnance Datum (OD) or relative depths]; bulk sample – perhaps including size/weight; context number; archaeological character; and brief description). Further sampling requirements can be noted under Discussion given below.

Results. The content should present information on the site's location, known geology, and soils and perhaps also modern land use (e.g. using Soil Survey of England and Wales for the United Kingdom, USDA guides for the United States); systematically present the field information, for example, logically describing the oldest rocks/parent material or earliest stratigraphy first; and presenting each geological/archaeological section/soil profile in turn – referring to tabulated sample lists and descriptions, section drawings, field drawings, photographs/digital images, and any other data (figures and tables) that have been gathered (e.g. pH, HCl carbonate reaction; survey and coring data).

Discussion. This can be organized as follows: first the scene is set by discussing the known information about the geoarchaeology of the site and what the fieldwork has revealed, perhaps broadly discussing systematically, the stratigraphic sequence in terms of oldest, or by area on the site. Second, discuss the geoarchaeological questions posed by the site – referring to *why the site was visited/aim of geoarchaeological study* and what are the broad aims and objectives of the investigation – and what information has been gained from the fieldwork. For example, this is the time to identify which samples have the potential (from a laboratory study) of answering the geoarchaeological questions. (It may be necessary here to identify where further samples should be taken.) Third, suggest the various techniques that can be employed to answer these questions. Here, a proper literature-supported track record for each technique in providing useful data and interpretations has to be argued. For example, areas of supposed occupation could be isolated by measuring the LOI, P and χ of samples taken through a grid survey (see Chapter 16). Or, in order to seek out possible locations of Palaeolithic hearths in caves, soil micromorphology should be combined with microprobe and FTIR analyses to identify ash transformations (see Chapter 8).

Future work and costs. The number of samples and techniques to be applied need to be listed, based upon the preceding Discussion and the

(cont.)

Box 17.1 *(cont.)*

List of Samples. Estimated costs of analyses and reporting (see below) should be given; and this should include the number of working days involved so that the work can scheduled properly. It is also useful here to indicate how long procedures take – for example, soil micromorphology requires lengthy processing times before thin sections can be made – several months (see Chapter 16). Also, a worker may have a backlog of work and may need to inform the "client" when the work can commence. Reports often have to fit a deadline, so that the archaeologist can write up his report with all contributions present. This is especially true of CRM/mitigation archaeology.
Bibliography. A reasonable list of cited publications is presented; more are normally required for the post-excavation report/publication (see below).
Professional responsibilities. Lastly, fieldwork may take a single day, and as such, the "client" may expect a fieldwork report the very next day (by email/fax) or very soon after. This may arise because the geoarchaeological information is required immediately in order to know: which important contexts require further sampling and which areas of the site should be chosen to be further trenched/excavated, etc. It may be that a number of questions were answered on the site during the visit, but still these answers need to be reinforced by a well-thought-out report. This is especially important when short or long field trips are carried out. During long periods of fieldwork on site, many "operational" questions can be resolved on the site itself.

Box 17.2 Reporting – London Guildhall

An example of fully integrated report format: ***The Guildhall of London: from an eleventh-century settlement to a civic and commercial capital*** (Report to Museum of London Archaeological Service).
Microstratigraphy: soil micromorphology, chemistry, magnetic susceptibility and pollen by Richard I Macphail, Institute of Archaeology, University College London, John Crowther, Department of Archaeology, University of Wales, Lampeter, G. M. Cruise (ex-Institute of Archaeology, University College London) May 2003

1 Introduction and background to the study
This presents the background to the medieval site study, from the fieldwork seasons through to the assessment of samples 1992–1997 (soil micromorphology, chemistry, and palynology), and highlights the archaeological research questions specific to the London Guildhall site, and the relationship of these issues to other sites in London, the United Kingdom, and Europe and the experimental base of the investigation. In short, the site is composed of (1) first century AD buried Romano-British Alfisols formed in brickearth and deposits associated with the Roman amphitheatre, such as arena floor deposits and sediments associated with the repair and abandonment of the site (all reported separately for a "Roman" monograph); and (2) post-Roman and early medieval dark earth (late third century to middle eleventh century AD) and mainly eleventh–twelfth century humic and moist occupation deposits associated a settlement and the remains of its wooden structures.

2 Samples and methods
Monolith and bulk samples, and subsampled monoliths (Fig. 17.1), provided 50 thin sections (Macphail: soil micromorphology,

(cont.)

Box 17.2 *(cont.)*

microprobe), 48 bulk samples (Crowther: chemistry) and 31 pollen samples (Cruise: palynology). Thin section and bulk samples are listed separately as data presentations (including counts of soil micromorphological features) and as integrated table that includes the pollen samples, and full archaeological context information.

2.1 Soil micromorphology
Specific methods for sample preparation, petrological, and microprobe analysis are given, alongside the ways the data is integrated into soil microfabric types (SMTs).

2.2 Chemistry
Techniques and their relevance to the site study are listed, together with the specific methods of analysis and statistical manipulations. Analyses comprised: of LOI; the heavy metals Pb, Zn, and Cu; and phosphate-P_i (inorganic phosphate) and phosphate-P_o (organic phosphate); magnetic susceptibility measurements, including χ_{max}, in order to calculate % χ_{conv}.

2.3 Pollen
This includes information on how monolith samples were subsampled for pollen ahead of subsampling for bulk analyses and resin impregnation, as well as the specific techniques involved in pollen sample preparation, identification, and counting.

3 Results

3.1 Soil micromorphology
Eighty seven contexts and layers were identified, described, and counted to produce microstratigraphic data that were later combined with bulk and palynological information to produce linked SMTs and microfacies types (MFTs) (Fig. 17.2). 12 MFTs and their variants are described. Data are presented as tables of counts, descriptions, selected thin section scans, and photomicrographs and microprobe elemental maps (CD-ROM archive of all thin section scans, and photomicrographs and microprobe elemental maps also supplied to the Museum of London Archaeological Service for their archive)

3.2 Chemistry
The results are presented under the following headings: general characteristics; key anthropogenic indicators – LOI, P, χ heavy metals; and main findings–summary of characteristics and statistical associations between different measurements for each of the soil MFT.

3.3 Pollen
The results are presented for the three polleniferous MFT, namely: MFT 4 (uppermost dark earth), MFT 5 (stabling deposits), and MFTs 7 (domestic floor deposits).

4 Discussion: MFT and chronological sequence

Figure 17.1 Example of early medieval stratigraphy at the London Guildhall; Kubiena box samples 980 and 981, and plastic downpipe sample 979; BCF = brickearth clay floor; DFD = domestic floor deposits.

(cont.)

Box 17.2 *(cont.)*

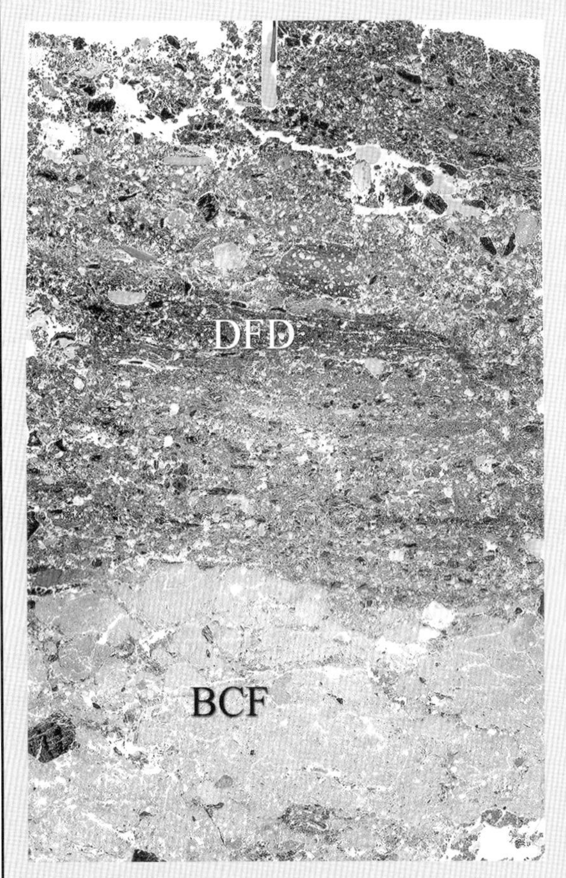

FIGURE 17.2 Thin section 979 (Fig. 17.1); BCF = brickearth clay floor (microfacies type – MFT 1d brickearth floor slab); DFD = domestic floor deposits (microfacies – MFT 7a domestic beaten floor deposits). 13 cm long scanned thin section.

This section takes the form of discussing the stratigraphic sequences of different periods (mainly the dark earth to AD 1050; AD 1050–1140, and AD 1140–1230) on a location-specific basis across the site using the integrated data of the three disciplines, soil micromorphology, chemistry, and palynology.

5 Discussion: research questions

This section addresses the archaeological research questions that were driving the investigation. It expands on these issues according to the findings under the following headings.

5.1 Settlement over the arena: origins and topography (research objective 7.4.1)

The ca 600-year-long formation of dark earth had essentially formed a "greenfield" site that progressively became occupied – first through stock concentrations and midden/latrine disposal.

5.2 Earlier medieval buildings, construction, use, and history (research objective 7.4.2.3–13)

This subsection discusses the buildings and constructional methods, and includes experimental data from the United Kingdom and Sweden; for example, the original interpretation of the site was for the use of turf in building in order to account for the accumulation of humic deposits. These, in fact, turned out to be mainly formed from stabling waste.

5.3 Local economy and diet (research objective 7.4.2.14)

The contribution of this integrated study was to show that domestic animals were probably of local origin (probably kept for dairy products) rather than the settlement acted as a market centre for nonlocal stock.

5.4 Microstratigraphic signatures (research objective 7.4.3.27)

The findings from the site-specific study and ongoing experiments were examined to produce microstratigraphic signatures based upon the three techniques employed. The results allowed the differentiation of domestic (Fig. 17.2) and stabling floor deposits. The investigation also showed that the concentrations of heavy metals in the early medieval deposits came from organic inputs (bone, cess, and stabling waste) and had no links to industrial processes.

6 Acknowledgements

These are quite lengthy because the work was both supported by funding (from various sources – e.g. Corporation of London, MoLAS, English Heritage, British Academy) and by the enthusiasm of friends and colleagues (e.g. Mike Allen, Marie-Agnès Courty, Roger Engelmark, Johan Linderholm, and Peter Reynolds).

7 References

sites from the New and Old Worlds, as well as much practical and worldwide experience of working with teams to produce reports and publications. A few of our insights are presented here.

In England, the ultimate authority in archaeology is English Heritage, and city and county councils employ their standards and working practices. One example can be cited. By 1990, English Heritage had produced a document called *Management of Archaeological Projects*, which laid out the ways in which archaeological sites were managed (English Heritage, 1991a,b). The chief concern here was the concept of the *assessment report*. This document is produced after a site has been fully excavated and sampled, and an agreed portion of the collected samples would have been processed and analyzed. Together with the fieldwork findings, the document reports the potential of further analyses and full reporting (Corporation of London, 2004; www.cityoflondon.gov.uk/plans). The geoarchaeology is, however, normally only part of a holistic approach to archaeological investigation, which in addition to the feature and artifact study will include a full arsenal of environmental techniques. This is also true of many European countries (e.g. Guyard, 2003; Maggi, 1997).

The assessment concept is very laudable as all spending needs to be justified. More often these days, as contract companies now excavate most sites, the assessment process has to be a rapid and inexpensive exercise. The assessment will require the findings from the fieldwork (see Box 17.1) and perhaps a preliminary laboratory study. In the United Kingdom, there may well be a separation of techniques designed to address broad landscape issues (geotechnical, geomorphological, and geological) from soils and microstratigraphic methods. In the context of the latter, issues at this stage can be: "is this context phosphate rich?" or "are there preserved microfeatures of occupation?" Landscape questions may include sea- or base-level fluctuations and how they affect sites in a regional context, and other long-term histories. The findings from both, will be fully integrated in the published report (Bell *et al.*, 2000, Collins, 1998; Roberts and Parfitt, 1999; Wilkinson and Murphy, 1995). Moreover, it is essential at this time that the research aims and objectives of the site be addressed, and that new information is fed back to augment and improve such research aims. As an example, data retrieved during the assessment of the early Medieval levels from London Guildhall have been usefully employed in a number of publications on methodologies and urban archaeology (Cruise and Macphail, 2000; Macphail and Cruise, 2000; Macphail *et al.*, 2004), As a result, the carefully argued assessment eventually led to the full analysis of this exceptional site (Macphail *et al.*, 2003; Bowsher *et al.*, in press)(see Box 17.2).

Other organizations and teams in the United Kingdom and elsewhere, however, are generally satisfied that during the fieldwork, which may have taken place during a day, or over a long season, or seasons, the selection of samples for laboratory analysis and reporting has already been thoughtfully carried out. Moreover, timetables and funding are finite, and the aim is to produce a report/publication to a deadline and to budget. Also, during long projects, samples are continually being processed and studied so that later sampling becomes more focused. This issue also comes under the heading *Management of Archaeological Sites* (Barham and Macphail 1995; English Heritage, 1991a,b), such as monitoring the Rose Theatre (Elizabethan and Shakespearean, London) and wetland sites in the Fens of East Anglia, England (French, 2003: 159; Sidell, 2003). In the United States, there is a government *Final Governing Standards* document for the management of both the environment and cultural heritage of sites. This document is formulated individually for each site, but includes protective requirements of the United States and applicable international agreements (e.g. UNESCO conventions, International Charters, and World Heritage Conventions; 1972 Paris Treaty on World Cultural and Natural Heritage).

17.3 Fieldwork and assessment/evaluation reporting

An example of a field report protocol and suggested contents is given in Box 17.1. Such reports end up as grey literature (McClellan *et al.*, 2003) and may be used by a variety of persons other than the archaeologist involved, (e.g. the "curator" for the town or district, or other planning officers), or be utilized in a review, many years later. Consequently, it is important to specify in the Introduction the details of the site visit; date and people involved; site details, period, specific finds, etc; reasons for visit/site questions being addressed, and tasks/methods carried out. The Results section should include a tabulated list of profiles/sections examined and samples taken, including type (e.g. undisturbed monolith, bulk sample, and associated samples for pollen and macrofossil analysis), including their provenience. This table may also contain descriptive field information, and any laboratory data, such as pH. Aerial photos, field photos, section drawings, survey results, etc., may also be included along with references to descriptions and desktop studies (e.g. data from topographic, geological, and soil maps). The Discussion obviously varies according to the task, but it should specifically cover: "reasons for visit/site questions being addressed", logically employing the field data gathered and any supportive laboratory information, either organized according to site chronology or by section/profile location. In support of these arguments, examples of analog sites, similar situations, and experimental results should be cited.

It is here in the discussion that follow-up work suggestions can be made. These should be argued on the basis of the field findings, the quality of the samples taken, the importance of the site, and the questions being addressed. Here, appropriate laboratory techniques can be discussed along with their track record in achieving answers to the questions specific to the site. This report section should also contain bibliographic support, for example, indicating why this suggested work would contribute to an established research agenda, for a region or topic.

Follow-up work needs to be costed: cost per sample preparation, analysis, and per diem (day rate) of study and reporting. It is also important to give a timescale, for the length of time this work will take, and state when the work can begin. It should not be forgotten that while a pollen sample can be processed in a day, soil studies, especially soil micromorphology, take a long time: sample processing takes weeks to months from monolith to thin section. Whereas a thin section may be described in a day, interpretation may not be instantaneous, and complementary microprobe studies have to be planned often well in advance because instrument time is limited.

17.4 Postexcavation reporting and publication

The format of a full postexcavation report (and publication) follows the same plan as field reporting described above, although publication in a journal is likely to be much more thematically based. It will be shorter and less detailed, but more focused. Unlike many reports that tend to be single authored, a publication may be multiauthored, and may contain inputs from a series of specialists. This type of publication will therefore need a series of *Methods, Results,* and *Discussion* sections (and supporting sample tables and data presentations; see Appendix 17A.2) for each specialist, before an overall Discussion and Conclusion can be written. Some examples of multi-authored reports and publications are "Microstratigraphy: soil micromorphology, chemistry and pollen" of the Mediaeval London Guildhall, Box 17.2), and "Archaeological soil and pollen analysis of experimental floor deposits; with special reference to Butser Ancient Farm, Hampshire, UK", and "Geology and Stratigraphy of the Wilson Leonard Site" (Collins, 1998; Goldberg and Holliday, 1998; Macphail *et al.*, 2003, 2004).

The London Guildhall study, recounts detailed findings from 50 thin sections (and selected microprobe studies), 48 bulk, and 31 pollen analyses; the results were presented chronologically, context-by-context. The final report on Butser summarized the results on experimental floors and viewed these findings in the light of the literature and three archaeological analogs. It is no easy task to bring together data from several independent disciplines (soil micromorphology, chemistry and magnetic susceptibility, and palynology), and produce a consensus interpretation that makes sense and is useful and clear to the archaeologist(s) (Macphail and Cruise, 2001)(see below for data correlation). Soil micromorphological and bulk data have therefore to be presented not only clearly, but transparently. The presentation of soil micromorphological information including data gathering and data "testing" protocols, has been discussed earlier (Chapter 16). Ways to present bulk data have also been presented (Chapters 2 and 16) as both tables and diagrams. In some cases however, special associations (e.g. field versus micromorphological findings) may need to be demonstrated through statistics (see Section 17.4.1). Equally, a site report might include the integration of finds, contexts, descriptions, photographs, altitudes, soils, and geology, within a landscape, and here GIS becomes a requirement (see Section 17.4.2). Large numbers of samples and data-particularly images-require that they be archived (see Section 17.4.3). Currently, there is also the possibility of publishing on the internet, where there is no limit to the amount of supportive data that can be presented.

Geoarchaeological publications include short and long thematic papers, and large-scale site-based and project-based reports, as tackled by various authors (Courty and Weiss, 1997; Courty *et al.*, 1998; Holliday and Johnson, 1989).

17.4.1 Statistical support

Statistical packages are widely available, and used throughout the biological, geographical, and social sciences, as well as in archaeology. Here, we simply provide some examples, where statistical analyses of geoarchaeological data have produced some geoarchaeological insights that help integrate soil micromorphological and bulk analyses, thus making interpretations more robust.

Early Mediaeval London Guildhall (Bowsher et al., in press). Here organic matter loss on-ignition-(LOI), P, magnetic susceptibility (χ), and heavy metals (Cu, Pb, and Zn) were analyzed ($n = 48$) by John Crowther (University of Wales, Lampeter, United Kingdom)(see Chapter 16, Laboratory Methods). Pearson product moment correlation coefficients were used to examine the relationships between these various properties, and analysis of variance (using the Scheffé procedure) was used to compare the mean values for individual groupings of MFTs, as identified through soil micromorphology (complemented by microprobe and palynology; Macphail *et al.*, 2003). An example of one of these MFTs, namely MFT 5a (*well-preserved highly organic stabling refuse*), is given in Table 16A.6, Analysis of variance was undertaken only on groupings with $\geq$4 samples. In cases, where the data for individual properties had a skewness value of ≥ 1.0, a log10 transformation was applied in order to increase the parametricity. Statistical significance was assessed at $p = 0.05$ (i.e. 95% confidence level).

Statistical analyses found that, while the majority of the phosphate is present in inorganic forms (phosphate-P_i : P ratio, 72.0–92.8%), there is a strong correlation between phosphate-P_i and LOI ($r = 0.705$, $p < 0.001$)(Tables 17.1 and 17.2). This suggests that organic sources, other than or in addition to bone, have made a significant contribution to phosphate enrichment (here dung of domestic stock and human excrement, coprolites and cess being the probable sources). The concentrations of three heavy metals (Cu, Pb, and Zn) show highly significant ($p < 0.001$) direct correlations. There are also strong direct correlations between phosphate and heavy metal concentrations. Magnetic susceptibility enhancement, however, is much less strongly correlated with the chemical properties, which suggests that the circumstances leading to chemical enrichment are not necessarily the ones associated with burning. It was concluded therefore, that in the early Medieval levels at the London Guildhall, heavy metal concentrations were linked to organic sources (bone, dung, and cess). This is unlike the situation of post-Medieval Tower of London (location of Royal Ordinance and Mint; Keevill, 2004; Macphail and Crowther, 2004) there were no links to industry or craft working of alloys, consistent with the archaeological finds recovery of these levels.

TABLE 17.1 Discriptive statistics of results from bulk samples from Early Medieval London Guildhall: summary of analytical data for all samples ($n = 48$)* (Bowsher *et al.*, in press; Macphail *et al.*, in press) (by kind permission of John Crowther, University of Wales, Lampeter)

Property	*n*	Mean	Standard deviation	Minimum	Maximum
LOI (%)	45	12.0	9.26	1.20	43.4
pH (1:2.5, water)	40	7.3	0.43	6.1	8.1
Carbonate (est) (%)	47	1.4	2.21	<0.1	>10.0
Phosphate-P_o (mg g^{-1})	48	1.26	0.892	0.190	4.74
Phosphate-P_i (mg g^{-1})	48	7.52	6.75	0.864	48.2
Phosphate-P (mg g^{-1})	48	8.78	7.44	1.20	52.6
Phosphate-P_0 : P (%)	48	15.2	4.94	7.2	28.0
Phosphate-P_i : P (%)	48	84.8	4.94	72.0	92.8
χ (10^{-8} m^3 kg^{-1})	48	7.4	36.2	5.5	196
χ_{max}(10^{-8} m^3 kg^{-1})	44	731	370	130	2080
χ_{conv}(%)	44	9.72	5.80	0.26	22.0
Pb ($\mu g\ g^{-1}$)	42	345	283	11.4	1280
Zn ($\mu g\ g^{-1}$)	42	101	76.0	10.4	468
Cu ($\mu g\ g^{-1}$)	42	127	77.0	7.30	333

* In certain cases the sample supplied was too small to undertake the full range of analyses. In these cases priority was given to the determination of phosphate and χ.

TABLE 17.2 Pearson product–moment correlation coefficients (r) for relationships between the various soil properties for all the bulk samples analysed** (n = 40–48, depending on combination of variables) (by kind permission of John Crowther, University of Wales, Lampeter)

	pH	Carb[a]	P_i[a]	P_o[a]	P[a]	P_i : P	χ[ab]	χ_{max}[ab]	χ_{conv}[ab]	Pb[a]	Zn[a]	Cu
LOI[a]	−0.456	n.s.	0.705*	0.822*	0.738*	n.s.	0.324	n.s.	n.s.	0.601*	0.902*	0.569*
pH		0.485	n.s.	−0.337	n.s.	n.s.	n.s.	n.s.	n.s.	n.s.	−0.424	n.s.
Carb[a]			n.s.	n.s.	n.s.	n.s.	n.s.	n.s.	n.s.	n.s.	n.s.	n.s.
P_i[a]				0.839*	0.997*	0.303	n.s.	0.347	n.s.	0.558*	0.858*	0.643*
P_o[a]					0.881*	n.s.	n.s.	n.s.	n.s.	0.524*	0.847*	0.575*
P[a]						n.s.	n.s.	0.346	n.s.	0.561*	0.874*	0.643*
P_i : P							n.s.	n.s.	n.s.	n.s.	n.s.	n.s.
χ[ab]								n.s.	0.720*	0.475	n.s.	n.s.
χ_{max}[ab]									−0.598*	n.s.	n.s.	n.s.
χ_{conv}[b]										0.363	n.s.	n.s.
Pb[a]											0.701*	0.660*
Zn[a]												0.691*

** Statistical significance: n.s. = not significant (i.e. $p = 0.05$), * = significant at $p < 0.001$.
[a] Indicates $\log_{10}$ transformation applied to the data set.
[b] In the case of the magnetic properties, correlations between the untransformed variables (used in assessing the relative importance of χ_{conv} and χ_{max} in affecting χ) are as follows: χ with χ_{conv} $r = 0.687$ ($p < 0.001$; see figure 1); and χ with χ_{max} $r = -0.134$ (n.s.).

Prehistoric Raunds, Northamptonshire, UK (Healy and Harding, forthcoming). The archaeological soil study focused upon a buried 3.5 k long prehistoric landscape that included the investigation of five treethrow holes, one dating to 4360–3980 cal BC and containing Late Mesolithic/Early Neolithic flints, two Neolithic barrows (e.g. 3940–3780 cal BC) and six Bronze Age barrows (e.g. 2470–1880 cal BC) (Macphail, forthcoming). The buried soils involved are Alfisols (argillic brown earths/luvisols). During 1985–99, 14 prehistoric sites that were totally (e.g. 0.80–1.30 m) or partially sealed by early medieval alluvium, were studied through 52 thin sections and 70 bulk analyses. Bulk methods included organic carbon, LOI, various extraction methods applied to P, carbonate, χ, and grain size (see Table 16A.5).

The study focused on the treethrow holes (Macphail and Goldberg, 1990), the barrow make-ups, and the buried soil profiles; one interpretation of the specific chemistry and micromorphology of the soils was animal management. Measurements on burned soil from treethrow holes found distinctively high values of χ ($753–894 \times 10^{-8}$ SI kg^{-1}). Textural features (six types of clay coatings counted at Raunds) are far more numerous in treethrow holes than outside (control soils), and mainly originate from soil disruption induced by treethrow, rather than through simple forest soil illuviation/*lessivage* (see Chapter 9).

Use of fire. Statistically significant (95% confidence level) correlation was found between χ and measures of organic matter, which probably reflects both the natural topsoil formation of magnetic maghaemite, as well as the presence of burned soil in the ancient topsoils. The use of fire would have been an important mechanism for opening up and maintaining open conditions for the pastoral Neolithic and Bronze Age landscape as interpreted from the soil data. The burning of *in situ* "fallen" trees at Raunds infers both purposeful killing and the pulling over of trees, or a remarkable number of blowdowns being burned.

Animal management. There are both strongly significant correlations (99.9%) between both P_{nitric} and P_{citric}, and measurements of organic matter, which support the identification of animal management (deposition of organic phosphate in the form of dung; Courty *et al.*, 1994; Macphail and Cruise, 2001). In addition, the counting of textural pedofeatures (e.g. Turf Mound and Barrow 5) and analysis of some these through microprobe, suggested a link between animal activity and topsoil slaking/poaching (internal crusting) and the concomitant anomalous deposition of dark red-colored clay coatings. The last are enriched in organic matter and phosphate, and only occur in monument-buried topsoils/ mound material (see Fig. 16.5). They are absent from the treethrow features. Arguments, based upon soil micromorphology, for a grazed landscape and occasional animal concentrations at some premonument locations, where there was both trampling and deposition of excreta, is supported by strongly significant correlations between clay and organic carbon, and clay and P_{nitric} for Raunds as a whole. These suggestions are consistent with environmental interpretations based upon widespread studies of plant and insect (e.g. dung beetles) macrofossils, faunal remains, and pollen (Healy and Harding, forthcoming).

17.4.2 Geographical information system

Geographical information system (GIS) studies are a valuable tool for integrating a variety of data sets not only from the geoarchaeologist but from the entire research team. Classically, GIS has been used for modeling past landscape and ancient settlement morphologies (Harris and Lock, 1995). It can, however, make the day-to-day running of a large or long-lived excavation more efficient. Increasingly, as sites are excavated, all finds, features, and samples are located in three-dimensions (by tape measure and level, or by infrared EDM techniques). As spot dates on finds are produced, archaeological features can be phased. Such GIS "maps" of the site, its contents and surroundings, can then be interrogated to show the distribution patterns of archaeological finds, as well as geoarchaeological data. Moreover, with the increased portability of computers and the use of total stations, the data can be interrogated while the site is being excavated. This allows for dynamic interpretations and hypothesis generation that can often be tested in the field, either immediately or by the end of the season.

These types of procedures have been practiced at several prehistoric sites in the Old World, such as at the Mossel Bay Archaeology Project, South Africa (C. Marean and P. Nilssen, directors), and Pech de l'Azé IV and Roc de Marsal, in the Dordogne of France (H. Dibble and S. McPherron, directors; http://www.oldstoneage.com/). Similarly, at West Heslerton, North Yorkshire, UK, this type of recording has been in operation since the late 1980s, by Dominic Powesland and his team, the most recent report being on the Anglian (Saxon) cemetery (Haughton and Powesland, 1999). A number of other organizations (Framework Archaeology, United Kingdom) are also employing this approach not only to aid their own recording and site interpretation, but also to allow specialists such as geoarchaeologists, to interrogate the site's database in order to locate their samples in relation to other contexts and features. The system also supports images of field photos and section drawings.

In landscape studies, and at the interpretation stage (see below), the background geology, relief, and soils can be used to aid the reconstruction of land use based upon the archaeological record and background archaeological models of a region, its terrain, and culture. The location and extent of inferred resources, whether from woodland, pasture, or good arable land, can be mapped for any period and country – for example, the Neolithic of Southern Italy (Robb and Van Hove, 2003).

It is essential, however, that such reconstructions are based on relevant data, both at the landscape and the site scale (Courty *et al.*, 1994). Detailed information may have been recorded over a region on soil type, slope type and angle, relief, precipitation, and vegetation, with the intention of relating them to site distribution or land use. The tacit assumption, of course, is that the present configurations of these parameters are close to those of ancient times. Past experiences in the field have shown that site distributions can be truncated or lost by wholesale erosion or burial, thus removing critical information from the database. Workers employing GIS need to be aware of such situations.

At the site level, data may well provide information on artifacts, crops, domestic animals, and wild resources that were being utilized during a specific period. As we have emphasized, however, the documentation and reconstruction of past soils is much more problematic. The soils of the Neolithic, for example, may have been quite different from those of the Iron Age, or those of today. Erosion, colluviation, alluviation, weathering and podzolization, manuring and plaggen soil, or *terra preta* formation, may have transformed the soil cover. The inferred fertility of Bronze Age fields, based upon the modern horticultural production of strawberries, did not convince one of the authors (RIM) at one conference. Therefore, it is the role of a geoarchaeologists to make sure that any geological and soil information employed in GIS modeling, is correct.

17.4.3 Archiving

It has already been mentioned that if undisturbed monolith samples are impregnated in resin (Chapter 15), they form a sample archive, as carried out for a number of Fenland sites in the United Kingdom (C. French, University of Cambridge, personal communication). Both (PG and RIM) have a store of impregnated blocks that can be used for future study. Wessex Archaeology (United Kingdom) also has a policy of archiving blocks (M. Allen, personal communication). Bulk samples can also be archived; the late Professor Dimbleby (Institute of Archaeology, UCL) recommended that these be air-dried to halt most biological activity thus allow for reasonable preservation. It should be remembered, however, that labile soil characteristics, such as pH, are liable to change. Hence it is best to analyze samples that still retain their field moisture content. A list of archived samples can be appended to the geoarchaeological report.

Workers normally retain and score their regional field notebooks, or copied or scanned versions, and if possible, also hard copies of datasheets although now most information is recorded and archived electronically. Nowadays, it is common to produce a report that is composed of text, bibliography, tables, figures, and photo-images. Generally, a report is supported by a CD-ROM archive, where, for example, a large number of images and raw data (e.g. field notes, spreadsheets) can be stored. The internet also allows the presentation of whole site reports, with supporting data, and images.

Such archiving is particularly important, as some sites are reopened and reexcavated during the lifetime of an individual researcher or over several generations (Bell *et al.*, 1996). It thus behoves any researcher to leave as detailed a "paper trail" as possible in order to facilitate the efforts of later researchers. Personal experience has demonstrated that "retrofitting" one's own research results with those produced only a decade earlier, at the same site, proved to be a difficult task. Uncertainties about past and present stratigraphic correlations, for example, were never satisfactorily removed.

17.5 Site interpretation

17.5.1 Introduction

In this book we have repeatedly stressed how to record field geoarchaeological and laboratory data. Less attention has been paid to interpretation, except for illustrative examples, and the presentation of type sites. In the view of one eminent and respected American archaeologist (Professor David Sanger), the easy part is description – the Devil is in the interpretation; and this applies to both *the site* and *the geoarchaeology*. He also suggests that the means to interpret a site takes *something else* which is hard to teach or acquire. This is not the place, however, to note instances where the temptation has been to forsake interpretations of "difficult situations, materials, or sites". Equally, an interpretation may not be based upon extant data, but rather the interpretation of the data has been fitted or rammed into some kind of preconceived, and often false, notion or model.

It should be quite clear from previous chapters that the first step must be the correct recording of data, because without this strategy no accurate interpretations are possible. We have also tried to provide examples of what questions need to be asked of a site before the correct recording and analyses can take place. If the archaeologist, through ignorance of geoarchaeology, or the geoarchaeologist, through ignorance of archaeology, does not know what questions to ask during a site investigation, the final interpretations are likely to fall well short of a site's potential. Most geoarchaeologists come from a geoscience, geographical, and/or pedological background, so that geoarchaeology may well be a closed book to many archaeologists. We therefore hope that the thematic studies presented throughout this book will be an aid to both these groups.

17.5.2 Correlating and integrating geoarchaeological data with those of other disciplines

It is important that geoarchaeological data are not collected, analyzed, and interpreted in isolation (good use of complementary geoarchaeological data are in Dimbleby, 1962; Evans, 1972, 1999; Sharples, 1999; Whittle, forthcoming). Geoarchaeological information is only one part in the study of a site. Holistic archaeological studies involve both cultural and environmental inputs. Geoarchaeology can provide information that is useful to environmental reconstruction and also add to debates concerning cultural activity. It is much more multifaceted than is often realized within the archaeological literature. It also has a role in theoretical studies (see Chapter 10). This section demonstrates how geoarchaeological investigation, if well thought through from the outset, can be successfully integrated with other data in archaeological projects. This approach, first influences the fieldwork and sampling agenda (Chapter 15), laboratory methods applied (Chapter 16), and the way the results are viewed and integrated into the site or project discussion.

17.5.2.1 Data gathering

It is crucial that all geoarchaeological sampling is correlated with any other environmental sampling or recovery of artifacts. The exact

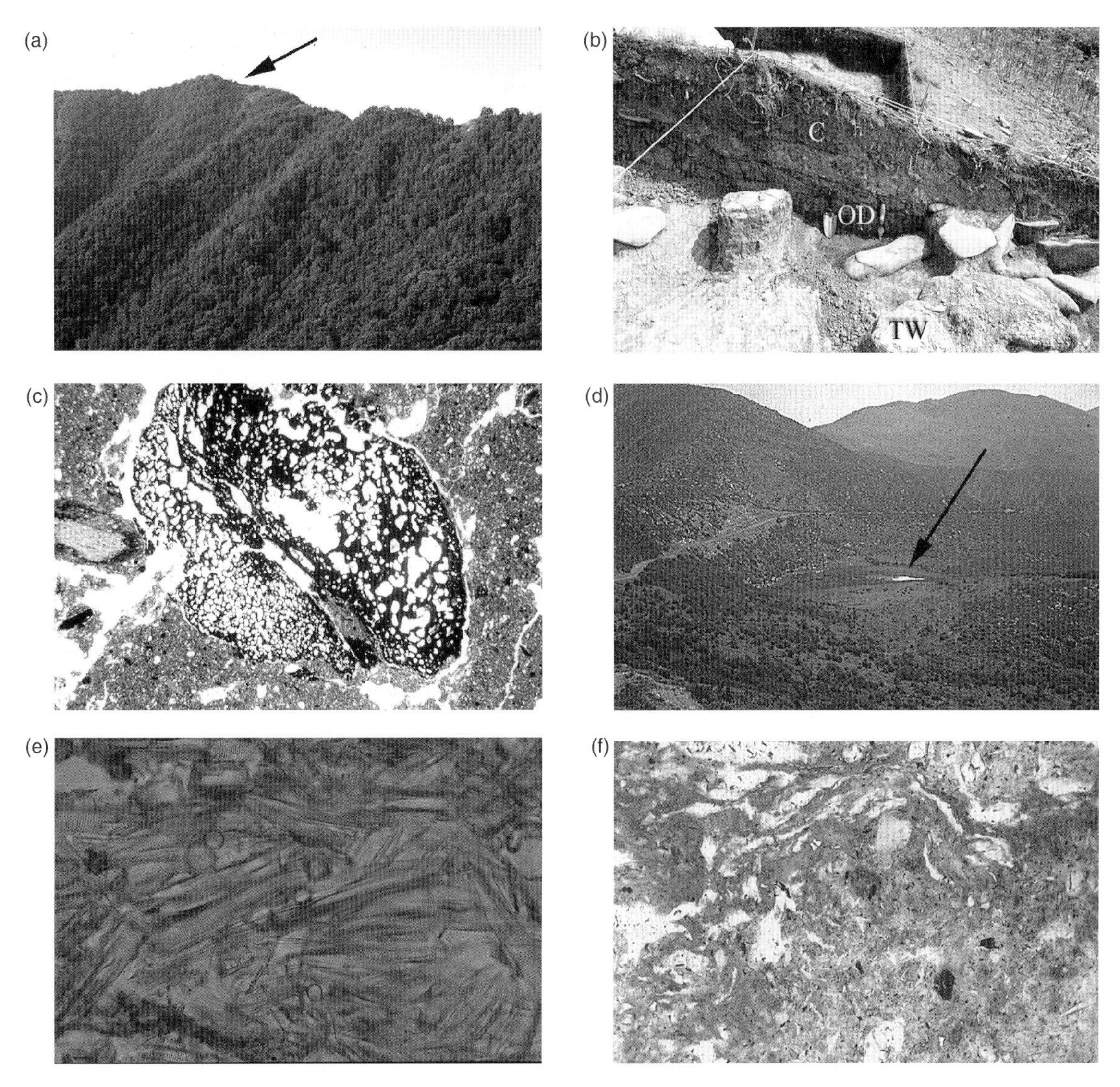

Figure 17.3 Site, microfossils and geoarchaeology – examples from the Ligurian Apennines, Italy. (a) The Chalcolithic, Bronze, and Iron Age ridge top site of Castellaro di Uscio (arrow) at 728 m above sea level (asl) in the western Apennines, where issues of steep slope, high relative relief, aspect, and ridge top route ways were reported (Maggi, 1990); also discussed was use of terraces and house platforms along an archaeological catena (Courty *et al.*, 1989); (b) Late Bronze Age terrace wall (TW), with occupation deposits (OD) sealed below more recent colluvium (C); (c) Iron Age occupation soil contains charred *Triticum* (wheat) grain (identified by S. College, Institute of Archaeology, University College London). Archaeobotanical studies by R. Nisbet, were an integral part of the site's environmental and economic reconstruction (local wood resources and food stuffs through time)(see also Maggi and Nisbet, 2000; Nisbet and Biagi, 1987); (d) the early Holocene to Neolithic to Medieval peat bog site (arrow) of Lago di Bargone (900 m asl, 9 km from the Mediterranean) (Cruise *et al.*, 1996; Cruise and Maggi, 2001) provides both a local and regional environmental record principally based upon palynology, in an area (eastern Apennines) where peat records are rare; because of slope and climate, (e) during the early Holocene, the site was fully aquatic, and hence the common presence of mid- to fully eutrophic diatoms

Figure 17.3 (*cont.*)
and diatomaceous silts at around 3.00 m, these remains together with the palynology are indicative of a lake surrounded by white fir (*Abies*) woodland; (f) pollen data and the development of weakly formed "bog iron" (also marked by a peak of phosphate and *x* measured after ignition at 550°C; Cruise and Macphail, 2000) replacing some organic materials, suggests fluctuating water tables and a grazed open woodland environment, likely contemporary, with increased upland pastoral activity during the Chalcolithic and Early Bronze Ages. Such interdisciplinary studies help to answer questions concerning the links, or simple time correlation, between past human activity, vegetation change, and climate.

relationships between the geoarchaeological samples and the archaeological contexts that they are investigating, need to be recorded. This would include the correlation of geoarchaeological findings with those from fossil analyses (Fig. 17.3).

Some examples of the analysis of fossils can be noted. Whole charred cereal grains such as *Triticum* (Fig. 17.3c) and *Einkorn*, and large pieces of hazelnut shells have been identified in thin section (Courty *et al.*, 1989). When these occur in deposits that contain mainly finely fragmented charcoal, this implies at least two phases of input with perhaps strong physical working in between (Saville, 1990). The identification of species from wood charcoal to species is more problematic, however, because wood charcoal is only identifiable if sectioned in exactly the right planes, and this does not happen very often (see Arene Candide below for a notable exception). Macrofossil analysis more often relies on flots (or recovered mineralized material) from large (e.g. 2 L bags) bulk samples, but sometimes the exact relationships between important layers and the macrofossils they contain can be crucial. The worst disputes between geoarchaeologists and other environmentalists often arise from the study of samples from different site locations, so that their stories do not tie up. Just as frequently, however, the geoarchaeologists and environmentalists (like the archaeologist) work with different agendas and try to tackle quite dissimilar questions. They also may not wish to fully confront the limitations of their own technique and the ways in which taphonomy and site formation processes affect the extent of recovered biological and geological/pedological materials and information. This is where a useful dialogue between specialists should take place, as every technique is likely to add some unique information to the consensus interpretation of a site. No one need feel defensive at such times.

If work is to be truly interdisciplinary (Fig. 17.3), geoarchaeologists should be aware of their roles in a project. Under the umbrella of a project, the geoarchaeologist may be:

- an equal partner amongst a team, providing environmental information to the archaeologist/project director, who will develop a consensus interpretational chapter (e.g. Bell *et al.*, 1996, 2000; Macphail *et al.*, 2003, 2004; Maggi, 1997; Saville, 1990; Sharples, 1991; Whittle, forthcoming);
- within the geoarchaeological component, interpretations are enhanced by a biologist(s) providing data on specific layers or contexts, and here the palaeoenvironmentalist plays a supportive role (French, 1998, 2001; Macphail and Goldberg, 1999); or
- Where the geoarchaeologist carries out focused analyses of particular contexts/layers, palaeobiological and cultural investigations of full sequences is aided by geoarchaeological information (Fig. 17.4); here the geoarchaeologist carries out a

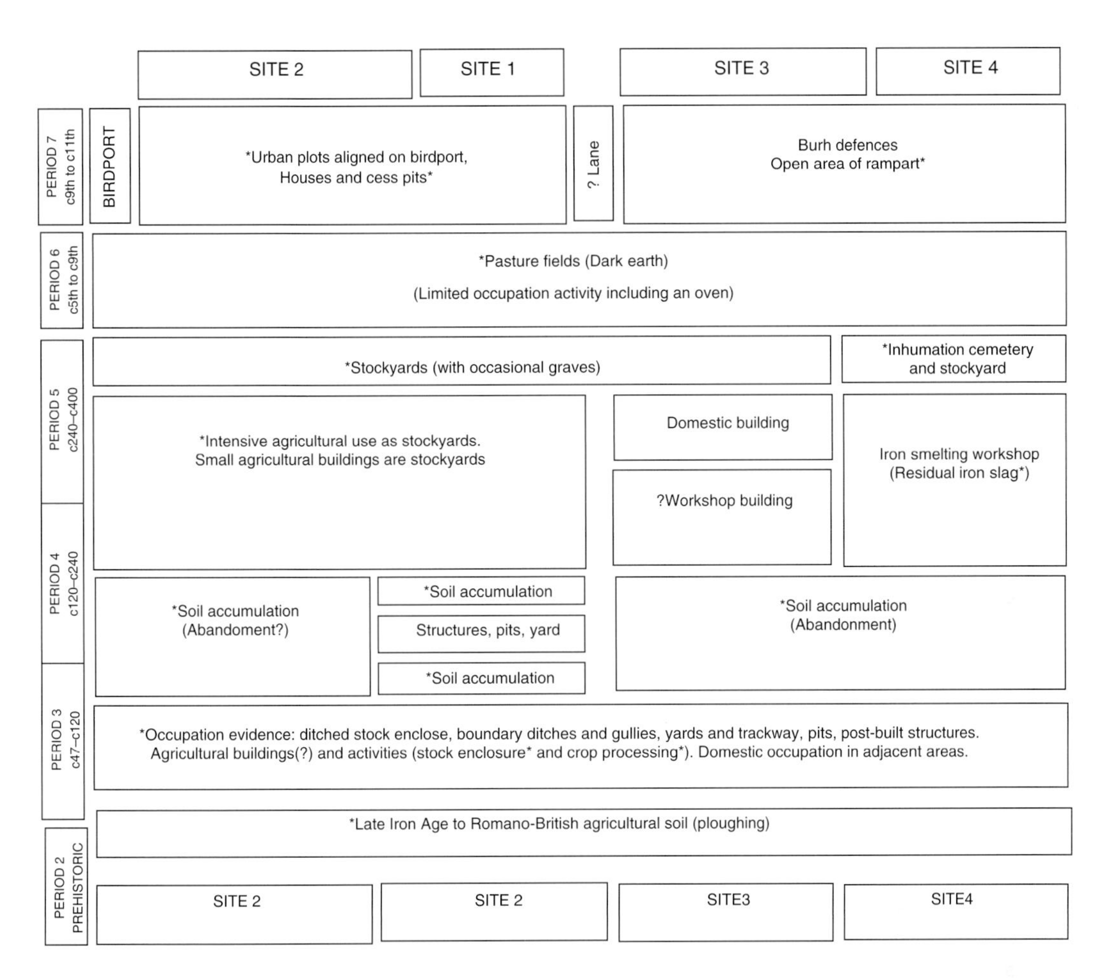

FIGURE 17.4 Deansway, Worcester, land use diagram (after Dalwood and Edwards, 2004): chronological development and spatial variability; soil micromorphology and bulk analyses contributed to the identification of Roman agricultural activity and processing, Late Roman middening and animal pounding, Dark Age pastoralism (see Greig in Dalwood and Edwards, 2004), Saxon cess disposal and domestic and craft activities. Geoarchaeological contributions to the site understanding are marked with an *.

secondary role (e.g. Ligurian Apennines, Italy)(Cruise, 1990; Cruise *et al.*, 1996).

It should not be forgotten that the geoarchaeological study may well be a team effort. The people providing chemical, magnetic, and soil micromorphological data, also need to communicate and discuss the questions that they are addressing; just as much, the geoarchaeologist needs to *connect* with the archaeologist and palaeoenvironmentalist.

17.5.2.2 Case study: Arene Candide, Liguria, Italy

This cave was analyzed and published in two parts, first on the Pleistocene deposits

(Bietti, 1994) and second for the Holocene sequence (Maggi, 1997). The Pleistocene geoarchaeology is presented here.

17.5.2.2.1 The Late Pleistocene deposits at Arene Candide

The study was predominantly a micromorphological one, composed of seven thin sections taken from profiles excavated in the 1940s and 1950s and reexposed in the 1990s (Macphail *et al.*, 1994). This multiauthor study also integrated new radiocarbon dating assays, and macrofossil observations of small mammal teeth and charcoal in the thin sections. The samples were correlated with proxy climatic sequence based upon the avian and mammal faunas that had already been studied. The chief findings related to the coincidence of different data, from the top to the base of the sequence:

- 10,000–11,700 BP (Final Upper Palaeolithic/ 'Late Epigravettian'): sealed by a rock fall containing a red deer skull dating to ca 10,000 BP (8,980 Cal BC) and correlating to the very end of the cool Younger Dryas of Northern Italy, there are:
  1. uppermost occupation deposits (two thin sections) containing a trampled surface of fine bone fragments and charcoal, with faunal indications of a temperate humid climate with major exploitation of boar, roe, and red deer;
  2. a series of convoluted hearths (two thin sections) containing both charred twig wood and coarse charcoal of *Arbutus, Quercus*, and *Prunus* (identified in thin section), the macro-convolutions and a platy, link-capping microfabric indicating a cool climate fluctuation(s) (freezing and thawing) that is/are corroborated by the presence of elk, marmot, and especially wolverine, within the Allerød;
  3. lowermost layers (three thin sections) that contain fewer artifacts, although charcoal of *Quercus* and softwoods such as *Juniperus* and *Taxus* are present (identified in thin section), alongside guano deposits, and associated phosphatized rock and numerous gastroliths (rounded 4–5 mm size crop stones commonly of nonlimestone "exotic" rocks). Other indicators of birds are the remains of regurgitation pellets (from raptors) and the teeth of possible marmot, lemming, pine mouse, and vole (identified in thin section). These layers also show an intermittent climatic progression from cool (elk, lynx, and mammoth) to eventually mainly temperate (e.g. prevalence of rabbit over hare) conditions reflecting the Older Dryas developing into the Allerod above;
- 18,500–25,600 BP (Upper Palaeolithic): highly biologically worked, mainly organic, remains that include guano, with remains of seed testa and faunal remains indicating predominant use of the cave by birds and carnivores such as hyenas, leopards, and bears (who were probably denning) during the Late Würm ("Tursac Interstadial").

In the case of the Holocene deposits at Arene Candide, the stratigraphic analysis here encompasses: the Early Neolithic; the Early, Middle, and Late Middle Neolithic; and the Late Neolithic (Chassey)(Maggi, 1997)(Table 17.3; see Fig. 10.1). The chief findings relate to the identification of sedimentary microfacies resulting from areas and times of domestic occupation and the stabling of herbivores (sheep/goat and cattle; Macphail *et al.*, 1997; Rowley-Conwy, 1997), a sediment-type repeated across the Mediterranean (Binder *et al.*, 1993; Boschian, and Montagnari-Kokelji, 2000; Cremaschi *et al.*, 1996; Sordoiliet, 1997). The interpretation of the main characteristics stemmed strongly from the research at Butser Ancient Farm and experiments carried out in France (Wattez, 1992)(Chapter 12). Roberto Maggi (1997) was strongly convinced that all data sources contributed to a holistic understanding of the cave sediments and the cultural activity that produced them. The work showed, that the cave fills were predominantly anthropogenic in

TABLE 17.3 Extract of summary table from Arene Candide (edited from Maggi, 1997: 636–7). Here two of nine archaeological phases (Late Pleistocene to Roman–post-Roman) are presented, where recovered artifacts, biofacts, and soil micromorphology of the cave sediments are compared. The consensus interpretation of the SMP 1 (square mouth pottery) is for year-round occupation of the cave, mainly with areas of either domestic occupation (with occupation surfaces; imports of grasses for human bedding; and a concentration of artifact discard), or stabling activities. Stabling became dominant in the SMP 2, with only stabling deposits present and dominated by leaf hay foddering practices that affected the local environment; the amounts of artifacts declined. Such stabling activities – associated with transhumance – continued into the Late Neolithic/Chassey Period.

Phases	Dates	Soil micro*	Pollen	Charcoal	Phytoliths	Land snails	Marine molluscs	Micro mammals	Animal bones
Middle Neolithic SMP 1	5,900–5,700 BP 4,800–4,500 BC	Major human occupation contemporaneous with stabling of herbivores – human occupation surfaces; human waste; imported plants for human bedding	Increase of cereals; decline of *Quercus;* Erica and wild grasses present	Dramatic decrease in forest broadleaved species	Several maxima of grasses	Increase in anthropogenic activity – evergreen oak communities	Decrease in patella; increase in species for ornaments	More humid terrain indicators	Increase in meat strategy; sheep size decrease; decrease of wild game; sheep killed all year round
Middle Neolithic SMP 2	5,600–5,400 BP 4,500–4,300 BC	Dominant cave use for herbivore stabling episodes; cattle coprolites > sheep/goat; periods of abandonment	Increase in cereals fluctuating; herbs and shrubs	Fluctuation in forest species = changes in anthropogenic pressure	Several maxima of leaves of trees and shrubs > grasses	Perhaps decline in human disturbance	As above	No change	Sheep size decrease; goats rise to 30% of sheep/goat assemblage

(cont.)

TABLE 17.3 (*cont.*)

Phases	Pottery style	Pottery temper	Chipped industry	Ground stones	Obsidian	Bone tools	Ochre	Burial	Anthropology
Middle Neolithic SMP 1	Scratched decoration; large sherds	Local cave calcite; coarse specimens from 15 to 20 km northwest	Rich SMP industry; sickles, arrowheads plant processing; resharpening	Maximum variety of activities and raw materials	Pieces from Sardinia, Palmari and Lipari; no cores; use-wear/retouch	240 tools, >50% total one tools here; awls; manufacture and use of hides	Abundant	Stone cist; also children and babies in pits	Shorter stature compared to Palaeolithic; plenty of cares; tuberculosis indicators of use of dairy products/ cattle domestication
Middle Neolithic SMP 2	Excised and spiral decoration	As above	Scarce – as above	No data	No data	58 tools; awls dominate; balance of wild versus domestic	Present	Probably no graves	No data

* Soil micromorphology

origin, and models of "natural" infilling just did not work. Maggi tabulated the summarized the results for both the cultural and environmental data including the sediments (see Table 17.2).

17.6 Conclusions

This chapter has attempted to demonstrate that the way results are reported and published. Both affect the way future site studies are organized and carried out, and the way geoarchaeological information is understood and valued – or not – by archaeologists. If geoarchaeology is to be useful and successful, it must be both focused and attentive to the needs of the site and the archaeological and environmental team. Some of the diverse ways of reporting have been discussed in this chapter. A field evaluation/sample assessment differs from a full postexcavation report. Equally, a publication in a journal will contrast strongly with the full archive "document" that can be supplied on a CD-ROM. Simply, reports are not articles either in terms of the quantity of data presented or how it is presented. At the reporting stage, various methods to improve the ways the geoarchaeological information is presented and supplied to the archaeologist, are suggested. These vary from tables, images, and figures, to the use of statistics and integration with GIS. What is emphasized, however, is the careful correlation of information with the archaeological and other environmental findings, in order to produce both independent and consensus interpretations within a team. This leads to sustainable and robust archaeological reconstructions of both past landscapes and cultural activity.

18

Concluding remarks and the geoarchaeological future

The preceding chapters illustrate that geoarchaeology is quite a broad field of enquiry, and many different kinds of people are carrying out a wide variety of research both in the field and in the laboratory. The spectrum of specialists and topics range from geoscientists (geologists *sensu lato*, geographers, pedologists) examining landscapes, soils and deposits, to archaeologists using aerial photos and soil survey data to infer ages of landscapes and understand site distributions, or geochemistry to evaluate bone diagenesis.

At the outset, we have tried to provide some of the basic characteristics of sediments and soils, which constitute the ultimate framework for archaeological deposits and sites. This is the material that is excavated in order to expose and capture the archaeological record. The spatio-temporal articulation of these objects and features with deposits and landscapes is covered in the chapter on stratigraphic concepts. In our opinion, an awareness of the concepts *facies and microfacies* is crucial for appreciating the changes in the localized nature of archaeological deposits within a site and its surroundings. Such awareness also enables the investigator to link these changes to processes of deposition or formation (usually subsumed under the heading of "site formation processes"). In the first instance, these processes can be of geological origin, and in this light, we have included a discussion of large-scale geological systems, such as coasts, streams and wetlands, and aeolian terrains. Equally, soils and soil formation are given equal prominence, and our examination of fluvial systems includes water movement on slopes.

Alternatively, many deposits in archaeological sites – particularly Late Pleistocene and Holocene ones where activities associated with hunter-gatherers and more complex societies (e.g. Neolithic, Bronze Age, Roman, Medieval) appear – are tied directly to human activities. In the Old World in particular, some deposits owe their very origin to human impact, and theoretically can be traced from hill slope clearance and cultivation, via colluviation to floodplain sediments. Anthropogenic aspects of deposition and site formation have previously been underplayed in many geoarchaeological publications, particularly those from North America. In view of this paucity, we devoted a relatively large proportion of the book toward the description, analysis, and interpretation of archaeological deposits, and the human activities associated with them. For example, we described archaeological deposits that are solely of human (anthropogenic) origin, and these can be studied and interpreted only if this provenance is fully recognized: such techniques as granulometry may well produce uninterpretable results without soil micromorphology, for instance. Although much can be, and has been, borrowed from the natural sciences, some archaeological sites have only begun to be understood through experiments designed by archaeologists, the chief examples being studies of buried soils and materials, and ethnoarchaeological reconstructions of settlements.

We have gone into a significant amount of detail describing geoarchaeological field strategies. No amount of laboratory work can resuscitate, incompletely or poorly documented or contextualized field observations, data, or samples. We have also provided a survey of many of the laboratory techniques that are currently employed in geoarchaeological studies, and emphasize that none of these can be employed blindly.

Finally, at the end, we have presented some practical tips and strategies on what and how to present geoarchaeological data in written reports. Field and laboratory information are critical to the unraveling of geoarchaeological problems, but what is included in a report and how it is presented, can make a large difference in how successfully the geoarchaeological results and interpretations are accepted by the geoarchaeologist's audience and clients. Everything must be made clear to archaeologists and other specialist team members of the project, so that the geoarchaeological results can be *usefully* incorporated into the overall understanding of the site and its history.

So, where is geoarchaeology heading, and what are its new directions? Before geoarchaeology was even formulated as such in the early part of the twentieth century, the major focus on geoarchaeological types of studies was clearly at the landscape or environmental scale, as can be seen by the works of Kirk Bryan, for example, in the American Southwest (see articles in Mandel, 2000). Emphasis was on reconstructing Quaternary landscape history and palaeoenvironments. Things are a lot different today. By the end of 2004, a perusal of the year's table of contents from the journal *Geoarchaeology* (www.interscience.wiley.com), reveals a wide variety of topics that are subsumed under the subject, Geoarchaeology:

- dating prehistoric canals in Arizona
- soil fertility and management and agriculture in Peru
- geochemistry of basalt grinding stones in Jordan
- obsidian sourcing in Mexico
- coastal settlement in Hawaii
- New Zealand horticulture
- Alluviation and Medieval and prehistoric tin mining in Cornwall
- Effects of bears on rockshelter sediments in Switzerland
- Grazing and land degradation in Iceland
- Differentiating between naturally and culturally flaked pebbles
- Temper and clay source of Helladic pottery, Greece

Editorial selection notwithstanding, this somewhat eclectic choice of articles on our part, nevertheless demonstrates how widely geoarchaeology is practiced in the last few years. This diversification of topics has come about since or because of the inception of the journal, *Geoarchaeology*, and the *Journal of Archaeological Science*, which is affiliated with the Society of Archaeological Science in the United States. We anticipate that this ever-increasing breadth of practice will continue into the future.

Nevertheless, a few broad subject areas/topics might be more extensively dealt with in the future. In Europe, for example, because archaeological soil micromorphology has developed exponentially since the publication of *Soils and Micromorphology in Archaeology* (Courty et al., 1989) with regular meetings of a working group that has provided extensive training, there has been much emphasis on the *detailed* microstratigraphic analysis of sites. This approach, which initially was practiced in a few confined hotspots (e.g. United Kingdom,

France) is now becoming more widespread and a generally accepted, and is almost a routine approach across Europe, complementing studies carried out by more traditional methods. What is needed, however, is the development and acceptance of standardized methods and approaches of recording and analysis (e.g. GIS) of everything from soil microfabrics to phosphate, so that results are readily presented and comprehended by all. For example, the analysis of phosphate varies from using plant available extraction methods to ICP where more complete extraction occurs, including lattice P. Clearly, chemists are not all trying to tackle archaeological questions sensibly, and obviously such different methods cannot produce comparable data.

In the eastern Mediterranean area, where tells and mounds abound, geoarchaeological participation is relatively meager (with the exception of Çatal Höyük; Matthews et al., 1996) and the subject unappreciated in comparison to the number of sites that are or have been excavated (Goldberg, 1988). This unwillingness to exploit geoarchaeologists' expertise is not unique to the Middle East, however, and this deficiency underscores a serious shortcoming in the way that archaeology is practiced, not only in this region, but over much of the globe. In fact, at the risk of overstating the case, not soliciting geoarchaeological inputs during an excavation or project is irresponsible, and effectively is the equivalent of saying, "let's ignore the pottery at this site because we don't have the resources to study it." Clearly, such actions result in biased/truncated datasets with incomplete or incorrect interpretations.

North American geoarchaeology still tends to be focused on large- to medium-scale geoarchaeology, with emphasis being given to landscape and soil histories. What appears to be very much lacking is attention to detailed studies of anthropogenic deposits, such as those found in large mounds in the Midwest, or in Pueblos in the southwest. Cultural deposits from Neolithic, Bronze Age, Roman and Medieval sites have been studied in considerable breadth/depth for more than two decades in the Old World and occur in "historical" sites in North America, from Colonial sites in New England to Spanish Mission sites in the southwest. Yet the number of geoarchaeological studies has been limited.

We can only hope that we've been successful in our goals and that geoarchaeology will gain in its application to all types of archaeological problems in all continents.

Appendices

15A Field methods

15A.1

Augering. Back strain or even back injury can result from augering. If several days of augering are contemplated do not forget to pull the auger out using a straight back. Whether working alone or with two people trying to pull the auger out of the ground, one of the best techniques is to grip the handle, then only using the legs "pull" the auger out of the ground. Never just use the back and arms.

16A Laboratory methods

16A.1

Bulk sample size. Logically, a 10 g sample can be employed for magnetic studies, which are non-destructive, and small subsamples can then be used for LOI, phosphate, and other extractions.

16A.2

Sulfate problems in grain size analysis. Sometimes the sample may contain SO_4, for example, marine and estuarine deposits, and particles settle very quickly leaving a clear supernatant, that can be tested for SO_4 content with $BaCl_2$, which forms a white powder (Avery and Bascomb, 1974). SO_4 is removed by drawing off the supernatant, and replacing this with deionized water, a process that may need to be repeated until all the SO_4 is removed.

16A.3

Grain size analysis. Grain size data is a useful comparison to grain size analysis in thin sections because grains are normally seen as *smaller* than their real size in thin section (Stoops, 2003: 14). On the other hand, the presence of ash crystals, clasts, mica, and phytoliths, for example, can be noted in thin section, and in fact clay translocation can only be truly identified through the presence of textural pedofeatures – clay coatings (e.g. Fig. 16.5a,b).

16A.4

Loss-on-ignition (LOI) and associated soil analyses. Typically, a temperature of 550°C is used for 2 h in geoarchaeology, although soil scientists are more likely to employ a lower temperature for longer, that is, 375°C for 6 h, because some carbonate can be lost above this temperature (Ball, 1964). In fact, the measurement of a sample at both 550°C and 800°C, is believed to yield estimates of both organic matter and an approximation of the carbonate content. Equally, the rubefied colors seen after ignition provide a qualitative measure of iron content, because all the iron has become oxidized. Exact amounts of calcium carbonate are calculated from the amount of CO_2 given off from a sample in a calcimeter when HCl is added (Avery and Bascomb, 1974). pH (soil reaction) is measured using a 2.5 proportion of deionized water (H_2O) to soil, although parallel testing is sometimes carried out using $CaCl_2$. In both cases, standard buffer solutions are used (e.g. pH 4.0 and pH 7.0) to calibrate the pH meter – which should also have a thermometer attachment, as temperature also affects pH (Gale and Hoare, 1991: 274). The fresher the soil/sediment sample the better, as drying out and/or microbiological activity in a stored sample, could well transform the pH.

Eh (in volts) – level of oxidation–reduction (redox potential) (Levy and Toutain, 1982) and saline content (Avery and Bascomb, 1974: 39) are other characteristics that can be measured and these affect electrical conductivity of deposits; they are important when attempting to preserve, *in situ*, deposits, biological materials, and metal objects (French, 2003, Fig. 11.3). For example, Tylecote found that acid soils were more aggressive in corroding buried tin-bronzes and copper, whereas alkaline soils were benign (Tylecote, 1979) – although corrosion still takes place (T. Rehren, personal communication; see Chapter 13 and Plates 13.2–13.3). Strangely, peats were also found to be benign, in spite of their acidity, probably due to the protective action of polyphenols.

16A.5

Thin section sample processing methods. Since it is paramount to use undisturbed samples (see Chapter 15) they must be impregnated with a resin in order to produce a hardened block that can then be cut up, mounted on a glass slide and ground to a thickness of 20–30 μm to produce a thin section (FitzPatrick, 1984; Murphy, 1986). Most important, however, is to produce a thin section that contains the fewest "processing artifacts" (e.g. incomplete impregnation, over-grinding, residues of grinding grits) as these may hinder description and thus cause interpretational problems (Stoops, 2003: 133).

Samples need to have soil moisture removed, as this inhibits curing of the resin; but any shrinkage can cause artifacts such as cracks. Air drying, acetone replacement (of water), and freeze drying have all been tested for a variety of materials, with acetone-replacement being reckoned the best method generally for soils (Murphy, 1986). On the other hand, acetone can dissolve organic matter, remove some minerals (e.g. halite), and cause some poorly cohesive samples to collapse (e.g. Pleistocene lagoonal silts). Impregnation is often recommended by using a vacuum to aid the deep penetration of resin into the sample, especially when dealing with clays with low interconnected porosity. Again, some poorly cohesive deposits can slump, for example poorly developed subsoils formed in volcanic ash. In addition, caution needs to be exercized when using a vacuum, as bubbling can occur if the vacuum process is rushed. It is best if a large vacuum chamber is available (as in a dedicated soil science laboratory) and a number of samples can be impregnated at one time (Courty *et al.*, 1989: 57–59). If laboratory facilities are not readily available, however, reasonable impregnation can often be achieved without a vacuum, if impregnation is not hurried (see below); this is particularly true for the impregnation of large samples that cannot fit into a standard 250 mm diameter vacuum chamber. In fact, for the many workers who do not have access to a large dedicated laboratory, it is be suggested that for most routine samples, air-drying and resin impregnation under normal atmospheric pressure is totally adequate. Use of a fume cupboard is absolutely essential on health grounds (see appropriate manufacturer's guide to toxicity and carcinogenic qualities of resin types; MSDS in the United States). Cracks/fissures relating to air drying can be recognized as artifacts and included in the description, but then "ignored" during interpretational process if appropriate.

Samples are typically impregnated with epoxy or polyester resin. Epoxy has been more traditionally used in soil science, whereas polyester has gained greater acceptance for archaeological studies. Generally, the latter is much cheaper, particularly if large samples are being impregnated, and easier to use, as proportions of resin to catalyst are less demanding with polyester. It should be pointed out that the use of polyester requires a low-viscosity product (although it can be diluted with styrene – see below), and one that is "*un*promoted," that is, one that lacks the cobalt accelerator, which is readily found in marine supply stores, for example: patching of boats and surfboards requires that the resin hardens relatively quickly (i.e. minutes). Impregnation of soils and sediments requires long soak times, in the order of days. Unpromoted resins can be purchased from wholesale suppliers of resins, and is normally available in 5 gallon containers, or in Europe in 5 kg plastic containers.

A simple approach to sample impregnation could employ the following impregnation routine, once an air-dried sample has been placed in a resin – and moderately heat-resistant and totally leak-proof container. At the beginning of the working day, pour resin into the container, around the sample, to a depth of approximately 30 mm. This must be done using a fume cupboard. If the resin has been diluted, and viscosity is reduced by adding 25–30% solvent such as acetone or styrene, natural capillarity will suck the resin up into the sample. If the sample has been warmed beforehand, at approximately 40°C, this will also help reduce the viscosity of the resin mixture. Never pour the resin *over* the sample, as this can trap air inside it, leading to poor impregnation of the center. Every hour or so, add more resin to the container. The sample will eventually show that it has absorbed resin by becoming darker and being "wetted" by the resin. At the end of the working day, it is usually possible to top up the container with resin, and the sample can be completely covered. As the solvent is fumed off in the fume cupboard, it can be topped up, but without adding any solvent. An example of a resin–styrene mixture is given in Table 16A.1.

Under normal temperatures (20°C), full impregnation and curing can take place between 1 and 3 months, so it is often necessary to accelerate curing by placing the container holding the sample in an oven (in fume cupboard), raising the temperature progressively from 40 to 60°C, and/or by adding additional catalyst to the resin mixture. It can be stated that the longer the impregnation time the better the impregnation, so curing should not be begun at least for two weeks. If the sample does not harden, it will have to be given longer time in the oven. Do not be tempted to raise the temperature too high to increase the rate of curing, as the whole sample can become brittle and crack up. Some plastic containers may also melt.

More rapid impregnations can be made using Canada balsam or low temperature thermoplastics (also used for mounting heavy minerals) but these may become unstable if subjected to X-rays in the microprobe. Methyl-acrylate, and similar chemicals can be used to impregnate sample in 24 h, but this material is expensive, highly toxic, and carcinogenic. In any case, all procedures should be carried out in a fume cupboard, including oven curing, and it is essential that proper protective rubber gloves, eye guards, and even a gas-mask should be worn, in addition to a lab coat (FitzPatrick, 1984).

Table 16A.1 Typical resin mixture for primary impregnation and topping up

	Polyester resin	**Styrene**	**Catalyst**
Primary impregnation resin mixture	ca 60–75%[a] (e.g. 600–750 ml)	ca 25–40%[a] (e.g. 250–400 ml)	10–20 ml[b]
Top up resin mixture	100% (e.g. 1000 ml)	None	10–20 ml[b]

[a]Extra styrene can be added if a particularly dense material, with few connecting voids, is being impregnated.
[b]According to manufacturer's recommendations. Normally, methyl-ethyl-ketone peroxide (MEKP).

The next stage is to cut up the impregnated block ahead of thin section manufacture. It is not common to have the expensive equipment and the expertise necessary to make good quality thin sections, and so many workers are happy to send off samples to a manufacturer; this is becoming routine procedure in earth science departments in the United States. If, however, a sample was carefully taken in the field, with its exact height above sea level (see Chapter 15), and is longer than the standard thin sections available, the impregnated block may need cutting into exactly measured and labeled subsamples before sending off for manufacture – it is not the responsibility of the manufacturer to maintain this control over the *exact* provenance of each thin section (Table 16A.2). Cutting can be done using an electric motor saw; inexpensive tile cutting saws are quite suitable and readily available in large hardware superstores. It is quite reasonable to cut thick slices from well-impregnated blocks, using water as a lubricant, but all later cutting and grinding should be done employing oil or paraffin; but be aware of the fire risk. Water entering the sample at a later stage can cause artifacts, such as swelling of the clays and buckling of the sample, which will crack it off its glass mount.

The motor saw is also useful when dealing with undisturbed monolith samples that are too big to put into a container for resin impregnation. These can be cut into suitable lengths to fit available containers (Table 16A.2), but this should normally be done without any water lubrication, as running water could well wash away sample juxtaposed to the saw cut.

It is not appropriate to discuss the various thin section manufacturing methods and machines here, as this is a specialist area, and will vary from country to country. Readers are referred to Courty *et al.* (1989); FitzPatrick (1984); Guilloré (1985); Murphy (1986) and micromorphology web sites (http://www.soils.org/divs/ s9/micromorph/micro.html). It can be noted here that if workers wish to carry out any kind of instrumentation analysis (e.g. EDAX, microprobe, FTIR) the thin section must be left uncovered – that is, without a glass cover slip – and the manufacturer must

TABLE 16A.2 Soil micromorphology – making thin sections from a long monolith. An example of producing a continuous series of thin sections from a 0.50 m long monolith taken from 10.00 to 9.50 m OD/height above sea level (asl)

Field monolith sample, e.g. sample [M212] taken at 10.00 m asl	Impregnation in two 160 mm long "resin-proof" containers	Separated blocks ready for cutting into blocks for 75 mm thin sections	Final thin sections (asl)
M212: 10.00–9.50 m (subsampled into 0–150, 150–300, 300–450, 450–500 mm lengths)	*Container A* Subsample A1: M212/0–150 mm Subsample A2: M212/150–300 mm *Container B* Subsample B1: M212/300–450 mm Subsample B2: M212/450–500 mm[a]	Subsample A1: M212/0–150 mm Subsample A2: M212/150–300 mm Subsample B1: M212/300–450 mm Subsample B2: M212/450–500 mm	M212a (10.00–9.925 m) M212b (9.925–9.850 m) M212c (9.850–9.775 m) 212d (9.775–9.700 m) 212f (9.700–9.625 m) 212g (9.625–9.550 m) 212h (9.550–9.500 m)

[a]Extra 100 mm long space in container B can be infilled with waste blocks in order not to waste resin.

be told to leave it off. In addition, a worker may wish to carry out a very fine polish of the thin section (using a 1000–1200 grade abrasive paper) to improve the optical qualities of the thin section. The manufacturer will always supply cover slips that can be glued down once any instrumentation or fine polishing has been carried out.

16A.6

Handling thin sections. There are some important points to remember when first handling thin sections and undertaking their study under the petrological microscope. Firstly, thin sections are expensive in terms of time taken to prepare them and the costs incurred through sample preparation and their manufacture. Also a particular slice of sample is essentially unique. It is therefore important to remember that thin sections are delicate, and very easily broken. If dropped – and it is remarkable how many laboratories have stone or tiled floors – they will smash to smitherines. Thin sections should always be held or contained in a box, and *never* balanced on sheets of paper or trays. When using the microscope, they should always be observed using the lowest magnification first (see Table 16A.2), as at high magnifications the distance between the thin section and the objective is miniscule. If the microscope is racked down too far, the slide will be broken. If in any doubt, only rack down while watching how close the objective is getting to the top of the thin section/cover slip. It should also be remembered that the manufacturer normally marks the slide on the *back*, and so sometimes the thin section can be accidentally observed upside down. When this mistake happens, it is impossible to rack down the high magnification objective far enough to focus, because of the thickness of the glass. Again, this can lead to accidental breakage of the thin section, so it is necessary to check that thin sections are observed the right way up – with the cover slip up.

17A.2

Presentation of data. Data should all be tabulated as the first form of presentation, with the first column providing information on the sample numbers and/or contexts studied. It is important to organize the sample list in a logical way, perhaps organized by stratigraphic position – from the highest to the lowest for soils; with each section being listed in numerical sequences, or by area of excavation. This is all to help both write-up and interpret the results, and to allow the reader to easily read the report/publication. It is therefore crucial that samples referred to in the text are exactly the same as in the tables. Also, a sample number may have a very complicated code that permits it to be understood exactly by the various people involved in the site. For example, a 400 mm long monolith labeled M1, which has the top O.D. (asl) of 10.00 m, is used for bulk samples for pollen and chemistry, and a continuous sequence of 4 × 100 mm long thin sections. The top subsample taken for pollen and chemistry is sample M1: 10.00–9.98 m and represents the topmost 20 mm. The topmost thin section is M1: 10.00–9.90 m, and chemical and pollen data from sample M1: 10.00–9.98 m are representative of the topmost 20 mm of thin section M1: 10.00–9.90. If this is simplified, and samples are called 1a, 1b, 1c, and so on, it can cause problems, because these numbers could be from the top down or the bottom up. Also, what thickness of the stratigraphic unit was used to provide subsample 1a? Some years may pass before results from several disciplines need to be correlated. If sample numbers do not match, or if there is any confusion, some results may not be able to be used.

It may not be possible to get all data onto the same table. It is therefore conventional to separate tables according to main techniques, with for example, grain size data being presented in one table and chemical and magnetic properties data in another. Equally, lists of x-ray fluorescence analysis of elements from bulk samples

16A.7 Recording and presenting soil micromorphology

TABLE 16A.3 Suggested outline for presenting soil micromorphological information (and identification of soil microfabric types – SMTs), and any associated data that aid the identification of a microfacies type (MFT), for one or multiple thin sections

Material SMT and MFT	Sample number	Sampling depth, Soil Micromorphology (SM), *in situ* thin section data (i.e. collected from the thin section itself), and *relevant* associated Bulk Data (BD) including macro- and micro-fossil (MFT) data	Context, *brief* comments and discussion, and interpretation
Soil Microfabric Type (SMT)/Microfacies Type (MFT)		*Depth*: Relative sampling depth and/or height above sea level (asl or O.D.) *SM*: • *Heterogeneity*: • *Structure and Voids*: • *Coarse Mineral*: Includes coarse : fine ratio (C : F) and size limit for C : F, e.g. 10 μm • *Coarse Organic*: • *Coarse Anthropogenic*: • *Fine Fabric*: In addition to color PPL and OIL, and interference colors [XPL] and autofluorescence; includes information on *related distribution*, and composition of fine material; individual materials can be referred to as soil microfabric types or SMTs • *Pedofeatures*: In the order of *textural, depletion, crystalline, amorphous, and cryptocrystalline, fabric* and *excrement* pedofeatures *In situ data* (from thin section and/or mirror-image polished block): (e.g. Image analysis (IA), microchemistry (microprobe), complementary macro- and micro-fossil data (identifications of *in situ* charcoal, diatoms, palynomorphs, phytoliths, nematode eggs) *BD*: e.g. chemistry – loss-on-ignition, phosphate, heavy metals; magnetic susceptibility; grain size; mineralogy (X-ray diffraction, heavy minerals)	*Context*: archaeological information – area, context, phase, and preliminary interpretation/identification *Comments and discussion:* Highlighted soil micromorphological data and other analytical information to produce basic first and second level identifications/interpretations *Interpretation*: Summarized second and third level interpretation

Table 16A.4 *Soil analysis*: summarized data for soil microfabric types/microfacies types from the mound and buried turf (MFT 3) and buried upper subsoil (MFT 4) from two Neolithic and six Early Bronze Age barrows at Raunds, Northamptonshire, United Kingdom (see Figs 16.5a–d and Fig. 16.6b)

Material (SMT/MFT)	**Sample number BD = bulk**	**Sampling depth, Soil Micromorphology (SM), *in situ* thin section data (i.e. collected from the thin section itself), and *relevant* associated Bulk Data (BD) including macro- and micro-fossil data**	**Context, comments and discussion, and *interpretation***
Soil Microfabric Type 3 / Microfacies Type 3	M8, M35, **M36, M39, M40**, BD5, 6, 10, 11, 24	e.g. 670–720 mm *SM*: moderately homogeneous; *Structure and Voids*: massive structured, compact; mainly frequent to dominant burrows and chambers, with very few to common channels and vughs (Total voids: mean 24% voids, range 10–30% voids; $n = 18$ – as above); C : F 65:35; *Coarse mineral*: very poorly sorted; very few gravel-size flint and quartzite; very dominant subangular silt-size to subrounded and rounded medium sand-size quartz; very few opaque minerals and silt-size mica; rare soil clasts; *Fine material*: heavily dotted, blackish, dark reddish brown (PPL), isotic to extremely low interference colours (XPL) (single spaced porphyric, speckled b-fabric); brownish orange with few black inclusions (OIL); many to abundant relict amorphous organic matter; occasional to many fine charcoal and occasional rubefied material; *Pedofeatures*: rare to occasional dusty clay, rare to many dark red clay, rare to many pan-like, occasional to many yellowish brown clay and rare to many compound textural features (see Macphail, 1999); rare biogenic calcite; rare to occasional blackish iron and manganese nodular impregnations (Fe/Mn) *BD*: sandy loam, with 9–16% clay, 25–31% silt, 56–65% sand; poorly humic 0.2–0.6% Organic C, 2.0–4.6% LOI; mean 330 ppm $P_2O_{5ignited}$, range 230–440 ppm $P_2O_{5ignited}$, 3.4–3.5 P ratio ($n = 3$); mean 1350 ppm P_{nitric}, range 1290–1410 ppm P_{nitric} ($n = 2$)	*Context*: Barrow mound and buried turf. *Comments and discussion*: humic, part-biologically worked soils with anomalous presence of textural pedofeatures (including dark red clay and many pan-like) – not oriented to way up in turf (also secondary/post depositional yellow brown clay). Presence of charcoal, fine burned mineral material and enhanced magnetic susceptibility also contribute to the anthropogenic signature. *Buried topsoil and local turf (of landscape open since earliest Neolithic times) used for barrow construction. Anomalous presence of organic phosphate-rich textural pedofeatures (e.g. red clay) and pans are evidence of soil poaching and a proxy indication of domestic stock concentrations. Postburial (Saxon) alluvial flooding of site led to yellow clay inwash*

TABLE 16A.4 *(Cont.)*

Soil Microfabric Type 4/ Microfacies Type 4	*M9* (M39, M40) BD7, 25, 26	e.g. 710–750 mm *SM*: massive structured, with common to dominant burrows and chambers, few to frequent channels and vughs (e.g. total voids: mean 37% voids, range 20–40% voids); C : F 85 : 15; *Coarse mineral* as SMF3; *Fine mineral*: heavily speckled, greyish yellow brown (PPL), very low interference colours (XPL) (single spaced porphyric, speckled b-fabric); pale yellow orange (OIL); occasional to many relict amorphous organic matter; occasional charcoal and rubefied material; *Pedofeatures*: rare to many dusty clay, occasional to abundant dark red clay, occasional to abundant pan-like, rare yellowish brown clay and rare to occasional compound textural features (see Key); rare to occasional blackish iron and manganese nodular impregnations (Fe/Mn) *BD*: sandy loam to loamy sand, with 6–11% clay, 22–28% silt, and 61–72% sand; poorly humic (0.1–0.3% Organic C, 2–3% LOI), non-calcareous (<0.1% $CaCO_3$); 270 ppm $P_2O_{5 ignited}$, range 240–330 ppm $P_2O_{5 ignited}$, 2.3–2.9 P ratio ($n = 3$) *Microprobe*: Textural features Clay void infill: mean 0.11% P, 1.05% Ca, 0.06% Na, 0.06% S, 7.19% Fe, 0.37% Mn, 0.65% Mg, 16.6% Si, 8.3% Al, and 1.56% K ($n = 19$); Coarse capping/pan-like textural feature (excluding large sand grains): mean 0.07% P, 0.48% Ca, 0.19% Na, 0.13% S, 3.29% Fe, 0.03% Mn, 0.21% Mg, 20.6% Si, 3.0% Al, and 0.88% K ($n = 6$); dark red clay void coating: mean 0.25% P, 1.03% Ca, 0.38% Na, 0.04% S, 20.6% Fe, 0.12% Mn, 1.09% Mg, 12.2% Si, 7.8% Al, and 0.97% K ($n = 9$).	Barrow-buried upper subsoil (Eb or A2 horizon): As above, with again anomalous presence of textural pedofeatures and microprobe data showing higher concentrations of P here compared to underlying Bt horizon

(Macphail, 2003a, Forthcoming; Macphail and Cruise, 2001).

TABLE 16A.5 *Freshwater sediment analysis*: full presentation of soil micromorphology and complementary data and its brief interpretation for two units at Lower Palaeolithic Boxgrove, West Sussex, United Kingdom. (Hundreds of Acheulean hand axes found in Unit 4*, along with 2 human teeth; a human tibia was found just above Unit 4 (Figs 16.6a–b)

Material (SMT/MFT)	Sample number	Sampling depth, Soil Micromorphology (SM), *in situ* thin section data (i.e. collected from the thin section itself; microprobe and image analysis, IA), and associated Bulk Data (BD)	Context, comments and discussion, and *interpretation*
Sediment Microfabric Type 4d1/1; Microfacies 4d1	*M14c/ Unit 4d1*	*Relative depth*: 270–370 mm *SM*: *Heterogeneity*: homogeneous *Structure and voids*: massive with dominant fine to medium, coarse meso to macro-channels, showing partial collapse, with polyconcave and closed vughs, and few complex packing voids; coarse voids also include few macro- to mega-vesicles; *Coarse mineral*: C : F, 45 : 55, dominant coarse silt to very fine sand-size quartz (and chalk clasts), common calcite/aragonite; very few chalk fossils, Cretaceous clay, glauconite, mica, opaques, and calcitic root pseudomorph fragments; *Coarse organic*: rare organic staining and amorphous remains of plant tissue lengths (few calcite root pseudomorphs); *Coarse anthropogenic(?)*:3 mm long bone/scat fragment; *Fine fabric*: cloudy, speckled grey (PPL), very high interference colors (close porphyric, crystallitic b-fabric, XPL), gray to white (OIL); *Pedofeatures: Textural*, very abundant intercalations (interlaced), void infills, and ped coatings of gray to pale brown to dark gray calcitic layered impure silt, micrite, and brown micritic clay coatings, infilling around collapsed peds; *Crystalline*: few calcite root pseudomorphs; *Fabric*: many fragments of microlaminated textural features; many coarse 15 mm wide burrows; *Excrement*: rare possible excrements associated with relict roots *IA*: 8% voids; dominant fine and medium (61.9% – 101–200 μm^2), frequent coarse (24.1% – 201–500 μm^2) meso voids, few fine (6.5% – 501–1000 μm^2) macro voids, and very few medium and coarse macro and mega-voids (2.5% – 1000–>2000 μm^2) *BD*: Sandy silt loam (17.1% clay, 59.9% silt, 22.1% fine sand), 46.2–56.6% $CaCO_3$ (mean 53.5%), 0.458–0.678 phosphate-P mg g^{-1} (mean 0.565 phosphate-P mg g^{-1}), 0.12% org. C, 0.17–0.21% Ext. Fe (mean 0.19%), χ 1.55–1.82 $\times 10^{-8}$ SI kg^{-1}, and 3.54–4.71% $\chi_{conv.}$ ($n = 5$)	*Context*: Quarry 1B, Trench 5; Unit IVd1 *Comments and discussion:* biologically (likely earthworms and roots) worked fresh water pond sediment with later influence of fluvial deposition/meltwater causing collapse of ped (soil structures) and as overlying Unit 8ac is deposited (also causing contemporary textural feature fragmentation), voids are infilled with clay to silt-size material that is washed in, drastically reducing porosity *Interpretation: continuing massive deposition with likely homogenization through biological/freeze-thaw, but with increasing amounts and influence of debris flow sedimentation through time*

Table 16A.5 (*Contd.*)

Sediment Microfabric Type 4*/1; Microfacies 4*/1	*M13c/Unit* 4*	*Relative depth*: 670–730 mm (Upper 4*); *SM*: *Heterogeneity*: homogeneous *Structure and voids*: massive, burrow, and channel microstructure; common very fine to fine (0.1–0.4 mm), frequent medium (0.8 mm), and coarse (1 mm), and few patches of very fine packing pores (15% voids); *Coarse mineral*: C : F, 60 : 40, very dominant very coarse silt-size and very fine sand-size subangular to subrounded quartz (very few feldspars), frequent fine "shell/fossil" fragments; very few rounded glauconite, limonite, and other opaques and mica (rare coarse mollusc shell; rare fossils and ostracods, and foraminifera); *Coarse organic*: (see fine fabric) rare organic staining and amorphous remains of plant tissue lengths (few calcite root pseudomorphs); *Coarse anthropogenic(?)*: absent *Fine fabric*: fine speckled, cloudy grayish brown (PPL), moderately high interference colors (dominantly interaggregate related distribution – gefuric, with coated grains to porphyric especially in lowermost 30 mm, XPL), grayish white (OIL) – high amounts of calcitic fine material; rare fine organic and tissue fragments and amorphous staining; *Pedofeatures: Textural*: many weak, thin (0.5–0.8 mm) chalky clay (micritic) stringers, textural concentrations associated with sub horizontal divisions (palaeosurfaces?) and burrow and channel margins; *Crystalline*: rare calcitic root pseudomorphs; *Amorphous and cryptocrystalline*: occasional iron staining, especially of root traces; *Fabric*: many 0.5–2 mm and very abundant >2 mm size burrows; *Excrement*: aggregated total excremental fabric? *IA*: 15% voids; very dominant fine and medium (78.5% – 101–200 μm^2), frequent coarse (20.3% – 201–500 μm^2) meso voids, very few fine (0.9% – 501–1000 μm^2) macro voids, and very few medium and coarse macro and mega-voids (0.3% – 1000–>2000 μm^2)	*Context*: Quarry 1B, Trench 5; Unit 4*. *Comments and discussion:* rather open, coarsely burrowed and finely channeled calcitic fine sandy silt loam sediment develops that is slightly more coarse and less calcareous and ferruginous compared to the underlying Unit 3 and basal Unit 4* deposits. Also calcitic separations/stringers along possible palaeosurfaces and in channels and along burrow margins indicates wash coinciding with biological (fauna and plants) activity. Lower Unit IV*, that little bit more compact, calcareous and iron stained – similarly calcareous as uppermost Unit 3 *Interpretation:* *Appears that primary deposition of Unit4* was a likely calcareous slurry, forming a compact basal 30 mm layer, with some calcareous sediment washing into voids in Unit 3 – possibly also leading to calcite root pseudomorphs. Upward, Unit 4* is less densely calcitic and more open, as presumably biological activity by roots and earthworms worked the accumulating sediment*

BD: Sandy silt loam (11.4% clay, 57.3% silt, 30.0% fine sand), 21.1–21.9% $CaCO_3$ (mean 21.6%), 0.500–0.634 phosphate-P mg g^{-1} (mean 0.558 phosphate-P mg g^{-1}), 0.05% Org. C, 0.18–0.21% Ext. Fe, χ 2.59–2.73 $\times 10^{-8}$ SI kg^{-1} and 2.54–3.02% $\chi_{conv.}$ ($n = 3$) *Microprobe*: 0.14% Na, 0.14% Mg, 1.41% Al, 29.08% Si, 0.03% P, 0.01% S, 0.65% K, 10.54% Ca, 0.02% Ti, 0.01% Mn, and 1.1% Fe ($n = 47$) Elemental map: Fe is widespread, with fine (0.1 mm) Fe punctuations (mottling) throughout. Very low P, low Al, and very high Si throughout. High Ca is ubiquitous, and commonly concentrated into laminae. Lower 4*: 730–760 mm *SM*: compact, massive, and channel microstructure; 10–15% voids, dominant fine (0.4 mm), and frequent medium (0.8 mm) channels; Mineral and organic matter as above, but embedded (porphyric; C : F, 45 : 55), coated and interaggregate related distribution (porphyric, chitonic, and gefuric) *BD*: 23.0–24.2% $CaCO_3$ (mean 23.6%), 0.722–804 P mg g^{-1} (mean 0.763), 0.06% org. C, 0.36–0.37% Ext. Fe, χ 2.91–3.39 $\times 10^{-8}$ SI kg^{-1} and 1.11–1.13% $\chi_{conv.}$ ($n = 2$) *Field*: pale yellow (2.5Y7/4), sharp sloping boundary to Unit 3 below	*that infills above truncated Unit 3*

(Macphail *et al.*, 2001; Roberts *et al.*, Forthcoming)

Table 16A.6 *Anthropogenic sediment analysis*: summarized data for microfacies types from the Early Medieval London Guildhall, United Kingdom; data summarized from individual sample descriptions and counts, microprobe bulk chemical and pollen analyses

Facies (MFT) (examples of samples and contexts)	Micromorphology and microprobe	Chemistry	Pollen	Identifications and interpretations
MFT 5a {251} {687} {755} {977} {978}	Very dominantly organic, with (a) intact horizontally layered Poaceae tissues, commonly with abundant phytoliths and long articulated phytoliths; (b) sometimes with dark brown humified plant material, (c) intercalated with silt, autofluorescent under blue light, layered Ca, P, and K distribution (probe); (d) sometimes inclusions of amorphous organic matter with calcium oxalate crystals and Ca/P/K chemistry (probe); commonly limpid brown, speckled brown to dark brown/blackish (PPL), isotic, with rare scatter of high interference colors (isotic or crystallitic, XPL), very dark brown to black (OIL); very abundant plant organs and tissues, with rare pollen, diatoms, and ash; wood may occur; very few burrows and excrements	MFT 5a/b/c: moderate to very high LOI (range, 10.3–43.4%, $n = 13$). Mostly non-calcareous (maximum, 2% carbonate). Clear to strong indications of phosphate (phosphate-P, range = 5.79–20.7 mg g^{-1}), and possible to strong evidence of heavy metal enrichment (especially Pb and Cu). No χ enhancement in some samples, but possible to strong enhancement in others	High concentrations of well-preserved, dominantly cereal t. and Poaceae pollen. Samples with >30% grass should be viewed as relatively undiluted stabling refuse, with major inputs from animal feed, bedding, and dung. Samples with <30% grass contain very high amounts of cereal t., possibly as special feed ("pig" coprolite association). Rare concentrations of woody pollen probably indicative of construction. Weed assemblage probably indicates "local" animal husbandry.	*Well-preserved highly organic stabling refuse* (Periods 10 and 11), with (a) fodder and bedding, (b) dung of herbivores (cattle better preserved than sheep/goat); stabling floor crust and (c) omnivore (pig) dung. These deposits have been very little biologically worked before or after dumping. Also this material includes "raw" dung being used for lining wattle walls or to support and level (?) floor foundations. (There is also a strong statistical correlation between LOI and phosphate-P, showing phosphate enrichment is due to animal husbandry represented by MFT 5. There is also strong evidence of the link between phosphate-P and heavy metal concentrations.)

(Macphail *et al.*, 2001; in press)

should also be given separately to element counts made by microprobe on thin sections.

In addition to tables, it is useful to explore various other ways of presenting data so that it is easily appreciated. Cumulative curves, bar-charts, scattergrams, including regression data, can all be utilized (Figs 16.1 and 16.4). It is more difficult to combine selective data from different sources and disciplines, to enable "consensus" findings to be graphically presented (Fig. 11.14). On the other hand, all these make the analysis of the results, their discussion and interpretation much easier, and also understandable to the non-specialist.

Bibliography

Acott, T.G., Cruise, G.M. and Macphail, R.I. (1997) Soil micromorphology and high resolution images. In: *Soil Micromorphology: Diversity, Diagnostics and Dynamics* (Eds S. Shoba, M. Gerasimova and R. Miedema), pp. 372–378. International Soil Science Society, Moscow–Wageningen.

Adams, R.M. (1981) Heartland of Cities: Surveys of Ancient Settlement and Land Use on the Central Floodplain of Euphrate. Chicago: University of Chicago Press.

Adovasio, J.M., Donahue, J., Carlisle, R.C., Cushman, R., Stuckenrath, R. and Wiegman, P. (1984) Meadowcroft Rockshelter and the Pleistocene/Holocene transition in southwestern Pennsylvania. In: *Contributions in Quaternary Vertebrate Paleontology* (Eds H.H. Genoways and M.R. Dawson), *Carnegie Museum of Natural History Special Publication 8*, pp. 347–369.

Ahlbrandt, T.S. and Fryberger, S.G. (1982) Eolian deposits. In: *Sandstone Depositional Environments* (Eds P.A. Scholle and D.R. Spearing), Memoir 31, pp. 11–47. The American Association of Petroleum Geologists, Tulsa.

Aitken, M.J. and Valladas, H. (1992) Luminescence dating relevant to human origins. *Philosophical Transactions of the Royal Society of London, Series B: Biological Sciences*, **337**: 139–144.

Akazawa, T. (1976) Middle Palaeolithic assemblages of the Douara Cave in Syria and their relationship to the Middle Palaeolithic sequence in the Levant and the adjacent regions. *Deuxiéme Colloque sur la Terminologie de la Préhistoire du Proche-Orient "Terminology of Prehistory of the Near East"*, 138–152.

Akazawa, T. (1979) Flint factory sites in Palmyra Basin. Paleolithic site of Douara Cave and paleography of Palmyra Basin in Syria. Part II: Prehistoric occurrences and chronology in Palmyra Basin. *Bulletin of the University Museum, University of Tokyo*, **16**: 159–200.

Akeret, Ö. and Rentzel, P. (2001) Micromorphology and plant macrofossil analysis of cattle dung from the Neolithic lake shore settlement of Arbon Bleiche 3. *Geoarchaeology*, **16**: 687–700.

Albanese, J. (1974) Geology of the Casper archeological site. In: *The Casper Site: a Hell Gap Bison Kill on the High Plains* (Ed G.C. Frison), pp. 266. Academic Press, New York.

Albanese, J.P. and Frison, G.C. (1995) Cultural and landscape change during the middle Holocene, Rocky Mountain area, Wyoming and Montana Archaeological geology of the Archaic period in North America. *Special Paper-Geological Society of America*, **297**: 1–19.

Albert, R.M., Lavi, O., Estroff, L. et al. (1999) Mode of occupation of Tabun Cave, Mt Carmel, Israel, during the Mousterian period: a study of the sediments and phytoliths. *Journal of Archaeological Science*, **26**: 1249–1260.

Albert, R.M., Weiner, S., Bar-Yosef, O. and Meignen, L. (2000) Phytoliths in the Middle Palaeolithic Deposits of Kebara Cave, Mt Carmel, Israel: study of the plant materials used for fuel and other purposes. *Journal of Archaeological Science*, **27**: 931–947.

Allen, J.R.L. (1970) *Physical Process of Sedimentation*. George Allen & Unwin Ltd, London, 248 pp.

Allen, J.R.L. (1971) *Physical Process of Sedimentation*. George Allen & Unwin Ltd, London, 248 pp.

Allen, M.J. (1988) Archaeological and environmental aspects of colluviation in south-east England. In: *Man-made Soils* (Eds W.G.-V. Waateringe and M. Robinson), BAR International Series 410, pp. 67–92. British Archaeological Reports, Oxford.

Allen, M.J., Godden, D. and Matthews, C. (2002) Mesolithic, Late Bronze Age and Medieval activity at Katherine Farm, Avonmouth, 1998. Archaeology in the Severn Estuary 2002. Bath, Severn Estuary Levels Research Committee, pp. 85–105.

Allen, M.J. (1990) Magnetic susceptibility. In: *Excavations at Brean Down, Somerset* (Ed M. Bell), Archaeological Report No. 15, pp. 197–202. English Heritage, London.

Allen, M.J. (1992) Products of erosion and the prehistoric land use of the Wessex Chalk. In: *Past and Present Soil Erosion* (Eds M. Bell and J. Boardman), Monograph 22, pp. 37–52. Oxbow, Oxford.

Allen, M.J. (1994) The land-use history of the Southern English Chalklands with an Evaluation of the Beaker period using Environmental data: colluvial deposits as environmental and cultural indicators: Southampton, Southampton University.

Allen, M.J. (1995b) Land molluscs. In: *Balksbury Camp, Hampshire: Excavations 1973 and 1981* (Eds G.J. Wainwright and S.M. Davis), pp. 92–100. English Heritage, London.

Allen, M.J. and Macphail, R.I. (1987) Micromorphology and magnetic susceptibility studies: their combined role in interpreting archaeological soils and sediments. In: *Soil Micromorphology* (Eds N. Fedoroff, L.M. Bresson and M.A. Courty), pp. 669–676. Association Française pour l'Étude du Sol, Plaisir.

Allen, T.G. (1995a) *Lithics and Landscape: Archaeological Discoveries on the Thames Water Pipeline at Gatehampton Farm, Goring, Oxfordshire 1985–92*. Oxford Archaeological Unit Thames Valley Landscapes Monograph No. 7. Oxford University Committee for Archaeology, Oxford.

Altemüller, H.J. and Van Vliet-Lanöe, B. (1990) Soil thin section fluorescence microscopy. In: *Soil Micromorphology: A Basic and Applied Science* (Ed L.A. Douglas), pp. 565–579. Elsevier, Amsterdam.

Amick, D.S. and Mauldin, R.P. (1989) *Experiments in Lithic Technology*. Bar International Series, Oxford.

Ammerman, A. and Polgase, C. (1997) Analyses and descriptions of the obsidian collections from Arene Candide. In: *Arene Candide: A Functional and Environmental Assessment of the Holocene Sequence (Excavations Bernabò Brea-Cardini 1940–50)*. (Ed R. Maggi), New Series Number 5, pp. 573–592. Memorie dell'Istituto Italiano di Paleontologia Umana, Roma.

Amos, F., Shimron, A. and Rosenbaum, J. (2003) Radiometric dating of the Siloam Tunnel, Jerusalem. *Nature*, **425**: 169–171.

Ampe, C. and Langohr, R. (1996) Distribution of circular structures and link with the soilscape in the sandy and loamy sandy area of NW Belgium. Fortuitous or a deliberate choice? In: *Paleoecology* (Eds L. Castelletti and M. Cremaschi), pp. 59–68. ABACO, Forli.

Andrews, P. (1990) *Owls, Caves and Fossils*. Predation, preservation and accumulation of small mammal bones in caves, with an analysis of the Pleistocene cave faunas from Westbury-sub-Mendip, Somerset, UK. Natural History Museum Publications. The Natural History Museum, London.

Andrews, P., Cook, J., Currant, A. and Stringer, C. (1999) *Westbury Cave. The Natural History Museum Excavations 1976–1984*. CHERUB (Centre for Human Evolutionary Research at the University of Bristol), Bristol.

Audouze, F. and Enloe, J.G. (1997) High resolution archaeology at Verberie: limits and interpretations. *World Archaeology*, **29**: 195–207.

Arensberg, B., Bar-Yosef, O., Chech, M., Goldberg, P., Laville, H., Meignen, L., Rak, Y., Tchernov, E., Tiller, A. and Vandermeersch, B. (1985) Human Paleontology: a Neanderthalian sepulture in the Kebara Cave (Israel). *C. R. Acad. Sci. Paris*, 300: Series II, **6**: 227–230.

Armitage, P.L., Locker, A. and Straker, V. (1987) Environmental archaeology in London: a review. In: *Environmental Archaeology: a Regional Review* Vol II (Ed H.C.M. Keeley), Occasional paper No. 1, pp. 252–331. English Heritage, London.

Armour-Chelu, M. and Andrews, P. (1994) Some effects of bioturbation by earthworms (oligochaeta) on archaeological sites. *Journal of Archaeological Science*, **21**: 433–443.

Arpin, T., Macphail, R.I. and Boschian, G. (1998) Summary of the spring 1998 meeting of the Working Group on archaeological soil micromorphology, February 27–March 1, 1998. *Geoarchaeology*, **13**: 645–647.

Arpin, T., Mallol, C. and Goldberg, P. (2002) A new method of analyzing and documenting micromorphological thin sections using flatbed scanners: applications in geoarchaeological studies. *Geoarchaeology*, **17**: 305–313.

Arrhenius, O. (1931) Die Bodenanalyse in dienst der Archäologie. *Zeitschrift für Pflanzenerhährung, Düngung und Bodenkunde Teil B*, **10**: 427–439.

Arrhenius, O. (1934) Fosfathalten i skånska jordar. *Sveriges Geologiska Undersökningar*, Ser. C, 393.

Arrhenius, O. (1955) *The Iron Age Settlements on Gotland and the Nature of the Soil*. Valhagan II, Munksgaard, Copenhagen, pp. 1053–1064.

Ashley, G.M. (2001) Archaeological sediments in springs and wetlands. In: *Sediments in Archaeological Context* (Eds J.K. Stein and W.R. Farrand), pp. 183–210. The University of Utah Press, Salt Lake City.

Ashley, G.M. and Driese, S.G. (2000) Paleopedology and paleohydrology of a volcaniclastic paleosol interval; implications for early Pleistocene stratigraphy and paleoclimate record, Olduvai Gorge, Tanzania. *Journal of Sedimentary Research*, **70**: 1065–1080.

Atkinson, M. and Preston, S.J. (1998) The Late Iron Age and Roman settlement at Elms Farm, Heybridge, Essex, excavations 1993–5: an interim report. *Britannia*, **XXIX**: 85–110.

Atkinson, R.J.C. (1957) Worms and weathering. *Antiquity*, **31**: 219–233.

Avery, B.W. (1964) *The Soils and Land-Use of the District around Aylesbury and Hemel Hempstead*. Her Majesty's Stationary Office, London.

Avery, B.W. (1980) *Soil Classification for England and Wales*, Soil Technical Monograph No. 13. Soil Survey of England and Wales, Harpenden.

Avery, B.W. (1990) *Soils of the British Isles*. CAB International, Wallingford, 463 pp.

Avery, B.W. and Bascomb, C.L. (1974) *Soil Survey Laboratory Techniques*. Soil Survey Technical Monograph, No. 14. Soil Survey of England and Wales, Harpenden.

Ayala, G. and French, C. (2005) Erosion modeling of past land-use practices in the Fiume di Sotto di Troina River Valley, North-Central Sicily. *Geoarchaeology*, **20**: 149–167.

Babel, U. (1975) Micromorphology of soil organic matter. In: *Soil Components: Organic Components* (Ed J.E. Giesking), Vol. 1, pp. 369–473. Springer-Verlag, New York.

Bagnold, R.A. (1941) *The Physics of Blown Sand and Desert Dunes*. Methuen and Co., London.

Bailloud, G. (1953) Note préliminarie sur l'industrie des niveaux supérieurs de la Grotte du Renne, à Arcy-sur-Cure (Yonne). *Bulletin de la Société Préhistorique Française*, **50**: 338–344.

Bakels, C.C. (1988) Pollen from plaggen soils in the province of North Brabant, the Netherlands. In: *Man-made Soils* (Eds W. Groenman-van Waateringe and M. Robinson), International Series 410, pp. 55–66. British Archaeological Reports, Oxford.

Baker, R.T. (1976) Changes in the chemical nature of soil phosphate during pedogenesis. *Journal of Soil Science*, **27**: 504–512.

Bakler, N. (1989) Regional geology. In: *Excavation at Tel Michal, Israel* (Eds Z. Herzog, J. George Rapp and O. Negbi), pp. 198–218. The University of Minnesota Press, Minneapolis.

Bal, L. (1982) *Zoological Ripening of Soils*. Agricultural Research Report. Centre for Agricultural Publishing and Documentation, Wageningen.

Balaam, N., Bell, M., David, A., Levitan, B., Macphail, R.I., Robinson, M. and Scaife, R.G. (1987) Prehistoric and Romano-British sites at Westward Ho! Devon: archaeological and palaeoenvironmental surveys 1983 and 1984. In: *Studies in Palaeoeconomy and Environment in South West England* (Eds N.D. Balaam, B. Levitan and V. Straker), British Series 181, pp. 163–264. British Archaeological Reports, Oxford.

Balaam, N., Corney, M., Dunn, C. and Porter, H. (1991) The surveys. In: *Maiden Castle. Excavations and field survey 1985–6* (Ed N.M. Sharples), Archaeological Report No. 19, pp. 37–42. English Heritage, London.

Balek, C.L. (2002) Buried artifacts in stable upland sites and the role of bioturbation: a review. *Geoarchaeology*, **17**: 41–51.

Ball, D.F. (1964) Loss-on-ignition as an estimate of organic matter and organic carbon in non-calcareous soils. *Journal of Soil Science*, **15**: 84–92.

Barclay, A., Lambrick, G., Moore, J. and Robinson, M. (2003) *Lines in the Landscape. Cursus Monuments in the Upper Thames Valley: Excavations at the Drayton and Lechlade Cursuses*, Thames Valley Landscapes Monograph Monograph No. 15. Oxford Archaeology, Oxford.

Barclay, G.T. (1983) Sites of the third millennium BC to the first millennium AD at North Mains, Strathallan, Perthshire. *Proceedings of the Society of Antiquities Scotland*, **113**: 122–281.

Barham, A.J. (1995) Methodological approaches to archaeological context recording: X-radiography as an example of a supportive recording, assessment and interpretive technique. In: *Archaeological Sediments and Soils: Analysis, Interpretation and Management* (Eds A.J. Barham and R.I. Macphail), pp. 145–182. Institute of Archaeology, University College London, London.

Barham, A.J. and Macphail, R.I. (Eds) (1995) Archaeological sediments and soils: analysis, interpretation and management: London, Institute of Archaeology – University College London.

Barker, G. (1985) *Prehistoric Farming in Europe*. Cambridge University Press, Cambridge.

Barnes, G.L. (1990) Paddy soils now and then. In: Soils and Early Agriculture (Ed K. Thomas), *World Archaeology*, **22** (1), pp. 1–17. Routledge, London.

Barton, R.N.E. (1992) *Hengistbury Head, Dorset. Volume 2: The Late Upper Palaeolithic & Early Mesolithic sites*, Monograph No. 34. Oxford University Committee for Archaeology, Oxford.

Barton, R.N.E. (1997) *English Heritage Book of Stone Age Britain*. B. T. Batsford/English Heritage, London.

Barton, R.N.E., Currant, A.P., Fernandez-Jalvo, Y. et al. (1999) Gibraltar Neanderthals and results of recent excavations in Gorham's, Vanguard and Ibex Caves. *Antiquity*, **73**: 13–23.

Bartov, Y., Stein, M., Enzel, Y., Agnon, A. and Reches, Z. (2002) Lake levels and sequence stratigraphy of Lake Lisan, the Late Pleistocene precursor of the Dead Sea. *Quaternary Research*, **57**: 9–21.

Baruch, U., Werker, E. and Bar-Yosef, O. (1992) Charred wood remains from Kebara Cave, Israel: preliminary results. *Actualites Botaniques*, **2**: 531–538.

Bar-Yosef, O. (1974) Late Quaternary stratigraphy and prehistory in Wadi Fazael, Jordan Valley: a preliminary report. *Paléorient*, **2**: 415–428.

Bar-Yosef, O. (1991) The history of excavations at Kebara Cave. In: *Le Squelette Moustérien de Kébara 2* (Eds O. Bar-Yosef and B. Vandermeersch), pp. 17–27. (Cahiers de Paléanthropologie), Éditions du C.N.R.S., Paris.

Bar-Yosef, O. (1994) The Lower Palaeolithic of the Near East. *Journal of World Prehistory*, **8**: 211–265.

Bar-Yosef, O. and Goren-Inbar, N. (1993) *The Lithic Assemblages of 'Ubeidiya: A Lower Palaeolithic Site in the Jordan Valley*. Qedem, 34. Institute of Archaeology – The Hebrew University of Jerusalem, Jerusalem.

Bar-Yosef, O. and Phillips, J.L. (Eds) (1977) *Prehistoric Investigations in Gebel Maghara, Northern Sinai, Qedem*. Hebrew University of Jerusalem, Jerusalem.

Bar-Yosef, O. and Sillen, A. (1994) Implications of the new accelerator date of the charred skeletons from Kebara Cave (Mt. Carmel). *Paléorient*, **19**: 205–208.

Bar-Yosef, O. and Tchernov, E. (1972) *On the palaeo-ecological history of the site of 'Ubeidiya*. The Israel Academy of Sciences and Humanities, Jerusalem.

Bar-Yosef, O., Vandermeersch, B., Arensburg, B. et al. (1992) The excavation in Kebara Cave, Mt. Carmel. *Current Anthropology*, **33**: 497–550.

Bar-Yosef, O., Arnold, M., Mercier, N. et al. (1996) The dating of the Upper Paleolithic layers in Kebara Cave, Mt. Carmel. *Journal of Archaeological Science*, **23**: 297–306.

Bascomb, C.L. (1968) Distribution of pyrophosphate-extractable iron and organic carbon in soils of various groups. *Journal of Soil Science*, **19**: 251–268.

Bateman, M.D., Holmes, P.J., Carr, A.S., Horton, B.P. and Jaiswal, M.K. (2004) Aeolianite and barrier dune construction spanning the last two glacial–interglacial cycles from the southern Cape coast, South Africa. *Quaternary Science Reviews*, **23**: 1681–1698.

Bateman, N. (1997) The London Amphitheatre: excavations 1987–1996. *Britannia*, **XXXVIII**: 51–85.

Bateman, N. (2000) *Gladiators at the Guildhall*. Museum of London Archaeology Service, London.

Begin, Z.B., Nathan, Y. and Ehrlich, A. (1980) Stratigraphy and facies distribution in the Lisan formation new evidence from the area south of the Dead Sea, Israel. *Israel Journal of Earth Sciences*, **29**: 182–189.

Behre, K.-E. (Ed) (1986) *Anthropogenic Indicators in Pollen Diagrams*. A. A. Balkema, Rotterdam.

Belfer-Cohen, A. and Goldberg, P. (1982) An Upper Paleolithic site in south central Sinai. *Israel Exploration Journal*, **32**: 185–189.

Bell, M. (1981) Seaweed as a prehistoric resource. In: *Environmental Aspects of Coasts and Islands* (Eds D. Brothwell and G.W. Dimbleby), pp. 117–126. British Archaeological Reports, Oxford.

Bell, M. (1983) Valley sediments as evidence of prehistoric land use on the South Downs. *Proceedings of the Prehistoric Society*, **49**: 118–150.

Bell, M. (1990) *Brean Down Excavations 1983–87*. English Heritage, London.

Bell, M. (1992) The prehistory of soil erosion. In: *Past and Present Soil Erosion* (Eds M. Bell and J. Boardman), Monograph 22, pp. 21–35. Oxbow, Oxford.

Bell, M. and Boardman, J. (Eds) (1992) *Past and Present Soil Erosion*. Oxbow, Oxford.

Bell, M., Fowler, M.J. and Hillson, S.W. (1996) *The Experimental Earthwork Project, 1960–1992*. Research Report, Research Report 100. Council for British Archaeology, York.

Bell, M., Caseldine, A. and Neumann, H. (2000) *Prehistoric Intertidal Archaeology in the Welsh Severn Estuary*, Research Report 120. Council for British Archaeology, York.

Bellhouse, R.L. (1982) Soils and archaeology. In: *North of England Soils Discussion Group, Proceedings* (Ed M.J. Alexander), Vol. 17, pp. 41–47.

Bertran, P. (1993) Deformation-induced microstructure in soils affected by mass movements. *Earth Surface Processes and Landforms*, **18**: 645–660.

Bertran, P. (1994) Degradation des niveaux d'occupation paléolithiques en contexte périglaciaire: Exemples et implications archéologiques. *Paléo*, **6**: 285–302.

Bertran, P. and Texier, J.-P. (1999) Facies and microfacies of slope deposits. *CATENA*, **35**: 99–121.

Bethell, P.H. and Máté, I. (1989) The use of soil phosphate analysis in archaeology: a critique. In: *Scientific Analysis in Archaeology*. (Ed J. Henderson), Monograph No. 19, pp. 1–29. Oxford University Committee, Oxford.

Bettis, E.A., III (1988) Pedogenesis in Late Prehistoric Indian Mounds, Upper Mississippi Valley. *Physical Geography*, **9**: 263–279.

Bevan, B.W. and Roosevelt, A.C. (2003) Geophysical exploration of Guajará, a Prehistoric Earth Mound in Brazil. *Geoarchaeology*, **18**: 287–331.

Beynon, D.E. (1981) *The Geology of Meadowcroft Rockshelter*. Ph.D. Thesis, University of Pittsburgh, Pittsburgh. 283 pp.

Beynon, D.E. and Donahue, J. (1982) The geology and geomorphology of Meadowcroft Rockshelter and the Cross Creek drainage. In: *Collected Papers on the Archaeology of Meadowcroft Rockshelter and the Cross Creek Drainage* (Eds R.C. Carlilse and J.M. Adovasio), pp. 31–52. University of Pittsburgh, Pittsburgh.

Beynon, D.E., Donahue, J., Schaub, R.T. and Johnston, R.A. (1986) Tempering types and sources for Early Bronze Age ceramics from Bab edh-Dhra' and Numeira, Jordan. *Journal of Field Archaeology*, **13**: 297–305.

Beyries, S. (1988) Functional variability of lithic sets in the Middle Paleolithic. In: *Upper Pleistocene Prehistory of Western Eurasia* (Eds H. Dibble and A. Montet-White), pp. 213–223. University Museum, University of Pennsylvania, Philadelphia.

Bibby, J.S. and Mackney, D. (1972) *Land Use Capability Classification*, Technical Monograph No. 1. The Soil Survey, Harpenden.

Biddle, M., Hudson, D. and Heighway, C. (1973) The future of London's past: a survey of the archaeological implications of planning and development in the nation's capital. In*: Rescue: A Trust for British Archaeology*, Rescue Publication 4.

Bierman, P.R. and Gillespie, A.R. (1994) Evidence suggesting that methods of rock-varnish cation-ratio dating are neither comparable nor consistently reliable. *Quaternary Research*, **41**: 82–90.

Bietti, A. (Ed) (1994) *The Upper Pleistocene Deposit of the Arene Candide Cave (Savona, Italy): New Studies on the 1940–42 Excavations*, Quaternaria Nova IV. Istituto Itlaliano di Paleontologia Umana, Roma, 375 pp.

Binder, D., Brochier, J.E., Duday, H. Helmer, D., Marinval, P., Thiebault, S. and Wattez, J. (1993) L'abri Pendimoun à Castellar (Alpes-Maritimes): nouvelles données sur le complex culturel de la imprimée dans son contexte stratigraphique. *Gallia Préhistoire*, **35**: 177–251.

Binford, L.R. (1981) Behavioral archaeology and the "Pompeii Premise." *Journal of Anthropological Research*, **37**: 195–208.

Bintliff, J. (1992) Erosion in the Mediterranean lands: a reconsideration of pattern, process and methodology. In: *Past and Present Soil Erosion* (Eds M. Bell and J. Boardman), Monograph 22, pp. 125–131. Oxbow, Oxford.

Birkeland, P.W. (1992) Quaternary soil chronosequences in various environments – extremely arid to humid tropical. In: *Weathering, Soils & Paleosols* (Eds I.P. Martini and W. Chesworth), Developments in Earth Surface Processes 2, pp. 261–281. Elsevier, Amsterdam.

Birkeland, P.W. (1999) *Soils and Geomorphology*. Oxford University Press, New York.

Bischoff, J.L., Garcia, J.F. and Straus, L.G. (1992) Uranium-series isochron dating at El Castillo Cave (Cantabria, Spain): The "Acheulean"/"Mousterian" question. *Journal of Archaeological Science*, **19**: 49–62.

Blake, M.E. (1947) *Ancient Roman Construction in Italy from the Prehistoric Period to Augustus*. Carnegie Institution of Washington, Washington.

Blum, M.D. and Valastro, S., Jr. (1992) Quaternary stratigraphy and geoarchaeology of the Colorado and Concho Rivers, West Texas. *Geoarchaeology*, **7**: 419–448.

Blum, M.D., Abbott, J.T. and Valastro, S., Jr. (1992) Evolution of landscapes on the Double Mountain Fork of the Brazos River, West Texas; implications for preservation and visibility of the archaeological record. *Geoarchaeology*, **7**: 339–370.

Boardman, J. (1992) Current erosion on the South Downs: implications for the past. In: *Past and Present Soil Erosion* (Eds M. Bell and J. Boardman), pp. 9–19. Oxbow, Oxford.

Boardman, J. and Evans, R. (1997) Soil erosion in Britain; a review. In: *The Human Impact Reader; Readings and Case Studies* (Eds A. Goudie, D.E. Alexander, A.B. Gomez, H.O. Slaymaker and S.W. Trimble), pp. 118–125. University of Oxford, Oxford.

Boardman, S. and Jones, G. (1990) Experiments on the effects of charring on cereal plant components. *Journal of Archaeological Science*, **17**: 1–11.

Bodziac, W.J. (2000) *Footwear Impression Evidence*. CRC Press, LLC.

Boggs, S. (2001) *Principles of Sedimentology and Stratigraphy*, 3rd edition. Prentice Hall, Upper Saddle River, NJ.

Boiffin, J. and Bresson L. M. (1987) Dynamique de formation des croutes superficielles: apport de l'analyse microscopique. Soil Micromorphology. (Eds N. Fedoroff, L. M. Bresson and M. A. Courty). Plaisir, Association Française pour l'Étude du Sol: 393–399.

Boivin, N.L. (1999) Life rhythms and floor sequences: excavating time in rural Rajasthan and Neolithic Çatalhöyük. *World Archaeology*, **31**: 367–388.

Bonifay, E. (1956) Les sédiments détritiques grossiers dans le remplissage des grottes – méthode d'étude morphologique et statistique. *L' Anthropologie*, **60**: 447–461.

Bordes, F. (1954a) Les gisements du Pech de l'Azé (Dordogne). I, Le Moustérien de tradition achuelèenne. *L'Anthropologie*, **58**: 401–432.

Bordes, F. (1954b) *Les limons quaternaires du Bassin de la Seine, stratigraphie et archéologie paléolithique*. Institut de Paléontologie Humaine, Paris. (Fondation Albert, 1er prince de Monaco). Archives; mémoire 26. Masson, Paris.

Bordes, F. (1955) Les gisements du Pech de l'Azé (Dordogne). I, Le Moustérien de tradition acheulèene (suite), avec une note paléontologique de J. Bouchud. *L'Anthropologie*, **59**: 1–38.

Bordes, F. (1972) *A Tale of Two Caves*. Harper and Row, New York.

Bordes, F. (1975) Sur la notion de sol d'habitat en Préhistoire Paléolithique. *Bulletin de la Société Préhistorique Française*, **72**: 139–144.

Bordes, F., Laville, H. and Paquereau, M.M. (1966) Observations sur le Pleistocène Supérieur du Gisement de Combe-Grenal. *Actes de la Societé Linneénne de Bordeaux*, **10**: 3–19.

Boschian, G. and Montagnari-Kokelji, E. (2000) Prehistoric shepherds and caves in the Trieste Karst (northeastern Italy). *Geoarchaeology*, **15**: 331–371.

Boschian, G., Langohr, R., Limbrey, S. and Macphail, R.I. (Eds) (2003) *Second International Conference on Soils and Archaeology, Pisa, 12th–15th May, 2003. Extended Abstracts*. Dipartimento di Scienze Archeologiche, Università di Pisa, Pisa.

Bouma, J., Fox, C.A. and Miedema, R. (1990) Micromorphology of hydromorphic soils: applications for soil genesis and land evaluation. In: *Soil Micromorphology* (Ed L.A. Douglas), pp. 257–278. A Basic and Applied Science. Elsevier, Amsterdam.

Bousman, C.B. and Goldberg, P. (2001) Rejecting complexity. *Discovering Archaeology*, **3**: 66–71.

Bousman, C.B., Collins, M.B., Goldberg, P., Stafford, T., Guy, J., Baker, B.W., Steele, D.G., Kay, M., Kerr, A., Fredlund, G., Dering, P., Holliday, V., Wilson, D., Gose, W., Dial, S., Takac, P., Balinsky, R., Masson, M. and Powell, J.F. (2002) The Palaeoindian-Archaic transition in North America: new evidence from Texas. *Antiquity*, **76**: 980–990.

Bowler, J.M., Johnston, H., Olley, J.M., Prescott, J.R., Roberts, R.G., Shawcross, W. and Spooner, N.A. (2003) New ages for human occupation and climatic change at Lake Mungo, Australia. *Nature* (London), **421**: 837–840.

Bowman, D. (1997) Geomorphology of the Dead Sea western margin. In: *The Dead Sea, the Lake and its Setting* (Eds T.M. Niemi, Z. Ben-Avraham and J.R. Gat), pp. 217–225. Oxford University Press, New York.

Bowman, D. and Giladi, Y. (1979) Pebble analysis for palaeoenvironmental recognition – the 'Ubeidiya living floors. *Israel Journal of Earth-Sciences*, **28**: 86–93.

Bowsher, D., Holder, N., Howell, I. and Dyson, T. (in press) *The London Guildhall: The Archaeology and History of the Guildhall Precinct from the Medieval Period to the 20th Century*. Museum of London Archaeological Service, London.

Bradford, J.S.P. and González, A.R. (1960) Photo interpretation in archaeology. In: *Manual of Photographic Interpretation*, pp. 717–733. American Society of Photogrammetry, Washington, D.C.

Brain, C.K. (1981) *The Hunters or the Hunted? An Introduction to African Cave Taphonomy*. The University of Chicago Press, Chicago.

Brammer, H. (1971) Coatings in seasonally flooded soils. *Geoderma*, **6**: 5–16.

Bresson, L.M. and Moran, C.J. (1998) High-resolution bulk density images using calibrated X-ray radiography of impregnated soil slices. *Soil Science Society of America Journal*, **62**: 299–305.

Breuning-Madsen, H., Holst, M.K. and Rasmussen, M. (2001) The chemical environment in a burial mound shortly after construction – an archaeological – pedological experiment. *Journal of Archaeological Science*, **28**: 691–697.

Breuning-Madsen, H., Holst, M.K., Rasmussen, M. and Elberling, B. (2003) Preserved within log coffins before and after barrow construction. *Journal of Archaeological Science*, **30**: 343–350.

Brewer, R. (1964) *Fabric and Mineral Analysis of Soils*. Wiley and Sons, New York.

Bridge, J.S. (2003) *Rivers and Floodplains: Forms, Processes, and Sedimentary Record*. Blackwell Science, Oxford.

Bridges, E.M. (1978) *World Soils*, 2nd edition. Cambridge University Press, Cambridge.

Bridges, E.M. (1990) *Soil Horizon Designations*. Technical paper 19. International Soil Reference and Information Centre, Wageningen.

Bridgland, D.R. (1999) Analysis of the raised beach gravel deposits at Boxgrove and related sites. In: *Boxgrove. A Middle Pleistocene Hominid Site at Eartham Quarry, Boxgrove, West Sussex* (Eds M.B. Roberts and S.A. Parfitt), pp. 100–110. English Heritage, London,

Brochier, J.E. (1983) Bergeries et feux néolithiques dans le Midi de la France, caractérisation et incidence sur la raisonnement sédimentologique. *Quartar*, **33/34**: 181–193.

Brochier, J.E., Villa, P. and Giacomarra, M. (1992) Shepherds and Sediments: geo-ethnoarchaeology of pastoral sites. *Journal of Anthropological Archaeology*, **11**: 47–102.

Brookes, I.A. (2001) Aeolian erosional lineation in the Libyan Desert, Dakhla region, Egypt. *Geomorphology*, **39**: 189–209.

Brookfield, M.E. (1984) Eolian sands. In: *Facies Models, 2nd edition* (Ed R.G. Walker), pp. 91–104. Geoscience Canada Reprint Series 1, Ottawa.

Brown, A.G. (1997) *Alluvial Geoarchaeology*. Cambridge Manuals in Archaeology. Cambridge University Press, Cambridge.

Brown, G. (1988) Whittington Ave (WIV88), Museum of London Archaeological Service, London.

Brown, G.E. (1990) Testing of concretes, mortars, plasters, and stuccos. *Archeomaterials*, **4**: 185–191.

Bruckert, S. (1982) Analysis of the organo-mineral complexes of soils. In: *Constituents and Properties of Soils* (Eds M. Bonneau and B. Souchier), pp. 214–237. Academic Press, London.

Bruins, H.J. and Yaalon, D.H. (1979) Stratigraphy of the Netivot section in the desert loess of the Negev (Israel). *Acta Geologica Hungarica*, **22**: 161–170.

Bull, P.A. (1981a) Some fine-grained sedimentation phenomena in caves. *Earth Surface Processes and Landforms*, **6**: 11–22.

Bull, P.A. (1981b) The scanning electron microscope as an adjunct to environmental reconstruction in archeological sites. In: *Eighth International Congress of Speleology* (Ed B.F. Beck), *Proceedings of the Eighth International Congress of Speleology*, pp. 340–342. Bowling Green, KY, United States.

Bull, P.A. and Goldberg, P. (1985) Scanning Electron Microscope analysis of sediments from Tabun Cave, Mount Carmel, Israel. *Journal of Archaeological Science*, **12**: 177–185.

Bull, P.A., Goudie, A.S. and Price Williams, D. (1987) Colluvium; a scanning electron microscope analysis of a neglected sediment type. In: *Clastic particles; scanning electron microscopy and shape analysis of sedimentary and volcanic clasts* (Ed J.R. Marshall), pp. 16–35. Van Nostrand Reinhold Co., New York.

Bullard, R.G. (1970) Geological studies in field archaeology: Tell Gezer, Israel. *Biblical Archaeologist*, **33**: 97–132.

Bullock, P. and Murphy, C.P. (Eds) (1983) *Soil Micromorphology* (Volumes 1 and 2). A B Academic Publishers, Berkhamsted.

Bullock, P., Fedoroff, N., Jongerius, A., Stoops, G., Tursina, T. and Babel, U. (1985) *Handbook for Soil Thin Section Description*. Waine Research, Wolverhampton, 152 pp.

Buol, S.W., Hole, F.D. and McCracken, R.J. (1973) *Soil Genesis and Classification*. The Iowa State University Press, Ames.

Burrin, P.J. and Scaife, R.G. (1984) Aspects of Holocene valley sedimentation and floodplain development in southern England. *Proceedings of the Geologists' Association*, **95**: 81–96.

Butzer, K.W. (1976a) *Geomorphology from the Earth*. Harper & Row, New York.

Butzer, K.W. (1976b) Lithostratigraphy of the Swartkrans formation. *South African Journal of Science*, **72**: 136–141.

Butzer, K.W. (1982) *Archaeology as Human Ecology*. Cambridge University Press, Cambridge.

Cammas, C. (1994) Approche micromorphologique de la stratigraphie urbaine à Lattes: premiers résultats. In: *Lattara 7*, Vol. 7. pp. 181–202, ARALO, Lattes.

Cammas, C., David, C. and Guyard, L. (1996a) La question des terre noires dans les sites tardo-antiques et médiéval: le cas du Collège de France (Paris, France). In: *Proceedings XIII International Congress of Prehistoric and Protohistoric Sciences*, Colloquium 14, pp. 89–93. ABACO, Forlì.

Cammas, C., Wattez, J. and Courty, M.-A. (1996b) L'enregistrement sédimentaire des modes d'occupation de l'espace. In: *Paleoecology; Colloquium 3 of XIII International Congress of Prehistoric and Protohistoric Sciences* (Eds L. Castelletti and M. Cremaschi), Vol. 3, pp. 81–86. ABACO, Forlì.

Campy, M. and Chaline, J. (1993) Missing records and depositional breaks in French late Pleistocene cave sediments. *Quaternary Research (New York)*, **40**: 318–331.

Canada, C.-N.R. (2004) Fundamentals of remote sensing. In: *http://www.ccrs.nrcan.gc.ca/ccrs/learn/tutorials/tutorials_e.html*. National Resources Canada, Ottawa.

Canti, M. (1995) A mixed approach to geoarchaeological analysis. In: *Archaeological Sediments and Soils: Analysis, Interpretation and Management* (Eds A.J. Barham and R.I. Macphail), pp. 183–190. Institute of Archaeology, London.

Canti, M. (1997) An investigation into microscopic calcareous spherulites from herbivore dung. *Journal of Archaeological Science*, **24**: 435–444.

Canti, M. (1998) The micromorphological identification of faecal spherulites from archaeological and modern materials. *Journal of Archaeological Science*, **25**: 435–444.

Canti, M. (1999) The production and preservation of faecal spherulites: animals, environment and taphonomy. *Journal of Archaeological Science*, **26**: 251–258.

Carter, S. P. (1990) The stratification and taphonomy of shells in calcareous soils: implications for landsnail analysis in archaeology. *Journal of Archaeological Science* 17: 495–507.

Carter, S. (1998a) Soil micromorphology. In: *St. Boniface Church, Orkney: Coastal Erosion and Archaeological Assessment* (Ed C. Lowe), pp. 172–186. Sutton Publishing/Historic Scotland, Stroud.

Carter, S. (1998b) The use of peat and other organic sediments as fuel in northern Scotland: identifications derived from soil thin sections. In: *Life on the Edge: Human Settlement and Marginality* (Eds C.M. Mills and G. Coles), Monograph 100, pp. 99–104. Oxbow, Oxford.

Carter, S. and Davidson, D. (1998) An evaluation of the contribution of soil micromorphology to the study of ancient arable cultivation. *Geoarchaeology*, **13**: 535–547.

Carver, R.E. (Ed) (1971) *Procedures in Sedimentary Petrology*. Wiley-Interscience, New York.

Catt, J.A. (1986) *Soils and Quaternary Geology. A Handbook for Field Scientists*. Clarendon Press, Oxford.

Catt, J.A. (1990) Paleopedology manual. *Quaternary International*, **6**: 1–95.

Catt, J.A. (1999) Particle size distribution and mineralogy of the deposits. In: *Boxgrove. A Middle Pleistocene Hominid Site at Eartham Quarry, Boxgrove, West Sussex* (Eds M.B. Roberts and S.A. Parfitt), Archaeological Report 17, pp. 111–118. English Heritage, London.

Catt, J.A. and Bronger, A. (Eds) (1998) *Reconstruction and Climatic Implications of Paleosols, Catena Special Issue*, Vol. 34. Elsevier, Amsterdam.

Catt, J.A. and Staines, S.J. (1998) Petrography of sediments and soils. In: *Late Quaternary Environmental Change in North-West Europe: Excavations at Holywell Coombe, South-East England*. (Eds R.C. Preece and D.R. Bridgland), pp. 69–85. Chapman & Hall, London.

Chase, P.G. (1986) *The Hunters of Combe Grenal: Approaches to Middle Paleolithic Subsistence in Europe*. BAR International Series; 286. B.A.R., Oxford, England, viii, 224

Chinzei, K. (1970) The Amud Cave site and its deposits. In: *The Amud Man and His Cave Site* (Eds H. Suzuki and F. Takai), pp. 21–52 and IV-1– IV-6. University of Tokyo, Tokyo.

Chinzei, K. and Kiso, T. (1970) Geology and geomorphology around the Amud Cave. In: *The Amud Man and His Cave Site* (Eds H. Suzuki and F. Takai), pp. 9–20 and III-1–III-4. University of Tokyo, Tokyo.

Chorley, R.J., Schumm, S.A. and Sugden, D.E. (1984) *Geomorphology*. Methuen, London.

Churchill, S.E. (2000) The Creswellian (Pleistocene) human axial skeletal remains from Gough's Cave (Somerset, England). *Bulletin of the Natural History Museum. Geology Series*, **56**: 141–154.

Ciezar, P., Gonzalez, V., Pieters, M., Rodet-Belarbi, I. and Van-Ossel, P. (1994) In suburbano – new data on the immediate surroundings of Roman and early medieval Paris. In: *Urban–Rural Connexions: Perspectives from Environmental Archaeology* (Eds A.R. Hall and H.K. Kenward), pp. 137–146. Oxbow Books, Oxford.

Clark, A. (1996) *Seeing Beneath the Soil, Prospecting Methods in Archaeology*, 2nd edition. Batsford, London.

Clark, K. (2000) Architect's specification: building analysis and conservation. In: *Interpreting Stratigraphy* (Ed S. Roskams), International Series 910, pp. 17–24. *British Archaeological Reports*, Oxford.

Clarke, D.L. (1973) Archaeology: the loss of innocence. *Antiquity*, **47**: 6–18.

Clayden, B. and Hollis, J.M. (1984) *Criteria for Differentiating Soil Series*. Soil Survey Technical Monograph, No. 17. Soil Survey of England and Wales, Harpenden.

Collcutt, S.N. (1999) Structural sedimentology at Boxgrove. In: *Boxgrove* (Eds M.B. Roberts and S.A. Parfitt), pp. 42–99. English Heritage, London.

Collins, M.B. (Ed) (1998) Wilson Leonard, An 11,000-year archeological record of hunter-gatherers in central Texas. *Studies in Archaeology*, **31**. Texas Archeological Research Laboratory, University of Texas at Austin, Austin.

Collins, M.B. and Mear, C.E. (1998) The site and its setting. In: *Wilson-Leonard: An 11,000-year Archaeologcal Record of Hunter-Gatherers in Central Texas* (Ed M.B. Collins), *Studies in Archaeology* **31**: 5–31. Texas Archaeological Research Laboratory, Austin.

Collins, M.B., Bousman, C.B., Goldberg, P., Takac, P.R., Guy, J.C., Lanata, J.L., Stafford, T., Jr. and Holliday, V.T. (1993) The Paleoindian sequence at the Wilson-Leonard site, Texas. *Current Research in the Pleistocene*, **10:** 10–12.

Collins, M.E., Carter, B.J., Gladfelter, B.G. and Southard, R.J. (Eds) (1995) *Pedological Perspectives in Archaeological Research, SSSA Special Publication Number 44*. Soil Science Society of America, Inc., Madison.

Collinson, J.D. (1996) Alluvial sediments. In: *Sedimentary Environments: Processes, Facies and Stratigraphy*. 3rd edition (Ed H.G. Reading), pp. 37–82. Blackwell Science, Oxford.

Collinson, J.D. and Thompson, D.B. (1989) *Sedimentary Structures*, 2nd edition. Chapman & Hall, London.

Conard, N.J. and Bolus, M. (2003) Radiocarbon dating the appearance of modern humans and timing of cultural innovations in Europe: new results and new challenges. *Journal of Human Evolution*, **44**: 331–371.

Conard, N.J., Dippon, G. and Goldberg, P. (2004) Chronostratigraphy and archeological context of the Aurignacian Deposits at Geißenklösterle. In: *The Chronology of the Aurignacian and of the Transitional Technocomplexes. Dating, Stratigraphies, Cultural Implications* (Eds J. Zilhão and F. d'Errico), pp. 165–176. Proceedings of Symposium 6.1 of the XIVth Congress of the UISPP, Liège.

Conry, M.J. (1971) Irish Plaggen soils, their distribution, origin and properties. *Journal of Soil Science*, **22**: 401–416.

Conway, J. (1983) An investigation of soil phosphorus distribution within occupation deposits from a Romano-British hut group. *Journal of Archaeological Science*, **10**: 117–128.

Conyers, L.B. (1995) The use of ground-penetrating radar to map the buried structures and landscape of the Ceren Site, El Salvador. *Geoarchaeology*, **10**: 275–299.

Conyers, L.B. and Goodman, D. (1997) *Ground-Penetrating Radar: An Introduction for Archaeologists*. AltaMira Press, Walnut Creek, CA, 232 pp.

Cooke, R., Warren, A. and Goudie, A. (1993) *Desert Geomorphology*. UCL Press, London.

Cornwall, I.W. (1958) *Soils for the Archaeologist*. The Macmillan Company, New York.

Cornwall, I.W. and Hodges, H.W.M. (1964) Thin sections of British Neolithic pottery: Windmill Hill-A Test-site. *Bulletin of Historical Archaeology*, **4**: 29–33.

Corporation of London (2004) *Archaeology in the City of London. Archaeology Guidance*, Planning Advice Note 3. Department of Planning & Transportation, London.

Courty, M.A. (1989) Analyse microscopique des sédiments du remplissage de la grotte de Vaufrey (Dordogne). In: *La Grotte de Vaufrey* (Ed J.-P. Rigaud), pp. 183–209. Mémoire de la Societé Préhistorique, Française.

Courty, M.A. (2001) Microfacies analysis assisting archaeological stratigraphy. In: *Earth Sciences and Archaeology* (Eds P. Goldberg, V.T. Holliday and C.R. Ferring), pp. 205–239. Kluwer, New York.

Courty, M.A. and Fedoroff, N. (1982) Micromorphology of a Holocene dwelling. In: *Proceedings Nordic Archaeometry*, PACT 7, pp. 257–277.

Courty, M.A. and Fedoroff, N. (1985) Micromorphology of recent and buried soils in a semiarid region of northwestern India. *Geoderma*, **35**: 287–332.

Courty, M.A. and Nornberg, P. (1985) Comparison between buried uncultivated and cultivated Iron Age soils on the west coast of Jutland, Denmark. In: *Proceedings of the Third Nordic Conference on the Application of Scientific Methods in Archaeology* (Eds T. Edgren and H. Jungner), pp. 57–69. The Finnish Antiquarian Society, Helsinki.

Courty, M.-A. and Weiss, H. (1997) The scenario of environmental degradation in the Tell Leilan region, NE Syria, during the late third millennium abrupt climate change. In: *NATO ASI Series, Vol. 149, Third Millennium BC Climate Change and Old World Collapse (Eds H.N. Dalfes, G. Kukla and H. Weiss)*, pp. 107–047. Springer-Verlag, Heidelberg.

Courty, M.A. and Vallverdú, J. (2001) The microstratigraphic record of abrupt climate changes in cave sediments of the western Mediterranean. *Geoarchaeology*, **16**: 467–500.

Courty, M.A., Dhir, P. and Raghavan, H. (1987) Microfabrics of calcium carbonate accumulation in arid soils of western India. In: *Soil Micromorphology* (Eds

N. Fedoroff, L.M. Bresson and M.A. Courty), pp. 227–234. AFES, Plaisir.

Courty, M.A., Goldberg, P. and Macphail, R.I. (1989) *Soils and Micromorphology in Archaeology*. Cambridge Manuals in Archaeology. Cambridge University Press, Cambridge.

Courty, M.A., Macphail, R.I. and Wattez, J. (1991) Soil micromorphological indicators of pastoralism: with special reference to Arene Candide, Finale Ligure, Italy. *Revista di Studi Liguri*, **LVII**: 127–150.

Courty, M.A., Fedoroff, N., Jones, M.K. and McGlade, J. (1994a) Environmental dynamics. In: *Temporalities and Desertification in the Vera Basin, Southeast Spain, Archaeomedes Project* (Ed S.E. van der Leeuw), Vol. 2, pp. 19–84, Brussels.

Courty, M.A., Goldberg, P. and Macphail, R.I. (1994b) Ancient people – lifestyles and cultural patterns. In: *Transactions of the 15th World Congress of Soil Science, International Society of Soil Science, Mexico*, Vol. 6a, pp. 250–269. International Society of Soil Science, Acapulco.

Courty, M.A., Marlin, C., Dever, L., Temblay, P. and Vachier, P. (1994c) The properties, genesis and environmental significance of calcitic pendents from the high Arctic (Spitsbergen). *Geoderma*, **61**: 71–102.

Courty, M.A., Cachier, H., Hardy, M. and Ruellan, S. (1998) Soil record of exceptional wild-fires linked to climatic anomalies (Inter-Tropical and Mediterranean regions). CD-ROM, Montpellier.

Cowan, C. (Ed) (2003) *Urban Development in North-West Roman Southwark: Excavations 1974–90*, Monograph 16. MOLAS, London.

Cowgill, J. (2003) The iron production industry and its extensive demand upon woodland resources: a case study from Creeton Quarry, Lincolnshire (Eds Murphy, P. and P.E.J. Wiltshire). The Environmental Archaeology of Industry, Symposia of the Association for Environmental Archaeology No. 20: Oxford, Oxbow, pp. 48–57.

Cremaschi, M. (1987) *Paleosols and Vetusols in the Central Po Plain (Northern Italy). A study in Quaternary Geology and Soil Development*. Unicopli, Milano.

Cremaschi, M., Di Lernia, S. and Trombino, L. (1996) From taming to pastoralism in a drying environment. Site formation processes in the shelter of the Tadrat Massif (Libya, Central Sahara). In: *Paleoecology* (Eds L. Castelletti and M. Cremaschi), Proceedings of the International Union of Prehistoric and Protohistoric Sciences, pp. 87–106. ABCO, Forlì.

Cremeens, D.L. (1995) Pedogenesis of Cotiga Mound, a 2100-year-old woodland mound in southwest West Virginia. *Soil Science Society of America Journal*, **59**: 1377–1388.

Cremeens, D.L. (2005) Micromorphology of Cotiga Mound, West Virginia. *Geoarchaeology*, **20**: 581–597.

Cremeens, D.L., Landers, D.B. and Frankenberg, S.R. (1997) Geomorphic setting and stratigraphy of Cotiga Mound, Mingo County, West Virginia. *Geoarchaeology*, **12**: 459–477.

Crombé, P. (1993) Tree-fall features on Final Palaeolithic and Mesolithic sites situated on sandy soils: how to deal with it. *Helenium*, **XXXIII**: 50–66.

Crowther, J. (1996a) Report on sediments from Building 5, Old Market Street "86," In: *Excavations at Usk 1986–1988 Brittania XXVII, 51–110* (Ed A.G. Marvell) *Britannia*, **XXVII**: 92–99.

Crowther, J. (1996b) Soil chemistry. In: *The Experimental Earthwork Project 1960–1992* (Eds M. Bell, P.J. Fowler and S.W. Hillson), Research Report 100, pp. 107–118. Council for British Archaeology, York.

Crowther, J. (1997) Soil phosphate surveys: critical approaches to sampling, analysis and interpretation. *Archaeological Prospection*, **4**: 93–102.

Crowther, J. (2000) Phosphate and magnetic susceptibility studies. In: *Prehistoric Intertidal Archaeology in the Welsh Severn Estuary* (Ed M. Bell, A. Caseldine and H. Neumann) pp. 57–58 (and CD). CBA Research Report, York.

Crowther, J. (2003) Potential magnetic susceptibility and fractional conversion studies of archaeological soils and sediments. *Archaeometry*, **45**: 685–701.

Crowther, J. and Barker, P. (1995) Magnetic susceptibility: distinguishing anthropogenic effects from the natural. *Archaeological Prospection*, **2**: 207–215.

Crowther, J., Macphail, R.I. and Cruise, G.M. (1996) Short-term burial change in a humic rendzina, Overton Down Experimental Earthwork, Wiltshire, England. *Geoarchaeology*, **11**: 95–117.

Cruise, G.M., (1990) Holocene peat initiation in the Ligurian Apennines, northern Italy. *Review of Palaeobotany and Palynology*, **63**: 173–182.

Cruise, G.M. and Macphail, R.I. (2000) Microstratigraphical signatures of experimental rural occupation deposits and archaeological sites. In: *Interpreting Stratigraphy* (Ed S. Roskams), Vol. 9, pp. 183–191. University of York, York.

Currant, A.P. (1991) A Late Glacial interstadial mammal fauna from Gough's Cave, Somerset, England. In: *The Late Glacial in North-West Europe; Human Adaptation and Environmental Change at the End of the Pleistocene* (Eds N. Barton, A.J. Roberts and D.A. Roe), Vol. 77, pp. 48–50, Oxford, England, United Kingdom.

Dalan, R.A. and Bevan, B.W. (2002) Geophysical indicators of culturally emplaced soils and sediments. *Geoarchaeology*, **17**: 779–810.

Dalrymple, J.B., Blong, R.J. and Conacher, A.J. (1968) A hypothetical nine unit landsurface model. *Zeitschrift für Geomorphologie*, **12**: 60–76.

Dalwood, H. and Edwards, R. (Eds) (2004) *Excavations at Deansway, Worcester, 1988–89: Romano-British Small Town to Late Medieval City*, CBA Research Report. Council for British Archaeology, York.

Danin, A. and Ganor, E. (1991) Trapping of airborne dust by mosses in the Negev Desert, Israel. *Earth Surface Processes and Landforms*, **16**: 153–162.

Darwin, C. (1888) *The Formation of Vegetable Mould through the Action of Worms, with Observations on their Habits*. Murray, London.

Davidson, D.A. (1973) Particle size and phosphate Analysis-Evidence for the Evolution of a Tell. *Archaeometry*, **15**: 143–152

Davidson, D.A. (1976) Processes of tell formation and erosion. In: *Geoarchaeology: Earth Science and the Past* (Eds D.A. Davidson and M.L. Shackley), pp. 255-266. Duckworth, London.

Davidson, D.A., Carter, S., Boag, B., Long, D., Tipping, R. and Tyler, A. (1999) Analysis of pollen in soils: processes of incorporation and redistribution of pollen in five soil profile types. *Soil Biology & Chemistry*, **31**: 643–653.

Davidzon, A. and Goring-Morris, A.N. (2003) Sealed in stone: the Upper Palaeolithic Early Ahmarian Knapping Method in the light of refitting studies at Nahal Nizzana XIII, Western Negev, Israel. *Journal of the Israel Prehistoric Society*, **33**: 75–205.

Deacon, H.J., Geleijnse, V.B., Thackeray, A.I., Thackeray, J.F. and Tursenius, M.L. (1986) Late Pleistocene cave deposits in the Southern Cape: current research at Klasies River. *Palaeoecology of Africa*, **17**: 31–37.

De Coninck, F. (1980) Major mechanisms in formation of spodic horizons. *Geoderma*, **24**: 101–128.

Delvigne, J. and Stoops, G. (1990) Morphology of mineral weathering and neoformation I. In: *Soil Micromorphology: A Basic and Applied Science* (Ed L.A. Douglas), pp. 471–482. Elsevier, Amsterdam.

Denny, C.S. and Goodlett, J.C. (1956) Microrelief resulting from fallen trees. In: *Surficial Geology and Geomorphology of Potter County* (Ed C.S. Denny), US Geological Survey Professional Paper 288, pp. 59–65.

Deocampo, D.M., Blumenschine, R.J. and Ashley, G.M. (2002) Wetland diagenesis and traces of early hominids, Olduvai Gorge, Tanzania. *Quaternary Research*, **57**, 271–281.

Devoy, R.J.N. (1982) Analysis of the geological evidence for the Holocene sea-level movements in south-east England. *Proceedings of the Geological Association*, **93**: 65–90.

Dick, W.A. and Tabatabai, M.A. (1977) An alkaline oxidation method for the determination of total phosphorus in soils. *Journal of the Soil Science Society of America*, **41**: 511–514.

Dijkmans, J.W.A. and Mücher, H.J. (1989) Niveo-aeolian sedimentation of loess and sand; an experimental and micromorphological approach. *Earth Surface Processes and Landforms*, **14**: 303–315.

Dillon, J., Jackson, S. and Jones, H. (1991) Excavations at the Courage Brewery and Park Street 1984–90. *London Archaeologist*, **6**: 255–262.

Dimbleby, G.W. (1962) *The Development of British Heathlands and their Soils*, Oxford Forestry Memoir No. 23. Clarendon Press, Oxford.

Dimbleby, G.W. (1985) *The Palynology of Archaeological Sites*. Academic Press, London.

Dimbleby, G.W. and Gill, J.M. (1955) The occurrence of podzols under deciduous woodland in the New Forest. *Forestry*, **28**: 95–106.

Dippon, G. (2003) *Die Taphonomie der Aurignacienhorizonte der Geissenklösterle-Höhle bei Blaubeuren*. MA, Eberhard-Karls-Universität Tübingen, Tübingen, 88 pp.

Dodonov, A.E. (1995) Geoarchaeology of Palaeolithic sites in loesses of Tajikistan (Central Asia). In: *Ancient Peoples and Landscapes* (Ed E. Johnson), pp. 127–136. Museum of Texas Tech University, Lubbock.

Dodonov, A.E. and Baiguzina, L.L. (1995) Loess stratigraphy of Central Asia; palaeoclimatic and palaeoenvironmental aspects. *Quaternary Science Reviews*, **14**: 707–720.

Donahue, J. and Adovasio, J.M. (1990) Evolution of sandstone rockshelters in eastern North America; a geoarchaeological perspective. In: *Archaeological Geology of North America* (Eds N.P. Lasca and J. Donahue), Centennial Volume No. 4, pp. 231–251. Geological Society of America, Boulder.

Doran, G.H. and Dickel, D.N. (1988) Multidisciplinary investigations at the Windover site. In: *Wet Site Archaeology* (Ed B.A. Purdy), pp. 263–289. Telford, Caldwell, NJ.

Dorn, R.I. (1983) Cation-ratio dating; a new rock varnish age-determination technique. *Quaternary Research (New York)*, **20**: 49–73.

Dorn, R.I. (1991) Rock varnish. *American Scientist*, **79**: 542–553.

Dorn, R.I. and Oberlander, T.M. (1981) Rock varnish origin, characteristics, and usage. *Zeitschrift für Geomorphologie*, **25**: 420–436.

Douglas, L.A. (Ed) (1990) *Soil Micromorphology: A Basic and Applied Science, Proceedings of the VIIIth International Working Meeting of Soil Micromorphology, San Antonio, Texas – July 1988*. Elsevier, Amsterdam.

Drewett, P. (1976) The excavation of four round barrows of the second millennium BC at West Heath, Harting, 1973–1975: *Sussex Archaeological Collections*, Vol. 14, pp. 126–150.

Drewett, P.L. (1989) Anthropogenic soil erosion in prehistoric Sussex: excavations at West Heath and Ferring, 1984. *Sussex Archaeological Collections*, **127**: 11–29.

Driessen, P., Deckers, J., Spaargaren, O. and Nachtergaele, F. (Eds) (2001) Lecture Notes on the Major Soils of the World, FAO.

Driskell, B.N. (1994) Stratigraphy and chronology at Dust Cave. *Journal of Alabama Archaeology*, **40**: 17–34.

Duchaufour, P. (1982) *Pedology*. Allen and Unwin, London.

Dunning, N., Rue, D.J., Beach, T., Covich, A. and Traverse, A. (1998) Human–environment interactions in a tropical watershed: the paleoecology of Laguna Tamarindito, El Petén, Guatemala. *Journal of Field Archaeology*, **25**: 139–151.

Easterbook, D.J. (1993) Surface processes and land forms. Macmillan, New York.

Edwards, C.A. and Lofty, J.R. (1972) *Biology of Earthworms*. Chapman & Hall Ltd, London.

Eidt, R.C. (1984) *Advances in Abandoned Settlement Analysis: Application to Prehistoric Anthrosols in Colombia, South America*, Vol. XIV. The Center for Latin America, University of Wisconsin-Milwaukee, Milwaukee.

Ellis, J.C. and Rawlings, M. (2001) Excavations at Balksbury Camp, Andover 1995–97. *Hampshire Field Club and Archaeological Society*, **56**: 21–93.

Ellwood, B.B., Petruso, K.M. and Harrold, F.B. (1997) High-resolution paleoclimatic trends for the Holocene identified using magnetic susceptibility data from archaeological excavations in caves. *Journal of Archaeological Science*, **24**: 569–573.

Ellwood, B.B., Harrold, F.B., Benoist, S.L., Thacker, P., Otte, M., Bonjean, D., Long, G.J., Shahin, A.M., Hermann, R.P., and Grandjean, F. (2004) Magnetic susceptibility applied as an age–depth–climate relative dating technique using sediments from Scladina Cave, a Late Pleistocene cave site in Belgium. *Journal of Archaeological Science*, **31**: 283–293.

Endo, K. (1973) Sedimentological analysis of the deposits in the Douara cave site. *Bulletin. University Museum Tokyo*, **5**: 89–111.

Endo, B. (1978) Excavation at the Douara Cave. *Bulletin. University Museum Tokyo*, **14**: 83–98.

Endo, K. (1978) Stratigraphy and paleoenvironments of the deposits in and around the Douara Cave site. *Bulletin. University Museum Tokyo*, **14**: 53–81.

Engelmark, R. (1985) Carbonized seeds in postholes – a reflection of human activity. *Proceedings of the Third Nordic Conference on the Application of Scientific Methods in Archaeology*. 57–69. (Eds T. Edgren and H. Jungner). *Helsinki, The Finnish Antiquarian Society*, 205–209.

Engelmark, R. (1992) A review of farming economy in South Scania based on botanical evidence. In: *The Archaeology of the Cultural Landscape*, Stockholm.

Engelmark, R. and Linderholm, J. (1996) Prehistoric land management and cultivation. A soil chemical study. *Arkaeologiske Rapporter fra Esbjerg Museum*, 315–322.

Engelmark, R. and Viklund, K. (1986) Järnålders-jordbruk in Norrland – Teori och praktik. *Populär Arkeologi*, **4**: 22–34.

English Heritage (1991a) *Management of Archaeological Projects*. English Heritage, London.

English Heritage (1991b) *Exploring Our Past. Strategies for the Archaeology of England*. English Heritage, London.

Entwhistle, J.A., Abrahams, P.W. and Dodgshon, R.A. (1998) Multi-element analysis of soils from Scottish Historical sites. Interpreting land-use history through physical and geochemical analysis of soil. *Journal of Archaeological Science*, **25**: 53–68.

Enzel, Y., Kadan, G. and Eyal, Y. (2000) Holocene earthquakes inferred from a fan-delta sequence in the Dead Sea graben. *Quaternary Research*, **53**: 34–48.

Enzel, Y., Bookman, R., Sharon, D. et al. (2003a) Late Holocene climates of the Near East deduced from Dead Sea level variations and modern regional winter rainfall. *Quaternary Research*, **60**: 263–273.

Enzel, Y., Wells, S.G. and Lancaster, N. (Eds) (2003b) *Paleoenvironments and paleohydrology of the Mojave and Southern Great Basin deserts, Special Paper – Geological Society of America*, 368.

Evans, E.E. (1957) *Irish Folk Ways*. Routledge and Kegan Paul, London.

Evans, J.G. (1971) Habitat changes on the calcareous soils of Britain: the impact of Neolithic man. In: *Economy and Settlement in Neolithic and Early Bronze Age Britain and Europe* (Ed D.D.A. Simpson), pp. 27–74. Leicester University Press, Leicester.

Evans, J.G. (1972) *Land Snails in Archaeology*. Seminar Press, London.

Evans, J.G. (1975) *The Environment of Early Man in the British Isles*. Elek, London.

Evans, J.G. (1999) *Land & Archaeology. Histories of Human Environment in the British Isles*. Tempus, Stroud.

Evans, J.G. and Limbrey, S. (1974) The experimental earthwork on Morden Bog, Wareham, Dorset, England: 1963–1972. *Proceedings of the Prehistoric Society*, **40**: 170–202.

Evans, R. (1992) Erosion in England and Wales – the present key to the past. In: *Past and Present Soil Erosion* (Eds M. Bell and J. Boardman), Monograph 22, pp. 53–66. Oxbow, Oxford.

Evans, R. and Jones, R.J.A. (1977) Crop marks and soils at two archaeological sites in Britain. *Journal of Archaeological Science*, **4**: 63–76.

Evershed, R.P., Bethell, P.H. and Walsh, N.J. (1997) 5ß-stigmastanol and related 5ß-stanols as biomarkers of manuring: analysis of modern experimental material and assessment of the archaeological potential. *Journal of Archaeological Science*, **24**: 485–495.

Eyre, S.R. (1968) *Vegetation and Soils: A World Picture.* Edward Arnold London, Ltd.

Fa, D.A. and Sheader, M. (2000) Zonation patterns and fossilization potential of the rocky-shore biota along the Atlantic-Mediterranean interface: a possible framework for environmental reconstruction. In: *Gibraltar During the Quaternary* (Eds C. Finlayson, G. Finlayson and D. Fa), pp. 237–251. Gibraltar Government Heritage Publications Monographs, Gibraltar.

FAO-UNESCO (1988) Soil Map of the World, World Resources Report, 60. FAO, Rome.

Faegri, K. and Iverson, J. (1989) *Textbook of Pollen Analysis, 4th edition.* John Wiley & Sons, Chichester.

Farrand, W.R. (1970) Geology, climate and chronology of Yabrud Rockshelter I. In: *Fruhe Menschheit und Umwelt* (Eds K.S. Gripp and R. Schwabedissen). Bohlau Verlag Köln, Vienna.

Farrand, W.R. (1979) Chronology and palaeo-environment of Levantine prehistoric sites as seen from sediment studies. *Journal of Archaeological Science,* **6**: 369–392.

Farrand, W.R. (1984) Stratigraphic classification: living within the law. *Quarterly Review of Archaeology,* **5**: 1–5.

Farrand, W.R. (1990) Origins of Quaternary-Pleistocene–Holocene stratigraphic terminology, Special Paper 242. In: *Establishment of a Geologic Framework for Paleoanthropology* (Ed L.F. Laporte), pp. 15–22. Geological Society of America, Boulder.

Farres, P.J., Wood, S.J. and Seeliger, S. (1992) A conceptual model of soil deposition and its implications for environmental reconstruction. In: *Past and Present Soil Erosion* (Eds M. Bell and J. Boardman), Monograph 22, pp. 217–226. Oxbow, Oxford.

Faul, M.L. and Smith, R.T. (1980) Phosphate analysis and three possible Dark Age ecclesiastical sites in Yorkshire. *Landscape History,* **2**: 21–38.

Feder, K.L. (1997) Data preservation: recording and collecting. In: *Field Methods in Archaeology* (Eds T.R. Hester, H.J. Shafer and K.L. Feder), 7th edition, pp. 113–142. Mayfield Publishing Co, Mountain View, CA.

Fedoroff, N. and Goldberg, P. (1982) Comparative micromorphology of two late Pleistocene palaeosols (in the Paris basin). *CATENA,* **9**: 227–51.

Fedoroff, N., Bresson, L.M. and Courty, M.A. (Eds) (1987) *Micromorphologie des Sols – Soil Micromorphology.* Association Française pour l'Étude du Sol, Plaisir.

Fedoroff, N., Courty, M.A. and Thompson, M.L. (1990) Micromorphological evidence of palaeoenvironmental change in Pleistocene and Holocene Paleosols. In: *Soil Micromorphology: A Basic and Applied Science* (Ed L.A. Douglas), pp. 653–666. Developments in Soil Science 19, Elsevier, Amsterdam.

Feibel, C.S. (2001) Archaeological sediments in lake margin environments. In: *Sediments in Archaeological Context* (Eds J.K. Stein and W.R. Farrand), pp. 127–148. The University of Utah Press, Salt Lake City.

Fenton, A. (1968) Alternating stone and turf – an obsolete building practice. *Folk Life,* **6**: 94–103.

Ferraris, M. (1997) Ochre remains. In: *Arene Candide: A Functional and Environmental Assessment of the Holocene Sequence (Excavations Bernarbo' Brea-Cardini 1940–50)* (Ed R. Maggi), pp. 593–598. Istituto Italiano di paleontologia Umana, Roma.

Ferring, C.R. (1986) Rates of fluvial sedimentation: implications for archaeological variability. *Geoarchaeology,* **1**: 259–274.

Ferring, C.R. (1990) Late Quaternary geology and geoarchaeology of the upper Trinity River drainage basin, Texas. Geological Society of America, Annual Meeting Field trip, Dallas.

Ferring, C.R. (1992) Alluvial pedology and geoarchaeological research. In: *Soils in Archaeology: Landscape and Human Occupation* (Ed V.T. Holliday), pp. 1–40. Smithsonian Institution Press, Washington.

Ferring, C.R. (1995) Middle Holocene environments, geology and archaeology in the Southern Plains. In: *Archaeological Geology of the Archaic Period in North America* (Ed E.A. Bettis, III), Special Paper 297, pp. 21–35. The Geological Society of America, Boulder, CO.

Ferring, C.R. (2001) Geoarchaeology in alluvial landscapes. In: *Earth Sciences and Archaeology* (Eds P. Goldberg, V.T. Holliday and C.R. Ferring), pp. 77–106. Kluwer-Plenum, New York.

Fisher, P.F. and Macphail, R.I. (1985) Studies of archaeological soils and deposits by micromorphological techniques. In: *Palaeoenvironmental Investigations: Research Design, Methods and Data Analysis N.G.A. Ralph,* (Eds N.R.J. Fieller, D.D. Gilbertson), pp. 92–112. British Archaeological Reports International Series, Oxford.

Fisher, R.V. and Schmincke, H.U. (1984) *Pyroclastic Rocks.* Springer-Verlag, Berlin.

FitzPatrick, E.A. (1984) *Micromorphology of Soils.* Chapman and Hall, London.

FitzPatrick, E.A. (1986) *An Introduction to Soil Science,* 2nd edition. Longman Scientific and Technical, Harlow.

FitzPatrick, E.A. (1993) *Soil Microscopy and Micromorphology.* John Wiley & Sons, Chichester.

Ford, D.C. and Williams, P. (1989) *Karst Geomorphology and Hydrology.* Chapman & Hall, London.

Ford, T.D. and Cullingford, C.H.D. (1976) *The Science of Speleology.* Academic Press, London; New York.

Foreman, S., Hiller, J. and Petts, D. (Eds) (2002) *Gathering the People, Settling the Land. The Archaeology of a Middle Thames Landscape. Anglo-Saxon to Post-Medieval,* Thames Valley Landscapes monograph No. 14. Oxford Archaeology, Oxford.

Fowler, M.J. (2002) Satellite remote sensing and archaeology: a comparative study of satellite imagery of the environs of Figsbury Ring, Wiltshire. *Archaeological Prospection*, **9**: 55–69.

Fowler, P.J. and Evans, J.G. (1967) Plough-marks, lynchets and early fields. *Antiquity*, **41**: 289–301.

Foxhill, L. (2000) The running sands of time: archaeology and the short-term. *World Archaeology*, **31**: 484–498.

Francus, P. (2004) Image analysis, sediments and Paleoenvironments Developments in palaeoenvironmental research; Vol. 7. Kluwer Aademic Publisher, Dordrecht; Boston, xviii, 300 pp.

Frechen, M., Neber, A., Tsatskin, A., Boenigk, W. and Ronen, A. (2004) Chronology of Pleistocene sedimentary cycles in the Carmel Coastal Plain of Israel. *Quaternary International*, **121**: 41–52.

Frederick, C. (2001) Evaluating causality of landscape change. In: *Earth Sciences and Archaeology* (Eds P. Goldberg, V.T. Holliday and C.R. Ferring), pp. 55–76. Kluwer-Plenum, New York.

Frederick, C.D., Bateman, M.D. and Rogers, R. (2002) Evidence for aeolian deposition in the Sandy Uplands of East Texas and the implications for archaeological site integrity. *Geoarchaeology*, **17**: 191–217.

French, C. (1998) Soils and sediments. In: *Etton. Excavations at a Neolithic causewayed enclosure near Maxey, Cambridgeshire, 1982–7* (Ed F. Pryor), pp. 311–331. Archaeological Report 18, English Heritage, London.

French, C. (2001) The development of the prehistoric landscape in the Flag Fen Basin. In: *The Flag Fen Basin. Archaeology and Environment of a Fenland landscape* (Ed F. Prior), pp. 400–404. Archaeological Reports, English Heritage, London.

French, C. (2003) *Geoarchaeology In Action. Studies in Soil Micromorphology and Landscape Evolution*. Routledge, London.

French, C. and Whitelaw, T.M. (1999) Soil erosion, agricultural terracing and site formation processes at Markiani, Amorgos, Greece: the micromorphological perspective. *Geoarchaeology*, **14**: 151–189.

Frison, G.C. (1974) *The Casper Site: a Hell Gap Bison Kill on the High Plains*. Studies in archeology. Academic Press, New York, Vol. xviii, 266 pp.

Fritz, W.J. and Moore, J.N. (1988) *Basics of Physical Stratigraphy and Sedimentology*. Wiley, New York.

Frumkin, A. (1998) Salt cave cross-sections and their paleoenvironmental implications. In: *Geomorphic Response of Mediterranean and Arid Areas to Climate Change* (Eds S. Berkowicz, H. Lavee and A. Yair), Vol. 23, pp. 183–191. Hebrew University of Jerusalem, Institute of Earth Science, Jerusalem, Israel.

Frumkin, A., Carmi, I., Zak, I. and Magaritz, M. (1994) Middle Holocene environment change determined from the salt caves of Mount Sedom, Israel. In: *Late Quaternary Chronology and Paleoclimates of the Eastern Mediterranean* (Eds O. Bar-Yosef and R.S. Kra), pp. 315–332. Radiocarbon, Tucson.

Fuchs, M. and Lang, A. (2001) OSL dating of coarse-grain fluvial quartz using single-aliquot protocols on sediments from NE Peloponnese, Greece. *Quaternary Science Reviews*, **20**: 783–787.

Fulford, M. and Wallace-Hadrill, A. (1995–96) The House of *Amarantus* at Pompeii (I, 9, 11–12): an interim report on survey and excavations in 1995–96. *Revista di Studi Pompeiani*, **VII**: 77–113.

Gabunia, L., Vekua, A., Lordkipanidze, D., Swisher, C.C., III, Ferring, R., Justus, A., Nioradze, M., Tvalchrelidze, M., Antón, S.C., Bosinski, G., Jöris, O., de Lumley, M.-A., Majsuradze, G. and Mouskhelishvili, A. (2000) Earliest Pleistocene hominid cranial remains from Dmanisi, Republic of Georgia; taxonomy, geological setting, and age. *Science*, **288**: 1019–1025.

Gabunia, L., Vekua, A. and Lordkipanidze, D. (2001) New human fossils from Dmanisi, eastern Georgia. *Archaeology, Ethnology & Anthropology of Eurasia*, **2**: 128–139.

Gaffney, C. and Gater, J. (2003) *Revealing the Buried Past, Geophysics for Archaeologists*. Tempus, Stroud, Gloucestershire.

Gage, M.D. (1999) Vacuum chambers for use in resin impregnation of core samples. *Geoarchaeology*, **14**: 307–311.

Gage, M.D. (2000) Ground-Penetrating Radar and Core Sampling at the Moundville Site. Masters thesis, The University of Alabama, Tuscaloosa, 170 pp.

Gage, M.D. and Jones, V.S. (1999) Ground penetrating radar. *Journal of Alabama Archaeology*, **45**: 49–61.

Gale, S.J. and Hoare, P.G. (1991) *Quaternary Sediments. Petrographic Methods for the Study of Unlithified Rocks*. Belhaven Press (John Wiley), London.

Galinié, H. (Ed) (2000) *Terres Noires–1, Documents. Sciences de la Ville*, No. 6. La Maison des Sciences de la Ville, Tours.

Galinié, H. (in press) L'expression terres noires, un concept d'attente. In: *Terres noires et stratigraphies de transition de l'Antiquité Tardive et du haut Moyen Age – Dark Earth in the Dark Ages* (Ed L. Verslype). Université Catholique de Louvain., Louvain-La-Neuve.

Gardner, P.S. (1994) Carbonized plant remains from Dust Cave. *Journal of Alabama Archaeology*, **40**: 192–211.

Garfinkel, Y. (1987) Burnt lime products and social implications in the pre-pottery Neolithic B villages of the Near East. *Paléorient*, **13**: 69–76.

Garrison, E.G. (2003) *Techniques in Archaeological Geology. Natural Science in Archaeology*. Springer, Berlin; New York.

Garrod, D. and Bate, D.M.A. (1937) *The Stone Age of Mount Carmel, Vol. 1*. Clarendon Press, Oxford.

Gasche, H. and Tunca, O. (1983) Guide to archaeostratigraphic classification and terminology:

definitions and prinicples, *Journal of Field Archaeology*, **10**: 325–335.

Gé, T., Courty, M.A., Matthews, W. and Wattez, J. (1993) Sedimentary formation processes of occupation surfaces. In: *Formation Processes in Archaeological Contexts* (Eds P. Goldberg, D.T. Nash and M.D. Petraglia), Monographs in, World Archaeology No. 17, pp. 149–163. Prehistory Press, Madison, WI.

Gebhardt, A. (1990) *Evolution du Paleopaysage Agricole dans Le Nord-Ouest de la France: apport de la micromorphologie*, L'Université de Rennes I, Rennes.

Gebhardt, A. (1992) Micromorphological analysis of soil structural modification caused by different cultivation implements. In: *Préhistoire de l'Agriculture: nouvelles approaches expérimentales et ethnographiques* (Ed P.C. Anderson), Monogaphie de CRA No. 6, pp. 373–392. Centre Nationale de la Recherche Scientifique, Paris.

Gebhardt, A. (1993) Micromorphological evidence of soil deterioration since the mid-Holocene at archaeological sites in Brittany, France. *The Holocene*, **3**: 331–341.

Gebhardt, A. (1995) Soil micromorphological data from traditional and experimental agriculture. In: *Archaeological Sediments and Soils: Analysis, Interpretation and management* (Eds A.J. Barham and R.I. Macphail), pp. 25–40. Institute of Archaeology, London.

Gebhardt, A. and Langohr, R. (1999) Micromorphological study of construction materials and living floors in the medieval motte of Werken (West Flanders, Belgium). *Geoarchaeology*, **14**: 595–620.

Gerson, R. and Amit, R. (1987) Rates and modes of dust accretion and deposition in an arid region; the Negev, Israel. In: *Special Scientific Meeting of the Geological Society of London: Desert Sediments, Ancient and Modern* (Eds L.E. Frostick and I. Reid), Vol. 35, pp. 157–169, London, United Kingdom.

Gifford, D.P. (1978) Ethnoarchaeological observations of natural processes affecting cultural materials. In: *Explorations in Ethnoarchaeology* (Ed R. Guild), pp. 77–101.

Gifford-Gonzalez, D.P., Damrosch, D.B., Damrosch, D.R., Pryor, J. and Thunen, R. (1985) The third dimension in site structure: an experiment in trampling and vertical dispersal. *American Antiquity*, **50**: 803–818.

Gilbertson, D.D. (1995) Studies of lithostratigraphy and lithofacies: a selective review of research developments in the last decade and their applications to geoarchaeology. In: *Archaeological Sediments and Soils: Analysis, Interpretation and Management* (Eds A.J. Barham and R.I. Macphail), pp. 99–145. Institute of Archaeology, London.

Gillieson, D. (1996) *Caves: Processes, Development, Management*. Blackwell, Oxford.

Girard, C. (1978) *Les industries Moustériennes de la Grotte de l'Hyène à Arcy-sur-Cure (Yonne)*. 11th supplément à Gallia Préhistoire. Centre National de la Recherche Scientifique, Paris.

Girard, M., Miskovsky, J.C. and Evin, J. (1990) La fin du Wuerm moyen et le debut du Würm superieur a Arcy-sur-Cure (Yonne); précisions paléoclimatiques et chronostratigraphiques d'après les remplissages des grottes; The end of the middle Wurm and beginning of the upper Wurm at Arcy-sur-Cure, Yonne; paleoclimatic chronostratigraphic precision from cave filling. *Mémoires du Musée de Préhistoire d'Ile-de-France*, **3**: 295–303.

Giuntoli, S. (1994) *The Golden Book of Pompeii, Herculaneum, Mt. Vesuvius*. Casa Editrice Bonechi, Florence.

Gladfelter, B.G. (1992) Soil properties of sediments in Wadi Feiran, Sinai: a geoarchaeological interpretation. In: *Soils in Archaeology: Landscape Evolution and Human Occupation* (Ed V.T. Holliday), pp. 169–192. Smithsonian Institution Press, Washington.

Gladfelter, B.G. (2001) Archaeological sediments in humid alluvial environments. In: *Sediments in Archaeological Context* (Eds J.K. Stein and W.R. Farrand), pp. 93–125. The University of Utah Press, Salt Lake City.

Glaser, B. and Woods, W.I. (Eds) (2004) *Amazonian Dark Earths: Exploration in Space and Time*. Springer, New York.

Glob, P.V. (1970) *The Bog People*. Cornell University Press, Ithaca.

Goldberg, P. (1969) Analyses of sediments of Jerf 'Ajla and Yabrud Rockshelters, Syria. In: *Union Internationale pour L'Etude Du Quaternaire V111 Congres INQUA*, pp. 747–754. Textes Réunis par Mireille Ters, Paris.

Goldberg, P. (1973) *Sedimentology, Stratigraphy, and Paleoclimatology of et-Tabun Cave, Mount Carmel, Israel*. Doctoral dissertation, The University of Michigan, Ann Arbor, 183 pp.

Goldberg, P. (1976) The Geology of Boker Tachtit, Boker, and Their Surroundings. In: *Prehistory and Paleoenvironments of the Central Negev, Vol. I* (Ed A.E. Marks), pp. 25–55. Southern Methodist University Press, Dallas.

Goldberg, P. (1977a) Late quaternary stratigraphy of Gebel Maghara. In: *Prehistoric Investigations in Gebel Maghara, Northern Sinai, Qedem* (Eds O. Bar-Yosef and J.L. Phillips), Vol. 7, pp. 11–31. Hebrew University, Jerusalem.

Goldberg, P. (1977b) Nahal Aqev (D35) stratigraphy and environment of deposition. In: *Prehistory and Paleoenvironments in the Central Negev* (Ed A.E. Marks), Vol. II, pp. 56–60.

Goldberg, P. (1978) Granulométrie de sédiment de la Grotte de Taboun, Mont-Carmel, Israël. *Géologie Méditerranéenne*, **4**: 371–383.

Goldberg, P. (1979a) Geology of Late Bronze Age mudbrick from Tel Lachish. *Tel Aviv, Journal of the Tel Aviv Institute of Archaeology*, **6**: 60–71.

Goldberg, P. (1979b) Micromorphology of Pech-de-l'Azé II sediments. *Journal of Archaeological Science*, **6**: 1–31.

Goldberg, P. (1979c) Micromorphology of sediments from Hayonim Cave, Israel. *CATENA*, **6**: 167–181.

Goldberg, P. (1981) Applications of micromorphology in archaeology. In: *Soil Micromorphology. Volume 1. Techniques and Applications* (Eds P. Bullock and C.P. Murphy), pp. 139–150. A B Academic Publishers, Berkhamsted.

Goldberg, P. (1983) The geology of Boker Tachtit, Boker, and their surroundings. In: *Prehistory and Paleoenvironments in the Central Negev, Israel* (Ed A.E. Marks), Vol. III, pp. 39–62. Department of Anthropology, Institute for the Study of Earth and Man, Southern Methodist University, Dallas.

Goldberg, P. (1984) Late Quaternary history of Qadesh Barnea Northeastern Sinai. *Zeitschrift für Geomorphologie N.F.*, **28**: 193–217.

Goldberg, P. (1986) Late Quaternary environmental history of the southern Levant. *Geoarchaeology*, **1**: 225–244.

Goldberg, P. (1987) The geology and stratigraphy of Shiqmim. In: *Shiqmim 1: Studies Concerning Chalcolithic Societies in the Northern Negev Desert, Israel (1982–1984)* (Ed T.E. Levy), pp. 35–43 and 435–444. B.A.R., Oxford.

Goldberg, P. (1988) The archaeologist as viewed by the geologist. *Biblical Archaeologist*, December: 197–202.

Goldberg, P. (1994) Interpreting Late Quaternary continental sequences in Israel. In: *Late Quaternary Chronology and Paleoclimates of the Eastern Mediterranean* (Eds O. Bar-Yosef and R.S. Kra), pp. 89–102. Radiocarbon, Tucson.

Goldberg, P. (2000) Micromorphology and site formation at Die Kelders Cave 1, South Africa. *Journal of Human Evolution*, **38**: 43–90.

Goldberg, P. (2001a) Geoarchaeology. *Geotimes*, **46**: 38–39.

Goldberg, P. (2001b) Some micromorphological aspects of prehistoric cave deposits. *Cahiers d'archéologie du CELAT*, **10, série archéometrie 1**: 161–175.

Goldberg, P. and Arpin, T. (1999) Micromorphological analysis of sediments from Meadowcroft Rockshelter, Pennsylvania: implications for radiocarbon dating. *Journal of Field Archaeology*, **26**: 325–342.

Goldberg, P. and Arpin, T. (2003) Micromorphology, sediments, and artifacts. *Abstracts, Society for American Archaeology, Annual Meeting, Milwaukee, April 2003.*

Goldberg, P. and Bar-Yosef, O. (1995) Sedimentary environments of prehistoric sites in Israel and the Southern Levant. In: *Ancient Peoples and Landscapes* (Ed E. Johnson), pp. 29–49. Museum of Texas Tech University, Lubbock.

Goldberg, P. and Bar-Yosef, O. (1998) Site formation processes in Kebara and Hayonim Caves and their significance in Levantine prehistoric caves. In: *Neandertals and Modern Humans in Western Asia* (Eds T. Akazawa, K. Aoki and O. Bar-Yosef), pp. 107–125. Plenum, New York.

Goldberg, P. and Brimer, B. (1983) Late Pleistocene geomorphic surfaces and environmental history of Avdat/Havarim area, Nahal Zin. In: *Prehistory and Paleoenvironments in the Central Negev, Israel, Vol. III* (Ed A.E. Marks), pp. 1–13. Department of Anthropology, Institute for the Study of Earth and Man, Southern Methodist University, Dallas.

Goldberg, P. and Bull, P.A. (1982) Scanning electron microscope analysis of sediments from Tabun Cave, Mount Carmel, Israel. *Eleventh International Congress on Sedimentology*, **11**: 143–144.

Goldberg, P. and Guy, J. (1996) Micromorphological observations of selected rock ovens, Wilson–Leonard site, Central Texas. In: *Paleoecology; Colloquium 3 of XIII International Congress of Prehistoric and Protohistoric Sciences* (Eds L. Castelletti and M. Cremaschi), pp. 115–122. ABACO, Forlì.

Goldberg, P. and Holliday, V.T. (1998) Geology and stratigraphy of the Wilson–Leonard Site. In: *Wilson Leonard, An 11,000-year Archeological Record of Hunter-Gatherers in Central Texas, Studies in Archaeology* (Ed M.B. Collins), Vol. 31, pp. 77–121. Texas Archeological Research Laboratory, University of Texas at Austin, Austin.

Goldberg, P. and Laville, H. (1988) Le contexte stratigraphique des occupations paléolithiques de la grotte de Kebara (Israël). *Paléorient*, **14**: 117–123.

Goldberg, P. and Macphail, R.I. (2000) Micromorphology of sediments from Gibraltar Caves: some preliminary results from Gorham's Cave and Vanguard Cave. In: *Gibraltar During the Quaternary: The Southernmost Part of Europe in the Last Two Million Years* (Eds C. Finlayson, G. Finlayson and D. Fa), pp. 93–108. Gibraltar Government Heritage Publications, Gibraltar.

Goldberg, P. and Macphail, R.I. (2003) Strategies and techniques in collecting micromorphology samples. *Geoarchaeology*, **18**: 571–578.

Goldberg, P. and Macphail, R.I. (in press) Chapter 3. Geomorphology of the ancient landscape. In: *Hunter-Gatherers of the Southern California Coast* (Ed B.F. Byrd), New Approaches to Anthropological Archaeology. Equinox Publishers, Oxford.

Goldberg, P. and Nathan, Y. (1975) The phosphate mineralogy of et-Tabun Cave, Mount Carmel, Israel. *Mineralogical Magazine*, **40**: 253–258.

Goldberg, P. and Rosen, A. (1987) Early Holocene paleoenvironments. In: *Shiqmim 1: Studies Concerning Chalcolithic Societies in the Northern Negev Desert, Israel (1982–1984)* (Ed T.E. Levy), pp. 23–33 and 434. B.A.R., Oxford.

Goldberg, P. and Sherwood, S.C. (1994) Micromorphology of Dust Cave sediments: some preliminary results. *Journal of Alabama Archaeology*, **40**: 56–64.

Goldberg, P. and Whitbread, I. (1993) Micromorphological study of a Bedouin tent floor. In: *Formation Processes in Archaeological Context* (Eds P. Goldberg, D.T. Nash and M.D. Petraglia), Monographs in World Archaeology No. 17, pp. 165–188. Prehistory Press, Madison.

Goldberg, P., Gould, B., Killebrew, A. and Yellin, J. (1986) The provenience of Late Bronze Age ceramics from Deir el Balah: comparison of the results of neutron activation analysis and thin section analysis. In: *Proceedings of 24th International Archaeometry Symposium* (Eds J.S. Olin and M.J. Blackman), pp. 341–351, Washington, D.C., 1984.

Goldberg, P., Lev-Yadun, S. and Bar-Yosef, O. (1994) Petrographic thin sections of archaeological sediments: a new method for palaeobotanical studies. *Geoarchaeology*, **9**: 243–257.

Goldberg, P., Holliday, V.T. and Ferring, C.R. (Eds) (2001a) *Earth Sciences and Archaeology*. Kluwer Academic/ Plenum Publishers, New York.

Goldberg, P., Weiner, S., Bar-Yosef, O., Xu, Q. and Liu, J. (2001b) Site formation processes at Zhoukoudian, China. *Journal of Human Evolution*, **41**: 483–530.

Goldberg, P., Schiegl, S., Meligne, K., Dayton, C. and Conard, N.J. (2003) Micromorphology and site formation at Hohle Fels Cave, Swabian Jura, Germany. *Eiszeitalter und Gegenwart*, **53**: 1–25.

Goldman-Finn, N.S. (1994) Dust Cave in regional context. *Journal of Alabama Archaeology*, **40**: 212–231.

Goodfriend, G.A. (1990) Rainfall in the Negev Desert during the middle Holocene, based on ^{13}C of organic matter in land snail shells. *Quaternary Research*, **34**: 186–197.

Goodfriend, G.A. (1991) Holocene trends in ^{18}O in land snail shells from the Negev Desert and their implications for changes in rainfall source areas. *Quaternary Research*, **35**: 417–426.

Goren, Y. (1999) On determining use of pastoral cave sites: a critical assessment of spherulites in archaeology. *Mitekufat haeven: Journal of the Israel Prehistoric Society*, **29**: 123–128.

Goren, Y. and Goldberg, P. (1991) Petrographic thin sections and the development of Neolithic plaster production in Northern Israel. *Journal of Field Archaeology*, **18**: 131–138.

Goren, Y., Gopher, A. and Goldberg, P. (1993) The beginnings of pottery production in the southern Levant: technological and social aspects. In: *Biblical Archaeology Today, 1990: Proceedings of the Second International Congress on Biblical Archaeology, Supplement.*, pp. 32–40. Israel Exploration Society, Jerusalem.

Goren-Inbar, N., Feibel, C. S., Verosub, K. L., Melamed, Y., Kislev, M. E., Tchernov, E. and Saragusti, I. (2000) Pleistocene milestones on the Out-of-Africa Corridor at Gesher Benot Ya'aqov, Israel. *Science*, **289**: 944–947.

Goren-Inbar, N., Goren, Y., Rabinovich, R., Saragusti, I. and Belitzky, S. (1992) Gesher Benot Ya'qov – the "bar": an Acheulian assemblage. *Geoarchaeology*, **7**: 27–40.

Goren, Y., Goring-Morris, A.N. and Segal, I. (2001) The technology of skull modelling in the pre-pottery Neolithic B (PPNB): Regional variability, the relation of technology and iconography and their archaeological implications. *Journal of Archaeological Science*, **28**: 671–690.

Goring-Morris, A.N. (1987) At the Edge: Terminal Pleistocene Hunter-Gatherers in the Negev and Sinai. British Archaeological Reports, International Series, **361**. BAR, Oxford.

Goring-Morris, N. (1993) From foraging to herding in the Negev and Sinai: the Early to Late Neolithic transition. *Paléorient*, **19**: 65–89.

Goring-Morris, N. and Belfer-Cohen, A. (1997) The articulation of cultural processes and late Quaternary environmental changes in Cisjordan. *Paléorient*, **23**: 71–93.

Goring-Morris, A.N. and Goldberg, P. (1990) Late Quaternary Dune incursions in the Southern Levant: archaeology, chronology, and paleoenvironments. *Quaternary International*, **5**: 115–137.

Gosden, C. (1994) *Social Being and Time*. Blackwell, Oxford.

Goudie, A.S. and Middleton, N.J. (2001) Saharan dust storms; nature and consequences. *Earth-Science Reviews*, **56**: 179–204.

Graham, E. (1998) Metaphor and metamorphism: some thoughts on environmental metahistory. In: *Advances in Historical Ecology* (Ed W. Balée). Columbia University Press, New York.

Graham, I.D.G. and Scollar, I. (1976) Limitations on magnetic prospection in archaeology imposed by soil properties. *Archaeo-Physika*, **6**: 1–124.

Graham, J. (1988) Collection and analysis of field data. In: *Techniques in Sedimentology* (Ed M.E. Tucker), pp. 5–62. Blackwell Scientific Publications, Oxford.

Grimes, W.F. (1968) *The Excavations of Roman and Medieval London*. Routledge and Kegan Paul, London.

Grimm, P. (1968) *Tilleda. Eine Kö nigspfalz am Kyffhäusser. Teil 1: Die Hauptburg*, Bd. 24. Deutsche Akademie der Wissenschaften zu Berlin, Schriften der Sektion Ur- und Frühgeschichte, Berlin.

Grøn, O. (1989) General spatial behaviour in small dwellings: a preliminary study in ethnoarchaeology and social psychology. In: *Mesolithic Europe* (Ed C. Bonscell), pp. 99–105.

Grøn, O., Aurdal, L., Christensen, F., Solberg, R. Macphil, R., Lous, J. and Loska, A. (2005) Locating invisible cultural heritage sites in agricultural fields. Development of methods for satellite monitoring of cultural heritage sites – report 2004: Oslo, Norwegian Directorate for Cultural Heritage.

Guccione, M.J., Sierzchula, M.C., Lafferty, R.H. and Kelley, D. (1998) Site preservation along an active meandering and avulsing river: the Red River, Arkansas. *Geoarchaeology*, **13**: 475–500.

Guélat, M. and Federici-Schenardi, M. (1999) Develier-Courtételle (Jura) L'histoire d'une cabane en fosse reconstituée grâce à la micromorphologie. *Helvetica Archaeologica,* **118/119**: 58–63.

Guillet, B. (1982) Study of turnover of soil organic matter using radio-isotopes. In: *Constituents and Properties of Soils* (Eds M. Bonneau and B. Souchier), pp. 238–257. Academic Press, London.

Guilloré, P. (1985) *Méthode de Fabrication Méchanique et en Séries des Lames Minces.* Institut National Agronomique, Paris.

Gustavs, S. (1998) Spätkaiserzeitliche Baubefunde von Klein Köris, Lkr. Dahme-Spreewald. In: *Haus und Hof im östlichen Germanien (Berlin, 1994)* (Eds J. Henning and A. Leube), Universtätsforschungen zur prähistorischen Archäologie *Band* 50 and Schriften zur Archäologie der germanischen und slawischen Frühgeschichte *Band* 2, pp. 40–66. Dr Rudolf Habelt GmbH, Bonn.

Guyard, L. (Ed) (2003) Le Collège de France (Paris). Du quartier gallo-romain au Quartier latin (1er s. av. J.-CXIXe s.), Documents d'archéologie française. Série Archéologie préventive. Éditions de la Maison des sciences de l'Homme, Paris.

Gvirtzman, G. and Wieder, M. (2001) Climate of the last 53,000 years in the Eastern Mediterranean, based on soil-sequence stratigraphy in the coastal plain of Israel. *Quaternary Science Reviews,* **20**: 1827–1849.

Haesaerts, P. and Teyssandier, N. (2003) The early Upper Paleolithic occupations of Willendorf II (Lower Austria): a contribution to the chronostratigraphic and cultural context of the beginning of the Upper Paleolithic in Central Europe. In: *The Chronology of the Aurignacian and of the Transitional Technocomplexes: Dating, Stratigraphies, Cultural Implications. Proceedings of Symposium 6.1 of the XIVth Congress of the UISPP, Trabalhos de Arqueologia, Vol.* 33, pp. 33–51, Université of Liège, Belgium, September 2–8, 2001.

Haesaerts, P., Mestdagh, H. and Bosquet, D. (1999) The sequence of Remicourt (Hesbaye, Belgium): new insights on the ped- and chronostratigraphy of the Rocourt Soil. *Geologica Belgica,* **2/3–4**: 5–27.

Haesaerts, P., Borziak, I., Vhirica, V., Koulakovsa, L. and van der Plicht, J. (2003) The East Carpathian loess record: a reference for the Middle and Late Pleniglaical stratigraphy in Central Europe. *Quaternaire (Paris),* **14**: 163–188.

Hahn, J. (1988) *Die Geißenklösterle-Höhle im Achtal bei Blaubeuren I. Fundhorizontbildung und Besiedlung im Mittelpaläolithikum und im Aurignacien.* Forschungen und Berichte zur Vor- und Frühgeschichte in Baden-Württemberg, Band 26. Kommissionsverlag, Korad Theiss Verlag, Stuttgart.

Hahn, J. (1995) Ausgrabungen 1994 im Hohle Fels Schelklingen, Alb-Donau-Kreis.

Hamblin, W.K. (1996) *Atlas of Stereoscopic Aerial Photographs and Remote Sensing Imagery of North America.* Crystal Productions, Glenview, IL.

Harden, J.W. (1982) A quantitative index of soil development from field descriptions. Examples from a chronosequence in central California. *Geoderma,* **28**: 1–28.

Harris, T.M. and Lock, G.R. (1995) Towards an evaluation of GIS in European archaeology; the past, present and future of theory and applications, (Eds Lock, G., and Stancic, Z.), Archaeology and Geographical Information Systems: A European Perspective: London, Taylor and Francis, pp. 349–365.

Harrower, M., McCorriston, J. and Oches, E.A. (2002) Mapping the roots of agriculture in southern arabia: the application of satellite remote sensing, global positioning system and geographic information system technologies. *Archaeological Prospection,* **9**: 35–42.

Hassan, F. (1995) Late Quaternary geology and geomorphology of the area of the vicinity of Ras en Naqb. In: *Prehistoric Cultural Ecology and Evolution: Insights from Southern Jordan* (Ed D.O. Henry), pp. 23–41. Plenum, New York.

Hastings, C.M. and Moseley, M.E. (1975) The adobes of Huaca del Sol and Huaca de la Luna, *American Antiquity,* **40**: 196–203.

Haughton, C. and Powlesland, D. (1999) *West Heslerton. The Anglian Cemetery.* English Heritage, London.

Haynes, C.V., Jr. (1982) Great Sand Sea and Selima sand sheet, eastern Sahara; geochronology of desertification. *Science,* **217**: 629–633.

Haynes, C.V., Jr. (1991) Geoarchaeological and paleohydrological evidence for a Clovis-Age drought in North America and its bearing on extinction. *Journal of Quaternary Research,* **35**: 438–450.

Haynes, C.V., Jr. (1995) Geochronology of paleoenvironmental change, Clovis type site, Blackwater Draw, New Mexico. *Geoarchaeology,* **10**: 317–388.

Hazelden, J., Sturdy, R.G. and Lovelend, P.J. (1987) Saline soils in North Kent. In: *SEESOIL* (Ed M.G. Jarvis), pp. 2–23. South East Soils Discussion Group, Bedford.

Healy, F., and Harding, J. (forthcoming) Raunds Area Project. The Neolithic and Bronze Age landscapes of West Cotton, Stanwick and Irthlingborough, Northamptonshire., Volume English Heritage Archaeological Report, English Heritage, London.

Heathcote, J.L. (2002) *An Investigation of the Pedosedimentary Characteristics of Deposits Associated with Managed Livestock.* Ph.D. Thesis, University College London, London.

Hedberg, H.D. (Ed) (1976) *International Subcommission on Stratigraphic Classification.* Wiley, New York.

Hedges, R.E.M. and Millard, A.R. (1995) Bones and groundwater: towards the modelling of diagenetic processes. *Journal of Archaeological Science,* **22**: 155–164.

Helms, J.G., McGill, S.F. and Rockwell, T.K. (2003) Calibrated, late Quaternary age indices using clast rubification and soil development on alluvial surfaces in Pilot Knob Valley, Mojave Desert, southeastern California. *Quaternary Research*, **60**: 377–393.

Henning, J. (1996) Landwirtschaft der Franken. In: *Die Franken–Wegbereiter Europas. Vor 1500 Jahren: König Chlodwig und seine Erben*, pp. 174–185, Mainz.

Henning, J. and Macphail, R.I. (2004) Das karolingische Oppidum Büraburg: Archäologische und mikromorphologische Studien zur Funktion einer frümittelaterlichen Bergbefestigung in Nordhessen (The Carolingian times *oppidum* Bueraburg: archaeological and soil micromorphological investigations on the function of an early medieval hillfort in North Hesse). In: *Parerga Praehistorica. Jubiläumsschrift zur Prähistorischen Archäologie 15 Jahre UPA* (Ed B. Hänsel), Band 100, pp. 221–252. Verlag Dr Rudolf Habelt GmbH, Bonn.

Henshilwood, C.S., d'Errico, F., Yates, R., Jacobs, Z., Tribolo, C., Duller, G.A.T., Mercier, N., Sealy, J.C., Valladas, H., Watts, I. and Wintle, A.G. (2002) Emergence of modern human behavior: Middle Stone Age engravings from South Africa. *Science*, **295**: 1278–1280.

Herz, N. and Garrison, E. (1998) *Geological Methods for Archaeology*. Oxford University Press, New York.

Hester, T.R., Shafer, H.J. and Feder, K.L. (1997) *Field Methods in Archaeology, 7th edition*. Mayfield Publishing Company, Mountain View, CA.

Hill, C. and Forti, P. (1997) *Cave Minerals of the World, 2nd edition*. National Speleological Society, Huntsville.

Hill, J. and Rowsome, P. (In preparation) Excavations at 1 Poultry: the Roman sequence, *Monograph Series. Museum of London Archaeological Service*, London.

Hilton, M.R. (2002) *Evaluating Site Formation Processes at a Higher Resolution: An Archaeological Case Study in Alaska Using Micromorphology and Experimental Techniques*, Doctoral Dissertation, University of California, Los Angeles.

Hirai, Y. (1987) Stratigraphy and sedimentological analysis of the Douara Cave deposits, 1984 excavations. *The University of Tokyo Museum Bulletin*, **29**: 49–60.

Hjulström, F. (1939) Transportation of detritus by moving water. In: *Recent Marine Sediments: A Symposium* (Ed P.D. Trask), pp. 5–31. American Association of Petroleum Geologists, Tulsa.

Hodder, I. (1986) *Reading the Past*. Cambridge University Press, Cambridge.

Hodder, I. (1999) *The Archaeological Process: An Introduction*. Blackwell, Oxford.

Hodgson, J.M. (1997) *Soil Survey Field Handbook*, Technical Monograph No. 5. Soil Survey and Land Research Centre, Silsoe.

Hogue, S.H. (1994) Human skeletal remains from Dust Cave. *Journal of Alabama Archaeology*, **40**: 173–191.

Holliday, V. and Johnson, E. (1989) Lubbock Lake: Late Quaternary cultural and environmental change on the Southern High Plain, USA. *Journal of Quaternary Science*, **4**: 145–165.

Holliday, V.T. (1985) Early and Middle Holocene soils at the Lubbock Lake archaeological site, Texas. *Catena*, **12**: 61–78.

Holliday, V.T. (1990) Pedology in archaeology. In: *Archaeological Geology of North America* (Eds N.P. Lasca and J. Donahue), Centennial Special Vol. 4, pp. 525–540. Geological Society of America, Boulder.

Holliday, V.T. (Ed) (1992) *Soils in Archaeology. Landscape Evolution and Human Occupation*. Smithsonian Institution Press, Washington.

Holliday, V.T. (1997) *Paleoindian Geoarchaeology of the Southern High Plains*. Texas Archaeology and Ethnohistory Series. University of Texas Press, Austin.

Holliday, V.T. (2004) *Soils in Archaeological Research*. Oxford University Press, Oxford.

Homsey, L. (2003) "Geochemical characterization of archaeological sediments at Dust Cave, Alabama." Unpublished M.S. thesis, University of Pittsburgh.

Hopkins, D.W., Wiltshire, P.E.J. and Turner, B.D. (2000) Microbial characteristics of soils from graves: an investigation at the interface of soil microbiology and forensic science. *Applied Soil Ecology*, **14**: 283–288.

Horowitz, A. (1979) *The Quaternary of Israel*. Academic Press, New York.

Horrocks, M., Coulson, S.A. and Walsh, K.A.J. (1999) Forensic palynology: variation in the pollen content of soil on shoes and shoeprints in soil. *Journal of Forensic Science*, **44**: 119–122.

Horwitz, L.K. and Goldberg, P. (1989) A study of Pleistocene and Holocene Hyaena Coprolites. *Journal of Archaeological Science*, **16**: 71–94.

Hovers, E., Ilani, S., Bar-Yosef, O. and Vandermeersch, B. (2003) An early case of color symbolism: ochre use by modern humans in Qafzeh Cave. *Current Anthropology*, **44**: 491–522.

Hovers, E., Rak, Y., Lavi, R. and Kimbel, W.H. (1995) Hominid remains from Amud Cave in the context of the Levantine Middle Paleolithic. *Paléorient*, **21**: 47–61.

Howard, A.J., Macklin, M.G. and Passmore, D.G. (Eds) (2003) *Alluvial Archaeology in Europe*. Balkema Publishers, Lisse.

Huckleberry, G. (2001) Archaeological sediments in dryland alluvial environments. In: *Sediments in Archaeological Context* (Eds J.K. Stein and W.R. Farrand), pp. 67–92. The University of Utah Press, Salt Lake City.

Hughes, P.J. and Lampert, R.J. (1977) Occupational disturbance and types of archaeological deposit. *Journal of Archaeological Science*, **4**: 135–140.

Humphrey, J.D. and Ferring, C.R. (1994) Stable isotopic evidence for latest Pleistocene and Holocene climatic change in North-Central Texas. *Quaternary Research*, **41**: 200–213.

Hunter, J., Roberts, C. and Martin, A. (1997) *Studies in Crime: An Introduction to Forensic Archaeology*. Routledge, London.

Iceland, H.B. and Goldberg, P. (1999) Late-Terminal Classic Maya pottery in Northern Belize: a petrographic analysis of sherd samples from Colha and Kichpanha. *Journal of Archaeological Science*, **26**: 951–966.

Imeson, A.C. and Jungerius, P.D. (1976) Aggregate stability and colluviation in the Luxembourg Ardennes: an experimental and micromorphological study. *Earth Surface Processes*, **1**: 259–271.

Imeson, A.C., Kwaad, F.J.P.M. and Mücher, H.J. (1980) Hillslope processes and deposits in forested areas of Luxembourg. In: *Timescales in Geomorphology* (Eds R.A. Cullingford, D.A. Davidson and J. Lewin), pp. 31–42, Hull, United Kingdom.

Jackson, M.L., Levelt, T.W.M., Syers, J.K., Rex, R.W., Clayton, R.N., Clayton, G.D., Sherman, G.D. and Uehara, G. (1971) Geomorphological relationships of tropospherically-derived quartz in soils on the Hawaiian Islands. *Soil Science Society of America Proceedings*, **35**: 515–525.

Jacob, J.S. (1995) Archaeological pedology in the Maya lowlands. In: *Pedological Perspectives in Archaeological Research* (Eds M.E. Collins, B.J. Carter, B.G. Gladfelter and R.J. Southard), SSSA Special Publication Number 44, pp. 51–80. Soil Science Society of America, Inc., Madison.

Jarvis, K.E., Wilson, H.E. and James, S.L. (2004) Assessing element variability in small soil samples taken during forensic investigation. In: *Forensic Geoscience: Principles, Techniques and Applications* (Eds K. Pye and D.J. Croft), Geological Society, London, Special Publications 232, pp. 171–182. The Geological Society of London, London.

Jarvis, M.G., Allen, R.H., Fordham, S.J., Hazleden, J., Moffat, A.J. and Sturdy, R.G. (1983) Soils of England and Wales. *Sheet 6*. South East England. Ordnance Survey, Southampton.

Jarvis, M.G., Allen, R.H., Fordham, S.J., Hazleden, J., Moffat, A.J. and Sturdy, R.G. (1984) *Soils and Their Use in South-East England, Bulletin* No. 15. Soil Survey of England and Wales, Harpenden.

Jelinek, A. (1982) The Tabun cave and Paleolithic man in the Levant. *Science*, **216**: 1369–1375.

Jelinek, A.J., Farrand, W.R., Haas, H., Horowitz, A. and Goldberg, P. (1973) New excavations at the Tabun Cave, Mount Carmel, Israel, 1967–1972 – a preliminary report. *Paléorient*, **1**: 151–183.

Jenkins, D. (1994) Interpretation of interglacial cave sediments from a hominid site in North Wales: translocation of Ca–Fe-phosphates. In: *Proceedings of IX International Working-Meeting on Soil Micromorphology, Townsville, Australia, July 1992* (Ed A. Ringrose-Vaose), pp. 293–305. Elsevier, Amsterdam.

Jenny, H. (1941) *Factors of Soil Formation*. McGraw-Hill, New York.

Jia, L. (Ed) (1999) *Chronicle of Zhoukoudian (1927–1937)*. Shanghai Scientific and Technical Publishers, Shanghai.

Jia, L. and Huang, W. (1990) *The Story of Peking Man*. Foreign Languages Press, Oxford University Press, Beijing, Hong Kong.

Jing, Z., G., Rapp, Jr., and Gao, T. (1995) Holocene landscape evolution and its impact on the Neolithic and Bronze Age sites in the Shangqiu area, northern China. *Geoarchaeology*, **10**: 481–513.

Joffe, J. (1949) *Pedology*. Pedology Publications, New Brunswick, New Jersey.

Johnson, D.L. (2002) Darwin would be proud: Bioturbation, dynamic denudation, and the power of theory in science. *Geoarchaeology*, **17**: 7–40.

Jones, R.J.A. and Evans, R. (1975) Soil and crop marks in the recognition of archaeological sites by aerial photography. In: *Aerial Reconnaissance for Archaeology* (Ed D.R. Wilson), Vol. 12, pp. 1–12. Council for British Archaeology Research Reports, Oxford.

Jongerius, A. (1983) The role of micromorphology in agricultural research. In: *Soil Micromorphology* (Eds P. Bullock and C.P. Murphy), Volume 1: Techniques and Applications, pp. 111–138. A B Academic Publishers, Berkhamsted.

Kapur, S., Sakarya, N. and FitzPatrick, E.A. (1992) Mineralogy and micromorphology of Chalcolithic and early Bronze Age Ikiztepe ceramics. *Geoarchaeology*, **7**: 327–337.

Karkanas, P., Kyparissi-Apostolika, N., Bar-Yosef, O. and Weiner, S. (1999) Mineral assemblages in Theopetra, Greece: a framework for understanding diagenesis in a prehistoric cave. *Journal of Archaeological Science*, **26**: 1171–1180.

Karkanas, P., Bar-Yosef, O., Goldberg, P. and Weiner, S. (2000) Diagenesis in prehistoric caves: the use of minerals that form *in situ* to assess the completeness of the archaeological record. *Journal of Archaeological Science*, **27**: 915–929.

Karkanas, P., Rigaud, J.-P., Simek, J.F., Albert, R.M. and Weiner, S. (2002) Ash bones and guano: a study of the minerals and phytoliths in the sediments of Grotte XVI, Dordogne, France. *Journal of Archaeological Science*, **29**: 721–732.

Keeley, H.C.M., Hudson, G.E. and Evans, J. (1977) Trace element contents of human bones in various states of preservation. *Journal of Archaeological Science*, **4**: 19–24.

Keevill, G. (Ed) (2004) *The Tower of London Moat: Archaeological Excavations 1995–1999*. Oxford Archaeology with Historic Royal Palaces, Oxford.

Kelly, J. and Wiltshire, P.E.J. (1996) Microbiological report. In: *The Experimental Earthwork Project 1960–1992* (Eds M. Bell, P.J. Fowler and S.W. Hillson), CBA Research Report, pp. 148–155. Council for British Archaeology, York.

Kemp, R.A. (1985) The Valley Farm Soil in southern East Anglia. In: *Soils and Quaternary Landscape Evolution* (Ed J. Boardman), pp. 179–196. John Wiley and Sons, Chichester.

Kemp, R.A. (1986) Pre-Flandrian Quaternary soils and pedogenic processes in Britain. In: *Paleosols: Their Recognition and Interpretation* (Ed V.P. Wright), pp. 242–262. Princeton University Press, Princeton, NJ.

Kemp, R.A. (1999) Soil micromorphology as a technique for reconstructing palaeoenvironmental change. In: *Paleoenvironmental Reconstruction in Arid Lands* (Eds A.K. Singhvi and E. Derbyshire), pp. 41–71. Oxford and IBH Publishing Co. Pvt. Ltd., New Delhi.

Kemp, R.A., Jerz, H., Grottenthaler, W. and Preece, R.C. (1994) Pedosedimentary fabrics of soils within loess and colluvium in southern England and southern Germany. In: *Soil micromorphology; Developments in Soil Science* (Eds A.J. Ringrose-Voase and G.S. Humphreys), Vol. 22, pp. 207–219. Elsevier, Townsville, Queensland, Australia.

Kemp, R.A., McDaniel, P.A. and Busacca, A.J. (1998) Genesis and relationship of macromorphology and micromorphology to contemporary hydrological conditions of a welded Argixeroll from the Palouse in Idaho. *Geoderma*, **83**: 309–329.

Kerr, P.K. (1995) Phosphate imprinting within Mound A at the Huntsville site. In: *Pedological Perspectives in Archaeological Research* (Eds M.E. Collins, B.J. Carter, B.G. Gladfelter and R.J. Southard), SSSA Special Publication Number 44, pp. 133–149. Soil Science Society of America, Madison.

Kidson, C. (1982) Sea level changes in the Holocene. *Quaternary Science Reviews*, **1**: 121–151.

King, R.B., Baillie, I.C., Abell, T.M.B., Dunsmore, J.R., Gray, D.A., Pratt, J.H., Versey, H.R., Wright, A.C.S., and Zisman, S.A. (1992) *Land Resource Assessment of Northern Belize.* Vols 1 and 2. Kent, U.K.

Kingery, W.D., Vandiver, P.B. and Prickett, M. (1988) The beginnings of pyrotechnology, part II: production and use of lime and gypsum plaster in the pre-pottery Neolithic Near East. *Journal of Field Archaeology*, **15**: 219–244.

Klein, C. (1986) *Fluctuations of the Level of the Dead Sea and Climatic Fluctuations during Historical Times.* Ph.D. Dissertation, Hebrew University of Jerusalem, Jerusalem.

Klingebiel, A.A. and Montgomery, P.H. (1961) *Land Capability Classification.* Soil Conservation Service Agricultural handbook No. 210. USDA, Washington.

Knecht, H., Pike-Tay, A. and White, R. (1993) *Before Lascaux: the Complex Record of the Early Upper Paleolithic.* CRC Press, Boca Raton.

Koizumi, T. (1978) Climato-genetic landforms around Jabal ad Douara and its surroundings. *Bulletin. University Museum Tokyo*, **14**: 29–51.

Kooistra, M.J. (1978) Soil development in recent marine sediments of the Intertidal Zone in the Oosterschelde – the Netherlands: a soil micromorphological approach. *Soil Survey Papers*, No. 14. Soil Survey Institute, Wageningen.

Kresten, P. and Hjärthner-Holdar, E. (2001) Analyses of the Swedish ancient iron reference slag W-25:R. *Historical Metallurgy*, **35**: 48–51.

Krinsley, D.H. and Doornkamp, J.C. (1973) *Atlas of Quartz Sand Surface Textures*. Cambridge University Press, Cambridge.

Krumbein, W.C. and Pettijohn, F.J. (1938) *Manual of Sedimentary Petrography*. Appleton-Century-Crofts, Inc., New York.

Krumbein, W.C. and Sloss, L.L. (1963) *Stratigraphy and Sedimentation*, 2nd edition. W.H. Freeman and Company, San Francisco.

Kubiëna, W.L. (1938) *Micropedology*. Collegiate Press, Ames, Iowa.

Kubiëna, W.L. (1953) *The Soils of Europe*. Thomas Murby, London, 317 pp.

Kubiëna, W.L. (1970) *Micromorphological Features of Soil Geography*. Rutgers University Press, New Brunswick.

Kukla, G.J. (1977) Pleistocene land–sea correlations I. Europe. *Earth-Science Reviews*, **13**: 307–374.

Kvamme, K.L. (2001) Current practices in archaeogeophysics. In: *Earth Sciences and Archaeology* (Eds P. Goldberg, V.T. Holliday and C.R. Ferring), pp. 353–384. Kluwer Academic/Plenum, New York.

Kwaad, F.J.P.M. and Mücher, H.J. (1977) The evolution of soils and slope deposits in the Luxembourg Ardennes near Wilts. *Geoderma*, **17**: 1–37.

Kwaad, F.J.P.M. and Mücher, H.J. (1979) The formation and evolution of colluvium on arable land in northern Luxembourg. *Geoderma*, **22**: 173–92.

Lambrick, G. (1992) Alluvial archaeology of the Holocene in the Upper Thames Basin 1971–1991: a review. In: *Alluvial Archaeology in Britain* (Eds S. Needham and M.G. Macklin), Oxbow Monograph 27, pp. 209–228. Oxbow, Oxford.

Langohr, R. (1991) Soil characteristics of the Motte of Werken (West Flanders, Belgium). In: *Methoden und Perspektiven der Archaologie des Mittelalters, Tagungsberichte zum interdisziplinaren Kolloquium, September 1989, Liestal, Switzerland* (Ed J. Tauber), pp. 209–223. Berichte aus der Arbeit des Amtes fur Museen und Archaologie des Kantons Baseeland, Heft.

Langohr, R. (1993) Types of tree windthrow, their impact on the environment and their importance for the understanding of archaeological excavation data. *Helenium*, **XXXIII**: 36–49.

Lattman, L.H. and Ray, R.G. (1965) *Aerial Photographs in Field Geology*. Holt, Rinehart and Winston, New York.

Lautridou, J.P. and Ozouf, J.C. (1982) Experimental frost shattering; 15 years of research at the Centre de Géomorphologie du CNRS. *Progress in Physical Geography*, **6**: 215–232.

Laville, H. (1964) *Recherches sédimentologiques sur la paéoclimatologie du Wurm recent en Périgord*, L'Université de Bordeaux, Bordeaux, pp. 4–19.

Laville, H. and Goldberg, P. (1989) The collapse of the Mousterian regime and the beginnings of the Upper Palaeolithic in Kebara Cave, Mount Carmel. *British Archaeological Reports, Int'l Series*, **497**: 75–95.

Laville, H., Rigaud, J.-P. and Sackett, J. (1980) *Rock Shelters of the Perigord*. Academic Press, New York.

Lawson, A.J. (2000) Potterne 1982–5: *Animal Husbandry in Later Prehistoric Wiltshire*, Wessex Archaeological Report No. 17. Wessex Archaeology, Salisbury.

Lawson, T., Hopkins, D.W., Chudeck, J.A., Janaway, R.C. and Bell, M.G. (2000) The Experimental Earthwork at Wareham, Dorset after 33 years: interaction of soil organisms with buried materials. *Journal of Archaeological Science*, **22**: 273–285.

Le Ribault, L. (1977) *L'exoscopie des quartz*. Masson, Paris.

Lechtman, H.N. and Hobbs, L.W. (1983) Roman concrete and the Roman architectural revolution. *Journal of Archaeological Science*, **13**: 81–128.

Legros, J.P. (1992) Soils of Alpine Mountains. In: *Weathering, Soils and Paleosols* (Eds I.P. Martini and W. Chesworth), Developments in Earth Surface Processes 2. Elsevier, Amsterdam.

Lehmann, J., Kern, D.C., Glaser, B. and Woods, W.I. (Eds) (2004) *Amazonian Dark Earths: Origins, Properties, Management*. Kluwer Academic Publishers, New York.

Leigh, D.S. (2001) Buried artifacts in sandy soils; techniques for evaluating pedoturbation versus sedimentation. In: *Earth Sciences and Archaeology* (Eds P. Goldberg, V.T. Holliday and C.R. Ferring), pp. 269–293. Kluwer-Plenum, New York.

Lelong, R. and Souchier, B. (1982) Ecological significance of the weathering complex: relative importance of general and local factors. In: *Constituents and Properties of Soils* (Eds M. Bonneau and B. Souchier), pp. 82–108. Academic Press, London.

Leone, A. (1998) Change or no change? Revised perceptions of urban transformation in late Antiquity. In: *TRAC 98 (Proceedings of the Eighth Annual Theoretical Roman Archaeology Conference, Leicester 1998)*, pp. 121–130, Oxford.

Leroi-Gourhan, A. (1961) Les fouilles d'Arcy-sur-Cure (Yonne). *Gallia Préhistoire*, **4**: 3–16.

Leroi-Gourhan, A. (1984) *Pincevent: campement magdalénien de chasseurs de rennes*. Guides archaéologiques de la France, 3. Ministère de la culture Direction du Patrimoine Sous-direction de l'archéologie: Impr. nationale, Paris.

Leroi-Gourhan, A. (1987) The archaeology of Lascaux Cave.

Leroi-Gourhan, A. (1988) Le passage Moustérien-Châtelperronnien à Arcy-sur-Cure. *Bulletin de la Société Préhistorique Française*, **85**: 102–104.

Leroi-Gourhan, A. and Allain, J. (1979) *Lascaux inconnu*. XIIe Supplément à Gallia Préhistoire.

Leroi-Gourhan, A. and Brézillon, M.N. (1972) *Fouilles de Pincevent. Essai d'analyse éthnographique d'un habitat magdalénien (la section 36)*. Supplément à "Gallia Préhistoire"; 7. Centre National de la Recherche Scientifique, Paris, 331 pp.

Levy, G. and Toutain, F. (1982) Aeration and redox phenomena in soils. In: *Constituents and Properties of Soils* (Eds M. Bonneau and B. Souchier), pp. 355–366. Academic Press, London.

Levy, T.E. (1995) Cult, metallurgy and rank societies – Chalcolithic Period (ca 4500–3500 BCE). In: *The Archaeology of Society in the Holy Land* (Ed T.E. Levy), pp. 226–244. Leicester University Press, London.

Lewis, H. (1998) *The Characterisation and Interpretation of Ancient Tillage Practices Through Soil Micromorphology: A Methodological Study*. Doctoral Dissertation, Cambridge.

Lewis, J.S., Wiltshire, P. and Macphail, R.I. (1992) A Late Devensian/Early Flandrian site at Three Ways Wharf, Uxbridge: environmental implications. In: *Alluvial Archaeology in Britain* (Eds S. Needham and M.G. Macklin), Monograph 27, pp. 235–248. Oxbow, Oxford.

Lillesand, T.M. and Kiefer, R.W. (1994) *Remote Sensing and Image Interpretation, 3rd edition*. Wiley & Sons, New York.

Limbrey, S. (1975) *Soil Science and Archaeology*. Academic Press, London.

Limbrey, S. (1990) Edaphic opportunism? A discussion of soil factors in relation to the beginnings of plant husbandry in south-west Asia. In: *Soils and Early Agriculture* (Ed K. Thomas), World Archaeology 22, Vol. 1, pp. 45–52. Routledge, London.

Limbrey, S. (1992) Micromorphological studies of buried soils and alluvial deposits in a Wiltshire river valley. In: *Alluvial Archaeology in Britain* (Eds S. Needham and M.G. Macklin), Oxbow Monograph 27, pp. 53–64. Oxbow, Oxford.

Linderholm, J. (2003) Soil chemical surveying: a path to a different understanding of prehistoric sites and societies in Northern Sweden. In: *Second International Conference on Soils and Archaeology* (Ed G. Boschian), pp. 114–119. Pisa, 12th–15th May, 2003. Extended Abstracts. Dipartimento di Scienze Archeologiche, Università di Pisa, Pisa.

Linderholm, J. and Lundberg, E. (1994) Chemical characterisation of various archaeological soil samples using main and trace elements determined by inductively coupled plasma atomic emission spectrometry. *Journal of Archaeological Science*, **21**: 303–314.

Littmann, T. (1991) Recent African dust deposition in West Germany – sediment characteristics and climatological aspects. *CATENA Supplement*, **20**: 57–73.

Liu, Z. (1985) Sequence of sediments at Locality 1 in Zhoukoudian and correlation with loess stratigraphy in Northern China and with the chronology of deep-sea cores. *Quaternary Research (New York)*, **23**: 139–153.

Liu, Z. (1987) Development and filling of caves in Dragon Bone Hill at Zhoukoudian, Beijing. In: *International Geomorphology 1986; Proceedings of the First International Conference on Geomorphology; Part II* (Ed V. Gardiner), pp. 1125–1141. John Wiley & Sons, Chichester, UK.

Liversage, D., Munro M.A.R, Courty M.A and Nørnberg, P. (1987) Studies of a buried Iron Age field. *Acta Archaeologica*, **1**: 55–84.

Livingstone, I. and Warren, A. (1996) *Aeolian Geomorphology; An Introduction*. Addison Wesley Longman, Harlow, UK.

Longworth, G., Becker, L.W., Thompson, R., Oldfield, F., Dearing, J.A. and Rummery, T.A. (1979) Mössbauer and magnetic studies of secondary iron oxides in soils. *Journal of Soil Science*, **30**: 93–110.

Lucas, G. (2001) *Critical Approaches to Fieldwork: Contemporary and Historical Archaeological Practice*. Routledge, London.

Lutz, H.J. and Griswold, F.S. (1939) The influence of tree roots on soil morphology. *American Journal Science*, **237**: 389–400.

Mackney, D. (1961) A podzol development sequence in oakwoods and heath in central England. *Journal of Soil Science*, **12**: 23–40.

Macnab, J.W. (1965) British strip lynchets. *Antiquity*, **39**: 279–290.

Macphail, R.I. (1980) Report on a soil in a Romano–British context at Lloyds Merchant Bank, London (LLO78) London. *Ancient Monuments Laboratory Report* 3045, English Heritage, London.

Macphail, R.I. (1981) Soil and botanical studies of the "Dark Earth." In: *The Environment of Man: the Iron Age to the Anglo-Saxon Period* (Eds M. Jones and G.W. Dimbleby), British Series 87, pp. 309–331. British Archaeological Reports, Oxford.

Macphail, R.I. (1986) Paleosols in archaeology: their role in understanding Flandrian pedogenesis. In: *Paleosols: Their Recognition and Interpretation* (Ed V.P. Wright), pp. 263–290. Blackwell Scientific Publications, Oxford.

Macphail, R.I. (1987) A review of soil science in archaeology in England. In: *Environmental Archaeology: A Regional Review Vol. II* (Eds H.C.M. Keeley), Occasional paper No. 1, pp. 332–379. Historic Buildings & Monuments Commission for England, London.

Macphail, R.I. (1990a) Soil history and micromorphology. In: *Brean Down Excavations 1983–1987* (Ed M. Bell), Archaeological Report No. 15, pp. 187–196. English Heritage, London.

Macphail, R.I. (1990b) The soils. *In: Hazleton North, Gloucestershire, 1979–82: The Excavation of a Neolithic Long Cairn of the Cotswold-Severn Group* (Ed A. Saville), Archaeological Report No. 13, pp. 223–226. English Heritage, London.

Macphail, R.I. (1990c) Soil Report on Carn Brea, Redruth, Cornwall, with Some Reference to Similar Sites in Brittany, France. 55/90, Ancient Monuments Laboratory. English Heritage, London.

Macphail, R.I. (1991) The archaeological soils and sediments, (Eds Sharples, N.M.), Maiden Castle: Excavations and field survey 1985–1986, Archaeological Report no 19: Vol. pp. 106–118, London, English Heritage.

Macphail, R.I. (1992a) Soil micromorphological evidence of ancient soil erosion. In: *Past and Present Soil Erosion* (Eds M. Bell and J. Boardmand), Monograph 22, pp. 197–216. Oxbow, Oxford.

Macphail, R.I. (1992b) Late Devensian and Holocene soil formation. In: *Hengistbury Head, Dorset. Volume 2: The Late Upper Palaeolithic & Early Mesolithic Sites* (Ed R.N.E. Barton), Monograph No. 34, pp. 44–51. Oxford University Committee for Archaeology, Oxford.

Macphail, R.I. (1994a) Soil micromorphological investigations in archaeology, with special reference to drowned coastal sites in Essex. In: *SEESOIL* (Eds H.F. Cook and D.T. Favis-Mortlock), Vol. 10, pp. 13–28. South East Soils Discussion Group, Wye.

Macphail, R.I. (1994b) The reworking of urban stratigraphy by human and natural processes. In: *Urban–Rural Connexions: Perspectives from Environmental Archaeology* (Eds A.R. Hall and H.K. Kenward), Monograph 47, pp. 13–43. Oxbow, Oxford.

Macphail, R.I. (1995) Report on the soils at Westhampnett bypass, West Susses: with special reference to the micromorphology of the late-glacial soil and Marl at Area 3, Wessex Archaeology, Southampton.

Macphail, R.I. (1998) A reply to Carter and Davidson's "An evaluation of the contribution of soil micromorphology to the study of ancient arable agriculture." *Geoarchaeology*, **13**: 549–564.

Macphail, R.I. (1999) Sediment micromorphology. In: *Boxgrove, A Middle Pleistocene Hominid Site at Eartham Quarry, Boxgrove, West Sussex* (Eds M.B. Roberts and S.A. Parfitt),

Archaeological Reports No. 17, pp. 118–148. English Heritage, London.

Macphail, R.I. (2000) Soils and microstratigraphy: a soil micromorphological and micro-chemical approach. In: *Potterne 1982–5: Animal Husbandry in Later Prehistoric Wiltshire* (Ed A.J. Lawson), Archaeology Report No. 17, pp. 47–70. Wessex Archaeology, Salisbury.

Macphail, R. I. (2002) Pevensey Castle: soil micromorphology and chemistry of the Roman deposits and 'dark earth'. University of Reading, Reading.

Macphail, R.I. (2003a) Attaining robust interpretations of archaeological "soils": examples from rural and urban contexts. In: *Second International Conference on Soils and Archaeology, Pisa, 12th–15th May, 2003. Extended Abstracts* (Ed G. Boschian), pp. 60–63. Dipartimento di Scienze Archeologiche, Università di Pisa, Pisa.

Macphail, R.I. (2003b) Industrial activities – some suggested microstratigraphic signatures: ochre, building materials and iron-working. In: *The Environmental Archaeology of Industry* (Eds P.E.J. Wiltshire and P. Murphy), AEA Symposia No. 20, pp. 94–106. Oxbow, Oxford.

Macphail, R.I. (2003c) Soil microstratigraphy: a micromorphological and chemical approach. In: *Urban Development in North-West Roman Southwark Excavations 1974–90* (Ed C. Cowan), Monograph 16, pp. 89–105. MoLAS, London.

Macphail, R.I. (2004) *Land at Underdown Lane, Eddington, Herne Bay, Kent: Soil Micromorphology*, Wessex Archaeology, Salisbury.

Macphail, R.I. (forthcoming) Soil report on the Raunds Area Project: results from the prehistoric period. In: *Raunds Area Project.* (Eds F. Healy and J. Harding), The Neolithic and

Macphail, R.I. (in press) "Dark earth": recent studies of "dark earth" and "dark earth-like" microstratigraphy in England. In: *Dark Earth in the Dark Ages* (Ed L. Verslype). Université Catholique de Louvain, Louvain.

Macphail, R.I. and Courty, M.A. (1985) Interpretation and significance of urban deposits. In: *Proceedings of the Third Nordic Conference on the Application of Scientific Methods in Archaeology* (Eds T. Edgren and H. Jungner), pp. 71–83. The Finnish Antiquarian Society, Helsinki.

Macphail, R.I. and Crowther, J. (2002a) *Battlesbury, Hampshire: Soil Micromorphology and Chemistry (W4896)*, Wessex Archaeology, Salisbury.

Macphail, R.I. and Crowther, J. (2002b) *Canterbury CW12 (Cycling Centre – Roman Rampart Site): Soil Micromorphology and Chemistry,* Canterbury Archaeological Trust, Canterbury.

Macphail, R.I. and Crowther, J. (2003) *Büraberg, North Hessen: Soil Micromorphology, Chemistry and Magnetic Properties of a Core Sample,* Johann Wolfgang Goethe-Universität, Frankfurt am Main.

Macphail, R.I. and Crowther, J. (2004a) Tower of London Moat: sediment micromorphology, particle size, chemistry and magnetic properties. In: *Tower of London Moat Excavation* (Ed G. Keevil), Historic Royal Palaces Monograph 1, pp. 41–43, 48–50, 78–79, 82–83, 155, 183–186, 202–204 and 271–284. Oxford Archaeology, Oxford.

Macphail, R.I. and Crowther, J. (2004b) *White Horse Stone: Soil Micromorphology, Phosphate and Magnetic Susceptibility*. Oxford Archaeology, Oxford.

Macphail, R.I. and Crowther, J. (in press) Micromorphology and Post-Roman town research: the examples of London and Magdeburg. In: Post-Roman Towns and Trade in Europe, Byzantium and the Near-East. New methods of structural, comparative and scientific methods in archaeology (Ed J. Henning). Walter De Gruyter Gmbh & Co. KG, Berlin.

Macphail, R.I. and Cruise, G.M. (1996) Soil micromorphology. In: *A Moated Site in Tempsford Park, Tempsford* (Ed D. Shotliff). *Bedfordshire Archaeology,* **22**: 123–124.

Macphail, R.I. and Cruise, G.M. (1997) 7–11, Bishopsgate and Colchester House (PEP89), London: preliminary report on soil microstratigraphy and chemistry, MoLAS, London.

Macphail, R.I. and Cruise, G.M. (2000a) Rescuing our urban archaeological soil heritage: a multidisciplinary microstratigraphical approach. In: *Proceedings of the First International Conference on Soils of Urban, Industrial, Traffic and Mining Areas* (Eds W. Burghardt and C. Dornauf), Vol. 1, pp. 9–14. IUSS/IBU, Essen.

Macphail, R.I. and Cruise, G.M. (2000b) Soil micromorphology on the Mesolithic site. In: *Prehisotric Intertidal Archaeolegy in the Welsh Severn Estuary* (Eds M. Bell, A. Caseldine and H. Neumann), Council for British Archaeology, York, pp. 55–57 and CD-ROM.

Macphail, R.I. and Cruise, G.M. (2001) The soil micromorphologist as team player: a multianalytical approach to the study of European microstratigraphy. In: *Earth Science and Archaeology* (Eds P. Goldberg, V. Holliday and R. Ferring), pp. 241–267. Kluwer Academic/Plenum Publishers, New York.

Macphail, R. I., Cruise, G. M., Mellalieu, S. J. (1999a) Soil micromorphological and diatom analysis of thin sections and microchemical analysis of polished blocks. In: *The Excavation of a Ceremonial Site at Folly Lane, Verulamium* (Ed R. Niblett). pp. 365–384, Monograph No. 14. London, Britannia.

Macphail, R.I. and Goldberg, P. (1990) The micromorphology of tree subsoil hollows: their significance to soil science and archaeology. In: *Soil-Micromorphology: A Basic and Applied Science* (Ed L.A. Douglas). Developments in Soil Science 19, pp. 425–429. Elsevier, Amsterdam.

Macphail, R.I. and Goldberg, P. (1995) Recent advances in micromorphological interpretations of soils and sediments from archaeological sites. In: *Archaeological Sediments and Soils: Analysis, Interpretation and Management* (Eds A.J. Barham and R.I. Macphail), pp. 1–24e. Institute of Archaeology, London.

Macphail, R.I. and Goldberg, P. (1999) The soil micromorphological investigation of Westbury Cave. In: *Westbury Cave. The Natural History Museum Excavations 1976–1984* (Eds P. Andrews, J. Cook, A. Currant and C. Stringer), pp. 59–86. CHERUB (Centre for Human Evolutionary Research at the University of Bristol), Bristol.

Macphail, R.I. and Goldberg, P. (2000) Geoarchaeological investigations of sediments from Gorham's and Vanguard Caves, Gibraltar: microstratigraphical (soil micromorphological and chemical) signatures. In: *Neanderthals on the Edge* (Eds C.B. Stringer, R.N.E. Barton and C. Finlayson), pp. 183–200. Oxbow, Oxford.

Macphail, R.I. and Goldberg, P. (2003) Gough's Cave, Cheddar, Somerset: microstratigraphy of the Late Pleistocene/earliest Holocene–sediments. *Bulletin Natural History Museum, London (Geol.)*, **58:** 51–58.

Macphail, R.I. and Linderholm, J. (2004) Neolithic land use in south-east England: a brief review of the soil evidence. In: *Towards a New Stone Age* (Eds J. Cotton and D. Field), Research Report 137, pp. 29–37. CBA, York.

Macphail, R.I. and Linderholm, J. (in press (a). Interpreting fills of grubenhaüser: examples from England and Sweden. In: *Proceedings from the VII Conference on the Applications on Scientific Methods in Archaeology* (Eds R. Engelmark, and J. Linderholm). Department of Archaeology and Sami Studies, University of Umeå, Umeå.

Macphail, R.I. and Linderholm, J. (in press (b)) No. 1, Poultry (Roman): soil micromorphology. In: *Excavations at 1 Poultry: the Roman Sequences* (Eds J. Hill and P. Rowsome), Monograph Series. Museum of London Archaeological Service, London.

Macphail, R.I., Romans, J.C.C. and Robertson, L. (1987) The application of micromorphology to the understanding of Holocene soil development in the British Isles; with special reference to cultivation. In: *Soil Micromorphology* (Eds N. Fedoroff, L.M. Bresson and M.A. Courty), pp. 669–676. Association Française pour l'Étude du Sol, Plaisir.

Macphail, R.I., Courty, M.A. and Gebhardt, A. (1990a) Soil micromorphological evidence of early agriculture in north-west Europe. *World Archaeology*, **22**: 53–69.

Macphail, R.I., Courty, M.A. and Goldberg, P. (1990b) Soil micromorphology in archaeology. *Endeavour, New Series*, **14**: 163–171.

Macphail, R.I., Hather, J., Hillson, S.W. and Maggi, R. (1994) The Upper Pleistocene deposits at Arene Candide: soil micromorphology of some samples from the Cardini 1940–42 excavations. *Quaternaria Nova*, **IV**: 79–100.

Macphail, R.I., Crowther, J. and Cruise, G.M. (1995) The soils. In: *The King's Privy Garden at Hampton Court Palace 1689–1995* (Ed S. Thurley), pp. 116–118. Apollo, London.

Macphail, R.I., Courty, M.A., Hather, J. and Wattez, J. (1997) The soil micromorphological evidence of domestic occupation and stabling activities. In: *Arene Candide: A Functional and Environmental Assessment of the Holocene Sequence (Excavations Bernabò Brea-Cardini 1940–50)* (Ed R. Maggi), pp. 53–88. Memorie dell'Istituto Italiano di Paleontologia Umana, Roma.

Macphail, R.I., Cruise, G.M. and Linderholm, J. (1998b) Report on the soil micromorphology and chemistry at Balksbury Camp, Hampshire (1996–97), Wessex Archaeology, Salisbury.

Macphail, R.I., Cruise, G.M., Mellalieu, S.J. and Niblett, R. (1998c) Micromorphological interpretation of a "Turf-filled" funerary shaft at St. Albans, United Kingdom. *Geoarchaeology, 13*: 617–644.

Macphail, R.I., Crowther, J. and Cruise, G.M. (1999b) King Arthur's Cave: soils of the Allerød palaeosol, Oxford University, Oxford.

Macphail, R.I., Cruise, G.M., Engelmark, R. and Linderholm, J. (2000) Integrating soil micromorphology and rapid chemical survey methods: new developments in reconstructing past rural settlement and landscape organization. In: *Interpreting Stratigraphy* (Ed S. Roskams), Vol. 9, pp. 71–80. University of York, York.

Macphail, R.I., Acott, T.G. and Crowther, J. (2001) *Boxgrove: Sediment Microstratigraphy (Soil Micromorphology, Image Analysis and Chemistry)*, Institute of Archaeology, London.

Macphail, R.I., Crowther, J. and Cruise, G.M. (in press) Microstratigraphy: soil micromorphology, chemistry and pollen. In: *The London Guildhall: the archaeology and history of the Guildhall precinct from the medieval period to the 20th century* (Eds D. Bowsher, N. Holder, I. Howell, and T. Dyson) Museum of London Archaeological Service, London.

Macphail, R.I., Crowther, J., Acott, T.G., Bell, M.G. and Cruise, G.M. (2003b) The Experimental Earthwork at Wareham, Dorset after 33 years: changes to the buried LFH and Ah horizon. *Journal of Archaeological Science*, **30**: 77–93.

Macphail, R.I., Galinié, H. and Verhaeghe, F. (2003c) A future for dark earth? *Antiquity*, **77**: 349–358.

Macphail, R.I., Cruise, G.M., Allen, M.J., Linderholm, J. and Reynolds, P. (2004) Archaeological soil and pollen analysis of experimental floor deposits; with special reference to Butser Ancient Farm, Hampshire, UK. *Journal of Archaeological Science*, **31**: 175–191.

Macphail, R.I., Cruise, G.M., Gebhardt, A. and Linderholm, J. (forthcoming) West Heslerton: soil micromorphology and chemistry of the Roman and Saxon deposits. In: *West Heslerton Anglo-Saxon Settlment* (Ed J. Tipper). Landscape Research Centre, Yedingham.

Macphail, R.I., Cruise, G.M., and Mellalieu, S.J. (1999) Soil micromorphological and diatom analysis of thin sections and microchemical analysis of polished blocks, in Niblett, R., ed., The Excavation of a Ceremonial site at Folly Lane, Verulamium, Volume Britannia Monograph No. 14: London, Britannia, pp. 365–384.

Macumber, P.G., Head, M.J., Chivas, A.R. and De Deckker, P. (1991) Implications of the Wadi al-Hammeh sequences for the terminal drying of Lake Lisan, Jordan. *Palaeogeography, Palaeoclimatology, Palaeoecology*, **84**: 163–173.

Madella, M., Jones, M.K., Goldberg, P., Goren, Y. and Hovers, E. (2002) The exploitation of plant resources by Neanderthals in Amud Cave (Israel): the evidence from phytolith studies. *Journal of Archaeological Science*, **29**: 703–719.

Magaritz, M., Kaufman, A. and Yaalon, D.H. (1981) Calcium carbonate nodules in soils: 18O/16O and 13C/12C Ratios and 14C Contents. *Geoderma*, **25**: 157–172.

Magaritz, M., Goodfriend, G.A., Berger, W.H. and Labeyrie, L.D. (1987) Movement of the desert boundary in the Levant from latest Pleistocene to early Holocene. *Abrupt Climatic Changes; Evidence and Implications*, **216**: 173–183.

Maggi, R. (Ed) (1990) Archeologia Dell'Appennino Ligure. Gli scavi del Castellaro di Uscio: *un* insediamento di crinale occupato dal Neolitico alla conquista Romana, Istituto Internazionale Di Studi Liguri, Collezinoe Monographie preistoriche ed Archaeologiche, VIII. Bordighera, Chiavari.

Maggi, R. (Ed) (1997) *Arene Candide: a Functional and Environmental Assessment of the Holocene Sequence (Excavations Bernabò Brea-Cardini 1940–50)*. Memorie dell'Istituto Italiano di Paleontologia Umana, Roma.

Maggi, R. and Nisbet, R. (2000) Alberi da foraggio scalvatura Neolitica: Nuovi data dalle Arene Candide (Fodder trees and Neolithic shredding: new data from Arene Candide). In: *La Neolitizzazione Tra Oriente e Occidente* (Eds A. Pessina and G. Musicio), pp. 289–308. Comune di Udine; Museo Fruilano di Storia Naturale, Udine.

Maher, L.A. (2004) *The Epipalaeolithic in Context: Palaeolandscapes and Prehistoric Occupation of Wadi Ziqlab, Northern Jordan*. Doctoral Thesis, University of Toronto, Toronto.

Malinowski, R. (1979) Concretes and mortars in ancient aqueducts. *Concrete International*, **1**: 66–76.

Mallol, C. (2004) *Micromorphological Observations from the Archaeological Sediments of 'Ubeidiya (Israel), Dmanisi (Georgia) and Gran Dolina-Td10 (Spain) for the Reconstruction\of Hominid Occupation Contexts*. Doctoral Dissertation, Harvard University, Cambridge, 277 pp.

Mandel, R. (1995) Geomorphic controls of the Archaic record in the Central Plains of the United States. In: *Archaeological Geology of The Archaic Period in North America* (Ed E.A. Bettis, III), *Special Paper 297*, pp. 37–66. The Geological Society of America, Boulder, CO.

Mandel, R. and Bettis, E.A., III (2001) Use and analysis of soils by archaeologists and geoscientists; a North American perspective. In: *Earth Sciences and Archaeology* (Eds P. Goldberg, V.T. Holliday and C.R. Ferring), pp. 173–204. Kluwer Academic/Plenum Publishers, New York.

Mandel, R.D. (1992) Soils and Holocene landscape evolution in Central and Southwestern Kansas: implications for archaeological research. In: *Soils in Archaeology: Landscape Evolution and Human Occupation* (Ed V.T. Holliday), pp. 41–100. Smithsonian Institution Press, Washington.

Mandel, R.D. (2000a) The past, present, and future. In: *Geoarchaeology in the Great Plains* (Ed R.D. Mandel), pp. 286–295. University of Oklahoma Press, Norman.

Mandel, R.D. (2000b) *Geoarchaeology in the Great Plains*. University of Oklahoma Press, Norman, OK.

Mandel, R.D., Arpin, T. and Goldberg, P. (2003) *Stratigraphy, Lithology, and Pedology of the South Wall at the Hopeton Earthworks, South-Central Ohio*. Kansas Geological Survey Open File Report 2003-4649.

Mania, D. (1995) The earliest occupation of Europe: the Elbe-Saale region (Germany). In: *The Earliest Occupation of Europe* (Eds W. Roebroeks and T. van Kolfschoten), pp. 85–102. University of Leiden and European Science Foundation, Leiden.

Marcolongo, B. and Mantovani, F. (1997) *Photogeology: Remote Sensing Applications in Earth Science*. Science Publishers Inc., Enfield, NH.

Marean, C.W., Goldberg, P., Avery, G., Grine, F.E. and Klein, R.G. (2000) Middle Stone Age stratigraphy and excavations at Die Kelders Cave 1 (Western Cape Province, South Africa): the 1992, 1993, and 1995 field seasons. *Journal of Human Evolution*, **38**: 7–42.

Marean, C.W., Nillsen, P.J., Brown, K., Jerardino, A. and Stynder, D. (2004) Paleoanthropological investigations of Middle Stone Age sites at Pinnacle Point, Mossel Bay (South Africa): archaeology and hominid remains from the 2000 Field Season. *PaleoAnthropology*, **2004.05.02**: 14–83.

Marks, A.E. (1983) The sites of Boker Tachtit and Boker: a brief introduction. In: *Prehistory and Paleoenvironments in the Central Negev, Israel* (Ed A.E. Marks), Vol. III, pp. 15–37. Department of Anthropology, Institute for the Study of Earth and Man, Southern Methodist University, Dallas.

Marzo, M. and Puigdefábregas, C. (Eds) (1993) *Alluvial Sedimentation, Special Publication Number 17,*

International Association of Sedimentologists. Blackwell Scientific, Oxford.

Mason, J.A. and Kuzila, M.S. (2000) Episodic Holocene loess deposition in central Nebraska. *Quaternary International,* **67**, 119–131.

Matsatuni, A. (1973) Microscopic study on the "amorphous silica" in sediments from the Douara Cave. Bulletin. *The University of Tokyo Museum,* **5**: 127–131.

Matsui, A., Hiraya, R., Mijaji, A. and Macphail, R.I. (1996) Availability of soil micromorphology in archaeology in Japan (in Japanese). pp. 149–152, Archaeology Society of Japan.

Matthews, W. (1995) Micromorphological characterisation and interpretation of occupation deposits and microstratigraphic sequences at Abu Salabikh, Iraq. In: *Archaeological Sediments and Soils, Analysis, Interpretation and Management* (Eds T. Barham and R.I. Macphail), pp. 41–74. Archetype Books, London.

Matthews, W. and Postgate, J.N. (1994) The imprint of living in a Mesopotamian City: questions and answers. In: *Whither Environmental Archaeology* (Eds R. Luff and P. Rowley Conwy), Monograph 38, pp. 171–212. Oxbow Books, Oxford.

Matthews, W., French, C.A.I., Lawrence, T., Cutler, D.F. and Jones, M.K. (1996) Multiple surfaces: the micromorphology. In: *On the Surface: Çatalhöyük 1993–1995* (Ed I. Hodder), pp. 301–342. The McDonald Institute for Research and British Institute of Archaeology at Ankara, Cambridge.

Matthews, W., French, C.A.I., Lawrence, T., Cutler, D.F. and Jones, M.K. (1997) Microstratigraphic traces of site formation processes and human activities. *World Archaeology,* **29**: 281–308.

Matthews, W., Hastorf, C.A. and Begums, E. (2000) Ethnoarchaeology: studies in local villages aimed at understanding aspects of the Neolithic site. In: *Towards Reflexive Method in Archaeology: the Example at Çatalhöyük* (Ed I. Hodder), pp. 177–188. McDonald Institute for Archaeological Research and British Institute of Archaeology at Ankara, Cambridge.

Mazar, A. (1990) Archaeology of the land of the Bible. Doubleday, New York, 576 pp.

McBratney, A.B., Moran, C.J., Stewart, J.B., Cattle, S.R. and Koppi, A.J. (1992) Modifications to a method of rapid assessment of soil macropore structure by image analysis. *Geoderma,* **53**: 255–274.

McBratney, A.B., Bishop, T.F.A. and Teliatnikov, I.S. (2000) Two soil profile reconstruction techniques. *Geoderma,* **97**: 209–221.

McCann, J.M., Woods, W.I. and Meyer, D.W. (2001) Organic matter and anthrosols in Amazonia: interpreting the Amerindian legacy. In: *Sustainable Management of Soil Organic Matter* (Eds R.M. Rees, B.C. Ball, C.D. Campbell and C.A. Watson), pp. 180–189. CAB International, Wallingford.

McClellan, M.C., Goldberg, P., Mallol, C. and Larson, B.J. (2003) *An Archaeological Investigation on Naval Station Rota. Technical Addendum: Geomorphological Analysis.* The Environmental Company, Inc., Newport News.

McClure, H. (1976) Radiocarbon chronology of late-Quaternary lakes in the Arabian desert. *Nature (London),* **263**: 755–756.

McIntyre, D.S. (1958a) Permeability measurements of soil crusts formed by raindrop impact. *Soil Science,* **85**: 185–189.

McIntyre, D.S. (1958b) Soil splash and the formation of surface crusts by raindrop impact. *Soil Science,* **85**: 261–266.

McKeague, A., Macdougall, J.I. and Miles, N.M. (1973) Micromorphological, physical and mineralogical properties of a catena of soils from Prince Edward Island, in relation to their classification and genesis. *Canadian Journal of Soil Science,* **53**: 281–295.

McKeague, J.A. (1983) Clay skins and argillic horizons. In: *Soil Micromorphology* (Eds P. Bullock and C.P. Murphy), Volume 2: Soil Genesis, pp. 367–388. AB Academic Publishers, Berkhamsted.

McKee, E.D. (1979) Introduction to a study of global sand seas. In: *A Study of Global Sand Seas; U. S. Geological Survey Professional Paper* (Ed E.D. McKee), Vol. 1052, pp. 1–19.

Meeks, S.C. (1994) Lithic artifacts from Dust Cave. *Journal of Alabama Archaeology,* **40**: 79–106.

Meignen, L., Bar-Yosef, O. and Goldberg, P. (1989) Les structures de combustion moustériennes de la grotte de Kébara (Mont Carmel, Israël). *Nature et Fonction des Foyers Préhistoriques, Mémoires du Musée de Préhistoire d'Ille de France,* **2**: 141–146.

Meignen, L., Bar-Yosef, O., Goldberg, P. and Weiner, S. (2001) Le feu au Paléolithique moyen: Recherches sur les structures de combustion et le statut des foyers. L'exemple du Proche-Orient. *Paléorient,* **26**: 9–22.

Mekel, J.F.M. (1978) The use of aerial photographs and other images in geological mapping. ITC textbook; Chapter 8.1. *International Institute for Aerial Survey and Earth Sciences* (ITC), Enschede, The Netherlands.

Mercer, R. and Midgley, M.S. (1997) The Early Bronze Age Cairn at Sketewan, Balnaguard, Perthshire. *Proceedings of the Society of Antiquaries of Scotland,* **127**: 281–338.

Mercier, N., Valladas, H., Valladas, G. and Reyss, J.-L. (1995) TL dates of burnt flints from Jelinek's excavations at Tabun and their implications. *Journal of Archaeological Science,* **22**: 495–509.

Mercier, N., Valladas, H., Froget, L., Joron, J.L. and Ronen, A. (2000) Datation par thermoluminescence de la base du gisement paléolithique de Tabun (Mont Carmel, Israël); thermoluminescence dating of the lowest Paleolithic of Tabun, Mount Carmel, Israel. *Comptes Rendus de l'Académie des Sciences, Série II. Sciences de la Terre et des Planètes*, **330**: 731–738.

Metcalfe, D. and Heath, K.M. (1990) Microrefuse and site structure: the hearths and floors of the Heartbreak Hotel. *American Antiquity*, **55**: 781–813.

Miall, A.D. (1996) *The Geology of Fluvial Deposits*. Springer Verlag, New York.

Middleton, W.D. and Douglas-Price, T. (1996) Identification of activity areas by multi-element characterization of sediments from modern and archaeological house floors using inductively coupled plasma-atomic emission spectroscopy. *Journal of Archaeological Science*, **23**: 673–687.

Miedema, R. and Mermut, A.R. (1990) *Soil Micromorphology: an Annotated Bibliography 1968–1986*. C.A.B. International, Wallingford.

Miedema, R., Jongmans, A.G. and Slager, S. (1974) Micromorphological observations on pyrite and its oxidation products in four Holocene alluvial soils in the Netherlands. In: *Soil Microscopy* (Ed G.K. Rutherford), pp. 772–794. The Limestone Press, Kingston, Ontario.

Mikkelsen, J.H. and Langohr, R. (1996) A pedological characterisation of the Aubechies soil, a well preserved soil sequence dated to the earliest Neolithic agriculture in Belgium. In: *Paleoecology; Colloquium 3 of XIII International Congress of Prehistoric and Protohistoric Sciences* (Eds L. Castelletti and M. Cremaschi), pp. 143–149. ABACO, Forli.

Miksa, E.J. and Heidke, J.M. (2001) It all comes out in the wash: actualistic petrofacies modeling of temper provenance, Tonto Basin, Arizona, USA. *Geoarchaeology*, **16**: 177–222.

Miyaji, A. (forthcoming) Analysis of cultivated fields and their soils in Japanese archaeology: soil micromorphology. In: *Proceedings of Second International Conference on Soils and Archaeology* (Ed G. Boschian), Pisa, 12th–15th May 2003, Pisa.

Montufo, A.M. (1997) The use of satellite imagery and digital image processing in landscape archaeology. A case study from the Island of Mallorca, Spain. *Geoarchaeology*, **12**: 71–85.

Moore, D. (2005) The Roman Pantheon: the triumph of concrete. http://romanconcrete.com/.

Mora, P., Mora, L. and Philipott, P. (1984) *Conservation of Wall Paintings*. Butterworths, London.

Morisawa, M. (1985) *Rivers*. Longman, New York.

Morris, E.H. (1944) Adobe bricks in a pre-Spanish wall near Aztec, New Mexico. *American Antiquity*, **9**: 434–438.

Mücher, H.J. (1974) Micromorphology of slope deposits: the necessity of a classification. In: *Soil Microscopy* (Ed G.K. Rutherford), pp. 553–556. The Limestone Press, Kingston, Ontario.

Mücher, H.J. (1997) The response of a soil ecosystem to mowing and sod removal: a micromorphological study in The Netherlands. In: *Soil Micromorphology: Studies on Soil Diversity, Diagnostics and Dynamics* (Eds S. Shoba, M. Gerasimova and R. Miedema), pp. 271–281. International Soil Science Society, Moscow–Wageningen.

Mücher, H.J. and de Ploey, J. (1977) Sedimentary structures formed in eolian-deposited silt loams under simulated conditions on dry, moist and wet surfaces. In: *Soil Micromorphology: A Basic and Applied Science* (Ed L.A. Douglas), pp. 155–160. Elsevier, Amsterdam.

Mücher, H.J. and Vreeken, W.J. (1981) (Re)deposition of loess in southern Limbourg, The Netherlands. II. Micromorphology of the lower silt loam complex and comparison with deposits produced under laboratory conditions. *Earth Surface Processes and Landforms*, **6**: 355–363.

Mücher, H.J., de Ploey, J. and Savat, J. (1981) Response of loess materials to simulated translocation by water: micromorphological observations. *Earth Surface Processes and Landforms*, 6: 331–336.

Mücher, H.J., Slotboom, R.T. and ten Veen, W.J. (1990) Palynology and micromorphology of a man-made soil. A reconstruction of the agricultural history since Late-medieval times of the Posteles in the Netherlands. *Catena*, **17**: 55–67.

Mueller, G. (1967) *Methods in Sedimentary Petrology*. Hafner Publishing Company, Stuttgart.

Murphy, C.P. (1986) *Thin Section Preparation of Soils and Sediments*. A B Academic Publishers, Berkhamsted.

Murphy, C.P., Bullock, P. and Turner, R.H. (1977) The measurement and characterisation of voids in thin sections by image analysis. Part 1. Principles and techniques. *Journal of Soil Science*, **29**: 498–508.

Murphy, P. and Fryer, V. (1999) The plant macrofossils. In: *The Excavation of a Ceremonial site at Folly Lane, Verulamium*. (Ed R. Niblett), *Britannia Monograph* No. 14, pp. 384–388. Society for the Promotion of Roman Studies, London.

NASCN (1983) North American Stratigraphic Code. *American Association of Petroleum Geologists Bulletin*, **67**: 841–875.

Needham, S. and Macklin, M.G. (Eds) (1992) *Alluvial Archaeology in Britain*. Oxbow, Oxford.

Neev, D., Bakler, N. and Emory, K.O. (1987) *Mediterranean Coasts of Israel and Sinai*. Taylor & Francis, New York.

Newcomer, M.H. and Sieveking, G.D.G. (1980) Experimental flake scatter patterns: a new interpretive technique. *Journal of Field Archaeology*, **7**: 345–352.

Newell, R.R. (1980) Mesolithic dwelling structures: fact and fantasy. *Veröffentlichungen des Museums für Ur- und Frühgeschichte Potsdam*, Band 14/15: 235–284.

Niblett, R. (1999) *The Excavation of a Ceremonial Site at* Folly lane, Verulamium. Britannia Monograph Series 14. Society for the Promotion of Roman Studies, London.

Nicholas, P. and Thomas, K. (2004) Color atlas and manual of Microscopy for Criminalists, Chemists, and Conservators. Boca Raton, CRC Press, 310

Nichols, G. (1999) *Sedimentology & Stratigraphy.* Blackwell Science, Oxford.

Nielsen, A.E. (1991) Trampling the archaeological record: an experimental study. *American Antiquity,* **56**: 483–503.

Niemi, T.M. (1997) Fluctuations of Late Pleistocene Lake Lisan in the Dead Sea Rift. In: *The Dead Sea: The Lake and its Setting* (Eds T.M. Niemi, Z. Ben-Avraham and J.R. Gat), pp. 226–235. Oxford University Press, New York.

Nisbet, R. and Biagi, P. (1987) *Balm'Chanto: un Riparo sottoroccia dell'età del rame nelle Alpi Cozie.* New Press, Como.

Nordt, L.C. (1995) Geoarchaeological investigations of Henson Creek; a low-order tributary in central Texas. *Geoarchaeology,* **10**: 205–221.

Nordt, L.C. (2004) Late Quaternary alluvial stratigraphy of a low-order tributary in central Texas, USA and its response to climate and sediment supply. *Quaternary Research,* **62**: 289–300.

Norman, P. and Reader, F.W. (1912) Further discoveries relating to Roman London, 1906–12. *Archaeologia,* **LXIII**: 257–344.

Nunn, P.D. (2000) Environmental catastrophe in the Pacific islands around AD 1300. *Geoarchaeology,* **15**: 715–740.

Oldfield, F., Krawiecki, A., Maher, B., Taylor, J.J. and Twigger, S. (1985) The role of mineral magnetic measurements in archaeology. In: *Palaeoenvironmental Investigations: Research Design, Methods and Data Analysis* (Eds N.R.J. Fieller, D.D. Gilbertson and N.G.A. Ralph), International Series 266, pp. 29–43. British Archaeological Reports, Oxford.

Pape, J.C. (1970) Plaggen soils in the Netherlands. *Geoderma,* **4**: 229–255.

Parker Pearson, M. and Richards, C. (Eds) (1994) *Architecture and Order: Approaches to Social Space.* Routledge, New York.

Parkin, R.A., Rowley-Conwy, P. and Serjeantson, D. (1986) *Late Palaeolithic Exploitation of Horse and Red Deer at Gough's Cave,* Cheddar, Somerset.

Parnell, J.J., Terry, R.E. and Golden, C. (2001) Using in-field phosphate testing to rapidly identify middens at Piedras Negras, Guatemala. *Geoarchaeology,* **16**: 855–873.

Parsons, R.B., Scholtes, W.H. and Riecken, F.F. (1962) Soils of Indian mounds in northeastern Iowa as benchmarks for studies of soil genesis. *Soil Science Society of America Proceedings,* **26**: 32–35.

Peacock, D.P.S. (1969) Neolithic pottery production in Cornwall. *Antiquity,* **43**: 145–149.

Peacock, D.P.S. (1982) *Pottery in the Roman World: An Ethnoarchaeological Approach.* Longmans, London.

Peterken, G.F. (1996) *Natural Woodland. Ecology and Conservation in Northern Regions.* Cambridge University Press, Cambridge.

Pettijohn, F.J. (1975) *Sedimentary Rocks, Third Edition.* Harper & Row Publishers, New York.

Pettijohn, F.J., Potter, E.P. and Siever, R. (1973) *Sand and Sandstone.* Springer-Verlag, New York.

Pettitt, P. (1997) High resolution Neanderthals? Interpreting Middle Palaeolithic intrasite spatial data. *World Archaeology,* **29**: 208–224.

Peyrony, D. (1930) Le Moustier, ses gisements, ses industries, ses couches géologiques. *Revue Anthropologique,* 1–3 and 4–6.

Phillips, J.L. (1988) The Upper Paleolithic of the Wadi Feiran, southern Sinai. *Paléorient,* **14**: 183–200.

Picard, L. and Baida, U. (1966) *Geological Report on the Lower Pleistocene of the 'Ubeidiya Excavations.* Israel Academy of Sciences, Jerusalem.

Pirrie, D., Butcher, A.R., Power, M.R., Gottlieb, P. and Miller, G.L. (2004) Rapid quantitative mineral and phase analysis using automated scanning electron microscopy (QemScan); potential applications in forensic geoscience. In: *Forensic Geoscience: Principles, Techniques and Applications* (Eds K. Pye and D.J. Croft), Geological, London, Special Publications 232, pp. 123–136. The Geological Society of London, London.

Pohl, M.D., Bloom, P.R. and Pope, K.O. (1990) Interpretation of wetland farming in northern Belize: excavations at San Antonio Rio Hondo. In: *Ancient Maya Wetland Agriculture. Excavations on Albion Island, Northern Belize* (Ed M.D. Pohl), pp. 187–254. Westview, Boulder, CO.

Pollard, A.M. and Heron, C. (1996) *Archaeological Chemistry.* The Royal Society of Chemistry, Cambridge, UK.

Polunin, O. and Smythies, B.E. (1973) *Flowers of South-West Europe. A Field Guide.* Oxford University Press, London.

Ponting, M. (2004) The scanning electron microscope and the archaeologist. *Physics Education,* **39**: 166–170.

Pope, K.O. and Dahlin, B.H. (1989) Ancient Maya wetland agriculture: new insights from ecological and remote sensing research. *Journal of Field Archaeology,* **16**: 87–106.

Porat, N., Valladas, H., Bar-Yosef, O., Vandermeersch, B. and Schwarcz, H.P. (1994) Electron spin resonance dating of burned flint from Kebara Cave, Israel. *Geoarchaeology,* **9**: 393–407.

Porter, G. (1997) An early medieval settlement at Guildhall, City of London. In: *Urbanism in Medieval Europe (Papers of the 'Medieval Europe Brugge 1997' Conference Volume 1)*

(Eds G. De Boe and F. Verhaeghe), pp. 147–152. I. A. P. Rapporten 1, Zellick.

Potts, R., Behrensmeyer, A.K. and Ditchfield, P. (1999) Paleolandscape variation and Early Pleistocene hominid activities: Members 1 and 7, Olorgesailie formation, Kenya. *Journal of Human Evolution*, **37**: 747–788.

Preece, R.C. and Bridgland, D.R. (Eds) (1998) *Late Quaternary Environmental Change in North-West Europe*: Excavations at Holywell Coombe, South-East England. Chapman & Hall, London.

Preece, R.C., Kemp, R.A. and Hutchinson, J.N. (1995) A Late-glacial colluvial sequence at Watcombe Bottom, Ventnor, Isle of Wight, England. *Journal of Quaternary Science*, **10**: 107–121.

Price, T.D. and Feinman, G.M. (2005) *Images of the Past, 4th Edition*. McGraw-Hill, New York, 562 pp.

Proudfoot, V.B. (1976) The analysis and interpretation of soil phosphorous in archaeological contexts. In: *Geoarchaeology: Earth Science and the Past* (Eds D.A. Davidson and M.L. Shackley), pp. 93–113. Duckworth, London.

Pye, E. (2000/2001) Wall painting in the Roman empire: colour, design and technology. *Archaeology International*, 24–27.

Pye, K. (1987) *Aeolian Dust and Dust Deposits*. Academic Press, London.

Pye, K. (1995) The nature, origin and accumulation of loess. *Quaternary Science Reviews*, **14**: 653–666.

Pye, K. and Croft, D.J. (Eds) (2004) *Forensic Geoscience: Principles, Techniques and Applications*, Geological, London, Special Publications 232. The Geological Society of London, London.

Qian, F., Zhang, J. and Yin, W. (1982) Magnetostratigraphic study on the cave deposits containing fossil Peking Man at Zhoukoudian. In: *Quaternary Geological Research*, pp. 15.1. Reg. Cent. Quat. Geol. Beijing, China.

Rabenhorst, M.C., Wilding, L.P. and Girdner, C.L. (1984) Airborne dusts in the Edwards Plateau region of Texas. *Soil Science Society of America Journal*, **48**: 621–627.

Rabinovich, R. and Tchernov, E. (1995) Chronological, paleoecological and taphonomical aspects of the Middle Paleolithic site of Qafzeh, Israel. In: *Archaeozoology of the Near East II* (Eds H. Buitenhuis and H.-P. Uerpmann), pp. 5–44. Backhuys Publishers, Leiden.

Raghavan, H., Gaillard, C. and Rajaguru, S.N. (1991) Genesis of calcretes from the calc-pan site of Singi Talav near Didwana, Rajasthan, India; a micromorphological approach. *Geoarchaeology*, **6**: 151–168.

Ranov, V. (1995) The 'Loessic Palaeolithic' in South Tadjikistan, Central Asia: its industries, chronology and correlation. *Quaternary Science Reviews*, **14**: 731–745.

Rapp, A. and Nihlén, T. (1991) Desert dust-storms and loess deposits in North Africa and South Europe. *CATENA Supplement*, **20**: 43–55.

Rapp, G., Jr. (1975) The archaeological field staff: the geologist. *Journal of Field Archaeology*, **2**: 229–237.

Rapp, G., Jr. and Gifford, J. (1985) *Archaeological Geology*. Yale University Press, New Haven.

Rapp, G., Jr. and Hill, C.L. (1998) *Geoarchaeology. The Earth-Science Approach to Archaeological Interpretation*. Yale University Press, New Haven.

Rasmussen, P. (1993) Analysis of goat/sheep faeces from Egolzwil 3, Switzerland: evidence for branch and twig foddering of livestock in the Neolithic. *Journal of Archaeological Science*, **20**: 479–502.

Reading, H.G. (Ed) (1996) Sedimentary Environments, third edition. Blackwell Science, Oxford, 688, pp.

Reineck, H.E. and Singh, I.B. (1980) *Depositional Sedimentary Environments*. Springer-Verlag, Berlin.

Reitz, E.J. and Wing, E.S. (1999) *Zooarchaeology. Cambridge Manuals in Archaeology*. Cambridge University Press, Cambridge.

Renfrew, C. (1976) Archaeology and the earth sciences. In: *Geoarchaeology: Earth Science and the Past* (Eds D.A. Davidson and M.L. Shackley), pp. 1–5. Duckworth, London.

Renfrew, C. and Bahn, P. (2001) *Archaeology: Theories, Methods and Practice, 3rd Edition*. Thames and Hudson, London.

Rentzel, P. (1998) Ausgewähite Grubenstrukturen aus spätlatènezeitlichen Fundstelle Basel-Gasfabrik: Geoarchäologische interpretation der Grubenfüllungen. In: *Jahresbericht der archäologischen Bodenforschung des Kantons Basel-Stadt 1995*, pp. 35–79, Basel.

Rentzel, P. and Narten, G.-B. (2000) Zur Entstehung von Gehniveaus in sandig-lehmigen Ablagerungen – Experimente und archäologische Befunde (Activity surfaces in sandy-loamy deposits – experiments and archaeological examples). In: *Jahresbericht 1999*, pp. 107–127. Archäologische Bodenforschung des Kantons Basel-Stadt, Basel.

Retallack, G.J. (2001) *Soils of the Past, 2nd Edition*. Blackwell Science, Oxford.

Reynolds, P. (1979) *Iron Age Farm. The Butser Experiment*. British Museum Publications Ltd., London.

Reynolds, P. (1981) Deadstock and livestock. In: *Farming Practice in British Prehistory* (Ed R. Mercer), pp. 97–122. Edinburgh University Press, Edinburgh.

Reynolds, P. (1987) *Ancient Farming*. Shire Archaeology. Shire Publications Ltd., Aylesbury.

Reynolds, P. (1994) The life and death of a post-hole. In: *Interpreting Stratigraphy* (Ed E. Shepherd), Vol. 5, pp. 21–25. Norfolk Archaeological Unit, Norwich.

Reynolds, P. and Shaw, C. (2000) *Butser Ancient Farm. The Open Air Laboratory for Archaeology*. Butser Ancient Farm, Waterlooville.

Rice, P.M. (1987) *Pottery Analysis: A Sourcebook*. The University of Chicago Press, Chicago.

Rick, J.W. (1976) Downslope movement and archaeological intrasite spatial analysis. *American Antiquity*, **41**: 133–144.

Riley, D.N. (1987) *Air Photography and Archaeology*. University of Pennsylvania Press, Philadelphia.

Ringberg, B. (1994) The Swedish Varve chronology. *Pact*, **41-I.2**: 25–34.

Ringberg, B. and Erlstrom, M. (1999) Micromorphology and petrography of Late Weichselian glaciolacustrine varves in southeastern Sweden. *CATENA*, **35**: 147–177.

Rink, W.J. (2001) Beyond ^{14}C dating. In: *Earth Sciences and Archaeology* (Eds P. Goldberg, V.T. Holliday and C.R. Ferring), pp. 385–417. Kluwer Academic/Plenum, New York.

Rink, W.J., Bartoll, J., Goldberg, P. and Ronen, A. (2003) ESR dating of archaeologically relevant authigenic terrestrial apatite veins from Tabun Cave, Israel. *Journal of Archaeological Science*, **30**: 1127–1138.

Ritchie, J.C. and Haynes, C.V. (1987) Holocene vegetation zonation in the eastern Sahara. *Nature (London)*, **330**: 645–647.

Ritter, D.F., Kochel, R.C. and Miller, J.R. (2002) *Process Geomorphology, 4th Edition*. McGraw-Hill, New York.

Robb, J. and Van Hove, D. (2003) Gardening, foraging and herding: Neolithic land use and social territories in Southern Italy. *Antiquity*, **77**: 241–254.

Roberts, M.B. and Parfitt, S.A. (1999) *Boxgrove. A Middle Pleistocene Hominid Site at Eartham Quarry, Boxgrove, West Sussex*, Archaeological Report 17. English Heritage, London.

Roberts, M.B., Stringer, C.B. and Parfitt, S.A. (1994) A hominid tibia from Middle Pleistocene sediments at Boxgrove, UK. *Nature*, **369**: 311–313.

Roberts, M.B., Pope, M.I. and Parfitt, S.A. (Eds) (in preparation) Boxgrove: An Early Middle Pleistocene Hominid Site at Eartham Quarry, Boxgrove, West Sussex. *Excavations 1990–1996*. UCL Press, London.

Robinson, D. and Rasmussen, P. (1989) Botanical investigations at the Neolithic lake village at Weier, N. E. Switzerland: leaf hay and cereals as animal fodder. In: *The Beginnings of Agriculture* (Eds A. Milles, D. Williams and N. Gardner), International Series 496, pp. 137–148. British Archaeological Reports, Oxford.

Robinson, M. (1991) The Neolithic and Late Bronze Age insect assemblages. In: *Excavation and Salvage at Runnymede Bridge, 1978* (Ed S. Needham), pp. 277–326. British Museum, London.

Robinson, M. (1992) Environment, archaeology and alluvium on the river gravels of the South Midlands floodplains. In: *Alluvial Archaeology in Britain*, pp. 197–208. Oxbow, Oxford.

Rodwell, J.S. (Ed) (1992) *British Plant Communities. Volume 3. Grasslands and Montane Communities*. Cambridge University Press, Cambridge.

Rogers, M.J., Harris, J.W.K. and Feibel, C.S. (1994) Changing patterns of land use by Plio-Pleistocene hominins in the Lake Turkana Basin. *Journal of Human Evolution*, **27**: 139–158.

Rognon, P., Coudé-Gaussen, G., Fedoroff, N. and Goldberg, P. (1987) Micromorphology of loess in the northern Negev (Israel). In: *Micromorphologie des Sols – Soil Micromorphology* (Eds N. Fedoroff, L.M. Bresson and M.-A. Courty), pp. 631–638. Association Française pour l'Étude du Sol, Plaisir.

Rollefson, G.O. (1990) The uses of plaster at Neolithic 'Ain Ghazal Jordan. *Archeomaterials*, **4**: 33–54.

Romans, J.C.C. and Roberston, L. (1975a) Soils and archaeology in Scotland. In: *The Effect of Man on the Landscape: the Highland Zone* (Eds J.G. Evans, S. Limbrey and H. Cleere), Research Report No. 11, pp. 37–39. The Council for British Archaeology, York.

Romans, J.C.C. and Robertson, L. (1975b) Some genetic characteristics of the freely drained soils of the Ettrick Association in east Scotland. *Geoderma*, **14**: 297–317.

Romans, J.C.C. and Robertson, L. (1983a) The environment of north Britain: soils. In: *Settlement in North Britain 1000 BC to AD 1000* (Eds J.C. Chapman and H.C. Mytum), Vol. 118, pp. 55–80. British Archaeological Reports, British Series, Oxford.

Romans, J.C.C. and Robertson, L. (1983b) The general effects of early agriculture on soil. In: *The Impact of Aerial Reconnaissance on Archaeology* (Ed G.S. Maxwell), Research Report No. 49, pp. 136–41. Council for British Archaeology, London.

Rose, J., Boardman, J., Kemp, R.A. and Whitman, C.A. (1985) Palaeosols and the interpretation of the British Quaternary stratigraphy. In: *Geomorphology and Soils* (Eds K.S. Richards, R.R. Arnett and S. Ellis), pp. 348–375. George Allen & Unwin, London.

Rosen, A. (1986) Environmental change and settlement at Tel Lachish, Israel. *BASOR*, **266**: 45–58.

Rosen, A.M. (1986) *Cities of Clay. The Geoarchaeology of Tells*. University Press of Chicago, Chicago.

Roux, V. and Courty, M.-A. (1998) Identification of wheel-fashioning methods: technological analysis of 4th–3rd millennium BC oriental ceramics. *Journal of Archaeological Science*, **25**: 747–763.

Rowan, L.C. and Mars, J.C. (2003) Lithologic mapping in the Mountain Pass, California area using Advanced Spaceborne Thermal Emission and Reflection Radiometer (ASTER) data. *Remote Sensing of Environment*, **84**: 350–366.

Rowell, D.L. (1994) *Soil Science: Methods and Applications*. John Wiley & Sons, New York.

Rowley-Conwy, P. (1997) The animal bones from Arene Candide. *Final Report. Arene Candide: A Functional and Environmental Assessment of the Holocene Sequence (Excavations

Bernabò Brea-Cardini 1940–50). R. Maggi. Roma, Istituto Italiano di Paleontologia Umana: pp. 153–277.

Rowsome, P. (2000) *Heart of the City. Roman, Medieval and Modern London Revealed by Archaeology at 1 Poultry*. Museum of London Archaeology Service, London.

Runia, L.T. (1988) So-called secondary podzolisation in barrows. In: *Man-made Soils* (Eds W. Groenman-van Waateringe and M. Robinson), *British Archaeological Reports*, Oxford, pp. 129–142.

Rust, A. (1950) *Die Hohlenfunde von Jabrud (Syrien)*. Neumunster.

Sabins, F.F. (1997) *Remote Sensing: Principles and Interpretation, 3rd Edition*. W.H. Freeman and Co., New York.

Saito, Y. and Tiba, T. (1978) Petrological study of flints from the Douara Basin, northeast of Palmyra. *Bulletin. University of Tokyo Museum*, **14**: 111–121.

Salway, P. (1993) *The Oxford Illustrated History of Roman Britain*. Oxford University Press, Oxford.

Sandor, J.A. (1992) Long term effects of prehistoric agriculture on soils: examples from New Mexico and Peru. In: *Soils in Archaeology: Landscape Evolution and Human Occupation* (Ed V.T. Holliday), pp. 217–245. Smithsonian Institution Press, Washington.

Sankey, D. (1998) Cathedrals, granaries and urban vitality in late Roman London. In: *Roman London. Recent Archaeological Work* (Ed B. Watson), Supplementary Series No. 24, pp. 78–82. *Journal of Roman Archaeology*, Portsmouth, Rhode Island.

Sankey, D. and McKenzie, M. (1998) 7–11 Bishopsgate: a hole in the heart of London's business district. *The London Archaeologist*, **16**: 171–179.

Sarnthein, M. and Koopman, B. (1980) Late Quaternary deep-sea record of northwest African dust supply and wind circulation. *Palaeoecology of Africa*, **12**: 239–253.

Savage, A. (1995) *The Anglo-Saxon Chronicles*. Tiger Books International, London.

Saville, A. (1990) *Hazleton North. The Excavation of a Neolithic Long Cairn of the Cotswold-Severn Group*, Archaeological Report No. 13, English Heritage, London.

Scaife, R.G. and Macphail, R.I. (1983) The post-Devensian development of heathland soils and vegetation. In: *Soils of the Heathlands and Chalklands* (Ed P. Burnham), Vol. 1, pp. 70–99. South-East Soils Discussion Group, Wye.

Schick, T. and Stekelis, M. (1977) Mousterian assemblages in Kebara Cave, Mount Carmel. *Eretz Israel*, **13**: 97–149.

Schiegl, S., Lev-Yadun, S., Bar-Yosef, O., Goresy, A.E. and Weiner, S. (1994) Siliceous aggregates from prehistoric wood ash: a major component of sediments in Kebara and Hayonim caves (Israel). *Israel Journal of Earth Sciences*, **43**: 267–278.

Schiegl, S., Goldberg, P., Bar-Yosef, O. and Weiner, S. (1996) Ash deposits in Hayonim and Kebara Caves, Israel: macroscopic, microscopic and mineralogical observations, and their archaeological implications. *Journal of Archaeological Science*, **23**: 763–781.

Schiegl, S., Goldberg, P., Pfretzschner, H.-U. and Conard, N.J. (2003) Paleolithic burnt bone horizons from the Swabian Jura: distinguishing between*in situ* fire places and dumping areas. *Geoarchaeology*, **18**: 541–565.

Schiffer, M.B. (1987) *Formation Processes of the Archaeological Record*. University of New Mexico Press, Albuquerque, NM.

Schiffer, M.B., Skibo, J.M., Boelke, T.C., Neupert, M.A. and Aronson, M. (1994) New perspectives on experimental archaeology: surface treatments and thermal response of the clay cooking pot. *American Antiquity*, **59**: 197–217.

Schmid, E. (1963) Cave sediments and prehistory. In: *Science in Archaeology* (Ed D. Brothwell), pp. 123–138.

Schoch, R.M. (1989) *Stratigraphy: Principles and Methods*. Van Nostrand Reinhold, New York, 375 pp.

Schreiner, A. (1997) *Einführing in die Quartärgeologie, 2. Auflage*. E. Schweizerbartische Verlagsbuchhandlung, Stuttgart.

Schuldenrein, J. (1991) Coring and the identity of cultural resource environments. *American Antiquity* **56**: 131–137.

Schuldenrein, J. (1995) Geochemistry, phosphate fractionation, and the detection of activity areas at prehistoric North American sites. In: *Pedological Perspectives in Archaeological Research* (Eds M.E. Collins, B.J. Carter, B.G. Gladfelter and R.J. Southard), SSSA Special Publication Number 44, pp. 107–132. Soil Science Society of America, Madison.

Schuldenrein, J. and Clark, G.A. (1994) Landscape and prehistoric chronology of west-central Jordan. *Geoarchaeology*, **9**: 31–55.

Schuldenrein, J. and Clark, G.A. (2001) Prehistoric landscapes and settlement geography along the Wadi Hasa, West-Central Jordan. Part I: geoarchaeology, human palaeoecology and ethnographic modeling. *Environmental Archaeology*, **6**: 23–38.

Schuldenrein, J. and Goldberg, P. (1981) Late Quaternary paleoenvironments and prehistoric site distributions in the Lower Jordan Valley: a preliminary report. *Paléorient*, **7**: 57–81.

Schuldenrein, J., Wright, R.P., Mughal, M.R. and Khan, M.A. (2004) Landscapes, soils, and mound histories of the Upper Indus Valley, Pakistan: new insights on the Holocene environments near ancient Harappa. *Journal of Archaeological Science*, **31**: 777–797.

Schwarcz, H. and Blackwell, B. (1983) $^{230}Th/^{234}U$ age of a Mousterian site in France. *Science*, **301**: 236–237.

Schwarcz, H.P., Grün, R., Vandermeersch, B., Bar-Yosef, O., Valladas, H. and Tchernov, E. (1988) ESR dates for the hominid burial site of Qafzeh in Israel. *Journal of Human Evolution*, **17**: 733–737.

Schweitzer, F.R. (1979) Excavations at Die Kelders, Cape Province, South Africa; the Holocene deposits. *Annals of the South African Museum*, **78, Part 10**: 233.

Scollar, I. (1990) *Archaeological Prospecting and Remote Sensing*. Topics in remote sensing; 2. Cambridge University Press, Cambridge (England); New York.

Second Benelux Colloquium, Earth surface processes and geomorphology. *Second Benelux Colloquium; Earth Surface Processes and Geomorphology*, **6**: 319–330.

Segerström, U. (1991) Soil pollen analysis – an application for tracing ancient arable patches. *Journal of Archaeological Science*, **18**: 165–175.

Shackley, M.L. (1978) The behaviour of artifacts as sedimentary particles in a fluviatile environment. *Archaeometry*, **20**: 55–61.

Shahack-Gross, R., Berna, F., Karkanas, P. and Weiner, S. (2004) Bat guano and preservation of archaeological remains in cave sites. *Journal of Archaeological Science*, **31**: 1259–1272.

Shahack-Gross, R., Marshall, F. and Weiner, S. (2003) Geo-ethnoarchaeology of pastoral sites: the identification of livestock enclosures in abandoned Maasai settlements. *Journal of Archaeological Science*, **30**: 439–459.

Shahack-Gross, R., Marshall, F., Ryan, K. and Weiner, S. (2004) Reconstruction of spatial organization in abandoned Maasai settlements: implications for site structure in the Pastoral Neolithic of East Africa. *Journal of Archaeological Science*, **31**; 1395–1411.

Shanks, M. and Tilley, C. (1987) *Social Theory and Archaeology*. Polity Press, Cambridge.

Sharer, R.J. and Ashmore, W. (2003) *Archaeology, Discovering our Past*. McGraw-Hill, New York.

Sharples, N.M. (Ed) (1991) *Maiden Castle. Excavations and Field Survey 1985–6*, Archaeological Report No 19. English Heritage, London.

Sheets, P.D.(1992) *The Ceren Site: A Prehistoric Village Buried by Volcanic Ash in Central America*. Case Studies in Archaeology Series. Harcourt Brace College Publishers, Fort Worth.

Shepard, A.O. (1939) Technological notes on the pottery of San Jose. In: *Excavations at San Jose, British Honduras* (Ed J.E. Thompson), Publication 506. Carnegie Institution, Washington, D.C.

Shepard, A.O. (1985) *Ceramics for the Archaeologist, 5th Edition*. Carnegie Institution of Washington, Washington, DC.

Sherwood, S.C. (2001) Microartifacts. In: *Earth Sciences and Archaeology* (Eds P. Goldberg, V.T. Holliday and C.R. Ferring), pp. 327–352. Kluwer Academic/Plenum Publishers, New York.

Sherwood, S.C. (2001) *The Geoarchaeology of Dust Cave: A Late Paleoindian Through Middle Archaic Site in the Middle Tennessee River Valley*. Doctoral Dissertation, University of Tennessee, Knoxville, TN.

Sherwood, S.C. (2005) The Geoarchaeology of the whitesburg Bridge site. Technical Report (submitted). office of Archaeological research, Albama State Museums, University of Albama Tuscaloosa.

Sherwood, S.C. and Goldberg, P. (2001) A geoarchaeological framework for the study of karstic cave sites in the eastern woodlands. *Midcontinental Journal of Archaeology*, **26**: 145–167.

Sidell, E.J. (2000) Dark earth and obscured stratigraphy. In: *Taphonomy and Interpretation* (Eds J.P. Huntley and S. Stallibrass), Symposia of the Association for Environmental Archaeology No. 14, pp. 35–42. Oxbow, Oxford.

Sidell, E.J. (2003) The London Thames: a decade of research into the river and its floodplain. In: *Alluvial Archaeology in Europe* (Eds A.J. Howard, M.G. Macklin and D.G. Passmore), pp. 133–143. A. A. Balkema Publishers, Lisse.

Simmons, I.G. (1975) The ecological setting of mesolithic man in the highland zone. In: *The Effect of Man on the Landscape: the Highland Zone* (Eds J.G. Evans, S. Limbrey and H. Cleere), pp. 57–63, CBA, York.

Simpson, I.A. (1997) Relict properties of anthropogenic deep top soils as indicators of infield land management in Marwick, West Mainland, Orkney. *Journal of Archaeological Science*, **24**: 365–380.

Simpson, I.A. (1998) Early land management at Tofts Ness, Sanday, Orkney: the evidence of thin section micromorphology. In: *Life on the Edge* (Eds C.M. Mills and G. Coles), pp. 91–98, Human Settlement and Marginality. Oxbow, Oxford.

Simpson, I.A., Barrett, J.H. and Milek, K.B. (2005) Interpreting the Viking Age to Medieval Period Transition in Norse Orkney through cultural soil and sediment analyses. *Geoarchaeology*, **20**: 355–377.

Simpson, I.A., Perdikaris, S., Cook, G., Campbell, J.L. and Teesdale, W.J. (2000) Cultural sediment analyses and transitions in early fishing activity at Langenesvæet, Vesterålen, Northern Norway. *Geoarchaeology*, **15**: 743–763.

Singer, R. and Wymer, J. (1982) *The Middle Stone Age at Klasies River Mouth in South Africa*. The University of Chicago Press, Chicago.

Smart, P. and Tovey, N.K. (1981) *Electron Microscopy of Soils and Sediments: Examples*. Clarendon Press, Oxford University Press, Oxford, V0l. viii, 177 pp.

Smart, P. and Tovey, N.K. (1982) *Electron Microscopy of Soils and Sediments: Techniques*. Oxford Science Publications. Clarendon Press, Oxford University Press, Oxford.

Smith, G., Macphail, R.I., Mays, S.A., Nowakowki, J., Rose, P., Scaife, R.G., Sharpe, A Tomalin, D.J. and Williams, D.F. (1996) Archaeology and environment of a Bronze Age Cairn and Romano-British field system at Chysauster, Gulval, near Penzance, Cornwall: *Proceedings of the Prehistoric Society*, Vol. 62, pp. 167–219.

Smith, N.J.H. (1980) Anthrosols and human carrying capacity in Amazonia. *Annals of the Association of American Geographers*, **70**: 553–566.

Smyth, M.P., Dunning, N.P. and Dore, C.D. (1995) Interpreting prehistoric settlement patterns: lessons from the Maya center of Sayil, Yucatan. *Journal of Field Archaeology*, **22**: 321–347.

Sneh, A. (1983) Redeposited loess from the Quaternary Besor Basin, Israel. *Israel Journal of Earth-Sciences*, **32**: 63–69.

Soil Conservation Service (1994) *Keys to Soil Taxonomy*. U.S. Department of Agriculture, Washington, D.C.

Soil Survey Staff (1999) *Soil Taxonomy*. U.S. Department of Agriculture, Agriculture Handbook 436. U.S. Government Printing Office, Washington, D.C.

Sombroek, W.G. (1966) *Amazon Soils. A Reconnaissance of the Soils of the Brazilian Amazon Region*. Centre for Agricultural Publication and Documentation, Wageningen.

Sordoillet, D. (1997) Formation des dépôts archéologiques en grotte: la Grotte du Gardon (Ain) durant le Néolithique. In: *Dynamique du Paysage. Entretiens de Géoarchéologie* (Eds J.-F. Bravard and M. Prestreau), Documents d'Archéologiques en Rhône-Alpes No. 15, pp. 39–57. Ministère de la Culture, Lyon.

Spataro, M. (2002) *The First Farming Communities of the Adriatic: Pottery Production and Circulation in the Early and Middle Neolithic*, Quaderno-9. Società per la Preistoria e Protostoria della Regione Friuli-Venezia Giulla, Trieste.

Stafford, B.D. (1977) Burin manufacture and utilization: an experimental study. *Journal of Field Archaeology*, **4**: 235–246.

Stafford, M.D., Frison, G.C., Stanford, D. and Zeimans, G. (2003) Digging for the color of life: paleoindian red ochre mining at the Powars II Site, Platte County, Wyoming, U.S.A. *Geoarchaeology*, **18**: 71–90.

Stanley, D.J., Chen, Z. and Song, J. (1999) Inundation, sea-level rise and transition from Neolithic to Bronze Age cultures, Yangtze delta, China. *Geoarchaeology*, **14**: 15–26.

Stead, I.M., Bourke, J.B. and Brothwell, D. (Eds) (1986) *The Lindow Man, the Body in the Bog*. British Museum Publications, London.

Stein, J.K. (1986) Coring archaeological sites. *American Antiquity*, **51**: 505–527.

Stein, J.K. (1987) Deposits for archaeologists. *Archaeological Method and Theory*, **11**: 337–395.

Stein, J.K. (1990) Archaeological stratigraphy. In: *Archaeological Geology of North America* (Eds N.P. Lasca and J. Donahue), Centennial Special Volume 4, pp. 513–523. Geological Society of America, Boulder.

Stein, J.K. (1992) *Deciphering a shell midden*. Academic Press, San Diego.

Stein, J.K. and Linse, A. (Eds) (1993) *Effects of Scale on Archaeological and Geoscientific Perspectives*, Special paper, 283. Geological Society of America, Boulder.

Stekelis, M. (1954) Nouvelles fouilles dans la grotte de Kebarah. In: *Cronica del IV Congreso Internacional de Ciendceas Prehistoricas y Prothistoricas.*, pp. 385–389, Marid.

Stephens, E.P. (1956) The uprooting of trees: a forest process. *Soil Science Society of America Proceedings*, **20**: 113–116.

Stern, N., Porch, N. and McDougall, I. (2002) FxJj43: a window into a 1.5 million-year-old palaeolandscape in the Okote Member of the Koobi Fora Formation, Northern Kenya. *Geoarchaeology*, **17**: 349–392.

Stewart, H. (1995) Cedar: Tree of life to the northwest coast Indians: *Seattle*, University of Washington Press.

Stewart, D.J. (1999) Formation processes affecting submerged archaeological sites: an overview. *Geoarchaeology*, **14**: 565–587.

Stiner, M., Kuhn, S.L., Surovell, T.A. et al. (2001) Bone preservation in Hayonim Cave (Israel): a macroscopic and mineralogical study. *Journal of Archaeological Science*, **28**: 643–659.

Stiner, M.C., Kuhn, S.L., Weiner, S. and Bar-Yosef, O. (1995) Differential burning, recrystallization, and fragmentation of archaeological bone. *Journal of Archaeological Science*, **22**: 223–237.

Stoltman, J.B. (2001) The role of petrology in the study of archaeological ceramics. In: *Earth Sciences and Archaeology* (Eds P. Goldberg, V.T. Holliday and C.R. Ferring), pp. 297–326. Kluwer Academic/Plenum Publishers, New York.

Stoops, G. (1984) The environmental physiography of Pessinus in function of the study of the archaeological stratigraphy and natural building materials. In *'Les Fouilles de la Rijksuniversiteit te Gent a Pessiononte 1967–73* (Eds J. Devreker and M. Waelkens), Vol. XXII, pp. 38–50. Dissertationes Archaeologicae Gardenses.

Stoops, G. (Ed) (2003a) *Achievements in Micromorphology, CATENA*, **54** (3). Elsevier, Amsterdam.

Stoops, G. (2003b) *Guidelines for Analysis and Description of Soil and Regolith Thin Sections*. Soil Science Society of America, Inc., Madison, Wisconsin.

Stringer, C. (2000) The Gough's Cave human fossils; an introduction. *Bulletin of the Natural History Museum. Geology Series*, **56**: 135–139.

Stringer, C.B., Trinkhaus, E., Roberts, M.B., Macphail, R.I. and Parfitt, S.A. (1998) The Middle Pleistocene human tibia from Boxgrove. *Journal of Human Evolution*, **34**: 509–547.

Stringer, C.B., Barton, R.N.E. and Finlayson, J.C. (Eds) (2000) *Neanderthals on the Edge*. Papers from a conference marking the 150th anniversary of the Forbes' Quarry discovery, Gibraltar. Oxbow, Oxford.

Sukopp, H., Blume, H.-P. and Kunick, W. (1979) The soil, flora and vegetation of Berlin's waste lands. In: *Nature in Cities* (Ed I.C. Laurie), pp. 115–132. John Wiley, Chichester.

Summerfield, M.A. (1991) *Global Geomorphology: An Introduction to the Study of Landforms*. John Wiley & Sons, New York.

Sunaga, K., Sakagami, K. and Seki, T. (2003) Chemical and physical properties of buried cultivated soils of Edo Periods under volcanic mudflow deposits of 1783 eruption of Asama Volcano (tables and abstract in English). *Pedologist*, **47**: 14–28.

Suzuki, H. and Takai, F. (Eds) (1970) *The Amud Man and His Cave Site*. The University of Tokyo, Tokyo.

Tan, K.H. (Ed) (1984) *Andosols*. Van Nostrand Reinhold Company, New York.

Tang, Y., Jia, J. and Xie, X. (2003) Records of magnetic properties in Quaternary loess and its paleoclimatic significance. *Quaternary International*, **108**: 33–50.

Tankard, A.J. and Schweitzer, F.R. (1974) The geology of Die Kelders Cave and environs: a palaeoenvironmental study. *South African Journal of Science*, **70**: 365–369.

Tankard, A.J. and Schweitzer, F.R. (1976) Textural analysis of cave sediments: Die Kelders, Cape Province, South Africa. In: *Geoarchaeology: Earth Science and the Past* (Eds D.A. Davison and M.L. Shackley), pp. 289–316. Duckworth, London.

Tansley, A.G. (1939) *The British Islands and their Vegetation*. Cambridge University Press, Cambridge.

Tchernov, E. (1987) The age of 'Ubeidiya Formation, and early Pleistocene hominid site in the Jordan Valley, Israel. *Israel Journal of Earth Sciences*, **36**: 3–30.

Teilhard de Chardin, P. (1941) Early man in China. *Institut de Géo-Biologie, Pékin (Peiping)*, **7**: 99.

Teilhard de Chardin, P. and Pei, W.C. (1934) New discoveries in Choukoutien 1933–1934. *Geological Socicty of China, B*, **13**: 369–394.

Terrible, F., Wright, R. and FitzPatrick, E.A. (1997) Image analysis in soil micromorphology: from univariate approach to multivariate solution. In: *Soil Micromorphology: Studies on Soil Diversity, Diagnostics, Dynamics* (Eds S. Shoba, M. Gerasimova and R. Miedema), pp. 397–417. International Society of Soil Science, Moscow–Wageningen.

Tesch, S. (1992) House, farm and village in the Kopinge area from the early Neolithic to the early middle ages. In: *The Archaeology of the Cultural Landscape. Fieldwork and Research in a South Swedish Rural Region* (Eds L. Larsson, J. Callmer and B. Stjarnquist). Acta Archaeologica Ludensia, Series 4, No. 19, Lund.

Théry-Parisot, I. (2002) Fuel management (bone and wood) during the Lower Aurignacian in the Pataud Rock Shelter (Lower Palaeolithic, Les Eyzies de Tayac, Dordogne, France). Contribution of experimentation. *Journal of Archaeological Science*, **29**: 1415–1421.

Thieme, H. (1997) Lower Palaeolithic hunting spears from Germany. *Nature (London)*, **385**: 807–810.

Thomas, D.S.G. (1989) Aeolian sand deposits. In: *Arid Zone Geomorphology* (Ed D.S.G. Thomas), pp. 232–261. Halsted Press (John Wiley & Sons), New York.

Thomas, K. (Ed) (1990) *Soils and Early Agriculture*. World Archaeology 22, Vol. 1. Routledge, London.

Thompson, R. and Oldfield, F. (1986) *Environmental Magnetism*. Allen & Unwin, London.

Thoms, A.V. (2000) Environmental background, land-use history, and site integrity. In: *Uncovering Camp Ford: Archaeological Interpretations of a Confederate Prisoner-of-War Camp in East Texas* (Ed A.V. Thoms), pp. 8–25. Reports of Investigations No. 1, Center for Ecological Archaeology, Texas A&M University, College Station.

Thomas, D.H. (1998) *Archaeology, Third Edition*. Hartcourt Brace College Publishers, Fort Worth.

Thorson, R.M. (1990) Archaeological geology. *Geotimes*, **35**: 32–33.

Thorson, R.M. and Hamilton, T.D. (1977) geology of the Dry Creek site: a stratified early man site in interior Alaska. *Quaternary Research*, **7**: 172–188.

Tipper, J. (2001) *Grubenhäuser: Pitfalls and Pitfills*, Ph. D Thesis, Cambridge University, Cambridge.

Tipper, J. (Ed) (forthcoming) *West Heslerton Anglo-Saxon Settlement*. Landscape Research Centre, Yeddingham.

Tite, M.S. (1972) The influence of geology on the magnetic susceptibility of soils on archaeological sites. *Archaeometry*, **14**: 229–236.

Tite, M.S. and Mullins, C.E. (1971) Enhancement of magnetic susceptibility of soils on archaeological sites. *Archaeometry*, **13**: 209–219.

Tovey, N.K. and Sokolov, V.N. (1997) Image analysis applications in soil micromorphology. In: *Soil Micromorphology: Studies on Soil Diversity, Diagnostics, Dynamics* (Eds S. Shoba, M. Gerasimova and R. Miedema), pp. 345–356. International Society of Soil Science, Moscow–Wageningen.

Tsatskin, A. and Nadel, D. (2003) Formation processes at the Ohalo II submerged prehistoric campsite, Israel, inferred from soil micromorphology and magnetic susceptibility studies. *Geoarchaeology*, **18**: 409–432.

Tsatskin, A., Weinstein-Evron, M. and Ronen, A. (1992) Preliminary paleoenvironmental studies of the lowest layers (G and F) of Tabun Cave, Mt. Carmel. *Israel Geological Society Annual Meeting*, **1992**: 153.

Tsoar, H. (1982) Internal structure and surface geometry of longitudinal (seif) dunes. *Journal of Sedimentary Petrology*, **52**: 823–831.

Tsoar, H. and Goodfriend, G.A. (1994) Chronology and palaeoenvironmental interpretation of Holocene aeolian sands at the inland edge of the Sinai-Negev erg. *The Holocene*, **4**: 244–250.

Tucker, M.E. (Ed) (1988) *Techniques in Sedimentology*. Blackwell Scientific Publications, Oxford.

Tucker, M.E. (2001) Sedimentary petrology: an introduction to the origin of sedimentary rocks. 3rd edition. Blackwell Science, Oxford, Maiden, MA. 262 pp.

Turner, B.D. (1987) Forensic entomology: insects against crime. *Science Progress*, **71**: 169–180.

Turner, B.D. and Wiltshire, P. (1999) Experimental validation of forensic evidence: a study of the decomposition of buried pigs in a heavy clay soil. *Forensic Science International*, **101**: 113–12.

Turville-Petre, F. (1932) The excavations in the Mugharet et-Kebarah. *Journal of the Royal Anthropological Institute of Great Britain and Ireland*, **62**: 271–276.

Tylecote, R.F. (1979) The effect of soil conditions on the long term corrosion of buried tin-bronzes and copper. *Journal of Archaeological Science*, **6**: 345–368.

Tylecote, R.F. (1986) *The Early History of metallurgy in Europe*. Longman, London.

Uesugi, Y. (1973) Grain size distribution characteristics of the sands around the Douara cave. *Bulletin. University Museum Tokyo*, **5**: 113-32.

Valladas, H., Geneste, J.M., Joron, J.L. and Chadelle, J.P. (1986) Thermoluminescence dating of Le Moustier (Dordogne, France). *Nature (London)*, **322**: 452–454.

Valladas, H., Joron, J.L., Valladas, G. et al. (1987) Thermoluminescence dates for the Neanderthal burial site at Kebara in Israel. *Nature*, **330**: 159–160.

Valladas, H., Reyss, J.L., Joron, J.L., Valladas, G., Bar-Yosef, O. and Vandermeersch, B. (1988) Thermoluminescence dating of Mousterian "Proto-Cro-Magnon" remains from Israel and the origin of modern man. *Nature*, **331**: 614–616.

van Andel, T.H., Zanger, E. and Demitrack, S. (1990) Land use and soil erosion in prehistoric and historic Greece. *Journal of Field Archaeology*, **17**: 379–396.

Vandermeersch, B. (1972) Ce Que Révèlent les Sépultures Moustériennes de Qafzeh en Israël. *Archaeologia*, **45**: 7–15.

van der Meer, J.J.M. (1982) *The Fribourg Area, Switzerland. A Study in Quaternary Geology and Soil Development*, Ph.D Thesis, University of Amsterdam, Amsterdam.

van de Westeringh, W. (1988) Man-made soils in the Netherlands, especially in sandy areas ("Plaggen Soils"). In: *Man-Made Soils* (Eds W. Groenman-van Waateringe and M. Robinson), International Series 410, pp. 5–19. British Archaeological Reports, Oxford.

Van Nest, J. (2002) The good earthworm: how natural processes preserve upland Archaic archaeological sites of western Illinois, U.S.A. *Geoarchaeology*, **17**: 53–90.

Van Nest, J., Charles, D.K., Buikstra, J.E. and Asch, D.L. (2001) Sod blocks in Illinois Hopewell mounds. *American Antiquity*, **66**: 633–650.

Van Vliet, B. (1980) Approche des conditions physicochimiques favorisant l'autofluorescence des minéraux argileux. *Pédologie*, **30**: 369–390.

Van Vliet, B., Faivre, P., Andreux, F., Robin, A.M. and Portal, J.M. (1983) Behaviour of some organic components in blue and ultra-violet light: application to the micromorphology of podzols, andosols and planosols. In: *Soil Micromorphology* (Eds P. Bullock and C.P. Murphy), Volume 1: Techniques and Applications, pp. 91–99. A B Academic Publishers, Berkhamsted.

Van Vliet-Lanöe, B. (1985) Frost effects in soils. In: *Soils and Quaternary Landscape Evolution* (Ed J. Boardman), pp. 117–158. John Wiley & Sons, Chichester.

Van Vliet-Lanöe, B. (1998) Frost and soils: implications for paleosols, paleoclimates and stratigraphy. *CATENA*, **34**: 157–183.

Van Vliet-Lanoë, B., Helluin, M., Pellerin, J. and Valadas, B. (1992a) Soil erosion in Western Europe: from the last interglacial to the present. In: *Past and Present Soil Erosion* (Eds M. Bell and J. Boardman), Monograph 22, pp. 101–114. Oxbow, Oxford.

Van Vliet-Lanoë, B., Fagnart, J.P., Langohr, R. and Munaut, A. (1992b) Importance de la succession des phases écologiques anciennes et actuelles dans la différenciation des sols lessivés de la couverture loessiques d'Europe occidentale: argumentation stratigraphique et archéologique. *Science du Sol*, **30**: 75–93.

Van Zeist, W. and Bottema, S. (1982) Vegetational history of the eastern Mediterranean and the Near East during the last 20,000 years. In: *Palaeoclimates, Palaeoenvironments and Human Communities in the Eastern Mediterranean Region in Later Prehistory Part ii*. (Eds J.L. Bintliff and W. van Zeist), pp. 277–321. British Archaeological Reports, Oxford,

Vekua, A., Lordkipanidze, D., Rightmire, G.P. et al. (2002) A new skull of early Homo from Dmanisi, Georgia. *Science*, **297**: 85–89.

Veneman, P.I.M., Jacke, P.V. and Bodine, S.M. (1984) Soil formation as affected by pit and mound relief in Massachsetts, USA. *Geoderma*, **33**: 89–99.

Viklund, K. (1998) *Cereals, Weeds and Crop Processing in Iron Age Sweden. Methodological and Interpretive Aspects of Archaeobotanical Evidence*. Archaeology and Environment 14, Umea.

Viklund, K., Engelmark, R. and Linderholm, J. (Eds) (1998) *Fåhus från bronsålder till idag, Skrifter om skogs- och lantbrukshistoria 12*. Nordiska Museet, Lund.

Villa, P. (1982) Conjoinable pieces and site formation processes. *American Antiquity*, **47**: 276–290.

Villa, P. and Courtin, J. (1983) The interpretation of stratified sites: a view from underground. *Journal of Archaeological Science*, **10**: 267–281.

Vita-Finzi, C. (1969) *The Mediterranean Valleys*. Cambridge University Press, Cambridge.

Vita-Finzi, C. (1973). *Recent Earth History*. Macmillian, London.

Waechter, J.d.A. (1951) Excavations at Gorham's Cave, Gibraltar. *The Prehistoric Society*, **15**: 83–92.

Wagstaff, M. (1992) Agricultural teraces: the Vailikos Valley, Cyprus. In: *Past and Present Soil Erosion* (Eds M. Bell and J. Boardman), Monograph 22, pp. 155–161. Oxbow, Oxford.

Wainwright, G.J. and Davies, S.M. (1995) *Balksbury Camp, Hampshire: Excavations 1973 and 1981*. Archaeological Report 4. English Heritage, London.

Walker, R.B. (1997) Late-Paleoindian faunal remains from Dust Cave, Alabama. *Current Research in the Pleistocene*, **14**: 85–87.

Walker, R.B., Detwiler, K.R., Meeks, S.C. and Driskell, B.N. (2001) Berries, bones, and blades: reconstructing late Paleoindian subsistence economy at Dust Cave, Alabama. *MCJA: Midcontinental Journal of Archaeology*, **26**: 169–197.

Walker, R.G. and Cant, D.J. (1984) Sandy fluvial systems. In: *Facies Models: Geoscience Canada Reprint Series 1* (Ed R.G. Walker), pp. 71–89. Geological Association of Canada.

Waters, M.R. (1986) *The Geoarchaeology of Whitewater Draw, Arizona*. Anthropological Papers of the University of Arizona, No. 45. University of Arizona Press, Tucson.

Waters, M.R. (1992) *Principles of Geoarchaeology: A North American Perspective*. University of Arizona Press, Tucson .

Waters, M.R. (2000) Alluvial stratigraphy and geoarchaeology in the American Southwest. *Geoarchaeology*, **15**: 537–557.

Waters, M.R. and Nordt, L.C. (1995) Late Quaternary floodplain history of the Brazos River in east-central Texas. *Quaternary Research*, **43**: 311–319.

Waters, M.R. and Ravesloot, J.C. (2000) Late Quaternary geology of the Middle Gila River, Gila River Indian Reservation, Arizona. *Quaternary Research*, **54**: 49–57.

Watson, A. (1989a) Desert crusts and varnishes. In: *Arid Zone Geomorphology* (Ed D.S.G. Thomas), pp. 25–55. Belhaven Press, London.

Watson, A. (1989b) Windflow characteristics and aeolian entrainment. In: *Arid Zone Geomorphology* (Ed D.S.G. Thomas), pp. 209–231. Belknap Press, London.

Watson, B. (Ed) (1998) Late Roman London. Roman London. Recent Archaeological Work. Portsmouth, Rhode Island, *Journal of Roman Archaeology*. Supplementary series No. 24.

Wattez, J. (1992) *Dynamique de formation des structures de combustion de la fin du Paléolithique au Néolithique Moyen*. Approche méthodologiques et implications culturelles, Thesis, Univeristé de Paris Sud (Sorbonne), Paris.

Wattez, J. and Courty, M.A. (1987) Morphology of ash of some plant materials. In: *Soil Micromorphology* (Eds N. Fedoroff, L.M. Bresson and M.A. Courty), pp. 677–683. AFES, Plaisir.

Wattez, J., Courty, M.A. and Macphail, R.I. (1990) Burnt organo-mineral deposits related to animal and human activities in prehistoric caves. In: *Soil Micromorphology: A Basic and Applied Science* (Ed L.A. Douglas), Developments in Soil Science 19, pp. 431–439. Elsevier, Amsterdam.

Wauchope, R. (1938) *Modern Maya Houses, a Study of Their Archaeological Significance, Publication 502*. Carnegie Institution of Washington, Washington 181 pp.

Weiner, S., Goldberg, P. and Bar-Yosef, O. (1993) Bone preservation in Kebara Cave, Israel using on-site Fourier Transform Infrared Spectrometry. *Journal of Archaeological Science*, **20**: 613–627.

Weiner, S., Schiegl, S., Goldberg, P. and Bar-Yosef, O. (1995) Mineral assemblages in Kebara and Hayonim Caves, Israel: excavation strategies, bone preservation, and wood ash remnants. *Israel Journal of Chemistry*, **35**: 143–154.

Weiner, S., Xu, Q., Goldberg, P., Liu, J. and Bar, Y.O. (1998) Evidence for the use of fire at Zhoukoudian, China. *Science*, **281**: 251–253.

Weiner, S., Bar-Yosef, O., Goldberg, P., Xu, Q. and Liu, J. (2000) Evidence for the use of fire at Zhoukoudian. *Acta Anthropologica Sinica*, **Supplement to Vol. 19:** 218–223.

Weiner, S., Goldberg, P. and Bar-Yosef, O. (2002) Three-dimensional distribution of minerals in the sediments of Hayonim Cave, Israel: diagenetic processes and archaeological implications. *Journal of Archaeological Science*, **29**: 1289–1308.

Weir, A.H., Catt, J.A. and Madgett, P. (1971) Postglacial soil formation in the loess of Pegwell Bay, Kent (England). *Geoderma*, **5**: 131–149.

Weiss, E., Wetterstrom, W., Nadel, D. and Bar-Yosef, O. (2004) The broad spectrum revisited: evidence from plant remains. *Proceedings of the National Academy of Sciences*, **101**: 9551–9555.

Wendorf, F. and Schild, R. (1980) *Prehistory of the eastern Sahara*. Academic Press, New York.

Wendorf, F., Schild, R. and Close, A.E. (Eds) (1993) *Egypt During the Last Interglacial: The Middle Paleolithic of Bir Tarfawi and Bir Sahara East*. Plenum Press, New York.

Wendorf, F., Schild, R., Close, A.E. Schwarcz, H.P., Miller, G.H., Grün, R., Bluszcz, A., Stokes, S., Morawska, L., Huxtable, J., Lundberg, ., Hill, C.L. and McKinney, C. (1994) A chronology for the Middle and Late Pleistocene wet episodes in the Eastern Sahara. In: *Late Quaternary Chronology and Paleoclimates of the Eastern Mediterranean* (Eds O. Bar-Yosef and R.S. Kra), pp. 147–168. *Radiocarbon*, Tucson.

Wentworth, C.K. (1922) A scale of grade and class terms for clastic sediments. *Journal of Geology*, **30**: 377–392.

West, L.T., Bradford, J.M. and Norton, L.D. (1990) Crust morphology and infiltrability in surface soils from the southeast and midwest U.S.A. In: *Soil Micromorphology: A Basic and Applied Science* (Ed L.A. Douglas), pp. 107–113. Elsevier, Amsterdam.

West, S.E. (1985) West Stow: the Anglo-Saxon village. *East Anglian Archaeology*, **24**.

Wheatley, D. and Gillings, M. (2002) *Spatial Technology and Archaeology: The archaeological Applications of GIS*. Taylor & Francis, London.

Whitbread, I.K. (1995) *Greek Transport Amphorae: a Petrological and Archaeological Study*. Fitch laboratory occasional paper 4. British School at Athens, [Athens].

Whittle, A. (2000) New research on the Hungarian Early Neolithic. *Antiquity*, **74**: 13–14.

Whittle, A. (Ed) Forthcoming. *The Early Neolithic on the Great Hungarian Plain: Investigations of the Körös Culture Site of Ecsegfalva 23, Co. Békés*. Institute of Archaeology, Budapest.

Whittle, A., Rouse, A.J. and Evans, J.G. (1993) A Neolithic downland monument in its environment: excavations at the Easton Down Long Barrow, Bishops Canning, North Wiltshire. *Proceedings of the Prehistoric Society*, **59**: 197–239.

Whittlesey, J.H., Myers, J.W. and Allen, C.C. (1977) The Whittlesey foundation 1976 field season. *Journal of Field Archaeology*, **4**: 181–196.

Wieder, M. and Adan-Bayewitz, D. (2002) Soil parent materials and the pottery of Roman Galilee: a comparative study. *Geoarchaeology*, **17**: 393–415.

Wieder, M. and Gvirtzman, G. (1999) Micromorphological indications on the nature of the late Quaternary Paleosols in the southern coastal plain of Israel. *CATENA (Giessen)*, **35**: 219–237.

Wilkinson, T.J. (1990) Soil development and early land use in the Jazira region, Upper Mesopotamia. In: *Soils and Early Agriculture* (Ed K. Thomas), World Archaeology 22, No. 1, pp. 87–103. Routledge, London.

Wilkinson, T.J. (1997) Holocene environments of the High Plateau, Yemen. Recent geoarchaeological investigations. *Geoarchaeology*, **12**: 833–864.

Wilkinson, T.J., French, C.A.I., Matthews, W. and Oates, J. (2002) Geoarchaeology, landscape and region. In: Excavation at tell Brak Vol. 2. Nagar in the Millennium BC (Eds D. Oats, J. Oats and H. McDonald), Cambridge: The McDonald Institute for Archaeological Research, pp. 1–14.

Wilkinson, T.J. (2003) *Archaeological Landscapes of the Near East*. The University of Arizona Press, Tucson.

Wilkinson, T.J. and Murphy, P. (1986) Archaeological survey of an intertidal zone: the submerged landscape of the Essex coast, England. *Journal of Field Archaeology*, **13**: 177–194.

Wilkinson, T.J. and Murphy, P.L. (1995) *The Archaeology of the Essex Coast, Volume I: The Hullbridge Survey*. East Anglian Archaeology Report No. 71. Essex County Council, Chelmsford.

Williams, J.W. and Jenkins, D.A. (1976) The use of petrographic, heavy mineral and arc spectrographic techniques in assessing the provenance of sediments used in ceramics. In: *Geoarchaeology: Earth Science and the Past* (Eds D.S. Davidson and M.L. Shackley), pp. 115–135. Duckworth, London.

Wilson, H.E. (2004) *Chemical and Textural Analysis of Soils and Sediments in Forensic Investigations*, PhD, Thesis, Royal Holloway University of London, London.

Wiltshire, P.E.J. (1999) Palynological analysis of filling in the funerary shaft. In: *The Excavation of a Ceremonial site at Folly Lane, Verulamium* (Ed R. Niblett), Britannia Monograph No. 14, pp. 347–365. Society for the Promotion of Roman Studies, London.

Wiltshire, P.E.J. (in press) Environmental profiling and forensic palynology: Old techniques and new applications. *Spotlight (National Crime Facility)*.

Wiltshire, P.E.J., Edwards, K.J. and Bond, S. (1994) Microbially-derived metallic sulphide spherules, pollen, and the waterlogging of archaeological sites. *Proceedings of the American Association of Sedimentary Palynologists*, **29**: 207–221.

Woods, W.I. and McCann, J.M. (1999) The Anthropogenic origin and persistence of Amazonian Dark Earths. *Conference of Latin Americanist Geographers*, **25**: 7–14.

Woodward, J.C. and Goldberg, P. (2001) The sedimentary records in Mediterranean rockshelters and caves: archives of environmental change. *Geoarchaeology*, **16**: 465–466.

Wright, V.P. (Ed) (1986) *Paleosols. Their Recognition and Interpretation*. Blackwell Scientific Publications, Oxford.

Wright, V.P. (1992) Problems in detecting environmental changes in pre-Quaternary paleosols. In: *Seesoil*. South East Soils Discussion Group, Wye, pp. 5–12.

Xu, Q., Tian, M., Li, L. and Liu, J. (1996) A brief introduction to the Quaternary geology and paleoanthropology of the Zhoukoudian site, Beijing. In: *Field Trip Guide; Volume 6, Beijing and its Adjacent Areas*. (Ed N. Deng), pp. T203.1–T203.10. Geological Publishing House, Beijing, China.

Yaalon, D.H. and Dan, J. (1974) Accumulation and distribution of loess-derived deposits in the semi-desert and desert fringe areas of Israel. *Zeitschrift für Geomorphologie Supplement Band*, **20**: 91–105.

Yaalon, D.H. and Ganor, E. (1975) The influence of dust on soils during the Quaternary. *Soil Science*, **116**: 146–155.

Yaalon, D.H. and Kalmar, D. (1978) Dynamics of cracking and swelling clay soils: displacement of skeletal grains, optimum depth of slickensides, and rate of intra-pedonic turbation. *Earth Surface Processes*, **3**: 31–42.

Yule, B. (1990) The "dark earth" and late Roman London. *Antiquity*, **64**: 620–628.

Zangger, E. (1992) Neolithic to present soil erosion in Greece. In: *Past and Present Soil Erosion* (Eds M. Bell and J. Boardman), Monograph 22, pp. 133–147. Oxbow, Oxford.

Zeuner, F.E. (1946) *Dating the Past. An Introduction to Geochronology*. Methuen & Co. Ltd., London.

Zeuner, F.E. (1953) The chronology of the Mousterian at Gorham's Cave Gibraltar. *The Prehistoric Society*, **19**: 180–188.

Zeuner, F.E. (1959) *The Pleistocene Period*. Hutchinson Scientific & Technical, London.

Zilberman, E. (1993) The late Pleistocene sequence of the northwestern Negev flood plains; a key to reconstructing the paleoclimate of southern Israel in the last glacial. *Israel Journal of Earth-Sciences*, **41**: 155–167.

Index

Page numbers in *italics* to figures; those in **bold** to tables

Sites and Place Names

Archaeological, Geological and Chronological Periods and Cultures